Series on Advances in Statistical Mechanics – Vol. 18

# Statistical Mechanics of Magnetic Excitations

## From Spin Waves to Stripes and Checkerboards

## SERIES ON ADVANCES IN STATISTICAL MECHANICS*

**Editor-in-Chief:** M. Rasetti *(Politecnico di Torino, Italy)*

---

*Published*

Vol. 7: The Hubbard Model – Recent Results
*edited by M. Rasetti*

Vol. 8: Statistical Thermodynamics and Stochastic Theory of Nonlinear Systems Far From Equilibrium
*by W. Ebeling & L. Schimansky-Geier*

Vol. 9: Disorder and Competition in Soluble Lattice Models
*by W. F. Wreszinski & S. R. A. Salinas*

Vol. 10: An Introduction to Stochastic Processes and Nonequilibrium Statistical Physics
*by H. S. Wio*

Vol. 12: Quantum Many-Body Systems in One Dimension
*by Zachary N. C. Ha*

Vol. 13: Exactly Soluble Models in Statistical Mechanics: Historical Perspectives and Current Status
*edited by C. King & F. Y. Wu*

Vol. 14: Statistical Physics on the Eve of the 21st Century: In Honour of J. B. McGuire on the Occasion of his 65th Birthday
*edited by M. T. Batchelor & L. T. Wille*

Vol. 15: Lattice Statistics and Mathematical Physics: Festschrift Dedicated to Professor Fa-Yueh Wu on the Occasion of his 70th Birthday
*edited by J. H. H. Perk & M.-L. Ge*

Vol. 16: Non-Equilibrium Thermodynamics of Heterogeneous Systems
*by S. Kjelstrup & D. Bedeaux*

Vol. 17: Chaos: From Simple Models to Complex Systems
*by M. Cencini, F. Cecconi & A. Vulpiani*

Vol. 18 Statistical Mechanics of Magnetic Excitations: From Spin Waves to Stripes and Checkerboards
*by E. Rastelli*

Vol. 19: An Introduction to Stochastic Processes and Nonequilibrium Statistical Physics
*by H. S. Wio, R. R. Deza & J. M. López*

*For the complete list of titles in this series, please go to
http://www.worldscientific.com/series/sasm

Series on Advances in Statistical Mechanics – Vol. 18

# Statistical Mechanics of Magnetic Excitations

## From Spin Waves to Stripes and Checkerboards

Enrico Rastelli
Institute of Materials for Electronics and Magnetism, CNR
and Department of Physics, University of Parma, Italy

World Scientific

NEW JERSEY • LONDON • SINGAPORE • BEIJING • SHANGHAI • HONG KONG • TAIPEI • CHENNAI

*Published by*

World Scientific Publishing Co. Pte. Ltd.
5 Toh Tuck Link, Singapore 596224
*USA office:* 27 Warren Street, Suite 401-402, Hackensack, NJ 07601
*UK office:* 57 Shelton Street, Covent Garden, London WC2H 9HE

**Library of Congress Cataloging-in-Publication Data**
Rastelli, Enrico.
Statistical mechanics of magnetic excitations : from spin waves to stripes and checkerboards / Enrico Rastelli, National Council of Research, Italy.
pages cm. -- (Series on advances in statistical mechanics ; volume 18)
Includes bibliographical references and index.
ISBN 978-9814355506 (hardcover : alk. paper)
1. Spin excitations. 2. Nuclear spin. 3. Magnetic resonance. 4. Statistical mechanics. 5. Spin waves. I. Title.
QC794.6.E9R37 2013
539.7'25--dc23

2012047961

**British Library Cataloguing-in-Publication Data**
A catalogue record for this book is available from the British Library.



Typeset by Stallion Press
Email: enquiries@stallionpress.com

Printed in Singapore.

# PREFACE

Elementary excitations in solid state physics are the key to understand the low temperature properties of the matter. Indeed, their energy and lifetime are deduced from the real and imaginary part of the complex pole of the related Green function that can be experimentally investigated through its relationship with the correlation function on the basis of the linear response theory.

This book is devoted to the elementary excitations in magnetic periodic structures, the spin waves or magnons, which are shown to behave like an ideal Bose gas at low temperature and an interacting Bose system at higher temperature.

The theoretical research in this field has been keeping up during the time since the pioneer investigation on spin oscillations in ferromagnets by Bloch and twenty years later in antiferromagnets by Anderson. The well-established model Hamiltonian used to describe a variety of insulating magnetic materials is the Heisenberg Hamiltonian. A fundamental contribution to the statistical treatment of the Heisenberg Hamiltonian was given by Dyson who established a connection between spin and Bose operators obtaining exact results for the isotropic ferromagnet in the low temperature limit.

The discovery of non-collinear magnetic order like helices and spirals stimulated theoreticians to understand the origin of such intriguing magnetic configurations. The basic underlying mechanism is the frustration entered by competing exchange interactions or lattice structure.

In the 1980's, the interest shifted towards the low-dimensional magnets in connection with the discovery of the high-$T_c$ superconductors (quasi-2D magnetic systems), the availability of excellent magnetic surfaces and ultrathin magnetic films. Theoretical interest in 2D magnetic models dates back to the exact evaluation of the partition function of the $S = \frac{1}{2}$, nearest neighbour (NN) 2D Ising model (Onsager, 1944) and the discovery of a new type of transition in 2D planar rotator (sometimes referred improperly as XY) model (Berezinskii, Kosterlitz and Thouless, 1970). Subsequently, the experimental investigation of quasi-1D magnetic systems like $ABX_3$ (where A is an alkali element, B is a transition metal and X is a halogen) and the theoretical discovery that antiferromagnetic chains with integer-spin show statistical properties different from the half-integer-spin chains (Haldane, 1980) stimulated the study of 1D magnets.

From the experimental point of view, the use of neutrons as a probe to investigate both the kind of magnetic order and the elementary excitations spectrum looking at the elastic and inelastic part of the neutron scattering cross-section was a milestone in the comprehension of the magnetism.

In Chap. 1, we deduce the Heisenberg Hamiltonian starting from the hydrogen molecule and we present some exact results for small ring of spin $S = \frac{1}{2}$. The low lying energy states of a ferromagnet are obtained.

In Chap. 2, we discuss the spin-boson transformations and we give the low temperature thermodynamic functions for non-interacting spin waves.

In Chap. 3, the interaction between the spin waves is accounted for using the method of the equation of motion of the Green function.

In Chap. 4, the systematic perturbation theory for the Green function is introduced making use of the Feynman diagrams.

In Chap. 5, the two-magnon bound states in ferromagnets are investigated and their dependence on the dimensionality and anisotropy is discussed.

In Chap. 6, we present explicit calculations of the spin waves in ferromagnets with planar anisotropy and show the way to overcome the problem of the kinematical consistency entered by the spin-boson transformation in systems with planar anisotropy.

In Chap. 7, we obtain the ground states and the spin waves of magnetic systems with non-collinear order. The exchange competition or the lattice structure leads to helix-configurations that can be checked in neutron scattering experiments.

In Chap. 8, we obtain the ground-state configurations and the spin waves in multilayers. The explicit calculation for ultrathin films and for a semi-infinite medium is explicitly performed.

In Chap. 9, we study the ground–state configuration and spin waves excitations in presence of long range dipole–dipole interaction. The explicit calculation for the high-$T_c$ superconductor $ErBa_2Cu_3O_{6+x}$ is performed including a crystalline electric field calculation to fit the model with the experimental data.

In Chap. 10, we study the effect of long range dipole–dipole interaction in 2D systems and we show how unusual stripe and checkerboard configurations originate in anisotropic systems as the Ising model. Then, we present results at finite temperature obtained by Monte Carlo simulations.

Chapters 1, 2, 5, 7 and 8 are appropriate for an upper level of undergraduate course of magnetism. Chapters 3, 4 and 6 are suitable for a postgraduate course. Finally, Chapters 9 and 10 have to be considered as a stimulus for candidates to the theoretical research in magnetism.

Let me conclude this introduction by thanking the people who contributed to my scientific formation. First of all, I would thank Prof. Luciano Reatto who introduced me to the study of the statistical mechanics and of the non-collinear magnetic systems. Then, I would thank Prof. Per-Anker Lindgård who addressed me towards a thorough study of the properties of the different spin-boson transformations and Prof. Steven Lovesey who pointed out the importance of the neutron scattering as a tool of investigation of the spin waves. I am particularly grateful to Prof. Brooks Harris for having introduced me to the many-body theory of magnetic insulators.

A very special thank goes to Prof. Armando Tassi who assisted me in writing this book and was my fundamental collaborator in my research activities in the past 40 years. Let me thank other collaborators with whom I was in touch in different periods of my research activity: Prof. Alberto Pimpinelli, Dr. Silvia Sedazzari, Dr. Albino Carbognani and, particularly, Dr. Sofia Regina who gave me a relevant contribution in Monte Carlo simulations.

Finally, I would like to thank the people who convinced me to write this book: Prof. Davide Cassi and my family. Their insistence and stimulus played a key role in pushing me to undertake this venture.

# CONTENTS

# Chapter 1

# MAGNETIC HAMILTONIANS

## 1.1. Hydrogen Molecule Hamiltonian

The magnetism is a typical quantum effect due to the Pauli "exclusion principle". To understand this, let us consider one of the most simple quantum system: the hydrogen molecule containing two protons (nuclei) and two electrons. In the non-relativistic approximation and neglecting the nuclear kinetic energy (adiabatic approximation), the Hamiltonian of a hydrogen molecule reads

$$\mathcal{H} = -\frac{\hbar^2}{2m}\nabla_1^2 - \frac{\hbar^2}{2m}\nabla_2^2 + e^2\left(\frac{1}{R} + \frac{1}{r_{12}} - \frac{1}{r_{1a}} - \frac{1}{r_{2a}} - \frac{1}{r_{1b}} - \frac{1}{r_{2b}}\right) \tag{1.1.1}$$

where $R$ is the distance between the nuclei; $r_{i\alpha}$ (with $i = 1,2$ and $\alpha = a,b$) is the distance between the $i$-th electron and the $\alpha$-th nucleus, $r_{12}$ is the distance between the two electrons; $\nabla_i^2$ (with $i = 1,2$) is the Laplacian corresponding to the $i$-electron; $e = -4.803 \times 10^{-10}$ statCoulomb is the charge of the electron and $m = 9.11 \times 10^{-28}$ g is its mass. Note that any magnetic interaction has been neglected in the Hamiltonian (1.1.1). The wave function $\psi$ of the two electrons is given by the product of a spatial function times a spin function like

$$\psi(\boldsymbol{r}_1\sigma_1, \boldsymbol{r}_2\sigma_2) = \phi(\boldsymbol{r}_1, \boldsymbol{r}_2)\chi(\sigma_1, \sigma_2) \tag{1.1.2}$$

where $\sigma_1$ and $\sigma_2$ are the components of the electronic spin along the (arbitrary) quantization axis. The wave function (1.1.2) has to be chosen antisymmetric under the simultaneous exchange of the coordinates and spin variables of the two electrons. This means that a spatial antisymmetric function $\phi_{\mathrm{A}}(\boldsymbol{r}_1, \boldsymbol{r}_2)$ has to be associated with a symmetric spin function $\chi_{\mathrm{S}}(\sigma_1, \sigma_2)$ and a symmetric space function $\phi_{\mathrm{S}}(\boldsymbol{r}_1, \boldsymbol{r}_2)$ has to be associated with an antisymmetric spin function $\chi_{\mathrm{A}}(\sigma_1, \sigma_2)$. The spin function $\chi$ will be chosen symmetric ($\chi_{\mathrm{S}}$) when the total spin of the two electons $\boldsymbol{S} = \boldsymbol{\sigma}_1 + \boldsymbol{\sigma}_2$ is characterized by a quantum number $S = 1$ (triplet) and it will be chosen antisymmetric ($\chi_{\mathrm{A}}$) when the total spin quantum number is $S = 0$ (singlet). As a consequence, the spatial wave function has to be antisymmetric ($\phi_{\mathrm{A}}$) for $S = 1$ and symmetric ($\phi_{\mathrm{S}}$) for $S = 0$. Assuming that the interaction between the hydrogen atoms making up the molecule is weak and that both hydrogen atoms are in the ground state, the space functions $\phi_S$ and $\phi_A$ can be written in terms of the spatial wave functions of each electron in the Coulomb field of each nucleus, that is

$$\phi_A(\boldsymbol{r}_1, \boldsymbol{r}_2) = \frac{1}{\sqrt{2(1-\Delta^2)}}[\phi_{1s}(r_{1a})\phi_{1s}(r_{2b}) - \phi_{1s}(r_{2a})\phi_{1s}(r_{1b})] \tag{1.1.3}$$

and

$$\phi_S(\boldsymbol{r}_1, \boldsymbol{r}_2) = \frac{1}{\sqrt{2(1+\Delta^2)}}[\phi_{1s}(r_{1a})\phi_{1s}(r_{2b}) + \phi_{1s}(r_{2a})\phi_{1s}(r_{1b})] \tag{1.1.4}$$

where $\phi_{1s}(r_{i\alpha})$ is the normalized wave function of the ground state of the hydrogen atom consisting on an electron $i$ orbiting around a nucleus $\alpha$, that is

$$\phi_{1s}(r_{1a}) = \frac{1}{\sqrt{\pi a_0^3}} e^{-\frac{r_{1a}}{a_0}}, \quad \phi_{1s}(r_{2a}) = \frac{1}{\sqrt{\pi a_0^3}} e^{-\frac{r_{2a}}{a_0}},$$
$$\phi_{1s}(r_{1b}) = \frac{1}{\sqrt{\pi a_0^3}} e^{-\frac{r_{1b}}{a_0}}, \quad \phi_{1s}(r_{2b}) = \frac{1}{\sqrt{\pi a_0^3}} e^{-\frac{r_{2b}}{a_0}} \tag{1.1.5}$$

where $a_0 = \hbar^2/(me^2) = 0.5292$ Å is the atomic unit length. The "overlap integral" $\Delta$ in Eqs. (1.1.3) and (1.1.4) assures the normalization of the wavefunctions $\phi_A$ and $\phi_S$ and it is given by

$$\Delta = \int d^3\boldsymbol{r}_1\, \phi_{1s}(r_{1a})\phi_{1s}(r_{1b}) = \frac{1}{\pi a_0^3}\int d^3\boldsymbol{r}_1\, e^{-\frac{r_{1a}+r_{1b}}{a_0}} = e^{-\rho}\left(1+\rho+\frac{1}{3}\rho^2\right) \tag{1.1.6}$$

with $\rho = R/a_0$. The explicit calculations may be found in the original papers.[1,2]

The mean values of the energy of the hydrogen molecule in the states corresponding to $S = 1$ ($\phi_A$) and $S = 0$ ($\phi_S$) are given by

$$E_{\uparrow\uparrow} = \int d\boldsymbol{r}_1\, d\boldsymbol{r}_2\, \phi_A(\boldsymbol{r}_1, \boldsymbol{r}_2)\, \mathcal{H}\, \phi_A(\boldsymbol{r}_1, \boldsymbol{r}_2) = 2E_{1s} + \frac{Q-A}{1-\Delta^2} \tag{1.1.7}$$

and

$$E_{\uparrow\downarrow} = \int d\boldsymbol{r}_1\, d\boldsymbol{r}_1\, \phi_S(\boldsymbol{r}_1, \boldsymbol{r}_2)\, \mathcal{H}\, \phi_S(\boldsymbol{r}_1, \boldsymbol{r}_2) = 2E_{1s} + \frac{Q+A}{1+\Delta^2} \tag{1.1.8}$$

where $E_{1s} = -\frac{e^2}{2a_0}$ is the ground state energy of a hydrogen atom consisting of electron 1 orbiting around the nucleus $a$ satisfying the Schrödinger equation

$$\left(-\frac{\hbar^2}{2m}\nabla_1^2 - \frac{e^2}{r_{1a}}\right)\phi_{1s}(r_{1a}) = E_{1s}\phi_{1s}(r_{1a}) \tag{1.1.9}$$

or electron 2 orbiting around the nucleus $b$ satisfying the Schrödinger equation

$$\left(-\frac{\hbar^2}{2m}\nabla_2^2 - \frac{e^2}{r_{2b}}\right)\phi_{1s}(r_{2b}) = E_{1s}\phi_{1s}(r_{2b}). \tag{1.1.10}$$

Obviously, $2E_{1s}$ is the energy of the system when the two hydrogen atoms are far away ($R \to \infty$). The quantity $Q$ appearing in Eqs. (1.1.7) and (1.1.8) is called the

*Coulomb integral* and it is given by

$$Q = e^2 \int d^3\boldsymbol{r}_1 \int d^3\boldsymbol{r}_2\ \phi_{1s}^2(r_{1a})\phi_{1s}^2(r_{2b}) \left(\frac{1}{r_{12}} - \frac{1}{r_{1b}} - \frac{1}{r_{2a}}\right) + \frac{e^2}{R} = I_{12} + 2I_1 + \frac{e^2}{R} \tag{1.1.11}$$

where

$$I_{12} = \int d^3\boldsymbol{r}_1 \int d^3\boldsymbol{r}_2\ \phi_{1s}^2(r_{1a})\phi_{1s}^2(r_{2b})\frac{e^2}{r_{12}} = \frac{e^2}{a_0\rho}\left[1 - e^{-2\rho}\left(1 + \frac{11}{8}\rho + \frac{3}{4}\rho^2 + \frac{1}{6}\rho^3\right)\right] \tag{1.1.12}$$

is the average Coulomb repulsion between the two electrons;

$$I_1 = -\int d^3\boldsymbol{r}_1\ \phi_{1s}^2(r_{1a})\frac{e^2}{r_{1b}} = -\frac{e^2}{a_0\rho}[1 - e^{-2\rho}(1+\rho)] \tag{1.1.13}$$

is the average Coulomb attraction between the electron 1 and the nucleus $b$; an indentical contribution comes from the Coulomb interaction between electron 2 and the nucleus $a$: this explains the factor 2 in Eq. (1.1.11). The last term of Eq. (1.1.11) represents the Coulomb repulsion between the two nuclei. The quantity $A$ appearing in Eqs. (1.1.7) and (1.1.8) is called *exchange integral* and it is given by

$$A = e^2 \int d^3\boldsymbol{r}_1 \int d^3\boldsymbol{r}_2\ \phi_{1s}(r_{1a})\ \phi_{1s}(r_{2a})\ \phi_{1s}(r_{1b})\ \phi_{1s}(r_{2b}) \left(\frac{1}{r_{12}} - \frac{1}{r_{1b}} - \frac{1}{r_{2a}}\right) + \frac{e^2}{R}\ \Delta^2 = J_{12} + 2\ J_1\ \Delta + \frac{e^2}{R}\ \Delta^2 \tag{1.1.14}$$

where

$$J_{12} = \int d^3\boldsymbol{r}_1 \int d^3\boldsymbol{r}_2\ \phi_{1s}(r_{1a})\ \phi_{1s}(r_{2a})\ \phi_{1s}(r_{1b})\ \phi_{1s}(r_{2b})\frac{e^2}{r_{12}} = \frac{e^2}{a_0}\Big\{e^{-2\rho}\left(\frac{5}{8} - \frac{23}{20}\rho - \frac{3}{5}\rho^2 - \frac{1}{15}\rho^3\right) + \frac{6}{5\rho}\left[\Delta^2\,(C + \ln\rho) + M^2\ \mathrm{Ei}(-4\rho) - 2\ M\ \Delta\ \mathrm{Ei}(-2\rho)\right]\Big\} \tag{1.1.15}$$

with

$$M = e^{\rho}\left(1 - \rho + \frac{1}{3}\rho^2\right), \tag{1.1.16}$$

$C = 0.57722$ is the Euler's constant[3] and

$$\mathrm{Ei}(x) = -\int_{-x}^{\infty} dt\ \frac{e^{-t}}{t} \tag{1.1.17}$$

is the exponential integral function;[4]

$$J_1 = -\int d^3\boldsymbol{r}_1\ \phi_{1s}(r_{1a})\ \phi_{1s}(r_{1b})\frac{e^2}{r_{1b}} = -\frac{e^2}{a_0}e^{-\rho}\,(1+\rho). \tag{1.1.18}$$

An identical contribution comes from the integral in Eq. (1.1.14) containing $\frac{1}{r_{2a}}$ that justifies the factor 2 in the last side of Eq. (1.1.14), the last term of which comes from the repulsion between the nuclei. Notice that in each term of the exchange integral, the wave function of the same electron 1 or 2 centered on both nuclei simultaneously appears so that, in view of the structure of the ground state wave functions (1.1.5), one expects an exponential decay of the exchange contribution as the distance between the nuclei increases. Replacing Eqs. (1.1.11) and (1.1.14) into Eqs. (1.1.7) and (1.1.8), one obtains

$$E_{\uparrow\uparrow} = \frac{e^2}{a_0}\left[-1 + \frac{q(\rho) - a(\rho)}{1-\Delta^2}\right] \tag{1.1.19}$$

and

$$E_{\uparrow\downarrow} = \frac{e^2}{a_0}\left[-1 + \frac{q(\rho) + a(\rho)}{1+\Delta^2}\right] \tag{1.1.20}$$

with

$$q(\rho) \equiv \frac{a_0}{e^2}Q = \frac{1}{\rho}e^{-2\rho}\left(1 + \frac{5}{8}\rho - \frac{3}{4}\rho^2 - \frac{1}{6}\rho^3\right) \tag{1.1.21}$$

and

$$a(\rho) \equiv \frac{a_0}{e^2}A = \frac{1}{\rho}\Delta^2\left[1 + \frac{6}{5}(C + \ln\rho)\right] - e^{-2\rho}\left(\frac{11}{8} + \frac{103}{20}\rho + \frac{49}{15}\rho^2 + \frac{11}{15}\rho^3\right)$$
$$+ \frac{6}{5\rho}M\,[M\,\mathrm{Ei}(-4\rho) - 2\Delta\,\mathrm{Ei}(-2\rho)]. \tag{1.1.22}$$

Note that each term forming the Coulomb integral $q(\rho)$ decays as $\frac{1}{\rho}$ when the distance between the nuclei increases as one can see from Eqs. (1.1.12) and (1.1.13) according to the electrostatic Coulomb interaction. Their sum, however, leads to exponential decaying as shown by (1.1.21). On the contrary, each term forming the exchange energy $a(\rho)$ decays exponentially with the distance as one can see directly by Eq. (1.1.18) and Eq. (1.1.15), recalling the asymptotic expansion of the exponential integral function[3]

$$\mathrm{Ei}(-x) = -\frac{e^{-x}}{x}\left(1 - \frac{1}{x} + \frac{2}{x^2} - \frac{6}{x^3} + \dots\right). \tag{1.1.23}$$

In Fig. 1.1, the energy of the triplet $E_{\uparrow\uparrow}$ and of the singlet $E_{\uparrow\downarrow}$ are shown as function of the distance between the nuclei. It is evident from Fig. 1.1 that two hydrogen atoms in their ground state are able to form a molecule only if the spins of the electrons are antiparallel (singlet): this "bonding state" shows a minimum for $\rho_{\text{eq}} = 1.6425$ corresponding to the equilibrium distance between nuclei in the hydrogen molecule ground state $R_{\text{eq}} = 0.87$ Å, to be compared with the experimental value $R_{\text{exp}} = 0.74$ Å. The value of the binding energy is given by the difference between the energy of two far-away hydrogen atoms and the minimum of the singlet energy that is $2E_{1s} - E_{\uparrow\downarrow}(\rho_{\text{eq}}) = 0.11597\frac{e^2}{a_0} = 3.15\,\text{eV}$ that to be compared with the experimental value 4.45 eV. The state with parallel spins (triplet) is an "antibonding state" and cannot form a molecule.

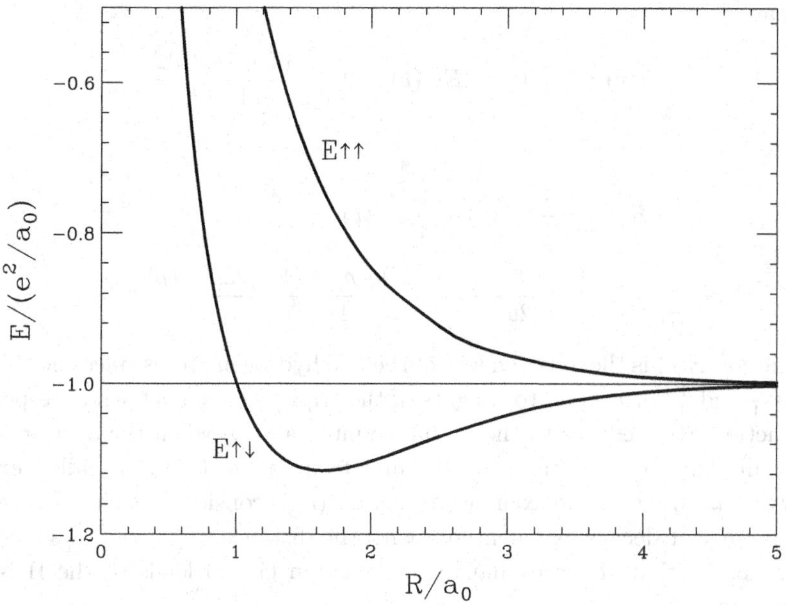


Fig. 1.1. Lowest energy levels (in units of $e^2/a_0$) of a hydrogen molecule for a triplet state ($E_{\uparrow\uparrow}$) and for a singlet state ($E_{\uparrow\downarrow}$) as function of the distance between the nuclei $\rho = R/a_0$.

## 1.2. Heisenberg Hamiltonian

The simple example of the hydrogen molecule in its ground state illustrates how the pure quantum effect of the electronic exchange leads to an effective interaction between the hydrogen atoms which is strongly dependent on the total spin of the two electrons even if the spin variables of the electrons do not appear explicitly in the Hamiltonian. Another way to write the Hamiltonian of a hydrogen molecule in its ground state is

$$\mathcal{H} = -J(r)\boldsymbol{\sigma}_1 \cdot \boldsymbol{\sigma}_2 + E(r) \tag{1.2.1}$$

where $\boldsymbol{\sigma}_1$ and $\boldsymbol{\sigma}_2$ are the electron spin operators and $J(r)$ and $E(r)$ are functions of $r$ chosen so that the eigenvalues of the operator $\mathcal{H}$ given by Eq. (1.2.1) are the same as $E_{\uparrow\uparrow}$ and $E_{\uparrow\downarrow}$ given by Eqs. (1.1.19) and (1.1.20), respectively. The eigenvalues of Hamiltonian (1.2.1) are easily obtained from the identity

$$\boldsymbol{\sigma}_1 \cdot \boldsymbol{\sigma}_2 = \frac{1}{2}\left(\boldsymbol{\sigma}_1 + \boldsymbol{\sigma}_2\right)^2 - \frac{1}{2}\left(\boldsymbol{\sigma}_1^2 + \boldsymbol{\sigma}_2^2\right) = \frac{1}{2}\boldsymbol{S}^2 - \frac{3}{4} \tag{1.2.2}$$

where the eigenvalues of $\boldsymbol{S}^2$ are $S(S+1)$ with $S = 0, 1$. Since the eigenvalues of the scalar product in Eq. (1.2.2) are $-\frac{3}{4}$ and $\frac{1}{4}$ for $S = 0$ and $S = 1$, respectively, the eigenvalues of Hamiltonian (1.2.1) coincide with $E_{\uparrow\uparrow}$ and $E_{\uparrow\downarrow}$ if $J(r)$ and $E(r)$ are chosen as

$$J(r) = E_{\uparrow\downarrow}(r) - E_{\uparrow\uparrow}(r) = 2\,\frac{e^2}{a_0}\,\frac{a(\rho) - q(\rho)\Delta^2}{1-\Delta^4} \tag{1.2.3}$$

and

$$E(r) = \frac{3}{4}E_{\uparrow\uparrow}(r) + \frac{1}{4}E_{\uparrow\downarrow}(r) = -\frac{e^2}{a_0} + \frac{e^2}{2a_0}\,\frac{(2+\Delta^2)q(\rho) - (1+2\Delta^2)a(\rho)}{1-\Delta^4}. \tag{1.2.4}$$

The function $E(r)$ is the mean energy of the two hydrogen atoms since the statistical weights $\frac{3}{4}$ and $\frac{1}{4}$ correspond to weights of the triplet and singlet state, respectively. The function $J(r)$ reduces to the exchange integral $a(\rho)$ when the overlap integral $\Delta$ is small. This fact explains why the first term of Eq. (1.2.1) is called *exchange Hamiltonian.* Note that the exchange integral $J(r)$ is considered a short-range interaction because it decays exponentially with the distance.

A generalization of the exchange Hamiltonian (1.2.1) leads to the Heisenberg Hamiltonian

$$\mathcal{H} = -\sum_{i,j} J_{ij}\boldsymbol{S}_i \cdot \boldsymbol{S}_j - h\sum_i S_i^z \tag{1.2.5}$$

where $i, j$ label the lattice sites; $2J_{ij} = 2J(|\boldsymbol{r}_i - \boldsymbol{r}_j|)$ is the exchange integral between the spins at $i$ and $j$ depending only on the distance between the spins; $\boldsymbol{S}_i$ is the spin associated to the site $i$ and $h = g\mu_B H$ where $g$ is the Landé factor, $\mu_B = 9.27\times10^{-20}$ erg $\times$ gauss$^{-1}$ is the Bohr magneton and $H$ the applied external magnetic field assumed to be directed along the $z$-axis. The simplest approximation neglects any interaction between the spins that are not nearest neighbour (NN) spins.

Let us give some examples of spin clusters for which the eigenvalues and eigenvectors of the Heisenberg Hamiltonian can be obtained exactly. Assuming that the spin quantum number is $\sigma = \frac{1}{2}$ let us consider the Hamiltonian given in Eq. (1.2.1) ignoring the term $E(r)$. Replacing $J(r)$ by $2J$ and adding the interaction with an external magnetic field, we obtain:

$$\mathcal{H} = -2J\boldsymbol{\sigma}_1 \cdot \boldsymbol{\sigma}_2 - h(\sigma_1^z + \sigma_2^z) = -J\left(\boldsymbol{S}^2 - \frac{3}{2}\right) - hS^z \tag{1.2.6}$$

where $\boldsymbol{S} = \boldsymbol{\sigma}_1 + \boldsymbol{\sigma}_2$ is the total spin of the complex and $S^z$ is the $z$-component of the total spin. The complete set of eigenvalues and eigenvectors of Hamiltonian (1.2.6) is easily obtained recalling the composition rule of the angular momenta $\boldsymbol{S}^2 = S(S+1)$ with $S = 1$ (triplet) and $S = 0$ (singlet). The result is illustrated in Fig. 1.2. For $J > 0$ the lowest energy level (ground state) and the corresponding eigenstate are given by

$$E_{\text{GS}} = -\frac{J}{2} - h, \quad |\uparrow\uparrow\rangle, \tag{1.2.7}$$

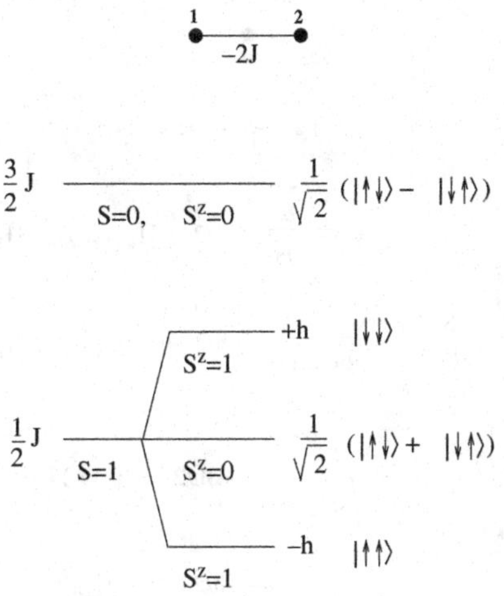


Fig. 1.2. Energy levels of a couple of spin interacting by an exchange integral $-2J$. The total spin number and its $z$-component associated to each state is shown.

respectively. The eigenstate $|\uparrow\uparrow\rangle$ is a contracted form of the state $|\sigma_1^z = \frac{1}{2}, \sigma_2^z = \frac{1}{2}\rangle$ where the symbols $\uparrow$ and $\downarrow$ mean $\sigma^z = \frac{1}{2}$ and $\sigma^z = -\frac{1}{2}$, respectively. The energies of the excited states and the corresponding eigenstates are given by

$$E_1 = -\frac{J}{2}, \quad \frac{1}{\sqrt{2}}(|\uparrow\downarrow\rangle + |\downarrow\uparrow\rangle), \tag{1.2.8}$$

$$E_2 = -\frac{J}{2} + h, \quad |\downarrow\downarrow\rangle \tag{1.2.9}$$

and

$$E_3 = \frac{3}{2}J, \quad \frac{1}{\sqrt{2}}(|\uparrow\downarrow\rangle - |\downarrow\uparrow\rangle). \tag{1.2.10}$$

In Fig. 1.3, we give the eigenvalues and eigenstates of a cluster of 3 spins forming a triangle described by the Heisenberg Hamiltonian

$$\begin{aligned}\mathcal{H} &= -2J(\boldsymbol{\sigma}_1 \cdot \boldsymbol{\sigma}_2 + \boldsymbol{\sigma}_2 \cdot \boldsymbol{\sigma}_3 + \boldsymbol{\sigma}_3 \cdot \boldsymbol{\sigma}_1) - hS^z \\ &= -J\left(\boldsymbol{S}^2 - \frac{9}{4}\right) - hS^z\end{aligned} \tag{1.2.11}$$

where the last step is obtained from the identity

$$\boldsymbol{S}^2 = \sum_{i=1}^{3}\sum_{j=1}^{3} \boldsymbol{\sigma}_i \cdot \boldsymbol{\sigma}_j = \frac{9}{4} + 2(\boldsymbol{\sigma}_1 \cdot \boldsymbol{\sigma}_2 + \boldsymbol{\sigma}_2 \cdot \boldsymbol{\sigma}_3 + \boldsymbol{\sigma}_3 \cdot \boldsymbol{\sigma}_1). \tag{1.2.12}$$

From Eq. (1.2.11), it is direct to write the complete set of eigenvalues and eigenstates ($2^3 = 8$) of the Hamiltonian using the composition rules of the angular momenta. In this case, $S = \frac{3}{2}$ and $S = \frac{1}{2}$ (twice degenerate). The result is shown in Fig. 1.3. Four eigenvalues belong to the quartet corresponding to $S = \frac{3}{2}$ and two eigenvalues (each of which twice degenerate) belong to the doublet corresponding to $S = \frac{1}{2}$.

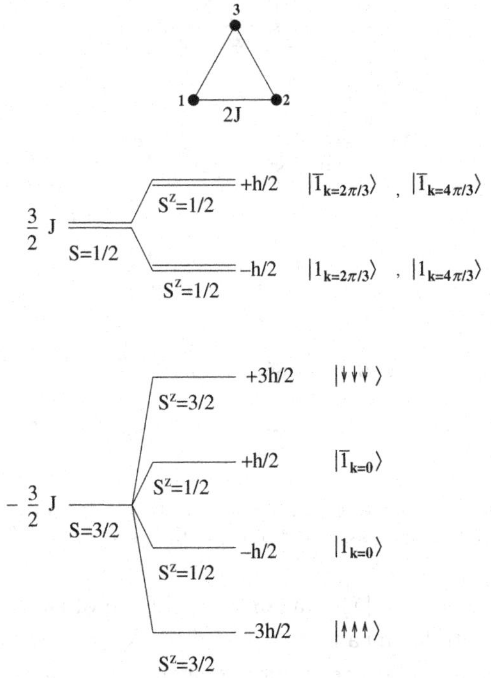


Fig. 1.3. Energy levels of a cluster of 3 spins forming a triangle. Each couple of spins interacts by an exchange integral $-2J$. The total spin number and its $z$-component associated to each state is shown. The double line in the highest doublet means that eigenvalue is twice degenerate.

There is no way to determine univocally the 2 states belonging to the eigenvalues $\frac{3}{2}J-\frac{h}{2}$ or $\frac{3}{2}J+\frac{h}{2}$. For a reason that will be clear later, we choose the two eigenstates belonging to the degenerate eigenvalue $\frac{3}{2}J-\frac{h}{2}$ as

$$|1_k>=\frac{1}{\sqrt{3}}(e^{ik}|\downarrow\uparrow\uparrow\rangle+e^{2ik}|\uparrow\downarrow\uparrow\rangle+|\uparrow\uparrow\downarrow\rangle) \tag{1.2.13}$$

with $k=\frac{2}{3}\pi$ and $k=\frac{4}{3}\pi$. The notation $|1_k\rangle$ is motivated by the fact that the state is a linear combination of states each containing 1 spin reversal with respect to the ground state $|0\rangle=|\uparrow\uparrow\uparrow\rangle$ localized in the different vertices of the triangle; the label $k$ reflects the $k$-dependence of the coefficients of the linear combination. Analogously, the two eigenstates belonging to the degenerate eigenvalue $\frac{3}{2}J+\frac{h}{2}$ are choosen as

$$|\bar{1}_k\rangle=\frac{1}{\sqrt{3}}(e^{ik}|\uparrow\downarrow\downarrow\rangle+e^{2ik}|\downarrow\uparrow\downarrow\rangle+|\downarrow\downarrow\uparrow\rangle) \tag{1.2.14}$$

with $k=\frac{2}{3}\pi$ and $k=\frac{4}{3}\pi$. Notice that the states of Eq. (1.2.14) are obtained by reversing the spins of the states of Eq. (1.2.13). They are a linear combination of states with a spin reversal with respect to the state $|\bar{0}\rangle=|\downarrow\downarrow\downarrow\rangle$. Note that the states given by Eq. (1.2.13) and the state belonging to the eigenvalue $-\frac{3}{2}-\frac{1}{2}h$ may be written in the more general form

$$|1_k\rangle = \frac{1}{\sqrt{3}}\sum_{n=1}^{3} e^{ikn}|1_n\rangle \tag{1.2.15}$$

with $k = \frac{2}{3}\pi$, $\frac{4}{3}\pi$ and 0, respectively, where $|1_n\rangle$ means a state characterized by a single deviation located at the site $n$. The three discrete values of $k$ reflect the periodic boundary conditions (PBC) of a periodic structure of 3 spins: $kN = 2\pi l$ with $N = 3$.

More work is required to obtain the complete set of eigenvalues and eigenstates of a cluster of 4 spins located on the vertices of a square as shown in Fig. 1.4. Assuming that the NN spins interact via an exchange integral $-2J_1$ and the next-nearest-neighbour (NNN) spins interact via an exchange integral $-2J_2$, the Hamiltonian reads

$$\begin{aligned}\mathcal{H} &= -2J_1(\boldsymbol{\sigma}_1\cdot\boldsymbol{\sigma}_2 + \boldsymbol{\sigma}_2\cdot\boldsymbol{\sigma}_3 + \boldsymbol{\sigma}_3\cdot\boldsymbol{\sigma}_4 + \boldsymbol{\sigma}_4\cdot\boldsymbol{\sigma}_1)\\ &\quad - 2J_2(\boldsymbol{\sigma}_1\cdot\boldsymbol{\sigma}_3 + \boldsymbol{\sigma}_2\cdot\boldsymbol{\sigma}_4) - hS^z\\ &= -J_1(\boldsymbol{S}^2 - 3) + 2(J_1 - J_2)(\boldsymbol{\sigma}_1\cdot\boldsymbol{\sigma}_3 + \boldsymbol{\sigma}_2\cdot\boldsymbol{\sigma}_4) - hS^z.\end{aligned} \tag{1.2.16}$$

The most direct way to obtain the eigenvalues of Hamiltonian (1.2.16) is to give the matrix elements of such Hamiltonian between the $2^4 = 16$ states $|\sigma_1^z\,\sigma_2^z\,\sigma_3^z\,\sigma_4^z\rangle$

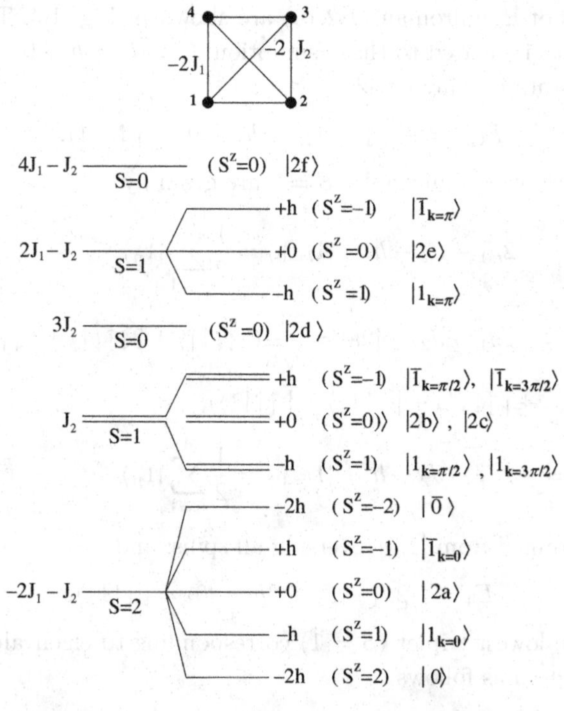


Fig. 1.4. Energy levels of a cluster of 4 spins forming a square with exchange interaction $-2J_1$ between spins along the edges of the square and $-2J_2$ between spins belonging to the diagonals of the square. The total spin number and its $z$-component associated to each state is shown.

with $\sigma_i^z = \pm\frac{1}{2}$ ($\uparrow$, $\downarrow$). The $16 \times 16$ matrix can be reduced to a matrix with blocks along the principal diagonal grouping the states according to $S^z$. Indeed, for $S^z = 2$ ($|\uparrow\uparrow\uparrow\uparrow\rangle$), the only matrix element is given by

$$\mathcal{H}^{(2)} = -2J_1 - J_2 - 2h. \tag{1.2.17}$$

For $S^z = 1$ ($|\downarrow\uparrow\uparrow\uparrow\rangle$, $|\uparrow\downarrow\uparrow\uparrow\rangle$, $|\uparrow\downarrow\downarrow\uparrow\rangle$ and $|\uparrow\uparrow\uparrow\downarrow\rangle$), the $4 \times 4$ matrix is given by

$$\mathcal{H}^{(1)} = \begin{pmatrix} -h & -J_1 & -J_2 & -J_1 \\ -J_1 & -h & -J_1 & -J_2 \\ -J_2 & -J_1 & -h & -J_1 \\ -J_1 & -J_2 & -J_1 & -h \end{pmatrix}. \tag{1.2.18}$$

For $S^z = 0$ ($|\downarrow\downarrow\uparrow\uparrow\rangle$, $|\downarrow\uparrow\downarrow\uparrow\rangle$, $|\downarrow\uparrow\uparrow\downarrow\rangle$, $|\uparrow\downarrow\downarrow\uparrow\rangle$, $|\uparrow\downarrow\uparrow\downarrow\rangle$, $|\uparrow\uparrow\downarrow\downarrow\rangle$), the $6 \times 6$ matrix is given by

$$\mathcal{H}^{(0)} = \begin{pmatrix} J_2 & -J_1 & -J_2 & -J_2 & -J_1 & 0 \\ -J_1 & (2J_1 - J_2) & -J_1 & -J_1 & 0 & -J_1 \\ -J_2 & -J_1 & J_2 & 0 & -J_1 & -J_2 \\ -J_2 & -J_1 & 0 & J_2 & -J_1 & -J_2 \\ -J_1 & 0 & -J_1 & -J_1 & (2J_1 - J_2) & -J_1 \\ 0 & -J_1 & -J_2 & -J_2 & -J_1 & J_2 \end{pmatrix}. \tag{1.2.19}$$

The blocks for $S^z = -1$ and $S^z = -2$ can be obtained directly from Eqs. (1.2.18) and (1.2.17) respectively, simply by changing the sign in front of $h$. The eigenvalues and eigenvectors of Hamiltonian (1.2.16) are shown in Fig. 1.4. Their ordering in increasing energies is related to the assumption $J_1 > J_2 > h > 0$.

The ground state is characterized by

$$E_{\text{GS}} = -2J_1 - J_2 - 2h, \quad |0\rangle = |\uparrow\uparrow\uparrow\uparrow\rangle. \tag{1.2.20}$$

The other 4 states of the quintuplet $S = 2$ are given by

$$E_1 = -2J_1 - J_2 - h, \quad |1_{k=0}\rangle = \frac{1}{2}\sum_{n=1}^{4} |1_n\rangle \tag{1.2.21}$$

$$E_2 = -2J_1 - J_2, \quad |2\mathrm{a}\rangle = \frac{1}{\sqrt{6}}(|\downarrow\downarrow\uparrow\uparrow\rangle + |\downarrow\uparrow\downarrow\uparrow\rangle + |\downarrow\uparrow\uparrow\downarrow\rangle$$
$$+ |\uparrow\downarrow\downarrow\uparrow\rangle + |\uparrow\downarrow\uparrow\downarrow\rangle + |\uparrow\uparrow\downarrow\downarrow\rangle), \tag{1.2.22}$$

$$E_3 = -2J_1 - J_2 + h, \quad |\bar{1}_{k=0}\rangle = \frac{1}{2}\sum_{n=1}^{4} |\bar{1}_n\rangle \tag{1.2.23}$$

where $|\bar{1}_n\rangle$ is obtained from $|1_n\rangle$ reversing all spins; and

$$E_4 = -2J_1 - J_2 + 2h, \quad |\bar{0}\rangle = |\downarrow\downarrow\downarrow\downarrow\rangle. \tag{1.2.24}$$

The states of the lowest triplet ($S = 1$) corresponding to eigenvalues with degeneracy 2 can be chosen as follows

$$E_5 = E_6 = J_2 - h, \quad |1_k\rangle = \frac{1}{2}\sum_{n=1}^{4} e^{ikn}|1_n\rangle \tag{1.2.25}$$

with $k = \frac{1}{2}\pi$ and $k = \frac{3}{2}\pi$;

$$E_7 = E_8 = J_2, \quad |2\mathrm{b}\rangle = \frac{1}{\sqrt{2}}(|\downarrow\downarrow\uparrow\uparrow\rangle + |\uparrow\uparrow\downarrow\downarrow\rangle), \quad |2\mathrm{c}\rangle = \frac{1}{\sqrt{2}}(|\downarrow\uparrow\uparrow\downarrow\rangle + |\uparrow\downarrow\downarrow\uparrow\rangle) \tag{1.2.26}$$

and

$$E_9 = E_{10} = J_2 + h, \quad |\overline{1}_k\rangle = \frac{1}{2}\sum_{n=1}^{4} e^{ikn}|\overline{1}_n\rangle \tag{1.2.27}$$

with $k = \frac{1}{2}\pi$ and $k = \frac{3}{2}\pi$. The lowest singlet ($S = 0$) is characterized by

$$E_{11} = 3J_2, \quad |2\mathrm{d}\rangle = \frac{1}{2}(|\downarrow\downarrow\uparrow\uparrow\rangle + |\uparrow\uparrow\downarrow\downarrow\rangle - |\downarrow\uparrow\uparrow\downarrow\rangle - |\uparrow\downarrow\downarrow\uparrow\rangle). \tag{1.2.28}$$

The next triplet ($S = 1$) is characterized by

$$E_{12} = 2J_1 - J_2 - h, \quad |1_\pi\rangle = \frac{1}{2}\sum_{n=1}^{4} e^{i\pi n}|1_n\rangle, \tag{1.2.29}$$

$$E_{13} = 2J_1 - J_2, \quad |2\mathrm{e}\rangle = \frac{1}{\sqrt{2}}(|\downarrow\uparrow\downarrow\uparrow\rangle + |\uparrow\downarrow\uparrow\downarrow\rangle) \tag{1.2.30}$$

and

$$E_{14} = 2J_1 - J_2 + h, \quad |\overline{1}_\pi\rangle = \frac{1}{2}\sum_{n=1}^{4} e^{i\pi n}|\overline{1}_n\rangle. \tag{1.2.31}$$

Finally, the highest singlet ($S = 0$) is characterized by

$$\begin{aligned} E_{15} &= 4J_1 - J_2, \\ |2\mathrm{f}\rangle &= \frac{1}{3\sqrt{3}}[|\downarrow\downarrow\uparrow\uparrow\rangle + |\uparrow\uparrow\downarrow\downarrow\rangle + |\downarrow\uparrow\uparrow\downarrow\rangle + |\uparrow\downarrow\downarrow\uparrow\rangle \\ &\quad - 2(|\downarrow\uparrow\downarrow\uparrow\rangle + |\uparrow\downarrow\uparrow\downarrow\rangle)]. \end{aligned} \tag{1.2.32}$$

Some general remarks can be deduced from the spectra of these small clusters. For $J > 0$, the ground state is the eigenstate corresponding to the saturated state with all spins up $|\uparrow \ldots \uparrow\rangle$ and it belongs to the maximum eigenvalue $NS(NS+1)$ of $\boldsymbol{S}^2$ and to the maximum eigenvalue $NS$ of $S^z$. The states with one deviation with respect to the ground state are given by the general equation

$$|1_k\rangle = \frac{1}{\sqrt{N}}\sum_{n=1}^{N} e^{ikn}|1_n\rangle \tag{1.2.33}$$

with $k = \frac{2\pi}{N}l$ and $l = 0, 1, 2, \ldots, N-1$. The state with $k = 0$, when acted upon by operators $S^z$ and $\boldsymbol{S}^2$, has eigenvalues $(NS - 1)$ and $NS(NS+1)$, respectively. In contrast, the states with $k \neq 0$, when acted upon by operators $S^z$ and $\boldsymbol{S}^2$, have eigenvalues $(NS-1)$ and $NS(NS-1)$, respectively. The energy of the states (1.2.33) for $J_2 = 0$ can be written

$$E_k = E_{\mathrm{GS}} + h + 2J_1(1 - \cos k). \tag{1.2.34}$$

As we will see in the next section, $E_k - E_{\mathrm{GS}}$ is the spin wave energy with wavevector $k$ in 1D with PBC.

## 1.3. Spin Wave Excitations

If we restrict the exchange interaction to NN spins, the Hamiltonian (1.2.5) becomes

$$\mathcal{H} = -J \sum_{i,\delta} \boldsymbol{S}_i \cdot \boldsymbol{S}_{i+\delta} - h \sum_i S_i^z. \tag{1.3.1}$$

The spins $\boldsymbol{S}_i$ occurring in (1.3.1) are angular momentum operators defined by the commutation rules[5]

$$[S_i^x, S_j^y] = iS_i^z \delta_{i,j}, \quad [S_i^z, S_j^x] = iS_i^y \delta_{i,j}, \quad [S_i^y, S_j^z] = iS_i^x \delta_{i,j}. \tag{1.3.2}$$

The commutation rules (1.3.2) imply that the components of the spins located at different lattice sites commute with each other so that the spins belonging to different lattice sites are treated as "distinguishable" particles. This is certainly true for the magnetic insulators in which the magnetic moments are localized on the lattice sites. It is useful to define the *raising* and *lowering* operators $S_i^+$ and $S_i^-$ as

$$S_i^+ = S_i^x + iS_i^y, \quad S_i^- = S_i^x - iS_i^y \tag{1.3.3}$$

whose commutation rules are obtained directly from Eq. (1.3.2)

$$[S_i^z, S_j^+] = S_i^+ \delta_{i,j}, \quad [S_i^z, S_j^-] = -S_i^- \delta_{i,j}, \quad [S_i^+, S_j^-] = 2S_i^z \delta_{i,j}. \tag{1.3.4}$$

From the commutation rules (1.3.4), it follows that in the representation where $\boldsymbol{S}_i^2 \equiv (S_i^x)^2 + (S_i^y)^2 + (S_i^z)^2$ and $S_i^z$ are diagonal, one has

$$\langle S'm'|\boldsymbol{S}_i^2|Sm\rangle = S(S+1)\,\delta_{S,S'}\,\delta_{m,m'}, \quad \langle S'm'|S_i^z|Sm\rangle = m\,\delta_{S,S'}\,\delta_{m,m'} \tag{1.3.5}$$

with $-S \leq m \leq S$. The matrix elements of the raising and lowering operators are

$$\langle S'm'|S_i^+|Sm\rangle = \sqrt{(S-m)(S+m+1)}\,\delta_{S,S'}\,\delta_{m',m+1} \tag{1.3.6}$$

and

$$\langle S'm'|S_i^-|Sm\rangle = \sqrt{(S+m)(S-m+1)}\,\delta_{S,S'}\,\delta_{m',m-1}. \tag{1.3.7}$$

By the use of the raising and lowering spin operators of Eq. (1.3.3), the Hamiltonian (1.3.1) becomes

$$\mathcal{H} = -J \sum_{i,\delta} \left[ S_i^z S_{i+\delta}^z + \frac{1}{2}(S_i^+ S_{i+\delta}^- + S_i^- S_{i+\delta}^+) \right] - h \sum_i S_i^z. \tag{1.3.8}$$

The constants of motion of the Hamiltonian (1.3.8), that is the operators that commute with it, are the square of the total spin of the system

$$\boldsymbol{S}_{\text{tot}}{}^2 = \left[ \sum_l \boldsymbol{S}_l \right]^2 = \sum_{l,m} \left[ S_l^z S_m^z + \frac{1}{2}(S_l^+ S_m^- + S_l^- S_m^+) \right] \tag{1.3.9}$$

and its $z$-component

$$S_{\text{tot}}^z = \sum_l S_l^z. \tag{1.3.10}$$

Note that the magnetization of the whole system is proportional to the total spin vector through the relationship $\boldsymbol{M}_{\text{tot}} = g\mu_B \boldsymbol{S}_{\text{tot}}$. The existence of constants of motion like those given in Eqs. (1.3.9) and (1.3.10) means that the Hamiltonian (1.3.8) cannot change the magnetization of the system or its $z$-component acting on a state of assigned $S_{\rm tot}$ and $S^z_{\rm tot}$. We have seen in the previous section that in the ground state $|0\rangle$, all spins are directed along the field assuming their maximum value. Acting with the operators (1.3.8)–(1.3.10) on the ground state, one obtains

$$\mathcal{H}|0\rangle = E_{\rm GS}|0\rangle, \quad E_{\rm GS} = -zJS^2N - hSN \tag{1.3.11}$$

and

$$\boldsymbol{S}_{\rm tot}{}^2|0\rangle = NS(NS+1)|0\rangle, \; S^z_{\rm tot}|0\rangle = NS|0\rangle \tag{1.3.12}$$

where $z$ is the number of NN (coordination number), $S$ is the spin quantum number of the magnetic moment located on each lattice site and $N$ is the number of lattice sites. As we have seen in the previous section, the lowest excitation energy levels are expected to correspond to states with only one spin deviation from the ground state. For this reason, we consider the state $S^-_j|0\rangle = \sqrt{2S}|1_j\rangle$ with a single spin deviation localized at the lattice site $\boldsymbol{r}_j$. The action of the operators (1.3.8)–(1.3.10) on the normalized state $|1_j\rangle$ gives

$$\mathcal{H}|1_j\rangle = (E_{\rm GS} + h + 2zJS)|1_j\rangle - 2JS\sum_{\boldsymbol{\delta}}|1_{j+\boldsymbol{\delta}}\rangle, \tag{1.3.14}$$

$$\boldsymbol{S}_{\rm tot}{}^2|1_j\rangle = NS(NS-1)|1_j\rangle + 2S\sum_m|1_m\rangle \tag{1.3.15}$$

and

$$S^z_{\rm tot}|1_j\rangle = (NS-1)|1_j\rangle, \tag{1.3.16}$$

showing that the state with one *localized* deviation is neither an eigenstate of the Hamiltonian (1.3.8) nor an eigenstate of the total spin operator $\boldsymbol{S}^2_{\rm tot}$ (1.3.9) even though it is an eigenstate of $S^z_{\rm tot}$ (1.3.10). In analogy with what we have seen in the case of small clusters, we define a *delocalized* state

$$|1_{\boldsymbol{k}}\rangle = \frac{1}{\sqrt{N}}\sum_j e^{i\boldsymbol{k}\cdot\boldsymbol{r}_j}|1_j\rangle \tag{1.3.17}$$

consisting of a linear combination of the "localized" states $|1_j\rangle$. The action of the operators (1.3.8)–(1.3.10) on the state (1.3.17) gives

$$\mathcal{H}|1_{\boldsymbol{k}}\rangle = \left[E_{\rm GS} + h + 2JS\sum_{\boldsymbol{\delta}}(1-\cos\boldsymbol{k}\cdot\boldsymbol{\delta})\right]|1_{\boldsymbol{k}}\rangle, \tag{1.3.18}$$

$$\boldsymbol{S}_{\rm tot}{}^2|1_{\boldsymbol{k}}\rangle = [NS(NS-1) + 2NS\;\delta_{\boldsymbol{k},0}]|1_{\boldsymbol{k}}\rangle \tag{1.3.19}$$

and

$$S^z_{\rm tot}|1_{\boldsymbol{k}}\rangle = (NS-1)|1_{\boldsymbol{k}}\rangle. \tag{1.3.20}$$

Eqs. (1.3.18)–(1.3.20) prove that $|1_{\boldsymbol{k}}\rangle$ is an eigenstate of the Hamiltonian, of the total spin operator and of the $z$-component of the total spin simultaneously. The cost of energy to create a delocalized spin deviation in the system is obtained by subtracting the ground-state energy (1.3.11) from Eq. (1.3.18), that is

$$\hbar\omega_{\boldsymbol{k}} = h + 2JS\sum_{\boldsymbol{\delta}}(1 - \cos \boldsymbol{k}\cdot\boldsymbol{\delta}). \tag{1.3.21}$$

For a ring of $N$ spins $S = \frac{1}{2}$ with NN exchange interaction ($\delta = \pm 1$ in units of lattice spacing), we recover the spin wave energy of small cluster given by Eq. (1.2.34). The energy (1.3.21) corresponding to a single spin deviation delocalized over the whole lattice is the *spin wave* energy and it is the lowest excited state of the Heisenberg Hamiltonian. Note that the mean value of the local spin operator $S_j^z$ on the localized state $|1_j\rangle$ is $S-1$ while its mean value on the delocalized state $|1_{\boldsymbol{k}}\rangle$ is $S - \frac{1}{N}$. Indeed, the single deviation is no more localized on a particular lattice site but distributed over the whole lattice, each spin component being reduced by $\frac{1}{N}$ on the average. Equation (1.3.19) shows that only the state corresponding to a spin wave with $\boldsymbol{k} = 0$ belongs to the maximum eigenvalue of $\boldsymbol{S}_{\mathrm{tot}}^2$ that is $NS(NS+1)$. On the contrary, the eigenstates of the spin waves with $\boldsymbol{k} \neq 0$ belong to the eigenvalue $(NS-1)NS$. In other words, the total spin quantum number is $NS$ for $|1_0\rangle$ and $NS-1$ for $|1_{\boldsymbol{k}}\rangle$, while the quantum number of the $z$-component of the total spin $S_{\mathrm{tot}}^z$ is $NS-1$ for any $|1_{\boldsymbol{k}}\rangle$ as shown by Eq. (1.3.20). This result is in agreement with the results obtained for small clusters in Section 1.2.

## 1.4. Two-Spin Deviation Excitations

In analogy with the choice done in the previous section for the states with a single spin deviation, we introduce the state with two spin deviations localized at the sites $\boldsymbol{r}_i$ and $\boldsymbol{r}_j$ $S_i^- S_j^- |0\rangle = 2S|1_i, 1_j\rangle$ for $i \neq j$ and $(S_i^-)^2|0\rangle = \sqrt{4S(2S-1)}|2_i\rangle$ for $i = j$. The states $|1_i, 1_j\rangle$ and $|2_i\rangle$ are normalized states and the coefficients in front of them are obtained from the matrix elements (1.3.7). The most general state with two spin deviations is given by

$$|\psi\rangle = \sum_{i,j} f_{ij} S_i^- S_j^- |0\rangle = \sum_{i\neq j} 2S f_{ij}|1_i, 1_j\rangle + \sum_i \sqrt{4S(2S-1)} f_{ii}|2_i\rangle \tag{1.4.1}$$

with the symmetry rules $f_{ij} = f_{ji}$ coming from the property $[S_i^-, S_j^-] = 0$. Notice that for $i = j$, the state $(S_i^-)^2|0\rangle$ is zero for $S = 1/2$. Using the commutation rules (1.3.4), the ground-state property $S_i^+|0\rangle = 0$ and the Heisenberg Hamiltonian (1.2.5)

$$\mathcal{H} = -\sum_{l,m} J_{lm}\left[S_l^z S_m^z + \frac{1}{2}(S_l^+ S_m^- + S_l^- S_m^+)\right] - h\sum_m S_m^z \tag{1.4.2}$$

where $J_{lm} = J_{ml}$, the Schrödinger equation can be written as

$$\mathcal{H}|\psi\rangle = E_{\mathrm{GS}}|\psi\rangle + \sum_{i,j} f_{ij}[\mathcal{H}, S_i^- S_j^-]|0\rangle = E|\psi\rangle \tag{1.4.3}$$

where $E_{\rm GS} = -\sum_{l,m} J_{lm}S^2 - hSN$ is the ground-state energy. Defining the two-spin wave excitation energy as $\hbar\omega = E - E_{\rm GS}$, Eq. (1.4.3) becomes

$$\hbar\omega \sum_{i,j} f_{ij} S_i^- S_j^- |0\rangle = \sum_{i,j} f_{ij} [\mathcal{H}, S_i^- S_j^-] |0\rangle. \tag{1.4.4}$$

The commutation rules (1.3.4) lead to the relationship

$$\begin{aligned}[\mathcal{H}, S_i^- S_j^-] = -2\sum_m [&J_{jm}(S_i^- S_m^- S_j^z - S_i^- S_j^- S_m^z) \\ &+ J_{im}(S_m^- S_j^- S_i^z - S_i^- S_j^- S_m^z) \\ &- J_{im} S_m^- S_j^- \delta_{i,j}] - 2J_{ij} S_i^- S_j^- + 2hS_i^- S_j^-.\end{aligned} \tag{1.4.5}$$

Using the identity $S_i^z|0\rangle = S|0\rangle$ from Eqs. (1.4.4) and (1.4.5), one obtains

$$\begin{aligned}&\hbar\omega \sum_{i,j} f_{ij} S_i^- S_j^- |0\rangle \\ &= \sum_{i,j} \left\{ \left[ 2h + 2S\sum_m (J_{im} + J_{jm}) - 2J_{ij} \right] f_{ij} \right. \\ &\quad \left. - 2S\sum_m (J_{jm} f_{im} + J_{im} f_{jm}) + J_{ij}(f_{ii} + f_{jj}) \right\} S_i^- S_j^- |0\rangle\end{aligned} \tag{1.4.6}$$

or

$$\begin{aligned}&\left(\hbar\omega - 2h - 4S\sum_m J_{im}\right) f_{ij} + 2S\sum_m (J_{jm} f_{im} + J_{im} f_{jm}) \\ &\quad = J_{ij}(f_{ii} + f_{jj} - 2f_{ij}).\end{aligned} \tag{1.4.7}$$

Let us introduce the Fourier transform[6]

$$f_{ij} = \frac{1}{N}\sum_{\boldsymbol{K}} e^{i\frac{\boldsymbol{K}}{2}\cdot(\boldsymbol{r}_i + \boldsymbol{r}_j)} F_{\boldsymbol{K}}(\boldsymbol{r}_{ij}) \tag{1.4.8}$$

where $\boldsymbol{r}_{ij} = \boldsymbol{r}_i - \boldsymbol{r}_j$ and $\boldsymbol{K} = 2\pi(\frac{l_x}{L_x}, \frac{l_y}{L_y}, \frac{l_z}{L_z})$ with $l_\alpha = 0, 1, 2, \ldots, L_\alpha - 1$ and $L_x L_y L_z = N$. Because of the symmetry rules $f_{ij} = f_{ji}$, one obtains the relationship $F_{\boldsymbol{K}}(\boldsymbol{r}_{ij}) = F_{\boldsymbol{K}}(\boldsymbol{r}_{ji})$ from Eq. (1.4.8). By replacing Eq. (1.4.8) into Eq. (1.4.7), one has

$$\begin{aligned}&\left(\hbar\omega - 2h - 4S\sum_m J_{im}\right) F_{\boldsymbol{K}}(\boldsymbol{r}_{ij}) \\ &\quad + 2S\sum_m [J_{jm} e^{i\frac{\boldsymbol{K}}{2}\cdot\boldsymbol{r}_{mj}} F_{\boldsymbol{K}}(\boldsymbol{r}_{ij} + \boldsymbol{r}_{jm}) + J_{im} e^{i\frac{\boldsymbol{K}}{2}\cdot\boldsymbol{r}_{mi}} F_{\boldsymbol{K}}(\boldsymbol{r}_{ij} + \boldsymbol{r}_{mi})] \\ &\quad = 2J_{ij}\left[\cos\left(\frac{\boldsymbol{K}}{2}\cdot\boldsymbol{r}_{ij}\right) F_{\boldsymbol{K}}(0) - F_{\boldsymbol{K}}(\boldsymbol{r}_{ij})\right].\end{aligned} \tag{1.4.9}$$

If we restrict the exchange interaction to NN, that is, $J_{ij} = J$ if $i$ and $j$ are NN sites and $J = 0$ otherwise, Eq. (1.4.9) becomes

$$(\hbar\omega - 2h - 4JzS)F_{\boldsymbol{K}}(\boldsymbol{r}) + 4JS\sum_{\boldsymbol{\delta}} e^{i\frac{\boldsymbol{K}}{2}\cdot\boldsymbol{\delta}} F_{\boldsymbol{K}}(\boldsymbol{r}+\boldsymbol{\delta})$$

$$= 2J(\boldsymbol{r})\left[\cos\left(\frac{\boldsymbol{K}}{2}\cdot\boldsymbol{r}\right)F_{\boldsymbol{K}}(0) - F_{\boldsymbol{K}}(\boldsymbol{r})\right] \tag{1.4.10}$$

where $\boldsymbol{\delta}$ is a vector connecting a spin with its $z$ NN. Let us introduce[6] the finite Fourier transforms

$$F_{\boldsymbol{K}}(\boldsymbol{r}) = \frac{1}{N}\sum_{\boldsymbol{q}} e^{i\boldsymbol{q}\cdot\boldsymbol{r}} f_{\boldsymbol{K}}(\boldsymbol{q}) \tag{1.4.11}$$

and

$$f_{\boldsymbol{K}}(\boldsymbol{q}) = \sum_{\boldsymbol{r}} e^{-i\boldsymbol{q}\cdot\boldsymbol{r}} F_{\boldsymbol{K}}(\boldsymbol{r}). \tag{1.4.12}$$

After multiplication of both sides of Eq. (1.4.10) by $e^{-i\boldsymbol{q}\cdot\boldsymbol{r}}$ and summation over $\boldsymbol{r}$, we obtain

$$[\hbar\omega - \hbar\omega_{\boldsymbol{K}}(\boldsymbol{q})]f_{\boldsymbol{K}}(\boldsymbol{q}) = 2J\sum_{\boldsymbol{\delta}}\cos\boldsymbol{q}\cdot\boldsymbol{\delta}\frac{1}{N}\sum_{\boldsymbol{q}'}\left(\cos\frac{\boldsymbol{K}}{2}\cdot\boldsymbol{\delta} - \cos\boldsymbol{q}'\cdot\boldsymbol{\delta}\right)f_{\boldsymbol{K}}(\boldsymbol{q}') \tag{1.4.13}$$

where

$$\hbar\omega_{\boldsymbol{K}}(\boldsymbol{q}) = 2h + 4JS\sum_{\boldsymbol{\delta}}\left(1 - \cos\frac{\boldsymbol{K}}{2}\cdot\boldsymbol{\delta}\ \cos\boldsymbol{q}\cdot\boldsymbol{\delta}\right) \tag{1.4.14}$$

is the energy of two *free* spin waves of momenta $\boldsymbol{k}_1 = \frac{\boldsymbol{K}}{2} + \boldsymbol{q}$ and $\boldsymbol{k}_2 = \frac{\boldsymbol{K}}{2} - \boldsymbol{q}$. As one can see from Eq. (1.4.13), the state corresponding to two *free* spin waves is *never* an eigenstate of the Heisenberg Hamiltonian because of the occurrence of the right-hand side of Eq. (1.4.13) that comes from the interaction potential between the spin waves. Let us write Eq. (1.4.13) for the "cubic lattices" and assume the lattice spacing $a = 1$:

$$[\hbar\omega - \hbar\omega_{\boldsymbol{K}}(\boldsymbol{q})]f_{\boldsymbol{K}}(\boldsymbol{q}) = 4J\sum_{\alpha}\cos q_\alpha\frac{1}{N}\sum_{\boldsymbol{q}'}\left(\cos\frac{K_\alpha}{2} - \cos q'_\alpha\right)f_{\boldsymbol{K}}(\boldsymbol{q}') \tag{1.4.15}$$

where $\alpha = x, y, z$ for a simple cubic (SC) lattice, $\alpha = x, y$ for a square (SQ) lattice and $\alpha = x$ for a linear chain (LC). We are able to prove that Eq. (1.4.15) has two kinds of solutions according to

i) the frequency $\omega$ is inside the two-spin wave band that is $\omega_{\text{bottom}} < \omega < \omega_{\text{top}}$ where $\hbar\omega_{\text{bottom}} = 8JS(d - \sum_\alpha \cos\frac{K_\alpha}{2})$ and $\hbar\omega_{\text{top}} = 8JS(d + \sum_\alpha \cos\frac{K_\alpha}{2})$ with $d = 1, 2, 3$ for LC, SQ and SC, respectively, or
ii) the frequency is below ($\omega < \omega_{\text{bottom}}$) or above ($\omega > \omega_{\text{top}}$) the two-spin wave band.

Let us begin considering the case i): we focus on a wavevector $\boldsymbol{q}_0$ and assume[7] that $f_{\boldsymbol{K}}(\pm\boldsymbol{q}_0) = N/2$. Then for $\boldsymbol{q} = \pm\boldsymbol{q}_0$, Eq. (1.4.15) becomes

$$\hbar\omega = \hbar\omega_{\boldsymbol{K}}(\boldsymbol{q}_0) + \frac{\Delta_{\boldsymbol{K}}}{N} \tag{1.4.16}$$

where

$$\Delta_{\boldsymbol{K}} = 8J\sum_{\alpha} \cos q_0^\alpha \left(\cos\frac{K_\alpha}{2} - \cos q_0^\alpha + V_{\boldsymbol{K}}^\alpha\right) \tag{1.4.17}$$

with

$$V_{\boldsymbol{K}}^\alpha = \frac{1}{N}\sum_{\boldsymbol{q}\neq\pm\boldsymbol{q}_0} \left(\cos\frac{K_\alpha}{2} - \cos q_\alpha\right) f_{\boldsymbol{K}}(\boldsymbol{q}). \tag{1.4.18}$$

For $\boldsymbol{q} \neq \pm\boldsymbol{q}_0$, Eq. (1.4.15) becomes

$$[\hbar\omega - \hbar\omega_{\boldsymbol{K}}(\boldsymbol{q})] f_{\boldsymbol{K}}(\boldsymbol{q}) = 4J\sum_{\alpha} \cos q_\alpha \left(\cos\frac{K_\alpha}{2} - \cos q_0^\alpha + V_{\boldsymbol{K}}^\alpha\right) \tag{1.4.19}$$

and using Eq. (1.4.16), one obtains

$$f_{\boldsymbol{K}}(\boldsymbol{q}) = 4J\sum_{\alpha} \frac{\cos q_\alpha \left(\cos\frac{K_\alpha}{2} - \cos q_0^\alpha + V_{\boldsymbol{K}}^\alpha\right)}{\hbar\omega_{\boldsymbol{K}}(\boldsymbol{q}_0) - \hbar\omega_{\boldsymbol{K}}(\boldsymbol{q}) + \frac{\Delta_{\boldsymbol{K}}}{N}}. \tag{1.4.20}$$

Multiplying both sides of Eq. (1.4.20) by $(\cos\frac{K_\beta}{2} - \cos q_\beta)$ with $\beta = x, y, z$ and summing over $\boldsymbol{q}$ with the restriction $\boldsymbol{q} \neq \pm\boldsymbol{q}_0$, we obtain

$$V_{\boldsymbol{K}}^\beta = \sum_{\alpha} \left(\cos\frac{K_\alpha}{2} - \cos q_0^\alpha + V_{\boldsymbol{K}}^\alpha\right) M_{\alpha\beta}(\boldsymbol{K}, \boldsymbol{q}_0) \tag{1.4.21}$$

where

$$M_{\alpha\beta}(\boldsymbol{K}, \boldsymbol{q}_0) = \frac{4J}{N}\sum_{\boldsymbol{q}\neq\pm\boldsymbol{q}_0} \frac{\cos q_\alpha \left(\cos\frac{K_\beta}{2} - \cos q_\beta\right)}{\hbar\omega_{\boldsymbol{K}}(\boldsymbol{q}_0) - \hbar\omega_{\boldsymbol{K}}(\boldsymbol{q}) + \frac{\Delta_{\boldsymbol{K}}}{N}}. \tag{1.4.22}$$

By the use of the matrix algebra, Eq. (1.4.21) becomes

$$V_{\boldsymbol{K}}^\alpha = \sum_{\beta\gamma} \left(\cos\frac{K_\beta}{2} - \cos q_0^\beta\right) M_{\beta\gamma} \left[\mathbf{1} - \mathbf{M}\right]^{-1}_{\gamma\alpha}. \tag{1.4.23}$$

with $\alpha, \beta, \gamma$ running over $x, y, z$. Replacing Eq. (1.4.23) into Eqs. (1.4.17) and (1.4.20), after some elementary matrix algebra, one obtains the complete solution of Eq. (1.4.15) in the case i), that is for $\omega_{\text{bottom}} < \omega < \omega_{\text{top}}$. We have

$$\hbar\omega = \hbar\omega_{\boldsymbol{K}}(\boldsymbol{q}_0) + \frac{8J}{N}\sum_{\alpha,\beta} \left(\cos\frac{K_\alpha}{2} - \cos q_0^\alpha\right) \left[\mathbf{1} - \mathbf{M}\right]^{-1}_{\alpha\beta} \cos q_0^\beta, \tag{1.4.24}$$

$$f_{\boldsymbol{K}}(\boldsymbol{q}) = 4J\sum_{\alpha\beta} \frac{\cos\frac{K_\alpha}{2} - \cos q_\alpha}{\hbar\omega_{\boldsymbol{K}}(\boldsymbol{q}_0) - \hbar\omega_{\boldsymbol{K}}(\boldsymbol{q}) + \frac{\Delta_{\boldsymbol{K}}}{N}} \left[\mathbf{1} - \mathbf{M}\right]^{-1}_{\alpha\beta} \cos q_\beta \tag{1.4.25}$$

for $\boldsymbol{q} \neq \pm\boldsymbol{q}_0$ and $f_{\boldsymbol{K}}(\pm\boldsymbol{q}_0) = N/2$ as assumed from the beginning. Strictly speaking, Eq. (1.4.24) is a self-consistent equation since $\mathbf{M}$ is function of $\Delta_{\boldsymbol{K}}/N$ as shown by Eq. (1.4.22). However, a good approximation for large $N$ neglects $\Delta_{\boldsymbol{K}}/N$ in Eqs. (1.4.22) and (1.4.25). From Eq. (1.4.24), one sees that the energy eigenvalue of the two-spin wave complex never coincides with the sum of the energies of the two *free* spin waves as anticipated during the discussion of Eq. (1.4.15). However, the interaction energy between the two spin waves given by the second term on the right-hand side of Eq. (1.4.24) decreases as $N$ increases and vanishes in the thermodynamic limit ($N \to \infty$) where the energy of the two-spin wave complex coincides with the sum of the energies of the two individual spin waves.

Now, we investigate the energy eigenvalue in the case ii) where the energy of the couple is outside the two-magnon band: in this case, since $\omega \neq \omega_{\boldsymbol{K}}(\boldsymbol{q})$, Equation (1.4.15) may be written as

$$f_{\boldsymbol{K}}(\boldsymbol{q}) = 4J \sum_{\alpha} \frac{\cos q_\alpha}{\hbar\omega - \hbar\omega_{\boldsymbol{K}}(\boldsymbol{q})} G_{\boldsymbol{K}}^{\alpha} \tag{1.4.26}$$

where

$$G_{\boldsymbol{K}}^{\alpha} = \frac{1}{N} \sum_{\boldsymbol{q}} \left( \cos\frac{K_\alpha}{2} - \cos q_\alpha \right) f_{\boldsymbol{K}}(\boldsymbol{q}). \tag{1.4.27}$$

Then multiplying both sides of Eq. (1.4.26) by $(\cos\frac{K_\beta}{2} - \cos q_\beta)$ and summing over $\boldsymbol{q}$, we obtain

$$G_{\boldsymbol{K}}^{\beta} = \sum_{\alpha} G_{\boldsymbol{K}}^{\alpha} M_{\alpha\beta}(\boldsymbol{K}, \omega) \tag{1.4.28}$$

where

$$M_{\alpha\beta}(\boldsymbol{K}, \omega) \equiv \frac{1}{2S} B_{\alpha\beta}(\boldsymbol{K}, \omega) = \frac{4J}{N} \sum_{\boldsymbol{q}} \frac{\cos q_\alpha \left( \cos\frac{K_\beta}{2} - \cos q_\beta \right)}{\hbar\omega - \hbar\omega_{\boldsymbol{K}}(\boldsymbol{q})}. \tag{1.4.29}$$

Equation (1.4.28) has a non-trivial solution for $G_{\boldsymbol{K}}^{\alpha}$ only if

$$\det\left[ \mathbf{1} - \frac{1}{2S} \mathbf{B}(\boldsymbol{K}, \omega) \right] = 0. \tag{1.4.30}$$

The roots $\omega_{\mathrm{BS}}$ of Eq. (1.4.30) are the bound states of two spin waves and they will be discussed in detail in Chapter 5.

For a two-magnon bound state of momentum $\boldsymbol{K}$, the probability of finding two spin deviations on sites apart from a lattice vector $\boldsymbol{r} \neq 0$ is

$$\frac{1}{2} \sum_{l} \frac{|\langle 1_l, 1_{l+r} | \psi \rangle|^2}{\langle \psi | \psi \rangle} = 8S^2 \sum_{l} \frac{|f_{l,l+r}|^2}{\langle \psi | \psi \rangle} = 8S^2 F_{\boldsymbol{K}}^2(\boldsymbol{r}) \frac{N}{\langle \psi | \psi \rangle} \tag{1.4.31}$$

and the probability of finding two spin deviations on the same site is

$$\sum_l \frac{|\langle 2_l|\psi\rangle|^2}{\langle\psi|\psi\rangle} = 4S(2S-1)\sum_l \frac{|f_{l,l}|^2}{\langle\psi|\psi\rangle} = 8S^2\left(1-\frac{1}{2S}\right)F_{\boldsymbol{K}}^2(0)\frac{N}{\langle\psi|\psi\rangle} \tag{1.4.32}$$

where

$$\langle\psi|\psi\rangle = 8S^2\left(1-\frac{1}{2S}\right)F_{\boldsymbol{K}}^2(0)N + N\sum_{\boldsymbol{r}\neq 0} 8S^2 F_{\boldsymbol{K}}^2(\boldsymbol{r}). \tag{1.4.33}$$

Equations (1.4.31) and (1.4.32) are obtained from Eqs. (1.4.1) and (1.4.8) in which the sum is restricted to the selected wavevector of the bound state $\boldsymbol{K}$. The factor $N$ comes from the sum over $l$ and from the property that in a lattice with PBC, the functions $|f_{l,l+r}|$ depend on $r$ but not on $l$, that is, they depend on the separation between the spin deviations but they are independent of the choice of the first deviation. The probability of finding two spin deviations on the same site in a bound state of momentum $\boldsymbol{K}$ is then

$$P_{\boldsymbol{K}}(0) = \frac{\left(1-\frac{1}{2S}\right)F_{\boldsymbol{K}}^2(0)}{\left(1-\frac{1}{2S}\right)F_{\boldsymbol{K}}^2(0) + \sum_{\boldsymbol{r}\neq 0} F_{\boldsymbol{K}}^2(\boldsymbol{r})} \tag{1.4.34}$$

and the probability of finding two spin deviations localized at distance $\boldsymbol{r}$ with $\boldsymbol{r}\neq 0$ is

$$P_{\boldsymbol{K}}(\boldsymbol{r}) = \frac{F_{\boldsymbol{K}}^2(\boldsymbol{r})}{\left(1-\frac{1}{2S}\right)F_{\boldsymbol{K}}^2(0) + \sum_{\boldsymbol{r}\neq 0} F_{\boldsymbol{K}}^2(\boldsymbol{r})}. \tag{1.4.35}$$

Note that $P_{\boldsymbol{K}}(0)$ vanishes for $S = 1/2$, as expected, since for $S = 1/2$ one spin deviation corresponds to a spin reversal so that any further deviation on the same site is excluded.

## 1.5. Two-Spin Deviation States in a Ring

In this section, we consider a ring of $N$ spins ($S = 1/2$) with the exchange interaction restricted to NN which is topologically equivalent to a LC with PBC. The aim of this section is to investigate the approach to thermodynamic limit ($N \to \infty$) starting from the exact solutions for finite clusters of Section 1.2. For a ring of $N$ spins Eq. (1.4.10) becomes

$$\begin{aligned}(\hbar\omega - 2h - 4J)F_K(r) &+ 2J\cos\frac{K}{2}[F_K(r+1) + F_K(r-1)] \\ &= 2J(r)\left[\cos\frac{K}{2}F_K(0) - F_K(r)\right]\end{aligned} \tag{1.5.1}$$

where $K = \frac{2\pi}{N}l$ with $l = 0, 1, \ldots, N-1$. The symmetry rules ($f_{ij} = f_{ji}$) and the PBC ($f_{i,j+N} = f_{ij}$) lead to the relations $F_K(r) = F_K(-r)$ and $F_K(N-r) = e^{-i\frac{K}{2}N}F_K(r)$, respectively. For $r = 0$, Eq. (1.5.1) becomes

$$(2h + 4J - \hbar\omega)F_K(0) - 4J\cos\frac{K}{2}\, F_K(1) = 0. \tag{1.5.2}$$

For $r = 1$, Equation (1.5.1) becomes

$$(2h + 2J - \hbar\omega)F_K(1) - 2J\cos\frac{K}{2}\,F_K(2) = 0. \tag{1.5.3}$$

Assuming $N$ is even, Eq. (1.5.1) becomes

$$(2h + 4J - \hbar\omega)F_K(r) - 2J\cos\frac{K}{2}[F_K(r-1) + F_K(r+1)] = 0 \tag{1.5.4}$$

for $2 \le r \le \frac{N}{2} - 1$ and

$$(2h + 4J - \hbar\omega)F_K\left(\frac{N}{2}\right) - 2J\cos\frac{K}{2}\left(1 + e^{i\frac{K}{2}N}\right)\,F_K\left(\frac{N}{2} - 1\right) = 0 \tag{1.5.5}$$

for $r = \frac{N}{2}$. The equations obtained from Eq. (1.5.1) for $r = \frac{N}{2} + 1, \ldots, N - 1$ can be reduced to Eqs. (1.5.3) and (1.5.4) by making use of the PBC. Equation (1.5.2) is of no physical interest because it simply gives the function $F(0)$ in terms of $F(1)$ but the probability of finding a doubly occupied site is zero when $S = 1/2$ as confirmed by Eq. (1.4.34). Notice that Eqs. (1.5.3)–(1.5.5) do not contain the unphysical amplitude $F(0)$. Then Eq. (1.5.5) gives

$$(2h + 4J - \hbar\omega)F_K\left(\frac{N}{2}\right) - 4J\cos\frac{K}{2}F_K\left(\frac{N}{2} - 1\right) = 0 \tag{1.5.6}$$

for $K = \frac{4\pi s}{N}$ with $s = 0, \pm 1, \ldots \pm(\frac{N}{4} - 1), \frac{N}{4}$ supposing $N$ is a multiple of 4 and $F_K(\frac{N}{2}) = 0$ for $K = \pm\frac{2\pi}{N}(2s + 1)$ with $s = 0, 1, \ldots, \frac{N}{4} - 1$. Eqs. (1.5.3)-(1.5.5) can be grouped in two matrix equations given by

$$\begin{pmatrix} 2h + 2J - \hbar\omega & -2J\cos\frac{K}{2} & \cdots & 0 \\ -2J\cos\frac{K}{2} & 2h + 4J - \hbar\omega & \cdots & 0 \\ 0 & -2J\cos\frac{K}{2} & \cdots & 0 \\ \vdots & \vdots & \ddots & \vdots \\ 0 & \cdots & 2h + 4J - \hbar\omega & -2J\cos\frac{K}{2} \\ 0 & \cdots & -4J\cos\frac{K}{2} & 2h + 4J - \hbar\omega \end{pmatrix}$$

$$\times \begin{pmatrix} F_K(1) \\ F_K(2) \\ F_K(3) \\ \vdots \\ F_K\left(\frac{N}{2} - 1\right) \\ F_K\left(\frac{N}{2}\right) \end{pmatrix} = 0 \tag{1.5.7}$$

with $F_K(\frac{N}{2}+r) = F_K(\frac{N}{2}-r)$ for $K = \frac{4\pi s}{N}$ and

$$\begin{pmatrix} 2h+2J-\hbar\omega & -2J\cos\frac{K}{2} & \cdots & 0 \\ -2J\cos\frac{K}{2} & 2h+4J-\hbar\omega & \cdots & 0 \\ 0 & -2J\cos\frac{K}{2} & \cdots & 0 \\ \vdots & \vdots & \ddots & \vdots \\ 0 & \cdots & -2J\cos\frac{K}{2} & 2h+4J-\hbar\omega \end{pmatrix}$$

$$\times \begin{pmatrix} F_K(1) \\ F_K(2) \\ F_K(3) \\ \vdots \\ F_K\left(\frac{N}{2}-1\right) \end{pmatrix} = 0 \tag{1.5.8}$$

with $F_K(\frac{N}{2}+r) = -F_K(\frac{N}{2}-r)$ for $K = \frac{2\pi}{N}(2s+1)$. Note that the square matrix occurring in Eq. (1.5.7) has dimension $\frac{N}{2}$ so that it has $\frac{N}{2}$ eigenvalues while the square matrix occurring in Eq. (1.5.8) has dimension $\frac{N}{2}-1$ so that it has $\frac{N}{2}-1$ eigenvalues. Since the wavevector $K$ can assume $\frac{N}{2}$ values in both Eqs. (1.5.7) and (1.5.8), the total number of eigenvalues for two-magnon excitations is $(\frac{N}{2})^2 + \frac{N}{2}(\frac{N}{2}-1) = \frac{N(N-1)}{2}$ corresponding to the number of states with two spin deviations. For comparison, let us recall that the ground state is unique, the one-magnon states are $N$ and the total number of states of the ring is $2^N$. We find explicitly the eigenvalues of (1.5.7) and (1.5.8) for $N = 8$ and 16. For $N = 8$, we give the 28 eigenvalues in Table 1.1. As one can see, the number of eigenvalues of the two-spin wave energy is 4 for $K = 0, \pm\frac{\pi}{2}$ and $\pi$ since the dimension of the square matrix occurring in Eq. (1.5.7) is 4 while the number of eigenvalues for $\pm\frac{\pi}{4}$ and $\pm\frac{3\pi}{4}$ is 3 since the dimension of the square matrix occurring in Eq. (1.5.8) is 3. The

Table 1.1. Two-spin deviation eigenvalues for $N = 8$ spins $S = 1/2$ on a ring.

| $K/\pi$ | $\hbar\omega/J$ |
|---|---|
| 0 | 0, 1.50604, 4.89008, 7.60388 |
| $\pm\frac{1}{4}$ | 0.58579, 3.12597, 6.28825 |
| $\pm\frac{1}{2}$ | 0.94863, 2, 4.51730, 6.53407 |
| $\pm\frac{3}{4}$ | 1.71175, 3.41421, 4.87403 |
| 1 | 2, 4, 4, 4 |

Table 1.2. Two-spin deviation eigenvalues for $N = 16$ spins $S = 1/2$ on a ring.

| $K/\pi$ | $\hbar\omega/J$ |
|---|---|
| 0 | 0, 0.34582, 1.32348, 2.76393, 4.41811, 6, 7.23607, 7.91259 |
| $\pm\frac{1}{8}$ | 0.15224, 0.81690, 2.03080, 3.58431, 5.20881, 6.62341, 7.58353 |
| $\pm\frac{1}{4}$ | 0.28039, 0.58579, 1.49441, 2.83236, 4.36875, 5.83797, 6.98600, 7.61433 |
| $\pm\frac{3}{8}$ | 0.63739, 1.23463, 2.27524, 3.60685, 4.99929, 6.21182, 7.03478 |
| $\pm\frac{1}{2}$ | 0.99626, 1.30412, 2, 3.04050, 4.23691, 5.38134, 6.27568, 6.76518 |
| $\pm\frac{5}{8}$ | 1.38294, 2.06999, 2.76537, 3.67369, 4.62711, 5.45827, 6.02263 |
| $\pm\frac{3}{4}$ | 1.70710, 2.51511, 2.85111, 3.41421, 4.07839, 4.71858, 5.22029, 5.49520 |
| $\pm\frac{7}{8}$ | 1.92388, 3.30209, 3.52881, 3.84776, 4.19298, 4.49803, 4.70645 |
| 1 | 2, 4, 4, 4, 4, 4, 4, 4 |

double degeneracy of the eigenvalues belonging to $\pm K \neq 0$ or $\pi$ is a consequence of the parity of the cosine function. At the zone boundary $K = \pi$, all eigenvalues are degenerate except one: indeed for $K = \pi$, the matrix occurring in Eq. (1.5.7) becomes diagonal with all elements equal to $2h+4J-\hbar\omega$ except one that is $2h+2J-\hbar\omega$. For N=16, we give the 120 eigenvalues in Table 1.2. As one can see for $K = \frac{\pi}{4}s$ with $s = 0, \pm1, \pm2, \pm3, 4$, one has 8 eigenvalues while for $K = \pm\frac{\pi}{8}(2s+1)$ with $s = 0, 1, 2, 3$, one has 7 eigenvalues. One can compare the eigenvalues of Table 1.2 for an assigned wavevector $K$ with the energy of two *free* spin waves of momenta $k_1 = \frac{K}{2} + q_0$ and $k_2 = \frac{K}{2} - q_0$, given by

$$\hbar\omega_K(q_0) = 4J\left(1 - \cos\frac{K}{2}\cos q_0\right) \tag{1.5.9}$$

in which $K$ is the momentum of the couple of the non-interacting magnons and $q_0 = \frac{2\pi}{16}l$ with $l = 0, \pm1, \pm2, \ldots, \pm7, 8$. The values $q_0 = 0$ and $q_0 = \pi$ correspond to the bottom and to the top of the two-magnon band, respectively. One can see that the energy of the two free spin waves differs from the exact eigenvalues given in Table 1.2 for two interacting spin waves except for $q_0 = \frac{K}{2}$ that occurs only for $K = \frac{\pi}{4}s$. This result is not surprising since we have seen that the interaction between spin waves prevents the states with two free spin waves to be eigenstates of the Heisenberg Hamiltonian.

The thermodynamic limit can be easily obtained from Eqs. (1.5.3) and (1.5.4). In the limit $N \to \infty$, Eq. (1.5.4) becomes an infinite set of equations with $r \geq 2$. Two kinds of solutions can be found that satisfy this infinite set. One is of the type

$$F_K(r) = f^{r-1}F_K(1) \tag{1.5.10}$$

with $r \geq 1$ and $f$ is a function of $\cos\frac{K}{2}$ to be determined. Note that the choice (1.5.10) reduces the infinite set (1.5.4) to the unique equation

$$\hbar\omega = 2h + 4J\left[1 - \frac{1+f^2}{2f}\cos\frac{K}{2}\right]. \tag{1.5.11}$$

Replacing Eq. (1.5.10) into the remaining Eq. (1.5.3), one obtains

$$\hbar\omega = 2h + 2J\left[1 - f\cos\frac{K}{2}\right]. \tag{1.5.12}$$

From Eqs. (1.5.11) and (1.5.12), one finds $f = \cos\frac{K}{2}$ and the complete solution of the infinite system (1.5.3) and (1.5.4) is

$$\hbar\omega = 2h + 2J\sin^2\frac{K}{2}, \tag{1.5.13}$$

$$F_K(r) = F_K(1)\left(\cos\frac{K}{2}\right)^{r-1} \tag{1.5.14}$$

and the probability of finding the two spin reversals at a distance $r$ becomes

$$P_K(r) = \sin^2\frac{K}{2}\left(\cos^2\frac{K}{2}\right)^{r-1} = \sin^2\frac{K}{2}\,e^{-\frac{r-1}{\xi}} \tag{1.5.15}$$

where $\xi = [-\ln(\cos^2\frac{K}{2})]^{-1}$. The probability (1.5.15) decays exponentially with the distance $r$ so that the maximum of probability corresponds to the case in which the two spin reversals are on NN sites. For $K \to 0$, the probability (1.5.15) reduces to $P_K(r) = \frac{1}{4}K^2 e^{-\frac{1}{4}(r-1)}$ that means the probability is spread over the whole ring. On the contrary, at the zone boundary ($K = \pi$), the probability is $P_K(1) = 1$ and $P_K(r) = 0$ for any $r \neq 1$, meaning that the spin reversals are located on NN sites.

The second solution is of the type

$$F_K(r) = A\cos(q_0 r)F_K(1) \tag{1.5.16}$$

with $r \geq 2$, where $A$ and and $q_0$ are functions to be determined. Once again, the choice (1.5.16) reduces the infinite set (1.5.4) to the unique equation

$$\hbar\omega = 2h + 4J\left(1 - \cos\frac{K}{2}\cos q_0\right) \tag{1.5.17}$$

that coincides with the energy of the two non-interacting spin waves given in Eq. (1.5.9). Replacing Eqs. (1.5.16) and (1.5.17) into the remaining Eq. (1.5.3), one obtains

$$A = \frac{2\cos\frac{K}{2}\cos q_0 - 1}{\cos\frac{K}{2}\cos 2q_0}. \tag{1.5.18}$$

The probability of finding two spin deviations at a distance $r$ is then

$$P_K(1) = \frac{1}{1 + A^2\sum_{r=2}^{\infty}\cos^2(q_0 r)} \to \frac{2}{A^2 N} \text{ for } N \to \infty \tag{1.5.19}$$

and

$$P_K(r) = \frac{A^2 \cos^2(q_0 r)}{1 + A^2 \sum_{r=2}^{\infty} \cos^2(q_0 r)} \to \frac{2}{N} \cos^2(q_0 r) \quad \text{for} \quad N \to \infty \tag{1.5.20}$$

with $r \geq 2$. The limit for $N \to \infty$ is obtained from the relationship

$$\frac{1}{N} \sum_{r=2}^{N} \cos^2(q_0 r) = \frac{1}{2} - \frac{1}{N} \cos^2 q_0. \tag{1.5.21}$$

As one can see from Eq. (1.5.20), the probability is ocsillating and no longer undergoes exponential decay like the first solution. It decreases as $N$ increases, indicating that the two deviations may be found at any distance with very small probability, in agremeent with a couple of delocalized deviations. The set of eigenvalues given in Eq. (1.5.17) with arbitrary $q_0$ confirms that this state corresponds to free spin wave whose energy is inside the two-magnon band $0 < q_0 < \pi$, in agreement with the expectation that in the thermodynamic limit, the eigenvalues with energy inside the two-magnon band are just the eigenvalues of two free spin waves.

In Fig. 1.5, we show the energy eigenvalues of a ring with $N = 8$ (crosses) corresponding to the numerical values of Table 1.1 and a ring with $N = 16$ (full circles) corresponding to the numerical values of 1.2. The continuous curve is the thermodynamic limit result given by Eq. (1.5.13) with $h = 0$. The scenario is clearer for $K$ near the zone boundary where the two-magnon band becomes narrower and reduces to a point at $K = \pi$. Indeed for a finite system, the eigenvalues belonging to a fixed wavevector $K$ group together within the two-magnon band delimited

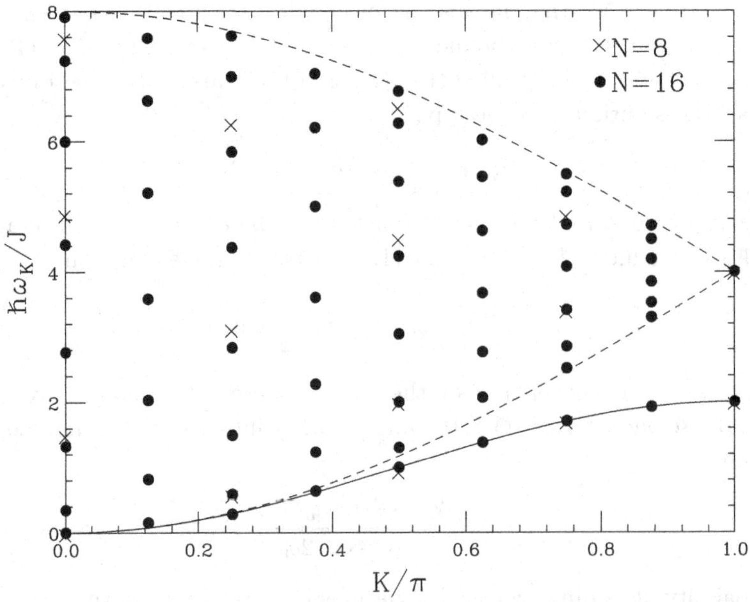


Fig. 1.5. Two spin wave energy $\hbar\omega/J$ versus $K/\pi$ for a ring of 8 (crosses) and 16 (full circle) spins $S = 1/2$. Dashed curves are the top and the bottom of the two-magnon band. Continuous curve is the bound state in the thermodynamic limit.

by the dashed lines shown in Fig. 1.5 except one eigenvalue that remains well below the band: this "isolated" eigenvalue approaches the two-magnon "bound state" in the thermodynamic limit (continuous curve).

It may be interesting to obtain the results shown in Fig. 1.5 which are given numerically in Tables 1.1 and 1.2 in a different way. Let us come back to Eq. (1.5.1): multiply both sides by $e^{-iqr}$ (with $q = \frac{2\pi}{N}l$ and $l = 0, 1, \ldots, N-1$) and sum over $r$ (with $r = 1, 2, \ldots, N-1$). We obtain

$$\left[\hbar\omega - 2h - 4J\left(1 - \cos\frac{K}{2}\cos q\right)\right]\sum_{r=1}^{N-1} F_K(r)e^{-iqr}$$
$$= 2J\left[\cos\frac{K}{2}\left(1 + e^{-i\frac{K}{2}N}\right) - \left(e^{-iq} + e^{iq}e^{-i\frac{K}{2}N}\right)\right] F_K(1). \quad (1.5.22)$$

The Fourier transforms (1.4.11) and (1.4.12) for the present 1D case reduce to

$$F_K(r) = \frac{1}{N}\sum_q e^{iqr} f_K(q) \quad (1.5.23)$$

and

$$f_K(q) = \sum_r e^{-iqr} F_K(r). \quad (1.5.24)$$

Equation (1.5.22) becomes

$$\left[\hbar\omega - 2h - 4J\left(1 - \cos\frac{K}{2}\cos q\right)\right] f_K(q)$$
$$= 4J\left(\cos\frac{K}{2} - \cos q\right)\frac{1}{N}\sum_{q'} e^{iq'} f_K(q') \quad (1.5.25)$$

for $K = \frac{4\pi}{N}s$ and

$$\left[\hbar\omega - 2h - 4J\left(1 - \cos\frac{K}{2}\cos q\right)\right] f_K(q) = 4iJ\sin q\frac{1}{N}\sum_{q'} e^{iq'} f_K(q') \quad (1.5.26)$$

for $K = \frac{2\pi}{N}(2s+1)$.

For $K = \frac{4\pi}{N}s$, one has the symmetry properties $F_K(N-r) = F_K(r)$ and $f_K(q) = f_K(2\pi - q)$ so that Eq. (1.5.25) becomes

$$\left(x + \cos\frac{K}{2}\cos\frac{2\pi}{N}l\right) f_K\left(\frac{2\pi}{N}l\right) = 4J\left(\cos\frac{K}{2} - \cos\frac{2\pi}{N}l\right) V_K \quad (1.5.27)$$

where

$$\hbar\omega = 2h + 4J(1 + x) \quad (1.5.28)$$

and

$$V_K = \frac{1}{N}\left[f_K(0) - f_K(\pi) + 2\sum_{l=1}^{\frac{N}{2}-1} f_K\left(\frac{2\pi}{N}l\right)\cos\frac{2\pi}{N}l\right]. \quad (1.5.29)$$

For $K = \frac{2\pi}{N}(2s+1)$, one has the symmetry properties $F_K(N-r) = -F_K(r)$ and $f_K(q) = -f_K(2\pi - q)$ so that Eq. (1.5.26) becomes

$$\left(x + \cos\frac{K}{2}\cos\frac{2\pi}{N}l\right) f_K\left(\frac{2\pi}{N}l\right) = 4J\, W_K \sin\frac{2\pi}{N}l \qquad (1.5.30)$$

where

$$W_K = \frac{2}{N}\sum_{l=1}^{\frac{N}{2}-1} f_K\left(\frac{2\pi}{N}l\right)\sin\frac{2\pi}{N}l. \qquad (1.5.31)$$

A "trivial" solution of Eq. (1.5.27) is obtained for $\frac{K}{2} = \frac{4\pi}{N}l$, that is, for $l = s$; the solution is

$$x = -\cos^2\frac{K}{2}. \qquad (1.5.32)$$

Note that Eq. (1.5.32) leads to the energy (1.5.9) of two non interacting spin waves with $q_0 = \frac{K}{2}$. This is a direct consequence of Eq. (1.5.25) where the interaction between the spin waves vanishes for $q = \frac{K}{2}$.

All the other solutions of Eq. (1.5.27) may be obtained as follows: for $x \neq -\cos\frac{K}{2}\cos\frac{2\pi}{N}l$, Equation (1.5.27) may be written as

$$f_K\left(\frac{2\pi}{N}l\right) = \frac{\cos\frac{K}{2} - \cos\frac{2\pi}{N}l}{x + \cos\frac{K}{2}\cos\frac{2\pi}{N}l} V_K. \qquad (1.5.33)$$

After having multiplied both sides of Eq. (1.5.33) by $\cos\frac{2\pi}{N}l$, performing the summation over $l$ and using the relationship (1.5.29), one obtains

$$[1 - M(K,x)]V_K = 0 \qquad (1.5.34)$$

where

$$M(K,x) = \frac{2}{N}\left[\frac{\cos^2\frac{K}{2} + x}{\cos^2\frac{K}{2} - x^2} + \sum_{l=1}^{\frac{N}{2}-1}\frac{\cos\frac{2\pi}{N}l\left(\cos\frac{K}{2} - \cos\frac{2\pi}{N}l\right)}{\cos\frac{K}{2}\cos\frac{2\pi}{N}l + x}\right]. \qquad (1.5.35)$$

The solutions of Eq. (1.5.34) are the zeros of the function $[1 - M(K,x)]$. For $K = \pi$ $(s = N/2)$, the solution (1.5.32) gives $x = 0$ while Eq. (1.5.35) reduces to $M(\pi, x) = -\frac{1}{2x}$ when making use of the identity[3] $\sum_{l=1}^{N/2-1}\cos^2\frac{2\pi}{N}l = \frac{N}{4} - 1$ so that Eq. (1.5.34) has only one zero for $x = -\frac{1}{2}$. Then, from Eq. (1.5.28) one has $\hbar\omega = 2h + 4J$ and $2h + 2J$ for $x = 0$ and $x = -\frac{1}{2}$, respectively, reflecting the property that at zone boundary, a bound state exists below the two-magnon band, the width of which is reduced to a single point as shown in Fig. 1.5. For $K \neq \pi$, the function $[1 - M(K,x)]$ is characterized by vertical asymptotes at $x = -\cos\frac{K}{2}\cos\frac{2\pi}{N}l$ for $l = 0, 1, \ldots, \frac{N}{2}$ with exclusion of $\frac{2\pi}{N}l = \frac{K}{2}$ and $l = \frac{N}{4}$ for which the vanishing of the numerators in the sum of Eq. (1.5.35) prevents any divergence. Fig. 1.6 shows the function $(1 - M)$ for $N = 16$ and $K = \frac{\pi}{4}$. Qualitatively, similar plots are obtained for $K = \frac{\pi}{4}s$ with $s = 0, 1, 2, 3$. On the contrary, for $s = 4$ $(K = \pi)$, the function $[1 - M(\pi, x)]$ reduces to the hyperbola $1 + \frac{1}{2x}$. To recover all the solutions of Eq. (1.5.27), the solution

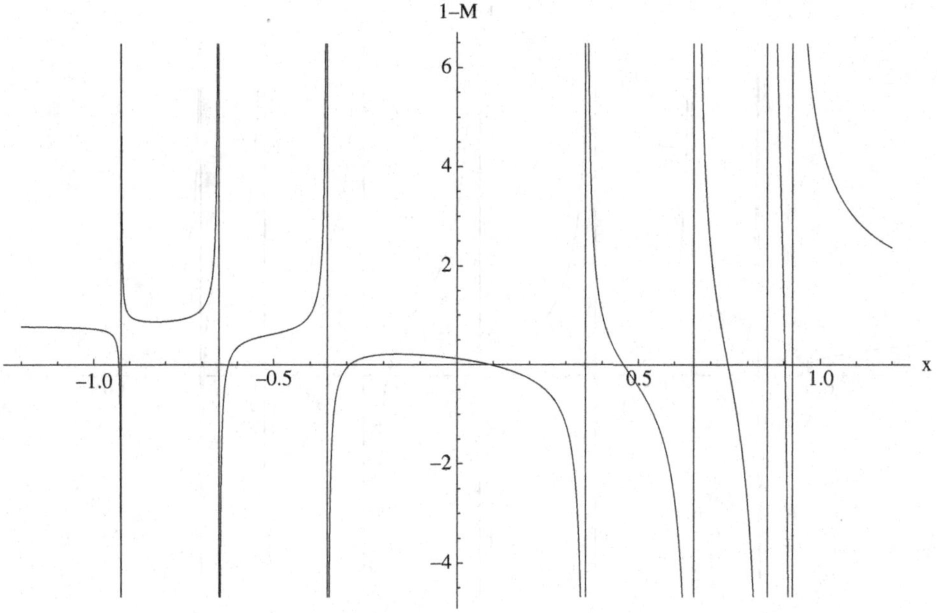


Fig. 1.6. $[1 - M(K,x)]$ vs $x$ for $N = 16$ and $K = \frac{\pi}{4}$.

$x = -\cos^2\frac{\pi}{8}$ obtained from Eq. (1.5.32) has to be taken into account. By means of similar steps, Eq. (1.5.30) leads to the equation

$$[1 + R(K,x)]W_K = 0 \tag{1.5.36}$$

where

$$R(K,x) = \frac{2}{N}\sum_{l=1}^{\frac{N}{2}-1}\frac{\sin^2\frac{2\pi}{N}l}{\cos\frac{K}{2}\cos\frac{2\pi}{N}l + x}. \tag{1.5.37}$$

The solutions of Eq. (1.5.36) are the zeros of the function $[1 + R(K,x)]$ shown in Fig. 1.7 for $N = 16$ and $K = \frac{3\pi}{8}$. Like $(1 - M)$ also $(1 + R)$ is characterized by vertical asymptotes at $x = -\cos\frac{K}{2}\cos\frac{\pi}{8}l$ with $l = 1, \ldots, 7$ and goes to 1 for $x \to \pm\infty$. The zeros of the function $(1 - M)$ shown in Fig. 1.6 and those of the function $(1 + R)$ shown in Fig. 1.7 lead to energies that coincide with the third $(K = \pm\frac{\pi}{4})$ and fourth line $(K = \pm\frac{3\pi}{8})$ of Table 1.2 for $h = 0$. Obviously, all the energy eigenvalues of Table 1.2 are recovered by evaluating the solutions of the equation $[1 - M(K,x)] = 0$ for $K = \frac{\pi}{4}s$ with $s = 0, 1, 2, 3, 4$ and of the equation $[1 + R(K,x)] = 0$ for $K = \frac{\pi}{8}(2s + 1)$ with $s = 0, 1, 2, 3$ and using the relationship $\hbar\omega = 4J(1 + x)$. Of course, for $K = \frac{\pi}{4}s$, one more solution given by $x = -\cos^2\frac{K}{2}$ has to be accounted for.

One can see that the number of zeros of Eqs. (1.5.34) and (1.5.36) increases as $N$ increases. Even though a numerical evaluation of the zeros is in order, we can deduce some important restrictions about the locations of the zeros of the functions

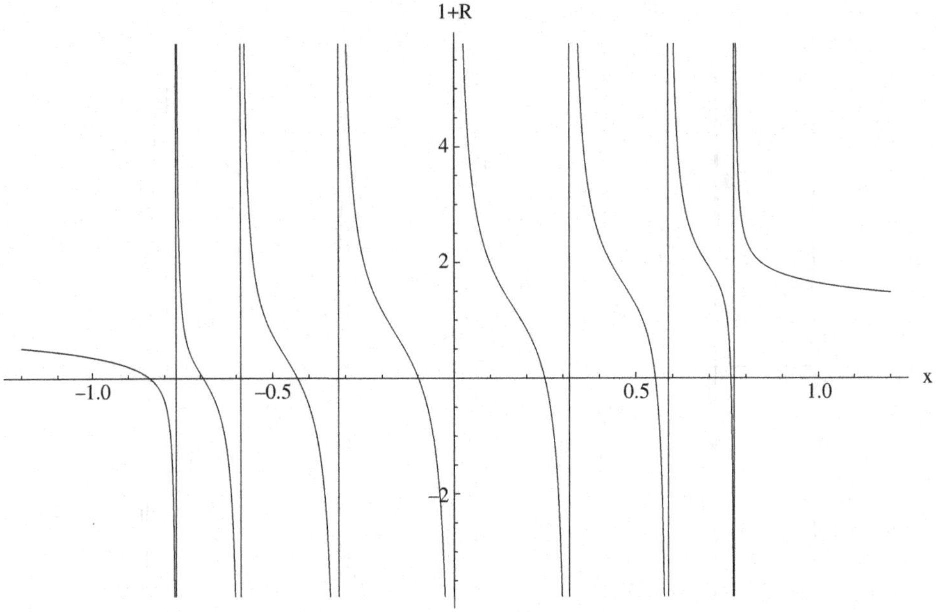


Fig. 1.7. $[1 + R(K, x)]$ vs $x$ for $N = 16$ and $K = \frac{3\pi}{8}$.

$(1 - M)$ and $(1 + R)$. For a fixed value of $K = \frac{4\pi}{N}s$, the solutions of Eq. (1.5.27) are subject to the following unequalities

$$x_0 \leq -\cos\frac{2\pi}{N}s, \tag{1.5.38}$$

$$-\cos\left(\frac{2\pi}{N}s\right)\cos\frac{2\pi}{N}(l-1) < x_l < -\cos\left(\frac{2\pi}{N}s\right)\cos\frac{2\pi}{N}l \tag{1.5.39}$$

for $0 < l < s$,

$$x_s = -\cos^2\frac{2\pi}{N}s \tag{1.5.40}$$

for $l = s$ and

$$-\cos\left(\frac{2\pi}{N}s\right)\cos\frac{2\pi}{N}l < x_l < -\cos\left(\frac{2\pi}{N}s\right)\cos\frac{2\pi}{N}(l+1) \tag{1.5.41}$$

for $s < l \leq \frac{N}{2} - 1$.

Analogously, for a fixed value of $K = \frac{2\pi}{N}(2s+1)$, the solutions of Eq. (1.5.30) are subject to the following unequalities

$$x_0 \leq -\cos\frac{2\pi}{N}(2s+1), \tag{1.5.42}$$

$$-\cos\left[\frac{2\pi}{N}(2s+1)\right]\cos\frac{2\pi}{N}l < x_l < \cos\left[\frac{2\pi}{N}(2s+1)\right]\cos\frac{2\pi}{N}(l+1) \tag{1.5.43}$$

for $0 < l \leq \frac{N}{2} - 2$.

The unequalities (1.5.38) and (1.5.42) show that the lowest solution ($x_0$) is always on the left of the first asymptote since $x_0 < -\cos\frac{K}{2}$ and corresponds to the bound state: $x_0 = x_{\rm BS}$. The location of $x_0$ is approximately independent of $N$ and it remains well apart from all the other zeros that accumulate as $N$ increases within the two-magnon band $-\cos\frac{K}{2} < x_l < \cos\frac{K}{2}$. More specifically, the unequalities (1.5.39), (1.5.41) and (1.5.43) imply that for large $N$,

$$x_l \to -\cos\frac{K}{2}\cos\frac{2\pi}{N}l \tag{1.5.44}$$

leading to the energy (1.5.17) corresponding to the energy of two free spin waves.

For comparison with the exact results obtained by solving Eqs. (1.5.34) and (1.5.36), we give the first-order results obtained by strictly following the steps of Section 1.4. The first-order solutions of Eqs. (1.5.27) and (1.5.30) are obtained after replacing $x$ by $-\cos q_0 \cos\frac{K}{2} + \frac{\Delta_K}{N}$ and neglecting $\frac{\Delta_K}{N}$ in the denominators of $f_K(q)$, $M(K, q_0)$ and $R(K, q_0)$. For $K = \frac{4\pi}{N}s$, one has $f_K(q_0) = \frac{N}{2}$ and

$$f_K^{(1)}(q) = \frac{\cos\frac{K}{2} - \cos q}{\cos\frac{K}{2}(\cos q - \cos q_0)} \frac{\cos q_0}{1 - M^{(1)}(K, q_0)} \tag{1.5.45}$$

for $q \neq q_0$, where

$$M^{(1)}(K, q_0) = \frac{2}{N}\left[\frac{\cos\frac{K}{2} - \cos q_0}{\cos\frac{K}{2}\sin^2 q_0} + \sum_{q\neq q_0} \frac{\cos q\,(\cos\frac{K}{2} - \cos q)}{\cos\frac{K}{2}(\cos q - \cos q_0)}\right]. \tag{1.5.46}$$

The first-order correction to the energy of the two free spin waves becomes

$$\Delta_K^{(1)} = 8J\left(\cos\frac{K}{2} - \cos q_0\right)\frac{\cos q_0}{1 - M^{(1)}(K, q_0)}. \tag{1.5.47}$$

For $K = \frac{2\pi}{N}(2s+1)$, one has $f_K(q_0) = \frac{N}{2}$ and

$$f_K^{(1)}(q) = -\frac{\sin q \sin q_0}{\cos\frac{K}{2}(\cos q - \cos q_0)} \frac{\cos q_0}{1 + R^{(1)}(K, q_0)} \tag{1.5.48}$$

for $q \neq q_0$, where

$$R^{(1)}(K, q_0) = \frac{2}{N}\sum_{q\neq q_0} \frac{\sin^2 q}{\cos\frac{K}{2}(\cos q - \cos q_0)}. \tag{1.5.49}$$

The first-order correction to the energy becomes

$$\Delta_K^{(1)} = -8J\frac{\sin^2 q_0}{1 + R^{(1)}(K, q_0)}. \tag{1.5.50}$$

Let us perform the sums over $q$ in Eqs. (1.5.46) and (1.5.49) for large $N$. Let us begin with the sum of Eq. (1.5.46). The first term on the right hand-side can be neglected because it is multiplied by a factor $\frac{2}{N}$ and the sum over $q$ with $q \neq q_0$

reduces to the principal value of the corresponding integral. Indeed, for large $N$, defining $\epsilon = \frac{2\pi}{N}$ and noticing that $\Delta q = \frac{2\pi}{N}$, Eq. (1.5.46) becomes

$$M^{(1)}(K,q_0) = \frac{1}{\pi}\int_0^{q_0-\epsilon} dq \frac{\cos q\left(\cos\frac{K}{2} - \cos q\right)}{\cos\frac{K}{2}(\cos q - \cos q_0)} + \frac{1}{\pi}\int_{q_0+\epsilon}^{\pi} dq \frac{\cos q\left(\cos\frac{K}{2} - \cos q\right)}{\cos\frac{K}{2}(\cos q - \cos q_0)}$$
$$\equiv \frac{\mathrm{P}}{\pi}\int_0^{\pi} dq \frac{\cos q\left(\cos\frac{K}{2} - \cos q\right)}{\cos\frac{K}{2}(\cos q - \cos q_0)} = 1 - \frac{\cos q_0}{\cos\frac{K}{2}} \tag{1.5.51}$$

where P (principal value) makes the integral in Eq. (1.5.51) convergent.[3] In a similar way, for large $N$, Eq. (1.5.49) reduces to

$$R^{(1)}(K,q_0) = \frac{\mathrm{P}}{\pi}\int_0^{\pi} dq \frac{\sin^2 q}{\cos\frac{K}{2}(\cos q - \cos q_0)} = -\frac{\cos q_0}{\cos\frac{K}{2}}. \tag{1.5.52}$$

Replacing Eqs. (1.5.51) and (1.5.52) into Eqs. (1.5.47) and (1.5.50) respectively, the first-order energy of the two interacting spin waves of momenta $k_1 = \frac{K}{2} + q_0$ and $k_2 = \frac{K}{2} - q_0$ is given by

$$\hbar\omega = 4J\left(1 - \cos\frac{K}{2}\cos q_0\right) + \frac{8J}{N}\cos\frac{K}{2}\left(\cos\frac{K}{2} - \cos q_0\right) \tag{1.5.53}$$

for $K = \frac{4\pi}{N}s$ and

$$\hbar\omega = 4J\left(1 - \cos\frac{K}{2}\cos q_0\right) - \frac{8J}{N}\frac{\cos\frac{K}{2}\sin^2 q_0}{\cos\frac{K}{2} - \cos q_0} \tag{1.5.54}$$

for $K = \frac{2\pi}{N}(2s+1)$. Note that Eq. (1.5.53) can be obtained directly from Eq. (1.4.24) while Eq. (1.5.54) has no analogous in Section 1.4. This comes from the finite nature of the ring: indeed, from Eq. (1.4.10), one obtains the two equations (1.5.25) and (1.5.26) according to $K = \frac{4\pi}{N}s$ or $K = \frac{2\pi}{N}(2s+1)$. In the same way, from Eq. (1.4.10), one should obtain different equations for finite $N$ according to $K_\alpha = \frac{4\pi}{L_\alpha}s_\alpha$ or $K_\alpha = \frac{2\pi}{L_\alpha}(2s_\alpha + 1)$ with $\alpha = x, y, z$ and $L_x L_y L_z = N$. However, in view of the thermodynamic limit, the effect of the distinction between the different choice of $\boldsymbol{K}$ becomes irrelevant so that in Section 1.4, we have performed the calculation for a generic $\boldsymbol{K} = 2\pi(\frac{l_x}{L_x}, \frac{l_y}{L_y}, \frac{l_z}{L_z})$ that for $l_\alpha = 0, 1, \ldots, L_\alpha - 1$, recovers all possible wavevectors leading to Eq. (1.4.15).

Another discrepancy appears when the first-order results given by Eqs. (1.5.53) and (1.5.54) are compared to the exact results obtained from the equations $[1 - M(\frac{4\pi}{N}s, x)] = 0$ and $[1 + R(\frac{2\pi}{N}(2s+1), x)] = 0$: we have seen that the number of the exact energy eigenvalues is $\frac{N(N-1)}{2}$ which correspond to the number of ways of putting a couple of spin deviations on two different sites. On the contrary, the first-order result leads to $\frac{N^2}{2}$ possible solutions. Also, this discrepancy disappears in the thermodynamic limit.

To conclude, in the study of the ring, we evaluate explicitly the bound state solution $x_0 < -\cos\frac{K}{2}$ for large $N$ where the sums can be replaced by integrals. For $K = \frac{4\pi}{N}s$, the bound state is obtained from the solution of the equation $[1 - M(K,x)] = 0$ where

$$M(K,x) = \frac{1}{\pi}\int_0^\pi \frac{\cos q\left(\cos\frac{K}{2} - \cos q\right)}{x + \cos\frac{K}{2}\cos q} = \left(1 + \frac{x}{\cos^2\frac{K}{2}}\right)[1 - xI_0(K,x)] \tag{1.5.55}$$

with

$$I_0(K,x) = \frac{1}{\pi}\int_0^\pi \frac{1}{x + \cos\frac{K}{2}\cos q}. \tag{1.5.56}$$

For $K = \frac{2\pi}{N}(2s+1)$, the bound state is obtained from the solution of the equation $[1 + R(K,x)] = 0$ where

$$R(K,x) = \frac{1}{\pi}\int_0^\pi \frac{\sin^2 q}{x + \cos\frac{K}{2}\cos q} = \left(1 - \frac{x^2}{\cos^2\frac{K}{2}}\right)I_0(K,x) + \frac{x}{\cos^2\frac{K}{2}}. \tag{1.5.57}$$

Replacing Eq. (1.5.55) into the equation $[1 - M(K,x)] = 0$ and Eq. (1.5.57) into the equation $[1 + R(K,x)] = 0$ and making use of the relationship[3]

$$I_0(K,x) = -\frac{1}{\sqrt{x^2 - \cos^2\frac{K}{2}}} \tag{1.5.58}$$

which is true for $x < -\cos\frac{K}{2}$, one finds the same solution for both equations, that is

$$x = -\frac{1}{2}\left(1 + \cos^2\frac{K}{2}\right) \tag{1.5.59}$$

leading to

$$\hbar\omega_{\mathrm{BS}} = 2h + 2J\sin^2\frac{K}{2} \tag{1.5.60}$$

that coincides with the bound state energy given by Eq. (1.5.13).

## 1.6. Spin Waves in Classical Mechanics

In this section, we will study the classical version of the spin waves that is the classical counterpart of the quantum spin waves obtained in the previous section. As for the theory of small oscillations in solids that leads to the classical normal modes or phonons in quantum mechanics, the classical spin waves may be seen as rotations of the spin vectors around the direction of the external magnetic field with a frequency which is the spin wave frequency. The spin vector at each lattice site sweeps the surface of a cone with a small apex angle with the same frequency and a phase difference constant, depending on the distance between the spins considered. These collective excitations are named spin waves or magnons and represent the magnetic normal modes of a ferromagnet while the phonons represent the vibrational normal modes of an elastic medium. First of all, let us replace the spin operators by usual angular momentum vectors $\hbar \boldsymbol{S}_j$ and assume that the magnetic moment localized on site $j$ is proportional to the angular momentum on that site, that is $\boldsymbol{\mu}_j = g\mu_B \boldsymbol{S}_j$. The effective magnetic field experienced by the magnetic moment on the site $j$ is the sum of the external magnetic field and the effective field due to the exchange interaction with the NN magnetic moments

$$\boldsymbol{H}_{\text{tot}} = \boldsymbol{H} + \boldsymbol{H}_j = \boldsymbol{H} + \frac{2J}{g\mu_B} \sum_{\boldsymbol{\delta}} \boldsymbol{S}_{j+\boldsymbol{\delta}}. \tag{1.6.1}$$

The Hamiltonian of a magnetic moment in the presence of a magnetic field is given by $-\boldsymbol{\mu}_j \cdot \boldsymbol{H}_{\text{tot}}$. If we extend the sum over all the lattice sites, we obtain

$$\mathcal{H} = -\sum_j \boldsymbol{\mu}_j \cdot \left[\boldsymbol{H} + \frac{1}{2}\boldsymbol{H}_j\right] = -g\mu_B \boldsymbol{H} \cdot \sum_j \boldsymbol{S}_j - J \sum_{i,\boldsymbol{\delta}} \boldsymbol{S}_j \cdot \boldsymbol{S}_{i+\boldsymbol{\delta}} \tag{1.6.2}$$

where the factor $\frac{1}{2}$ is necessary to avoid a double counting of the interaction with the effective field due to the NN moments. The torque equation of the classical mechanics is then

$$\frac{d}{dt}\left(\hbar \boldsymbol{S}_j\right) = \boldsymbol{\mu}_j \times \boldsymbol{H}_{\text{tot}}. \tag{1.6.3}$$

Using Eq. (1.6.1), the torque equation becomes

$$\frac{d\boldsymbol{S}_j}{dt} = \frac{g\mu_B}{\hbar} \boldsymbol{S}_j \times \boldsymbol{H} + \frac{2J}{\hbar} \sum_{\boldsymbol{\delta}} \boldsymbol{S}_j \times \boldsymbol{S}_{j+\boldsymbol{\delta}}. \tag{1.6.4}$$

Supposing that the external magnetic field is directed along the $z$-axis, the vector equation (1.6.4) reduces the three scalar equations

$$\frac{dS_j^x}{dt} = \frac{g\mu_B}{\hbar} S_j^y H + \frac{2J}{\hbar}\left[S_j^y \sum_{\boldsymbol{\delta}} S_{j+\boldsymbol{\delta}}^z - S_j^z \sum_{\boldsymbol{\delta}} S_{j+\boldsymbol{\delta}}^y\right], \tag{1.6.5}$$

$$\frac{dS_j^y}{dt} = -\frac{g\mu_B}{\hbar} S_j^x H - \frac{2J}{\hbar}\left[S_j^x \sum_{\boldsymbol{\delta}} S_{j+\boldsymbol{\delta}}^z - S_j^z \sum_{\boldsymbol{\delta}} S_{j+\boldsymbol{\delta}}^x\right], \tag{1.6.6}$$

$$\frac{dS_j^z}{dt} = \frac{2J}{\hbar}\left[S_j^x \sum_{\boldsymbol{\delta}} S_{j+\boldsymbol{\delta}}^y - S_j^y \sum_{\boldsymbol{\delta}} S_{j+\boldsymbol{\delta}}^x\right]. \tag{1.6.7}$$

Assuming that each spin makes a precession motion around the direction of the external magnetic field, one may writes its components as follows

$$S_j^x = u\ \cos(\boldsymbol{k} \cdot \boldsymbol{r}_j - \omega t), \tag{1.6.8}$$

$$S_j^y = u\ \sin(\boldsymbol{k} \cdot \boldsymbol{r}_j - \omega t) \tag{1.6.9}$$

and

$$S_j^z = \sqrt{S^2 - u^2}. \tag{1.6.10}$$

It is direct to prove that Eqs. (1.6.8)–(1.6.10) are solutions of Eqs. (1.6.5)–(1.6.7) for any $u$, provided that

$$\hbar\omega = g\mu_B H + 2J\sqrt{S^2 - u^2} \sum_{\boldsymbol{\delta}} (1 - \cos \boldsymbol{k} \cdot \boldsymbol{\delta}). \tag{1.6.11}$$

The solution (1.6.8)–(1.6.11) corresponds to a precession of each magnetic moment about the $z$-axis sweeping the surface of a cone of height $\sqrt{S^2 - u^2}$ and apex angle $\alpha = \arctan \frac{u}{S}$. The point of the vector $\boldsymbol{S}_i$ describes a uniform circular motion along the circumference of radius $u$ in a period $T = \frac{2\pi}{\omega}$. The magnetic moments located on the sites $\boldsymbol{r}_i$ and $\boldsymbol{r}_j$ have a constant phase difference given by $\theta = \boldsymbol{k} \cdot (\boldsymbol{r}_i - \boldsymbol{r}_j)$. The wavevectors $\boldsymbol{k}$ are determined by the PBC and by the lattice structure: their domain is extended over the first Brillouin zone (BZ) of the reciprocal lattice. In classical mechanics, the arbitrary value of $u$ is fixed by the initial conditions. As one can see from Eq. (1.6.11), the precession frequency coincides with the quantum spin wave frequency given in Eq. (1.3.21) if $u$ is small compared to $S$. A complete solution of the system of differential equations (1.6.5)–(1.6.7) may be obtained, neglecting terms proportional to $u^2$. This approximation implies that $S_i^z \simeq S$ and Eq. (1.6.7) is identically satisfied; Eqs. (1.6.5) and (1.6.6) become

$$\frac{dS_j^x}{dt} = \frac{g\mu_B}{\hbar} S_j^y H + \frac{2JS}{\hbar} \sum_{\boldsymbol{\delta}} (S_j^y - S_{j+\boldsymbol{\delta}}^y) \tag{1.6.12}$$

and

$$\frac{dS_j^y}{dt} = -\frac{g\mu_B}{\hbar} S_j^x H - \frac{2JS}{\hbar} \sum_{\boldsymbol{\delta}} (S_j^x - S_{j+\boldsymbol{\delta}}^x), \tag{1.6.13}$$

respectively. The general solution of Eqs. (1.6.12)–(1.6.13) is a superposition of normal modes given by

$$S_j^x = \sum_{\boldsymbol{k}} u_{\boldsymbol{k}} \cos(\boldsymbol{k} \cdot \boldsymbol{r}_j - \omega_{\boldsymbol{k}} t) \tag{1.6.14}$$

and

$$S_j^y = \sum_{\boldsymbol{k}} u_{\boldsymbol{k}} \sin(\boldsymbol{k} \cdot \boldsymbol{r}_j - \omega_{\boldsymbol{k}} t) \tag{1.6.15}$$

that looks like the superposition of the normal modes of a vibrating crystal in harmonic approximation.

## 1.7. Heisenberg Hamiltonian for Actual Compounds

The Hamiltonian (1.2.5) is the simplest Hamiltonian for an isotropic insulating ferromagnet. The most part of the theory of magnetism is based on the Hamiltonian (1.2.5) even though actual compounds cannot be described by a so simple Hamiltonian. For instance, in the transition metals like iron (Fe), cobalt (Co) or nikel (Ni) which are characterized by ions having an incomplete $3d$-shell, the effect of the conduction electrons is non-negligible. The interaction between the conduction electrons and the magnetic moments of the localized ions may be described by an effective exchange interaction not restricted to NN spins known as RKKY interaction[8] from the initials of the names of the proposers: Ruderman, Kittel, Kasuya, Yosida. These authors used a second-order perturbation theory to describe the "indirect exchange coupling". The mechanism consists of the polarization by a localized magnetic ion of the conduction electrons which in their turn, interact with other magnetic ions of the lattice, leading to an effective long-range interaction between the localized magnetic moments. An analogous approach is used to obtain the spin waves in rare earths like europium (Eu), gadolinium (Gd), terbium (Tb), dysprosium (Dy), holmium (Ho) and erbium (Er) where the magnetic ions are characterized by an incomplete $4f$-shell. For such elements, the Hamiltonian (1.2.5) is used with exchange integrals not restricted to NN but chosen to fit experimental data.[9]

The low temperature behaviour of some actual compounds is successfully described, adding anisotropic terms to the isotropic Hamiltonian (1.2.5). Typical anisotropic contributions to the spin Hamiltonian of a ferromagnet are the single-ion and exchange terms. The single-ion anisotropy due to the effect of a crystal field is described by the Hamiltonian

$$\mathcal{H}_D = -D\sum_i (S_i^z)^2. \tag{1.7.1}$$

This type of anisotropy forces the spins along the $z$ "easy-axis" for $D > 0$ or into the $xy$ "easy-plane" for $D < 0$. The simplest two-ion anisotropy is the exchange anisotropy described by the Hamiltonian

$$\mathcal{H}_K = -\sum_{i,\boldsymbol{\delta}} K\ S_i^z S_{i+\boldsymbol{\delta}}^z \tag{1.7.2}$$

that favours an alignement of the spins along $z$ for $K > 0$ (Ising-like symmetry) or a planar configuration for $K < 0$ (XY-like symmetry). The anisotropy contributions (1.7.1) and (1.7.2) may be grouped together into the anisotropic Hamiltonian

$$\begin{aligned}\mathcal{H} = &-\sum_{i,\boldsymbol{\delta}}\left[J^z\ S_i^z S_{i+\boldsymbol{\delta}}^z + \frac{1}{2}J^{\perp}(S_i^+ S_{i+\boldsymbol{\delta}}^- S_i^- S_{i+\boldsymbol{\delta}}^+)\right]\\ &-D\sum_i (S_i^z)^2 - h\sum_i S_i^z\end{aligned} \tag{1.7.3}$$

where $h = g\mu_B H$, $J^z = J + K$ and $J^{\perp} = J$. For easy-axis ferromagnets ($D > 0$ and $J^z > J^{\perp} > 0$), the spin wave spectrum (1.3.21) becomes

$$\hbar\omega_{\boldsymbol{k}} = h + D(2S-1) + 2J^z S\sum_{\boldsymbol{\delta}}\left(1 - \frac{J^{\perp}}{J^z}\cos\boldsymbol{k}\cdot\boldsymbol{\delta}\right). \tag{1.7.4}$$

As one can see from Eq. (1.7.4), the external magnetic field and the easy-axis single-ion anisotropy shift the spin wave spectrum upwards by an amount which is the same for any wavevector. On the contrary, the upwards shift caused by the exchange anisotropy is $\boldsymbol{k}$-dependent: its amount goes from a minimum $2J^z S\sum_{\boldsymbol{\delta}}(1 - J^{\perp}/J^z)$ at the zone centre to a maximum $2J^z S\sum_{\boldsymbol{\delta}}(1 + J^{\perp}/J^z)$ at the zone corner (ZC) $\boldsymbol{k}\cdot\boldsymbol{\delta} = \pi$. Replacing the Hamiltonian (1.3.8) by the anisotropic Hamiltonian (1.7.3),

Equation (1.4.10) for the two-spin waves energy, becomes

$$[\hbar\omega - 2h - 2D(2S-1) - 4zJ^z S]F_{\boldsymbol{K}}(\boldsymbol{r}) + 4J^\perp S \sum_{\boldsymbol{\delta}} e^{i\frac{\boldsymbol{K}}{2}\cdot\boldsymbol{\delta}} F_{\boldsymbol{K}}(\boldsymbol{r}+\boldsymbol{\delta})$$

$$= \left[2J^\perp(\boldsymbol{r})\cos\left(\frac{\boldsymbol{K}}{2}\cdot\boldsymbol{r}\right) - 2D\delta_{\boldsymbol{r},0}\right] F_{\boldsymbol{K}}(0) - 2J^z(\boldsymbol{r})F_{\boldsymbol{K}}(\boldsymbol{r}) \qquad (1.7.5)$$

and Eq. (1.4.15) for the "cubic lattices" becomes

$$[\hbar\omega - \hbar\omega_{\boldsymbol{K}}(\boldsymbol{q})]f_{\boldsymbol{K}}(\boldsymbol{q}) = 2D\tfrac{1}{N}\textstyle\sum_{\boldsymbol{q}'} f_{\boldsymbol{K}}(\boldsymbol{q}')$$

$$+4J^z \textstyle\sum_\alpha \cos q_\alpha \tfrac{1}{N}\sum_{\boldsymbol{q}'}\left(\cos\dfrac{K_\alpha}{2} - \dfrac{J^\perp}{J^z}\cos q'_\alpha\right) f_{\boldsymbol{K}}(\boldsymbol{q}') \qquad (1.7.6)$$

where

$$\hbar\omega_{\boldsymbol{K}}(\boldsymbol{q}) = 2h + 2D(2S-1) + 8J^z Sd - 8J^\perp S \sum_\alpha \cos\frac{K_\alpha}{2}\cos q_\alpha \qquad (1.7.7)$$

with $d = 1, 2, 3$ for LC, SQ, SC lattices, respectively. Using the definitions

$$G^\alpha_{\boldsymbol{K}} = \frac{1}{N}\sum_{\boldsymbol{q}}\left(\cos\frac{K_\alpha}{2} - \frac{J^\perp}{J^z}\cos q_\alpha\right) f_{\boldsymbol{K}}(\boldsymbol{q}) \qquad (1.7.8)$$

with $\alpha = x, y, z$ and

$$G^0_{\boldsymbol{K}} = \frac{1}{N}\sum_{\boldsymbol{q}} f_{\boldsymbol{K}}(\boldsymbol{q}), \qquad (1.7.9)$$

Equation (1.7.6) becomes

$$f_{\boldsymbol{K}}(\boldsymbol{q}) = -\left(\frac{D}{4J^\perp S}G^0_{\boldsymbol{K}} + \frac{J^z}{2J^\perp S}\sum_\alpha G^\alpha_{\boldsymbol{K}}\cos q_\alpha\right)\frac{1}{x + \sum_\sigma \cos\dfrac{K_\sigma}{2}\cos q_\sigma} \qquad (1.7.10)$$

where $\sigma = x, y, z$ and

$$x = \frac{\hbar\omega}{8J^\perp S} - \frac{h}{4J^\perp S} - \frac{D}{4J^\perp S} - d\frac{J^z}{J^\perp}. \qquad (1.7.11)$$

Multiplying both sides of Eq. (1.7.10) by $(\cos q_\beta - \frac{J^\perp}{J^z}\cos\frac{K_\beta}{2})$ where $\beta = x, y, z$ and summing over $\boldsymbol{q}$, one obtains the $d$ equations

$$\frac{D}{4J^\perp S}G^0_{\boldsymbol{K}}\left(I_\beta - \frac{J^\perp}{J^z}I_0\cos\frac{K_\beta}{2}\right) + \frac{J^z}{2J^\perp S}\sum_\alpha G^\alpha_{\boldsymbol{K}}\left(I_{\alpha\beta} - \frac{J^\perp}{J^z}I_\alpha\cos\frac{K_\beta}{2}\right) = 0 \qquad (1.7.12)$$

where

$$I_0 = \frac{1}{N}\sum_{\boldsymbol{q}}\frac{1}{x + \sum_\sigma \cos\frac{K_\sigma}{2}\cos q_\sigma}, \qquad (1.7.13)$$

$$I_\alpha = \frac{1}{N}\sum_{\boldsymbol{q}}\frac{\cos q_\alpha}{x + \sum_\sigma \cos\frac{K_\sigma}{2}\cos q_\sigma}. \qquad (1.7.14)$$

$$I_{\alpha\beta} = \frac{1}{N}\sum_{\boldsymbol{q}}\frac{\cos q_\alpha \cos q_\beta}{x + \sum_\sigma \cos\frac{K_\sigma}{2}\cos q_\sigma}. \qquad (1.7.15)$$

Moreover, summing over $\boldsymbol{q}$ on both sides of Eq. (1.7.10) one obtains

$$\left(1+\frac{D}{4J^{\perp}S}I_0\right)G^0_{\boldsymbol{K}}+\frac{J^z}{2J^{\perp}S}\sum_{\alpha}I_\alpha G^\alpha_{\boldsymbol{K}}=0. \tag{1.7.16}$$

The system made up of the $d$ equations (1.7.12) and the single equation (1.7.16) has non trivial solutions only if

$$\det\begin{pmatrix} 1+\frac{D}{4J^{\perp}S}I_0 & \frac{J^z}{2J^{\perp}S}I_x & \frac{J^z}{2J^{\perp}S}I_y & \frac{J^z}{2J^{\perp}S}I_z \\ \frac{D}{4J^{\perp}S}A_x & 1-\frac{1}{2S}B_{xx} & -\frac{1}{2S}B_{xy} & -\frac{1}{2S}B_{xz} \\ \frac{D}{4J^{\perp}S}A_y & -\frac{1}{2S}B_{yx} & 1-\frac{1}{2S}B_{yy} & -\frac{1}{2S}B_{yz} \\ \frac{D}{4J^{\perp}S}A_z & -\frac{1}{2S}B_{zx} & -\frac{1}{2S}B_{zy} & 1-\frac{1}{2S}B_{zz} \end{pmatrix}=0 \tag{1.7.17}$$

where

$$A_\alpha=I_\alpha-\frac{J^{\perp}}{J^z}I_0\cos\frac{K_\alpha}{2} \tag{1.7.18}$$

and

$$B_{\alpha\beta}=I_\alpha\cos\frac{K_\beta}{2}-\frac{J^z}{J^{\perp}}I_{\alpha\beta}. \tag{1.7.19}$$

The roots $x_{\rm BS}$ of the determinant equation (1.7.17) give the bound states of the anisotropic ferromagnet that we will study in more detail in Chapter 5. The probability of finding the two spin deviations at a distance $r$ is given by Eq. (1.4.34) and (1.4.35) with

$$F_{\boldsymbol{K}}(\boldsymbol{r})=\frac{1}{N}\sum_{\boldsymbol{q}}f_{\boldsymbol{K}}(\boldsymbol{q})\cos\boldsymbol{q}\cdot\boldsymbol{r} \tag{1.7.20}$$

where $f_{\boldsymbol{K}}(\boldsymbol{q})$ is given by Eq. (1.7.10).

For magnetic insulators like EuS, EuO, $BX_2$ and $ABX_3$, where A is an alkali metal like potassium (K), sodium (Na), cesium (Ce) or rubidium (Rb), B is a transition metal like vanadium (V), manganese (Mn) or copper (Cu) and X is a halogen like fluorine (F), chlorine (Cl), bromine (Br) or iodine (I), the anisotropic Hamiltonian (1.7.3) can be assumed to fit well with the magnetic excitations at low temperature.

In order to describe the ground state configuration of some heavy rare earths or some compounds of the family $BX_2$ or $ABX_3$, the extension of the exchange integral occurring in the Hamiltonian (1.2.5) to more distant neighbours with alternating sign can be used. In particular, the choice of an alternating sign of the exchange interactions moving from NN to more distant spins can give rise to non-collinear configurations like spiral or helix configurations that will be discussed in Chapter 7.

The sign of the exchange integral in Eq. (1.2.5) determines the kind of magnetic order in the ground state: for $J>0$, the ferromagnetic order is established and the ground state is characterized by an assembly of magnetic ions with all their angular momenta (spins) assuming the maximum value ($S$) and aligned in the same direction. For $J < 0$, an exact calculation of the ground state[10,11] and of the elementary excitations[12] is restricted to the LC with $S = 1/2$. Otherwise, approximate calculations are used: a reasonable approximation for the ground state is the "Neél state" in which all the magnetic moments point in the same direction but the spins located on the NN sites of a given spin are antiparallel to the spin in that site. Note that for $J < 0$, the order of the levels in Figs. 1.2–1.4 has to be reversed. For instance, the antiferromagnetic ground state of the cluster with 4 spins is given by Eq. (1.2.32): while the corresponding Neél state is given by the last two terms of Eq. (1.2.32). In any case, for an even number of spins, the antiferromagnetic ground state is always a singlet ($S_{\text{tot}} = 0$).

Note that the spin operators appearing in Hamiltonian (1.2.5) are "effective spin" operators that account for both the orbital and the spin contribution of the multi-electron ion. To determine the effective spin quantum number $S$ of the magnetic ion, one must evaluate the lowest energy levels of the ground-state multiplet of the "free" magnetic ion split by the crystalline field due to the NN non-magnetic ions by a perturbative method. The energy shift between the lowest levels so obtained, is of the order of the magnetic excitations, usually 0.01-0.1 eV. An example of such a calculation will be given in Chapter 9.

# Chapter 2

# SPIN WAVES IN FERROMAGNETS

## 2.1. Spin-Boson Transformation

The model Hamiltonian we will study in this chapter is the Heisenberg Hamiltonian for a ferromagnet, that is an exchange Hamiltonian with NN ferromagnetic interaction. The aim of the present section is to find a transformation from the spin operators to Bose "creation" and "destruction" operators which are essential to build a perturbative theory to infinite order. First of all, let us compare the matrix elements of the spin operators and the Bose operators. From the matrix elements given in Eqs. (1.3.6) and (1.3.7), one can write the spin operators $S_i^+$, $S_i^-$, $S_i^z$ and $\boldsymbol{S}_i^2$ as square matrices of dimension $(2S+1)$. For instance, for $S=1$, one obtains

$$S_i^+ = \begin{pmatrix} 0 & \sqrt{2} & 0 \\ 0 & 0 & \sqrt{2} \\ 0 & 0 & 0 \end{pmatrix}, \quad S_i^- = \begin{pmatrix} 0 & 0 & 0 \\ \sqrt{2} & 0 & 0 \\ 0 & \sqrt{2} & 0 \end{pmatrix}, \quad S_i^z = \begin{pmatrix} 1 & 0 & 0 \\ 0 & 0 & 0 \\ 0 & 0 & -1 \end{pmatrix}$$

$$\boldsymbol{S}_i^2 = \begin{pmatrix} 2 & 0 & 0 \\ 0 & 2 & 0 \\ 0 & 0 & 2 \end{pmatrix} = S(S+1)\begin{pmatrix} 1 & 0 & 0 \\ 0 & 1 & 0 \\ 0 & 0 & 1 \end{pmatrix}. \tag{2.1.1}$$

The Bose creation and destruction operators satisfy the commutation rules[5]

$$[a_i, a_j^+] = \delta_{i,j}, \quad [a_i, a_j] = [a_i^+, a_j^+] = 0. \tag{2.1.2}$$

In the *occupation number* representation where the operator number $n_i = a_i^+ a_i$ is diagonal, one has

$$\langle n'|a_i^+ a_i|n\rangle = n\delta_{n,n'}, \quad n \geq 0. \tag{2.1.3}$$

The matrix elements of $a_i$ e $a_i^+$ are given by

$$\langle n'|a_i|n\rangle = \sqrt{n}\delta_{n',n-1}, \quad \langle n'|a_i^+|n\rangle = \sqrt{n+1}\delta_{n',n+1}. \tag{2.1.4}$$

From the matrix elements (2.1.3) and (2.1.4), one sees that the Bose operators can be written as matrices of *infinite* dimension like

$$a_i = \begin{pmatrix} 0 & \sqrt{1} & 0 & \dots \\ 0 & 0 & \sqrt{2} & \dots \\ 0 & 0 & 0 & \dots \\ \vdots & \vdots & \vdots & \ddots \end{pmatrix}, \quad a_i^+ = \begin{pmatrix} 0 & 0 & 0 & \dots \\ \sqrt{1} & 0 & 0 & \dots \\ 0 & \sqrt{2} & 0 & \dots \\ \vdots & \vdots & \vdots & \ddots \end{pmatrix}$$

$$n_i = a_i^+ a_i = \begin{pmatrix} 0 & 0 & 0 & \dots \\ 0 & 1 & 0 & \dots \\ 0 & 0 & 2 & \dots \\ \vdots & \vdots & \vdots & \ddots \end{pmatrix}. \tag{2.1.5}$$

The great advantage of representing the spin operators $S_i^{\pm}$, $S_i^z$ in terms of Bose operators $a_i^+$, $a_i$ is that the commutation rules of the latter are *c*-numbers and this fact is crucial for building a series expansion of the Green function or propagator in a systematic way.[13] A spin-boson transformation has to satisfy the following requirements:

(1) The transformation must be *hermitian* such that the conjugation relations are preserved. This means that the raising and lowering spin operators written in terms of creation and destruction boson operators must be hermitian conjugate of each other.
(2) The transformation must be a *unitary* one in order to preserve the commutation rules between the spin operators given by Eq. (1.3.4) when the spin operators are expressed in terms of Bose operators for which the commutation rules are given by Eq. (2.1.2).
(3) The transformation must satisfy the equality between the matrix elements of the spin operators between the states $|S\,m\rangle$ and the matrix elements of their bosonic representations between the states $|n\rangle$.

The choice of such a transformation is not unique. The first spin-boson transformation was proposed long time ago by Holstein and Primakoff[14] (HP)

$$S_i^+ = \sqrt{2S}\left(1 - \frac{a_i^+ a_i}{2S}\right)^{\frac{1}{2}} a_i, \quad S_i^- = \sqrt{2S}a_i^+\left(1 - \frac{a_i^+ a_i}{2S}\right)^{\frac{1}{2}}, \quad S_i^z = S - a_i^+ a_i. \tag{2.1.6}$$

The HP transformation satisfies the properties (1)–(3) only if the restriction to the boson subspace with $n_i \le 2S$ is accounted for because of the presence of the square roots in Eq. (2.1.6). About twenty years later, a new transformation was proposed by Dyson[15] and Maleev[16] (DM) as follows:

$$S_i^+ = \sqrt{2S}\left(1 - \frac{a_i^+ a_i}{2S}\right) a_i, \quad S_i^- = \sqrt{2S}a_i^+ \quad \text{and} \quad S_i^z = S - a_i^+ a_i \tag{2.1.7}$$

with the restriction $n_i \leq 2S$ (physical states). The DM transformation does not satisfy requirement (1) since it is not hermitian as seen by Eq. (2.1.7). However, Dyson himself showed that even though the DM representation transforms the Heisenberg Hamiltonian into a non-hermitian boson Hamiltonian, the matrix elements of the Heisenberg Hamiltonian between the localized spin states are the same as the boson DM Hamiltonian between independent boson states, provided that only physical states are accounted for, that is $n_i \leq 2S$ (kinematical interaction). Moreover, Dyson showed that the release of the restriction $n_i \leq 2S$, that is the involvement of the *unphysical* states with $n_i \geq 2S$, does not affect the power law contributions in the low temperature results. Indeed, the effect of the unphysical states is proved to be lesser than $e^{-a\frac{T_c}{T}}$ where $T_c$ is the Curie temperature and $a$ is a numerical coefficient of the order of unity and independent of temperature.[15] A similar proof cannot be extended to the HP Hamiltonian.

## 2.2. Bosonic Approach to the Heisenberg Hamiltonian

The Heisenberg Hamiltonian in terms of the raising and lowering spin operators is given by Eq. (1.3.8). In the absence of the external magnetic field, Eq. (1.3.8) becomes

$$\mathcal{H} = -J\sum_{i,\delta}\left[S_i^z S_{i+\delta}^z + \frac{1}{2}(S_i^+ S_{i+\delta}^- + S_i^- S_{i+\delta}^+)\right]. \tag{2.2.1}$$

Now, let us replace the spin raising and lowering operators by the boson creation and destruction operators according to the HP transformation given by Eq. (2.1.6) or to the DM transformation given by Eq. (2.1.7). Let us begin with the HP transformation and expand the square roots in the HP transformation in powers of $a_i^+ a_i$ or even better, in a *normal ordered*[17] (NO) series expansion where all the creation operators $a_i^+$ are on the left of all the destruction operators $a_i$. So the HP transformation process can be written as

$$S_i^+ = \sqrt{2S}\left[1 - \frac{a_i^+ a_i}{4S} - \frac{(a_i^+ a_i)^2}{32S^2}\cdots\right] a_i = \sqrt{2S}\left\{1 - \left(1 - \sqrt{1 - \frac{1}{2S}}\right) a_i^+ a_i \right.$$
$$\left. + \left[1 - \sqrt{1 - \frac{1}{2S}} - \frac{1}{2}\left(1 - \sqrt{1 - \frac{1}{S}}\right)\right](a_i^+)^2 (a_i)^2 \cdots\right\} a_i, \tag{2.2.2}$$

$$S_i^- = \sqrt{2S} a_i^+\left[1 - \frac{a_i^+ a_i}{4S} - \frac{(a_i^+ a_i)^2}{32S^2}\cdots\right] = \sqrt{2S} a_i^+\left\{1 - \left(1 - \sqrt{1 - \frac{1}{2S}}\right) a_i^+ a_i \right.$$
$$\left. + \left[1 - \sqrt{1 - \frac{1}{2S}} - \frac{1}{2}\left(1 - \sqrt{1 - \frac{1}{S}}\right)\right](a_i^+)^2 (a_i)^2 \cdots\right\} \tag{2.2.3}$$

and

$$S_i^z = S - a_i^+ a_i. \tag{2.2.4}$$

The advantage of the NO expansion is that no terms with a smaller number of boson operators are produced by the higher order terms: this is not true for the usual expansion in powers of $(a_i^+ a_i)^n$. For instance, the term with 4 operators corresponding to $n = 2$ produces a NO term with 4 operators as well as a term with 2 operators: indeed, $a_i^+ a_i a_i^+ a_i = a_i^+ a_i^+ a_i a_i + a_i^+ [a_i, a_i^+] a_i = a_i^+ a_i^+ a_i a_i + a_i^+ a_i$. In the same way, the term with 6 operators ($n = 3$) produces NO terms with 6, 4 and 2 operators and so on. The truncation of the series (2.2.2) and (2.2.3) to the first few terms is reliable for large spin $S$ or most importantly, for small values of the average occupation number $\langle a_i^+ a_i \rangle$. The last condition is satisfied for any $S$ at low temperature where the number of spin deviations is small and the restriction $\langle n_i \rangle / 2S \leq 1$ may be neglected. Replacing the spin operators in the Hamiltonian (2.2.1) by Eqs. (2.2.2)–(2.2.4), one obtains

$$\mathcal{H}^{HP} = E_0^{HP} + \mathcal{H}_2^{HP} + \mathcal{H}_4^{HP} + \mathcal{H}_6^{HP} + \cdots \tag{2.2.5}$$

where

$$E_0^{HP} = -zJS^2 N, \tag{2.2.6}$$

$$\mathcal{H}_2^{HP} = JS \sum_{i,\delta} (a_i^+ a_i + a_{i+\delta}^+ a_{i+\delta} - a_{i+\delta}^+ a_i - a_i^+ a_{i+\delta}), \tag{2.2.7}$$

$$\mathcal{H}_4^{HP} = -J \sum_{i,\delta} \left[ \frac{1}{2} (a_i^+ a_{i+\delta}^+ a_i a_{i+\delta} + \text{h.c.}) - S \left( 1 - \sqrt{1 - \frac{1}{2S}} \right) \right.$$
$$\left. \times \, (a_{i+\delta}^+ a_i^+ a_i a_i + a_{i+\delta}^+ a_{i+\delta}^+ a_{i+\delta} a_i + \text{h.c.}) \right], \tag{2.2.8}$$

$$\mathcal{H}_6^{HP} = -JS \sum_{i,\delta} \left\{ \left[ 1 - \sqrt{1 - \frac{1}{2S}} - \frac{1}{2} \left( 1 - \sqrt{1 - \frac{1}{S}} \right) \right] (a_{i+\delta}^+ a_i^+ a_i^+ a_i a_i a_i \right.$$
$$+ a_{i+\delta}^+ a_{i+\delta}^+ a_{i+\delta}^+ a_{i+\delta} a_{i+\delta} a_i + \text{h.c.})$$
$$\left. + \left( 1 - \sqrt{1 - \frac{1}{2S}} \right)^2 (a_{i+\delta}^+ a_{i+\delta}^+ a_i^+ a_{i+\delta} a_i a_i + \text{h.c.}) \right\} \tag{2.2.9}$$

where " h.c." means "hermitian conjugate". Remember that the hermitian conjugate of a product of operators can be written as $(AB)^+ = B^+ A^+$. Notice that the usual HP expansion occurring in the textbooks is obtained from Eq. (2.2.8) replacing $S(1 - \sqrt{1 - \frac{1}{2S}})$ by the first few terms of its series expansion in powers of $\frac{1}{S}$ that is by $\frac{1}{4}(1 + \frac{1}{8S} + \frac{1}{32S^2})$ and from Eq. (2.2.9) replacing $S[1 - \sqrt{1 - \frac{1}{2S}} - \frac{1}{2}(1 - \sqrt{1 - \frac{1}{S}})]$ by $-\frac{1}{32S}(1 + \frac{3}{4S})$ and $S(1 - \sqrt{1 - \frac{1}{2S}})^2$ by $\frac{1}{16S}(1 + \frac{1}{4S})$. The main point to be noticed using the HP transformation is the generation of an *infinite* number of terms when the Heisenberg Hamiltonian is transformed into the equivalent boson Hamiltonian. On the contrary, the DM transformation (2.1.7) leads to an equivalent boson Hamiltonian consisting of only three terms given by

$$\mathcal{H}^{DM} = E_0^{DM} + \mathcal{H}_2^{DM} + \mathcal{H}_4^{DM} \tag{2.2.10}$$

where

$$E_0^{DM} = -zJS^2N, \tag{2.2.11}$$

$$\mathcal{H}_2^{DM} = JS\sum_{i,\boldsymbol{\delta}}(a_i^+a_i + a_{i+\boldsymbol{\delta}}^+a_{i+\boldsymbol{\delta}} - a_{i+\boldsymbol{\delta}}^+a_i - a_i^+a_{i+\boldsymbol{\delta}}) \tag{2.2.12}$$

and

$$\mathcal{H}_4^{DM} = -J\sum_{i,\boldsymbol{\delta}}\left[\frac{1}{2}\left(a_i^+a_{i+\boldsymbol{\delta}}^+a_ia_{i+\boldsymbol{\delta}} + \text{h.c.}\right) - \frac{1}{2}\left(a_{i+\boldsymbol{\delta}}^+a_i^+a_ia_i + a_i^+a_{i+\boldsymbol{\delta}}^+a_{i+\boldsymbol{\delta}}a_{i+\boldsymbol{\delta}}\right)\right]. \tag{2.2.13}$$

The ground-state energy $E_0^{DM}$ and the bilinear Hamiltonian $\mathcal{H}_2^{DM}$ coincide with those obtained by the HP transformation [compare Eq. (2.2.11) with (2.2.6) and Eq. (2.2.12) with (2.2.7)]. The interaction Hamiltonian is now reduced to the *single* term $\mathcal{H}_4^{DM}$ even though it is no longer hermitian as one can see from the second term of Eq. (2.2.13). In any case, this is not a serious problem because the DM boson Hamiltonian has been proven[15] to have the same matrix elements between the independent boson states as the original Heisenberg Hamiltonian between the spin states. Moreover, the kinematical interaction ($n_i \leq 2S$) can be ignored since it enters only exponentially small terms in the temperature expansion.

So far, we have considered localized spin deviations but the spin waves are delocalized excitations as we have seen in Section 1.3. For this reason, let us define the spin wave or *magnon* creation and destruction operators by using the Fourier transforms of the corresponding localized operators (2.1.2):

$$a_{\boldsymbol{k}} = \frac{1}{\sqrt{N}}\sum_i e^{-i\boldsymbol{k}\cdot\boldsymbol{r}_i}a_i, \quad a_{\boldsymbol{k}}^+ = \frac{1}{\sqrt{N}}\sum_i e^{i\boldsymbol{k}\cdot\boldsymbol{r}_i}a_i^+ \tag{2.2.14}$$

whose commutation rules, as a direct consequence of Eq. (2.1.2), are given by

$$[a_{\boldsymbol{k}}, a_{\boldsymbol{k}'}^+] = \delta_{\boldsymbol{k},\boldsymbol{k}'}, \quad [a_{\boldsymbol{k}}, a_{\boldsymbol{k}'}] = [a_{\boldsymbol{k}}^+, a_{\boldsymbol{k}'}^+] = 0. \tag{2.2.15}$$

Replacing the magnon operators (2.2.14) into the HP boson Hamiltonian (2.2.5), the spin wave or magnon Hamiltonian becomes

$$\mathcal{H}_2^{HP} = zJS\sum_{\boldsymbol{k}}[(1-\gamma_{\boldsymbol{k}}) + \text{c.c.}]a_{\boldsymbol{k}}^+a_{\boldsymbol{k}} \tag{2.2.16}$$

where "c.c." means "complex conjugate": for instance, the complex conjugate of $1-\gamma_{\boldsymbol{k}}$ is $1-\gamma_{\boldsymbol{k}}^*$ with

$$\gamma_{\boldsymbol{k}} = \frac{1}{z}\sum_{\boldsymbol{\delta}} e^{i\boldsymbol{k}\cdot\boldsymbol{\delta}} \tag{2.2.17}$$

where $z$ is the coordination number. The Hamiltonian (2.2.16) of non-interacting spin waves is called "harmonic" or "bilinear" Hamiltonian. The magnon-magnon interaction Hamiltonian is given by

$$\mathcal{H}_4^{HP} = -\frac{zJ}{N} \sum_{\boldsymbol{k}_1,\boldsymbol{k}_2,\boldsymbol{k}_3,\boldsymbol{k}_4} \delta_{\boldsymbol{k}_1+\boldsymbol{k}_2,\boldsymbol{k}_3+\boldsymbol{k}_4} v^{HP}_{\boldsymbol{k}_1,\boldsymbol{k}_2,\boldsymbol{k}_3,\boldsymbol{k}_4} a^+_{\boldsymbol{k}_1} a^+_{\boldsymbol{k}_2} a_{\boldsymbol{k}_3} a_{\boldsymbol{k}_4} \tag{2.2.18}$$

with

$$v^{HP}_{\boldsymbol{k}_1,\boldsymbol{k}_2,\boldsymbol{k}_3,\boldsymbol{k}_4} = \frac{1}{2}\gamma_{\boldsymbol{k}_1-\boldsymbol{k}_3} - S\left(1-\sqrt{1-\frac{1}{2S}}\right)(\gamma_{\boldsymbol{k}_1}+\gamma_{\boldsymbol{k}_2}) + \text{c.c.} \tag{2.2.19}$$

and

$$\mathcal{H}_6^{HP} = -\frac{zJ}{N^2} \sum_{\boldsymbol{k}_1,\boldsymbol{k}_2,\boldsymbol{k}_3,\boldsymbol{k}_4,\boldsymbol{k}_5,\boldsymbol{k}_6} \times \delta_{\boldsymbol{k}_1+\boldsymbol{k}_2+\boldsymbol{k}_3,\boldsymbol{k}_4+\boldsymbol{k}_5+\boldsymbol{k}_6} w^{HP}_{\boldsymbol{k}_1,\boldsymbol{k}_2,\boldsymbol{k}_3,\boldsymbol{k}_4,\boldsymbol{k}_5,\boldsymbol{k}_6} a^+_{\boldsymbol{k}_1} a^+_{\boldsymbol{k}_2} a^+_{\boldsymbol{k}_3} a_{\boldsymbol{k}_4} a_{\boldsymbol{k}_5} a_{\boldsymbol{k}_6} \tag{2.2.20}$$

where

$$w^{HP}_{\boldsymbol{k}_1,\boldsymbol{k}_2,\boldsymbol{k}_3,\boldsymbol{k}_4,\boldsymbol{k}_5,\boldsymbol{k}_6} = S\left(1-\sqrt{1-\frac{1}{2S}}\right)^2 \gamma_{\boldsymbol{k}_1+\boldsymbol{k}_2-\boldsymbol{k}_4} + S\left[1-\sqrt{1-\frac{1}{2S}}-\frac{1}{2}\left(1-\sqrt{1-\frac{1}{S}}\right)\right](\gamma_{\boldsymbol{k}_1}+\gamma_{\boldsymbol{k}_6}) + \text{c.c.} \tag{2.2.21}$$

To obtain Eqs. (2.2.16)–(2.2.21), the following relationship has been used

$$\frac{1}{N}\sum_i e^{i(\boldsymbol{k}_1+\boldsymbol{k}_2-\boldsymbol{k}_3-\boldsymbol{k}_4)\cdot \boldsymbol{r}_i} = \delta_{\boldsymbol{k}_1+\boldsymbol{k}_2,\boldsymbol{k}_3+\boldsymbol{k}_4}. \tag{2.2.22}$$

Replacing the magnon operators (2.2.14) into the DM Hamiltonian (2.2.10), one obtains

$$\mathcal{H}_2^{DM} = zJS\sum_{\boldsymbol{k}}[(1-\gamma_{\vec{k}}) + \text{c.c.}]a^+_{\boldsymbol{k}} a_{\boldsymbol{k}} \tag{2.2.23}$$

and

$$\mathcal{H}_4^{DM} = -\frac{zJ}{N} \sum_{\boldsymbol{k}_1,\boldsymbol{k}_2,\boldsymbol{k}_3,\boldsymbol{k}_4} \delta_{\boldsymbol{k}_1+\boldsymbol{k}_2-\boldsymbol{k}_3-\boldsymbol{k}_4} v^{DM}_{\boldsymbol{k}_1,\boldsymbol{k}_2,\boldsymbol{k}_3,\boldsymbol{k}_4} a^+_{\boldsymbol{k}_1} a^+_{\boldsymbol{k}_2} a_{\boldsymbol{k}_3} a_{\boldsymbol{k}_4} \tag{2.2.24}$$

where

$$v^{DM}_{\boldsymbol{k}_1,\boldsymbol{k}_2,\boldsymbol{k}_3,\boldsymbol{k}_4} = \frac{1}{2}[(\gamma_{\boldsymbol{k}_1-\boldsymbol{k}_3} - \gamma_{\boldsymbol{k}_1}) + \text{c.c.}]. \tag{2.2.25}$$

The magnon-magnon interaction potential $v^{DM}$ (dynamical interaction) goes to zero in the long wavelength limit [see Eq. (2.2.25)] so that the interaction between the spin waves of small wavevector becomes negligible. This is not the case for the interaction potential in the HP Hamiltonian as one can see from Eqs. (2.2.19) and (2.2.21). However, it was proven[18] that the results obtained from the boson HP Hamiltonian (2.2.16)–(2.2.21) reduce to those obtained from the DM Hamiltonian (2.2.23)–(2.2.25) if a grouping of all terms of the same order in $\frac{1}{S}$ is performed.

Note that the function $\gamma_{\boldsymbol{k}}$ is real for lattices with inversion symmetry where for each spin at the site $i+\boldsymbol{\delta}$ exists a spin located at $i-\boldsymbol{\delta}$. In particular, for all Bravais lattices we have $\gamma_{\boldsymbol{k}} = \gamma_{-\boldsymbol{k}} = \gamma_{\boldsymbol{k}}^*$ so that everywhere, the sign "c.c." can be neglected provided that the spin wave potentials are multiplied by a factor 2.

Let us conclude this section by writing the boson representation of the anisotropic Hamiltonians given by Eqs. (1.7.1)–(1.7.3):

$$\mathcal{H}_H = -h\sum_i S_i^z = -hSN + h\sum_{\vec{k}} a_{\boldsymbol{k}}^+ a_{\boldsymbol{k}}, \tag{2.2.26}$$

$$\begin{aligned}\mathcal{H}_D = -D\sum_i (S_i^z)^2 &= -DS^2N + D(2S-1)\sum_{\boldsymbol{k}} a_{\boldsymbol{k}}^+ a_{\boldsymbol{k}} \\ &\quad - \frac{D}{N}\sum_{\boldsymbol{k}_1,\boldsymbol{k}_2,\boldsymbol{k}_3,\boldsymbol{k}_4} \delta_{\boldsymbol{k}_1+\boldsymbol{k}_2,\boldsymbol{k}_3+\boldsymbol{k}_4} a_{\boldsymbol{k}_1}^+ a_{\boldsymbol{k}_2}^+ a_{\boldsymbol{k}_3} a_{\boldsymbol{k}_4}\end{aligned} \tag{2.2.27}$$

and

$$\begin{aligned}\mathcal{H}_K = -K\sum_{i,\boldsymbol{\delta}} S_i^z S_{i+\boldsymbol{\delta}}^z &= -zKS^2N + 2zKS\sum_{\boldsymbol{k}} a_{\boldsymbol{k}}^+ a_{\boldsymbol{k}} \\ &\quad - \frac{zK}{N}\sum_{\boldsymbol{k}_1,\boldsymbol{k}_2,\boldsymbol{k}_3,\boldsymbol{k}_4} \delta_{\boldsymbol{k}_1+\boldsymbol{k}_2,\boldsymbol{k}_3+\boldsymbol{k}_4} \gamma_{\boldsymbol{k}_1-\boldsymbol{k}_3} a_{\boldsymbol{k}_1}^+ a_{\boldsymbol{k}_2}^+ a_{\boldsymbol{k}_3} a_{\boldsymbol{k}_4}.\end{aligned} \tag{2.2.28}$$

Notice that the anisotropic Hamiltonians (2.2.26)–(2.2.28) are functions of the $z$ component of the spin operators so that they are the same in both HP and DM representations. This is no longer true for easy-plane anisotropic Hamiltonians ($D, K < 0$) for which the lowering and raising spin operators enter the Hamiltonians (2.2.27) and (2.2.28) since in this case, the magnetization lies in the $xy$ easy-plane.

As a summary, we stress that the DM transformation minimizes both the kinematical and the dynamical interaction in the limit of long wavelengths. The rigorous proof, however, is restricted to isotropic ferromagnets with NN exchange interaction even if it is reasonably believed that a similar conclusion can hold for other magnetic systems like antiferromagnets, ferrimagnets, helimagnets and anisotropic magnetic systems. In many actual calculations, both the kinematical and dynamical interactions are neglected (harmonic approximation) since only the low temperature region ($T \ll T_c$) is considered.

## 2.3. Harmonic Approximation

The harmonic approximation consists of keeping only the bilinear boson Hamiltonian: as one can see from Eqs. (2.2.16) and (2.2.23), the harmonic approximation is independent of the HP or DM transformation: in the harmonic approximation, the magnetic system is reduced to an ideal gas of non-interacting magnons whose Hamiltonian reads

$$\mathcal{H}_0 = \mathcal{H}_2^{HP} = \mathcal{H}_2^{DM} = \sum_{\boldsymbol{k}} \hbar\omega_{\boldsymbol{k}} a_{\boldsymbol{k}}^+ a_{\boldsymbol{k}} \tag{2.3.1}$$

where

$$\hbar\omega_{\boldsymbol{k}} = 2zJS(1-\gamma_{\boldsymbol{k}}) = 2JS\sum_{\boldsymbol{\delta}}(1-\cos\boldsymbol{k}\cdot\boldsymbol{\delta}). \tag{2.3.2}$$

Expanding Eq. (2.3.2) in powers of the wavevector $\boldsymbol{k}$, one sees that the energy spectrum (2.3.2) vanishes quadratically for $k \to 0$. For a SC lattice ($z = 6$) for which $\boldsymbol{\delta} = (\pm a, 0, 0)$, $(0, \pm a, 0)$ and $(0, 0, \pm a)$, where $a$ is the side of the cube, the energy spectrum Eq. (2.3.2) reads

$$\hbar\omega_{\boldsymbol{k}} = 12JS\left[1 - \frac{1}{3}(\cos k_x a + \cos k_y a + \cos k_z a)\right]. \tag{2.3.3}$$

For a body centered cubic (BCC) lattice ($z = 8$) for which $\boldsymbol{\delta} = (\pm\frac{a}{2}, \pm\frac{a}{2}, \pm\frac{a}{2})$, the energy spectrum reads

$$\hbar\omega_{\boldsymbol{k}} = 16JS\left(1 - \cos\frac{k_x a}{2}\cos\frac{k_y a}{2}\cos\frac{k_z a}{2}\right). \tag{2.3.4}$$

For a face centered cubic (FCC) lattice ($z = 12$) for which $\boldsymbol{\delta} = (\pm\frac{a}{2}, \pm\frac{a}{2}, 0)$, $(\pm\frac{a}{2}, 0, \pm\frac{a}{2})$ and $(0, \pm\frac{a}{2}, \pm\frac{a}{2})$, the energy spectrum reads

$$\hbar\omega_{\boldsymbol{k}} = 24JS\left[1 - \frac{1}{3}\left(\cos\frac{k_x a}{2}\cos\frac{k_y a}{2} + \cos\frac{k_x a}{2}\cos\frac{k_z a}{2} + \cos\frac{k_y a}{2}\cos\frac{k_z a}{2}\right)\right]. \tag{2.3.5}$$

In the limit of long wavelength, all cubic lattices yield energy spectra which are vanishing as $\hbar\omega_{\boldsymbol{k}} \to 2JS(ak)^2$. The energy of the uniform mode ($k = 0$) vanishes in agreement to the equivalent non-relativistic Goldstone theorem[19] that works for Hamiltonians with short-range interactions, continuous symmetry (the 3D rotation group in the present case) and a ground state with a broken symmetry (the ferromagnetic ground state). It is worthwhile noticing that the energy spectrum (2.3.2) is an *exact* eigenvalue of the ferromagnetic Heisenberg Hamiltonian (2.2.1) belonging to the eigenstate with a delocalized spin deviation with momentum $\boldsymbol{k}$ as one can see by comparing the spectrum (2.3.2) with the spectrum given by Eq. (1.3.21) obtained from an exact calculation on the Heisenberg Hamiltonian.

In presence of an external magnetic field, of a single-ion easy-axis anisotropy and of an exchange anisotropy, the energy spectrum becomes

$$\hbar\omega_{\boldsymbol{k}} = 2zJS(1-\gamma_{\boldsymbol{k}}) + h + D(2S-1) + 2zKS \tag{2.3.6}$$

as one can obtain from Eqs. (2.2.26)–(2.2.28). The main effect of these types of easy-axis anisotropy consists of moving the magnon spectrum upwards rigidly with the appearance of a gap at $\boldsymbol{k} = 0$. We stress that the harmonic approximation, which neglects any interaction between magnons, is expected to be a good representation of a ferromagnet at low temperature where the magnons can be treated as an ideal Bose gas. Obviously, since the number of magnons is not conserved (as well as in the case of a "phonon gas") but determined by the temperature of the thermal bath, the chemical potential of the magnons is zero and any Bose-Einstein condensation is prevented.

## 2.4. Low Temperature Thermodynamic Functions

The low temperature thermodynamic functions of a ferromagnet may be obtained from the harmonic Hamiltonian (2.3.1). The granpartition function[20] reads

$$\mathcal{Q} = e^{-\beta E_0}\mathrm{Tr}(e^{-\beta\mathcal{H}_0}) = e^{-\beta E_0}\sum_{\{n_{\boldsymbol{k}}\}} e^{-\beta\sum_{\boldsymbol{k}}\hbar\omega_{\boldsymbol{k}}n_{\boldsymbol{k}}}$$

$$= e^{-\beta E_0}\prod_{\boldsymbol{k}}\sum_{n_{\boldsymbol{k}}=0}^{\infty} e^{-\beta\hbar\omega_{\boldsymbol{k}}n_{\boldsymbol{k}}} = e^{-\beta E_0}\prod_{\boldsymbol{k}}\frac{1}{1-e^{-\beta\hbar\omega_{\boldsymbol{k}}}} \tag{2.4.1}$$

where $\beta = \frac{1}{k_BT}$, $E_0 = -zJS^2N$ is the ground-state energy and $\hbar\omega_{\boldsymbol{k}}$ is the energy spectrum of spin waves given by Eq. (2.3.2) in the isotropic case and by Eq. (2.3.6) in the anisotropic case. The trace (Tr) of the density matrix is performed over the states number $|\{n_{\boldsymbol{k}}\}\rangle$ that are eigenstates of $\mathcal{H}_0$ with eigenvalues $\sum_{\boldsymbol{k}}\hbar\omega_{\boldsymbol{k}}n_{\boldsymbol{k}}$. The free energy is directly connected to the grancanonical partition function

$$F = -k_BT\ln\mathcal{Q} = E_0 + k_BT\sum_{\boldsymbol{k}}\ln\left(1-e^{-\beta\hbar\omega_{\boldsymbol{k}}}\right)$$

$$= E_0 + k_BTN\frac{v_c}{(2\pi)^3}\int_{\mathrm{BZ}} d^3\boldsymbol{k}\ln(1-e^{-\beta\hbar\omega_{\boldsymbol{k}}})$$

$$= E_0 - k_BTN\sum_{n=1}^{\infty}\frac{1}{n}\frac{v_c}{(2\pi)^3}\int_{\mathrm{BZ}} d^3\boldsymbol{k}e^{-n\beta\hbar\omega_{\boldsymbol{k}}} \tag{2.4.2}$$

where $v_c$ is the volume of the unit cell (for instance, $v_c = \frac{a^3}{m}$ with $m = 1, 2, 4$ for SC, BCC, FCC lattices, respectively) and BZ indicates the first Brillouin zone. In Eq. (2.4.2), the series expansion of the logarithmic function $\ln(1-x) = -\sum_{n=1}^{\infty}\frac{x^n}{n}$ has been used. At low temperature, only magnons of low energy can be excited so that the magnon energy spectrum (2.3.2) can be replaced by its expansion for long wavelengths

$$\hbar\omega_{\boldsymbol{k}} = 2JS\sum_{\boldsymbol{\delta}}(\boldsymbol{k}\cdot\boldsymbol{\delta})^2$$

that reduces to $\hbar\omega_{\boldsymbol{k}} = 2JS(ak)^2$ for all cubic lattices. Then the free energy (2.4.2) becomes

$$F \simeq E_0 - k_BTN\sum_{n=1}^{\infty}\frac{1}{n}\frac{1}{8\pi^3m}\int_{BZ} d^3\boldsymbol{q}e^{-2n\beta JSq^2}$$

$$\simeq E_0 - k_BTN\frac{1}{8\pi^3m}\sum_{n=1}^{\infty}\frac{1}{n}\int_0^{\infty} q^2dq\int_0^{\pi}\sin\theta d\theta\int_0^{2\pi} d\phi e^{-2n\beta JSq^2}$$

$$= E_0 - k_BTN\frac{1}{2\pi^2m}\sum_{n=1}^{\infty}\frac{1}{n}\int_0^{\infty} q^2dqe^{-2n\beta JSq^2} \tag{2.4.3}$$

where $\boldsymbol{q}=a\boldsymbol{k}$. Due to the parabolic $q$-dependence of the magnon spectrum in Eq. (2.4.3), we have assumed spherical coordinates and replaced the original BZ with a sphere whose radius is sent to infinity. As we will see later, the higher powers in $q$ of the magnon spectrum and the actual lattice structure give rise to higher powers in the free energy temperature expansion. Making use of the relationship[3]

$$\int_0^\infty dx x^2 e^{-px^2} = \frac{1}{4p}\sqrt{\frac{\pi}{p}},$$

the free energy becomes

$$F = E_0 - k_B T N \zeta\left(\frac{5}{2}\right)\frac{1}{m}\left(\frac{k_B T}{8\pi JS}\right)^{\frac{3}{2}} \tag{2.4.4}$$

where $\zeta(\frac{5}{2}) = 1.34149$ is the Riemann function[4] $\zeta(p) = \sum_{n=1}^\infty n^{-p}$. As one can see, the first term in the temperature expansion of the free energy is proportional to $T^{\frac{5}{2}}$. In the presence of uniaxial anisotropy, the uniform mode has a gap $\hbar\omega_o = h+D(2S-1)+2zKS$ and the free energy is obtained from Eq. (2.4.4) after replacing $\zeta(\frac{5}{2})$ by the function $Z_{\frac{5}{2}}(\beta\hbar\omega_o)$ where $Z_p(x) = \sum_{n=1}^\infty n^{-p}e^{-nx}$. In this case, the free energy shows an exponential dependence on the temperature instead of a power law. Indeed, at low temperature ($k_B T \ll \hbar\omega_o$), the function $Z_{\frac{5}{2}}(\beta\hbar\omega_0)$ may be approximated by the first term of the series so that $Z_{\frac{5}{2}}(\beta\hbar\omega_0) \sim e^{-\beta\hbar\omega_o}$. From the free energy, all the other thermodynamic functions can be obtained. In particular, the entropy is given by

$$S = -\frac{\partial F}{\partial T} = 5k_B N\zeta\left(\frac{5}{2}\right)\frac{1}{2m}\left(\frac{k_B T}{8\pi JS}\right)^{\frac{3}{2}}, \tag{2.4.5}$$

the internal energy is then obtained from the relationship $U = F + TS$ or directly from the free energy derivative as

$$U = -T^2\frac{\partial}{\partial T}\left(\frac{F}{T}\right) = E_0 + 12\pi JSN\zeta\left(\frac{5}{2}\right)\frac{1}{m}\left(\frac{k_B T}{8\pi JS}\right)^{\frac{5}{2}}. \tag{2.4.6}$$

The same result could have been obtained from the equation $U = \sum_{\boldsymbol{k}} \hbar\omega_{\boldsymbol{k}}\langle a_{\boldsymbol{k}}^+ a_{\boldsymbol{k}}\rangle$ where the average number of magnons is evaluated as follows

$$\begin{aligned}
\langle a_{\boldsymbol{k}}^+ a_{\boldsymbol{k}}\rangle &= \frac{\mathrm{Tr}(a_{\boldsymbol{k}}^+ a_{\boldsymbol{k}} e^{-\beta\sum_{\boldsymbol{q}}\hbar\omega_{\boldsymbol{q}} a_{\boldsymbol{q}}^+ a_{\boldsymbol{q}}})}{\mathrm{Tr}(e^{-\beta\sum_{\boldsymbol{q}}\hbar\omega_{\boldsymbol{q}} a_{\boldsymbol{q}}^+ a_{\boldsymbol{q}}})} = \frac{\sum_{n_{\boldsymbol{k}}=0}^\infty n_{\boldsymbol{k}} e^{-\beta\hbar\omega_{\boldsymbol{k}} n_{\boldsymbol{k}}}}{\sum_{n_{\boldsymbol{k}}=0}^\infty e^{-\beta\hbar\omega_{\boldsymbol{k}} n_{\boldsymbol{k}}}} \\
&= -\frac{1}{\beta\hbar}\frac{\partial}{\partial\omega_{\boldsymbol{k}}}\ln\left(\sum_{n_{\boldsymbol{k}}=0}^\infty e^{-\beta\hbar\omega_{\boldsymbol{k}} n_{\boldsymbol{k}}}\right) \\
&= \frac{1}{\beta\hbar}\frac{\partial}{\partial\omega_{\boldsymbol{k}}}\ln\left(1 - e^{-\beta\hbar\omega_{\boldsymbol{k}}}\right) = \frac{1}{e^{\beta\hbar\omega_{\boldsymbol{k}}} - 1}
\end{aligned} \tag{2.4.7}$$

leading to the average occupation number of the Bose-Einstein statistics. The heat capacity is given by

$$C = \frac{\partial U}{\partial T} = 15 k_B N \zeta\left(\frac{5}{2}\right)\frac{1}{4m}\left(\frac{k_B T}{8\pi JS}\right)^{\frac{3}{2}}. \tag{2.4.8}$$

For an isotropic insulating ferromagnet, the heat capacity at low temperature is expected to behave like $C = A_{\text{mag}}T^{\frac{3}{2}} + B_{\text{vib}}T^3$ where the first term is the magnetic contribution and the second one is the vibrational contribution. A convenient way to separate the magnetic from the vibrational contribution is to draw the quantity $CT^{-\frac{3}{2}}$ as function of $T^{\frac{3}{2}}$. In this way, one obtains a straight line whose slope $B_{\text{vib}}$ gives the coefficient of the vibrational contribution while the intercept at $T = 0$ gives the coefficient $A_{\text{mag}}$ of the magnetic contribution.

The magnetization is given by

$$M = -\frac{\partial F}{\partial H} = g\mu_B\left\langle \sum_i S_i^z \right\rangle = g\mu_B\left(SN - \sum_{\boldsymbol{k}}\langle a_{\boldsymbol{k}}^+ a_{\boldsymbol{k}}\rangle\right) \tag{2.4.9}$$

or using Eq. (2.4.7),

$$\begin{aligned} M &= M_0\left[1 - \frac{1}{mS}\frac{v_c}{(2\pi)^3}\int_{\text{BZ}} d^3\boldsymbol{k}\frac{1}{e^{\beta\hbar\omega_{\boldsymbol{k}}} - 1}\right] \\ &= M_0\left[1 - \frac{1}{mS}\sum_{n=1}^{\infty}\frac{v_c}{(2\pi)^3}\int_{\text{BZ}} d^3\boldsymbol{k}e^{-n\beta\hbar\omega_{\boldsymbol{k}}}\right] \end{aligned} \tag{2.4.10}$$

where $M_0 = g\mu_B SN$ is the saturation magnetization at $T = 0$. In the low temperature limit, that is for $\hbar\omega_{\boldsymbol{k}} \simeq h + 2JS(ak)^2$, one obtains

$$\begin{aligned} M &\simeq M_0\left[1 - \frac{1}{mS}\sum_{n=1}^{\infty} e^{-n\frac{h}{k_B T}}\frac{1}{2\pi^2}\int_0^{\infty} q^2 dq e^{-2n\beta JSq^2}\right] \\ &\times M_0\left[1 - \frac{1}{mS}\left(\frac{k_B T}{8\pi JS}\right)^{\frac{3}{2}} Z_{\frac{3}{2}}\left(\frac{h}{k_B T}\right)\right] \end{aligned} \tag{2.4.11}$$

For $h = 0$, since $Z_{\frac{3}{2}}(0) = \zeta(\frac{3}{2}) = 2.61238$, the spontaneous magnetization $M_s = M(T, H = 0)$ deviates from its saturation value following the $T^{\frac{3}{2}}$-law found long time ago by Bloch.[21] The spontaneous magnetization is the order parameter of the ferromagnetic phase since it is non-zero in the ordered phase and vanishes in the (disordered) paramagnetic phase.

The susceptibility is given by

$$\chi = \frac{\partial M}{\partial H} \simeq \frac{(g\mu_B)^2 N}{m8\pi JS}\sqrt{\frac{k_B T}{8\pi JS}} Z_{\frac{1}{2}}\left(\frac{h}{k_B T}\right). \tag{2.4.12}$$

Equation (2.4.12) is obtained by taking advantage of the equality $Z'_{\frac{3}{2}}(x) = -Z_{\frac{1}{2}}(x)$. For weak magnetic field ($h \ll k_B T$), one obtains

$$\chi = \sqrt{\pi}\frac{(g\mu_B)^2 N}{m(8\pi JS)^{\frac{3}{2}}}\frac{k_B T}{\sqrt{h}}$$

since $Z_{\frac{1}{2}}(x) \simeq \sqrt{\frac{\pi}{x}}$, for $x \to 0$. For $h \to 0$, the susceptibility diverges at any finite temperature. This divergence is not surprising because in the ordered phase of the isotropic ferromagnet, any infinitesimal magnetic field is sufficient to direct the macroscopic magnetization along the field. In other words, the system reacts to the magnetic field with an infinite response. In a ferromagnet with an easy-axis anisotropy ($D, K > 0$), the divergence in the susceptibility disappears. Indeed, for small anisotropy ($\hbar\omega_o \ll k_B T$), the susceptibility becomes

$$\chi \simeq \sqrt{\pi}\frac{(g\mu_B)^2 N}{m(8\pi JS)^{\frac{3}{2}}}\frac{k_B T}{(\hbar\omega_o)^{\frac{1}{2}}} \tag{2.4.13}$$

while for low temperature but large anisotropy ($\hbar\omega_o \gg k_B T$), the susceptibility shows an exponential behavior

$$\chi \simeq \frac{(g\mu_B)^2 N}{m(8\pi JS)}\left(\frac{k_B T}{8\pi JS}\right)^{\frac{1}{2}} e^{-\beta\hbar\omega_o}. \tag{2.4.14}$$

Before concluding this section, we evaluate some of the higher order corrections in the temperature expansion of the thermodynamic functions entered by the actual $\boldsymbol{k}$-dependence of the magnon spectrum (2.3.2). To do this, we consider the SC magnon spectrum (2.3.3) and we replace it in the free energy (2.4.2) obtaining

$$\begin{aligned} F &= E_0 - k_B T N \sum_{n=1}^{\infty} \frac{e^{-12n\beta JS}}{n}\left(\frac{1}{2\pi}\int_{-\pi}^{\pi} dq_x e^{4n\beta JS\cos q_x}\right)^3 \\ &= E_0 - k_B T N \sum_{n=1}^{\infty} \frac{e^{-12n\beta JS}}{n}[I_0(4n\beta JS)]^3 \end{aligned} \tag{2.4.15}$$

where $E_0 = -6JS^2 N$ and $I_0(z) = \frac{1}{2\pi}\int_{-\pi}^{\pi} d\theta e^{z\cos\theta}$ is the modified Bessel function of the first kind and of order zero.[4] In the low temperature limit ($4\beta JS \gg 1$), we use the asymptotic expansion of $I_0(z)$

$$I_0(z) = \frac{e^z}{\sqrt{2\pi z}}\left(1 + \frac{1}{8z} + \frac{9}{128z^2} + \cdots\right) \tag{2.4.16}$$

leading to the free energy expansion

$$\begin{aligned} F = E_0 - k_B T N \Bigg[&\zeta\left(\frac{5}{2}\right)\left(\frac{k_B T}{8\pi JS}\right)^{\frac{3}{2}} + \frac{3\pi}{4}\zeta\left(\frac{7}{2}\right)\left(\frac{k_B T}{8\pi JS}\right)^{\frac{5}{2}} \\ &+ \frac{33\pi^2}{32}\zeta\left(\frac{9}{2}\right)\left(\frac{k_B T}{8\pi JS}\right)^{\frac{7}{2}}\Bigg] \end{aligned} \tag{2.4.17}$$

where[4] $\zeta(\frac{5}{2}) = 1.34149$, $\zeta(\frac{7}{2}) = 1.12673$, $\zeta(\frac{9}{2}) = 1.05471$. This result could also be obtained by expanding the magnon spectrum up to $k^6$. For the SC lattice, one obtains

$$\hbar\omega_{\boldsymbol{k}} = 2JS(ak)^2 - \frac{1}{6}JSa^4(k_x^4 + k_y^4 + k_z^4) + \frac{1}{180}JSa^6(k_x^6 + k_y^6 + k_z^6). \tag{2.4.18}$$

Performing an integration over a sphere instead of over the actual cubic cell and assuming the radius of the sphere goes to infinity, the half-integer powers of the series are recovered. However, the coefficients of the series are not the same. Indeed, the first two coefficients are unchanged whereas the third one gives $-\frac{\pi^2}{4}$ instead of the correct value $\frac{33\pi^2}{32}$. This means that the SC lattice BZ may be treated as a sphere up to the order $k^4$. The cubic structure of the BZ becomes important only at the order $k^6$. Even worse is the result for the BCC and FCC lattices where the actual structure affects the expansion coefficients at the order $T^{\frac{7}{2}}$ coming from the $k^4$-term of the spectrum. In any case, the coefficient of the main term of the expansion is correctly given by the parabolic approximation of the spectrum ($k^2$-term), ignoring any actual structure of the cell. Notice that the higher semi-integer powers of the temperature series expansion given in Eq. (2.4.17) are introduced as a result of the actual magnon spectrum, not by the magnon-magnon interaction which is neglected in the harmonic approximation. The series expansion for the internal energy and for the heat capacity of a SC lattice read

$$U = E_0 + k_B T N \left[ \frac{3}{2}\zeta\left(\frac{5}{2}\right)\left(\frac{k_B T}{8\pi J S}\right)^{\frac{3}{2}} + \frac{15\pi}{8}\zeta\left(\frac{7}{2}\right)\left(\frac{k_B T}{8\pi J S}\right)^{\frac{5}{2}} + \frac{231\pi^2}{64}\zeta\left(\frac{9}{2}\right)\left(\frac{k_B T}{8\pi J S}\right)^{\frac{7}{2}} \right] \tag{2.4.19}$$

and

$$\begin{aligned} C &= k_B N \left[ \frac{15}{4}\zeta\left(\frac{5}{2}\right)\left(\frac{k_B T}{8\pi J S}\right)^{\frac{3}{2}} + \frac{105\pi}{16}\zeta\left(\frac{7}{2}\right)\left(\frac{k_B T}{8\pi J S}\right)^{\frac{5}{2}} + \frac{2079\pi^2}{128}\zeta\left(\frac{9}{2}\right)\left(\frac{k_B T}{8\pi J S}\right)^{\frac{7}{2}} \right] \\ &= \frac{15}{4} k_B N \left[ 1.34149\left(\frac{k_B T}{8\pi J S}\right)^{\frac{3}{2}} + 6.19452\left(\frac{k_B T}{8\pi J S}\right)^{\frac{5}{2}} + 45.0865\left(\frac{k_B T}{8\pi J S}\right)^{\frac{7}{2}} \right], \end{aligned} \tag{2.4.20}$$

respectively. The series expansion for the magnetization can be obtained from Eq. (2.4.10), giving

$$\begin{aligned} M &= M_0 \left[ 1 - \frac{1}{S}\sum_{n=1}^{\infty} e^{-12n\beta J S}\left(\frac{1}{2\pi}\int_{-\pi}^{\pi} dq_x e^{4n\beta J S\cos q_x}\right)^3 \right] \\ &= M_0 \left\{ 1 - \frac{1}{S}\sum_{n=1}^{\infty} e^{-12n\beta J S}[I_0(4n\beta J S)]^3 \right\}. \end{aligned} \tag{2.4.21}$$

At low temperature ($k_BT \ll 4JS$), the asymptotic expansion (2.4.16) can be used to obtain

$$M = M_0\left\{1 - \frac{1}{S}\left[\zeta\left(\frac{3}{2}\right)\left(\frac{k_BT}{8\pi JS}\right)^{\frac{3}{2}} + \frac{3\pi}{4}\zeta\left(\frac{5}{2}\right)\left(\frac{k_BT}{8\pi JS}\right)^{\frac{5}{2}} + \frac{33\pi^2}{32}\zeta\left(\frac{7}{2}\right)\left(\frac{k_BT}{8\pi JS}\right)^{\frac{7}{2}}\right]\right\}$$

$$= M_0\left\{1 - \frac{1}{S}\left[2.61238\left(\frac{k_BT}{8\pi JS}\right)^{\frac{3}{2}} + 3.16081\left(\frac{k_BT}{8\pi JS}\right)^{\frac{5}{2}} + 11.4679\left(\frac{k_BT}{8\pi JS}\right)^{\frac{7}{2}}\right]\right\}. \quad (2.4.22)$$

In Eq. (2.4.22), the corrections to the $T^{\frac{3}{2}}$-law obtained by Bloch[21] are due to the actual dispersion relation of the magnon spectrum. The first correction entered by the magnon-magnon interaction is proportional to $T^4$ as it will be shown in Section 3.6.

## 2.5. Application to Quasi-2D and Quasi 1D-models

In order to understand how the thermodynamic functions are modified in low dimensional magnetic systems, we evaluate the thermodynamic quantities for a tetragonal (T) lattice where the $c$-edge of the unit cell may be chosen as $c \gg a$ or $c \ll a$. For a T ferromagnet, the exchange interaction between the 4 NN spins in the basal $ab$ plane is assumed to be $J > 0$ while the exchange interaction between the 2 NN spins along the $c$ axis is assumed to be $J' > 0$. According to $c \gg a$ or $c \ll a$, one takes $J \gg J'$ or $J \ll J'$, respectively. In the former case, we have a staking of ferromagnetic planes weakly coupled to each other (quasi-2D ferromagnet), in the latter case, we have weakly coupled ferromagnetic chains (quasi-1D ferromagnet). For a T lattice, the spectrum (2.3.3) becomes

$$\hbar\omega_{\mathbf{k}} = 8JS\left[1 - \frac{1}{2}(\cos k_xa + \cos k_ya)\right] + 4J'S(1 - \cos k_zc). \quad (2.5.1)$$

Obviously, for $c = a$ and $J = J'$, the T lattice reduces to the SC lattice and the results of the previous section are recovered. For a T lattice, the free energy can be obtained from Eq. (2.4.2), giving

$$F = E_0 - k_BTN\sum_{n=1}^{\infty}\frac{e^{-n(8J+4J')S\beta}}{n}[I_0(4n\beta JS)]^2 I_0(4n\beta J'S) \quad (2.5.2)$$

where $E_0 = -(4J+2J')S^2N$. At very low temperature $k_BT \ll 4JS, 4J'S$, one recovers the 3D result. Indeed, in this temperature range ($4\beta JS, 4\beta J'S \gg 1$), one can use the asymptotic expansion (2.4.16), obtaining

$$F = E_0 - k_B TN \left[ \zeta\left(\frac{5}{2}\right)\left(\frac{J}{J'}\right)^{\frac{1}{2}}\left(\frac{k_BT}{8\pi JS}\right)^{\frac{3}{2}} \cdots \right]. \tag{2.5.3}$$

As one can see, the free energy shows a $T^{\frac{5}{2}}$-dependence similar to that obtained for a SC lattice. The SC result is recovered for $J = J'$ as expected. The internal energy and the heat capacity of the T lattice become

$$U = E_0 + k_BTN \left[ \frac{3}{2}\zeta\left(\frac{5}{2}\right)\left(\frac{J}{J'}\right)^{\frac{1}{2}}\left(\frac{k_BT}{8\pi JS}\right)^{\frac{3}{2}} \cdots \right] \tag{2.5.4}$$

and

$$C = k_BN \left[ \frac{15}{4}\zeta\left(\frac{5}{2}\right)\left(\frac{J}{J'}\right)^{\frac{1}{2}}\left(\frac{k_BT}{8\pi JS}\right)^{\frac{3}{2}} \cdots \right], \tag{2.5.5}$$

respectively. The typical $T^{\frac{3}{2}}$ contribution to the heat capacity of a 3D ferromagnet is recovered at very low temperature.

For quasi-2D ferromagnets ($J' \ll J$) and for an intermediate temperature range $4J'S \ll k_BT \ll 4JS$ ($4\beta JS \gg 1 \gg 4\beta J'S$), the asymptotic expansion for $I_0(4\beta JS)$ is appropriate but the series expansion[4]

$$I_0(z) = 1 + \frac{1}{4}z^2 + \frac{1}{64}z^4 \cdots \tag{2.5.6}$$

has to be used for $I_0(4\beta J'S)$. Then the free energy (2.5.2) becomes

$$\begin{aligned} F &= E_0 - k_BTN \left[ Z_2(4\beta J'S)\left(\frac{k_BT}{8\pi JS}\right) \cdots \right] \\ &\simeq E_0 - k_BTN\zeta(2)\left(\frac{k_BT}{8\pi JS}\right) \cdots \end{aligned} \tag{2.5.7}$$

where[4] $\zeta(2) = \frac{\pi^2}{6} = 1.64493$. The internal energy and heat capacity read

$$U \simeq E_0 - k_BTN\zeta(2)\left(\frac{k_BT}{8\pi JS}\right) \cdots \tag{2.5.8}$$

and

$$C \simeq 2k_BN\zeta(2)\left(\frac{k_BT}{8\pi JS}\right) \cdots, \tag{2.5.9}$$

respectively. As one can see, the temperature dependence of the magnetic heat capacity of a quasi-2D ferromagnet is linear in $T$ so that one should expect a crossover between the $T^{3/2}$-law at very low temperature and a linear $T$-law at intermediate temperatures. The window of the linear contribution of the heat capacity vs temperature is related to the ratio $J'/J$. For typical quasi-2D magnets ($J'/J \sim 10^{-3}$–$10^{-5}$ and $J/k_B \sim 10$–$100\,\mathrm{K}$), the 3D-behaviour is restricted to temperatures $T \lesssim 1\,\mathrm{K}$ while the linear dependence may be extended from $T \lesssim 1\,\mathrm{K}$ to $T \lesssim 100\,\mathrm{K}$. Obviously, in the experiment, the elastic contribution (phonons) proportional to $T^{\frac{3}{2}}$ cannot be separated from the magnetic one. However, after drawing the function $T^{-1}C$ versus $T^{\frac{1}{2}}$, we may distinguish between the magnetic contribution (intercept at the origin) and the elastic contribution (slope of the straight line).

For quasi-1D ferromagnets ($J' \gg J$) in the intermediate temperature range $4JS \ll k_BT \ll 4J'S$ ($4\beta JS \ll 1 \ll 4\beta J'S$), the asymptotic expansion (2.4.16) is used for $I_0(4\beta J'S)$ and the series expansion (2.5.6) for $I_0(4\beta JS)$ so that one obtains

$$F = E_0 - k_BTNZ_{\frac{3}{2}}(8\beta JS)\left(\frac{k_BT}{8\pi J'S}\right)^{\frac{1}{2}}\cdots$$
$$\simeq E_0 - k_BTN\zeta\left(\frac{3}{2}\right)\left(\frac{k_BT}{8\pi J'S}\right)^{\frac{1}{2}}\cdots \tag{2.5.10}$$

The internal energy and heat capacity of a quasi-1D ferromagnet read

$$U \simeq E_0 + k_BTN\frac{1}{2}\zeta\left(\frac{3}{2}\right)\left(\frac{k_BT}{8\pi J'S}\right)^{\frac{1}{2}}\cdots \tag{2.5.11}$$

and

$$C \simeq \frac{3}{4}k_BN\zeta\left(\frac{3}{2}\right)\left(\frac{k_BT}{8\pi J'S}\right)^{\frac{1}{2}}\cdots, \tag{2.5.12}$$

respectively. A $T^{\frac{1}{2}}$-behaviour is expected for the heat capacity of a quasi-1D ferromagnet at intermediate temperatures.

Finally, let us consider the magnetization of the T ferromagnet. From Eq. (2.4.10), one has

$$M = M_0\left\{1 - \frac{1}{S}\sum_{n=1}^{\infty} e^{-n(8J+4J')\beta S}[I_0(4n\beta JS)]^2 I_0(4n\beta J'S)\right\}. \tag{2.5.13}$$

For temperatures such that $k_BT \ll 4JS$, $4J'S$ ($4\beta JS$, $4\beta J'S \gg 1$), the 3D-behaviour, that is the $T^{\frac{3}{2}}$-law of Bloch, is recovered

$$M = M_0\left[1 - \frac{1}{S}\zeta\left(\frac{3}{2}\right)\left(\frac{J}{J'}\right)^{\frac{1}{2}}\left(\frac{k_BT}{8\pi JS}\right)^{\frac{3}{2}}\cdots\right]. \tag{2.5.14}$$

For a quasi-2D ferromagnet ($J' \ll J$) at intermediate temperatures ($4J'S \ll k_BT \ll 4JS$), one obtains

$$M = M_0\left[1 + \frac{k_BT}{8\pi JS^2}\ln(1 - e^{-\frac{4J'S}{k_BT}})\right] \simeq M_0\left(1 - \frac{k_BT}{8\pi JS^2}\ln\frac{k_BT}{4J'S}\right). \tag{2.5.15}$$

where $\sum_{n=1}^{\infty} n^{-1}e^{-xn} = -\ln(1-e^{-x})$ has been used. In a quasi-2D ferromagnet, the magnetization is strongly depressed at intermediate temperature: this fact reflects the absence of long range order (LRO) at any finite temperature in a pure 2D Heisenberg ferromagnet ($J' = 0$). The absence of LRO in 2D systems with continuous symmetry and short-range interaction is rigorously proven by the Mermin-Wagner theorem.[22]

For a quasi-1D ferromagnet ($J' \gg J$) at intermediate temperatures ($4JS \ll k_B T \ll 4J'S$), one obtains

$$M = M_0 \left[ 1 - \frac{1}{S} Z_{\frac{1}{2}}(8\beta JS) \left( \frac{k_B T}{8\pi J'S} \right) \right] \simeq M_0 \left[ 1 - \frac{\pi}{S} \left( \frac{J'}{J} \right)^{\frac{1}{2}} \left( \frac{k_B T}{8\pi J'S} \right)^{\frac{3}{2}} \right] \tag{2.5.16}$$

since $Z_{\frac{1}{2}}(x) = \sum_{n=1}^{\infty} n^{-\frac{1}{2}} e^{-xn} \simeq \sqrt{\frac{\pi}{x}} + O(1)$ for $x \ll 1$. In the quasi-1D ferromagnet, the deviation from the saturation value of the magnetization may be very strong since for many actual compounds, one has $(J'/J)^{\frac{1}{2}} \sim 10^2$.

# Chapter 3

# INTERACTING SPIN WAVES IN FERROMAGNETS

## 3.1. Neutron Scattering Cross-Section

Thermal neutrons used in the experiments have a typical wavelength $\lambda \sim 1.81\,\text{Å}$ (wavevector $k_i = \frac{2\pi}{\lambda} \sim 3.47\,\text{Å}^{-1}$) and energy $E_i = \frac{\hbar^2 k_i^2}{2M_n} \sim 25\,\text{meV} = 290\,\text{K}$. The wavelength and the energy of the neutrons of the incident beam are comparable with the lattice spacing and with the energy of the elementary excitations of the sample, respectively, so that they are suitable probes to investigate the spin wave dispersion relation. Neutrons interact with matter through nuclear forces (neutron-ion interaction) and magnetic forces (neutron-electron interaction) due to the interaction between the neutron spin $\sigma = 1/2$ and the orbital and spin momentum of the electrons of an ion. The partial differential magnetic cross-section for unpolarized neutrons of incident energy $E_i$ scattered into an element of solid angle $\Omega$ with energy between $E_f$ and $E_f + dE_f$ is given by[23]

$$\frac{d^2\sigma}{d\Omega dE_f} = r_0^2 \frac{k_f}{k_i} \left[\frac{1}{2} g F(\boldsymbol{K})\right]^2 e^{-2W(\boldsymbol{K})} \sum_{\alpha,\beta} \left(\delta_{\alpha,\beta} - \frac{K_\alpha K_\beta}{K^2}\right) S^{\alpha\beta}(\boldsymbol{K},\omega) \tag{3.1.1}$$

where

$$S^{\alpha\beta}(\boldsymbol{K},\omega) = \sum_{i,j} e^{-i\boldsymbol{K}\cdot(\boldsymbol{r}_i - \boldsymbol{r}_j)} \frac{1}{2\pi\hbar} \int_{-\infty}^{+\infty} dt e^{-i\omega t} \langle S_i^\alpha S_j^\beta(t)\rangle \tag{3.1.2}$$

is called the *dynamical structure factor*; $r_0 = \frac{\gamma e^2}{m_e c^2} = -0.54\times 10^{-12}\,\text{cm}$; $g$ is the gyromagnetic ratio ($g = 2$ for spin-only scattering); $E_f$ is the final energy of the neutron, $\boldsymbol{k}_f$ and $\boldsymbol{k}_i$ are the wavevectors of the outgoing and incoming neutron, respectively: $\boldsymbol{K} = \boldsymbol{k}_i - \boldsymbol{k}_f$ is the scattering wavevector; $F(\boldsymbol{K})$ is the atomic form factor and $W(\boldsymbol{K}) = \frac{1}{2}\langle(\boldsymbol{K}\cdot\boldsymbol{u})^2\rangle$ is the Debye-Waller factor, in which $\langle \boldsymbol{u}^2\rangle$ is the mean square displacement of the ion. Labels $\alpha$ and $\beta$ refer to the cartesian components $x, y, z$; $\hbar\omega = E_i - E_f$ is the transferred energy between the neutron and the sample: if $\hbar\omega > 0$, the neutron *gives* energy to the sample while for $\hbar\omega < 0$, the neutron *gets* energy from the sample. The cross-section (3.1.1) is evaluated by treating the interaction between the magnetic moment of the neutron and the magnetic moment of the ion ($g\mu_B \boldsymbol{S}_i$) in the first Born approximation and assuming that the orbital angular momentum of the ion is zero (ion with half-filled shells) or quenched by the crystal field. If the $z$ component of the total spin of the system $\sum_i S_i^z$ is a constant of the motion, as it occurs for the isotropic or uniaxial ferromagnet, one has

$$\langle S_i^+ S_j^+(t)\rangle = \langle S_i^- S_j^-(t)\rangle = \langle S_i^+ S_j^z(t)\rangle = \langle S_i^- S_j^z(t)\rangle = 0 \tag{3.1.3}$$

since the raising and lowering spin operators $S_i^+$ and $S_i^-$ change the $z$-component of the total momentum by a unit. Equation (3.1.3) implies that all non-diagonal terms in Eq. (3.1.1) vanish and the cross-section (3.1.1) reduces to

$$\frac{d^2\sigma}{d\Omega dE_f} = r_0^2 \frac{k_f}{k_i}\left[\frac{1}{2}gF(\boldsymbol{K})\right]^2 e^{-2W(\boldsymbol{K})}\left[\left(1-\frac{K_z^2}{K^2}\right)S_\parallel(\boldsymbol{K},\omega) + \left(1+\frac{K_z^2}{K^2}\right)S_\perp(\boldsymbol{K},\omega)\right] \tag{3.1.4}$$

where

$$S_\parallel(\boldsymbol{K},\omega) \equiv S^{zz}(\boldsymbol{K},\omega) = \sum_{i,j} e^{-i\boldsymbol{K}\cdot(\boldsymbol{r}_i-\boldsymbol{r}_j)}\frac{1}{2\pi\hbar}\int_{-\infty}^{+\infty} dt e^{-i\omega t}\langle S_i^z S_j^z(t)\rangle \tag{3.1.5}$$

and

$$S_\perp(\boldsymbol{K},\omega) = S^{xx}(\boldsymbol{K},\omega) = S^{yy}(\boldsymbol{K},\omega) = \frac{1}{4}\sum_{i,j} e^{-i\boldsymbol{K}\cdot(\boldsymbol{r}_i-\boldsymbol{r}_j)}\frac{1}{2\pi\hbar}\int_{-\infty}^{+\infty} dt e^{-i\omega t}\langle S_i^+ S_j^-(t) + S_i^- S_j^+(t)\rangle. \tag{3.1.6}$$

Using the HP or DM spin-boson transformations (2.1.6) or (2.1.7), keeping only the higher order terms and using the Fourier transforms (2.2.14), one can write the correlation functions occurring in Eqs. (3.1.5) and (3.1.6) in terms of magnon creation and destruction operators:

$$\langle S_i^z S_j^z(t)\rangle \simeq S^2 - \frac{2S}{N}\sum_{\boldsymbol{q}}\langle a_{\boldsymbol{q}} a_{\boldsymbol{q}}^+\rangle \simeq \left\langle \frac{1}{N}\sum_i S_i^z\right\rangle^2 = \langle S_i^z\rangle^2 \tag{3.1.7}$$

and

$$\langle S_i^+ S_j^-(t) + S_i^- S_j^+(t)\rangle \simeq \frac{2S}{N}\sum_{\boldsymbol{q}}[e^{i\boldsymbol{q}\cdot(\boldsymbol{r}_i-\boldsymbol{r}_j)}\langle a_{\boldsymbol{q}} a_{\boldsymbol{q}}^+(t)\rangle + e^{-i\boldsymbol{q}\cdot(\boldsymbol{r}_i-\boldsymbol{r}_j)}\langle a_{\boldsymbol{q}}^+ a_{\boldsymbol{q}}(t)\rangle]. \tag{3.1.8}$$

Replacing Eqs. (3.1.7) and (3.1.8) into Eqs. (3.1.5) and (3.1.6), for a lattice with one atom per unit cell, one obtains

$$S_\parallel(\boldsymbol{K},\omega) = (N\langle S_i^z\rangle)^2 \sum_{\boldsymbol{G}} \delta_{\boldsymbol{K},\boldsymbol{G}}\delta(\hbar\omega) \tag{3.1.9}$$

and

$$S_\perp(\boldsymbol{K},\omega) = 2SN\sum_{\boldsymbol{q},\boldsymbol{G}}\delta_{\boldsymbol{K}+\boldsymbol{q},\boldsymbol{G}}\frac{1}{2\pi\hbar}\int_{-\infty}^{+\infty} dte^{-i\omega t}\langle a_{\boldsymbol{q}}a^+_{\boldsymbol{q}}(t)\rangle$$
$$+2SN\sum_{\boldsymbol{q},\boldsymbol{G}}\delta_{\boldsymbol{K}-\boldsymbol{q},\boldsymbol{G}}\frac{1}{2\pi\hbar}\int_{-\infty}^{+\infty} dte^{-i\omega t}\langle a^+_{\boldsymbol{q}}a_{\boldsymbol{q}}(t)\rangle \qquad (3.1.10)$$

where $\boldsymbol{G}$ is a reciprocal lattice vector. To obtain Eqs. (3.1.9) and (3.1.10), the following relations have been used:

$$\sum_{i,j} e^{-i\boldsymbol{K}\cdot(\boldsymbol{r}_i-\boldsymbol{r}_j)} = N^2\sum_{\boldsymbol{G}}\delta_{\boldsymbol{K},\boldsymbol{G}} \qquad (3.1.11)$$

and

$$\frac{1}{2\pi}\int_{-\infty}^{+\infty} dte^{-i\omega t} = \delta(\omega). \qquad (3.1.12)$$

Equation (3.1.11) takes into account that the vectors $\boldsymbol{q}$ of the Fourier transform (2.2.14) belong to the first Brilloin zone (BZ) while the scattering wavevector $\boldsymbol{K}$ is not restricted to the first BZ. Equation (3.1.12) is the well known representation of the Dirac $\delta$-function. For ferromagnets with more than one atom per unit cell (for instance, the BCC and FCC lattices in which the cubic cell is chosen as unit cell or the honeycomb lattice in 2D), the relationship (3.1.11) should be replaced by

$$\sum_{i,j} e^{-i\boldsymbol{K}\cdot(\boldsymbol{r}_i-\boldsymbol{r}_j)} = N_{\text{cell}}^2\sum_{\boldsymbol{G}}\delta_{\boldsymbol{K},\boldsymbol{G}}\sum_{l,l'=1,n} e^{-i\boldsymbol{K}\cdot(\boldsymbol{\rho}_l-\boldsymbol{\rho}_{l'})} \qquad (3.1.13)$$

where $\boldsymbol{\rho}_l$ is the position of the $l$-th atom in the unit cell. $N_{\text{cell}}$ is the number of unit cells and $N = nN_{\text{cell}}$ is the total number of atoms of the sample, $n$ being the number of atoms in the unit cell. Obviously, for $n = 1$ and $\boldsymbol{\rho}_1 = 0$, Eq. (3.1.13) reduces to Eq. (3.1.11).

In the harmonic approximation, the time evolution of the boson operators occurring in Eq. (3.1.10) is given by

$$a^+_{\boldsymbol{q}}(t) = a^+_{\boldsymbol{q}}e^{i\omega_{\boldsymbol{q}}t}, \quad a_{\boldsymbol{q}}(t) = a_{\boldsymbol{q}}e^{-i\omega_{\boldsymbol{q}}t} \qquad (3.1.14)$$

so that Eq. (3.1.10) becomes

$$S_\perp(\boldsymbol{K},\omega) = 2SN\sum_{\boldsymbol{q},\boldsymbol{G}}[(1+n_{\boldsymbol{q}})\delta_{\boldsymbol{K}+\boldsymbol{q},\boldsymbol{G}}\delta(\hbar\omega-\hbar\omega_{\boldsymbol{q}}) + n_{\boldsymbol{q}}\delta_{\boldsymbol{K}-\boldsymbol{q},\boldsymbol{G}}\delta(\hbar\omega+\hbar\omega_{\boldsymbol{q}})]. \qquad (3.1.15)$$

Then the first term of (3.1.4) gives the elastic contribution ($k_i = k_f$, $E_i = E_f$)

$$\left(\frac{d^2\sigma}{d\Omega dE_f}\right)_{\text{el}} = r_0^2\left[\frac{1}{2}gF(\boldsymbol{K})\right]^2 e^{-2W(\boldsymbol{K})}\left(1-\frac{K_z^2}{K^2}\right)(N\langle S_i^z\rangle)^2\delta(\hbar\omega)\sum_{\boldsymbol{G}}\delta_{\boldsymbol{K},\boldsymbol{G}} \qquad (3.1.16)$$

and the second term give the one-magnon inelastic contribution

$$\left(\frac{d^2\sigma}{d\Omega dE_f}\right)_{\text{inel}} = r_0^2 \frac{k_f}{k_i}\left[\frac{1}{2}gF(\boldsymbol{K})\right]^2 e^{-2W(\boldsymbol{K})}\left(1+\frac{K_z^2}{K^2}\right)$$
$$\times \frac{NS}{2}\sum_{\boldsymbol{q},\boldsymbol{G}}[(1+n_{\boldsymbol{q}})\delta_{\boldsymbol{K}+\boldsymbol{q},\boldsymbol{G}}\delta(\hbar\omega-\hbar\omega_{\boldsymbol{q}})$$
$$+ n_{\boldsymbol{q}}\delta_{\boldsymbol{K}-\boldsymbol{q},\boldsymbol{G}}\delta(\hbar\omega+\hbar\omega_{\boldsymbol{q}})]. \quad (3.1.17)$$

The elastic scattering cross-section (3.1.16) is proportional to the square of the magnetization leading to "Bragg peaks" located at each reciprocal lattice vector $\boldsymbol{G}$. The presence of the factor $\delta(\hbar\omega)$ implies that there is no energy transfer between the neutron and the sample so that the scattering is *elastic*. The elastic neutron scattering geometry is shown in Fig. 3.1 by the circumference that represents the locus of points such that $k_f' = k_i$. When the wavevctor $\boldsymbol{k'}_f$ falls on the reciprocal lattice point $\boldsymbol{G}'$, a Bragg peak occurs since $\boldsymbol{K} = \boldsymbol{G}'$. However, the Bragg peaks corresponding to scattering wavevectors $\boldsymbol{K}$ parallel to the magnetization direction disappear because the geometrical factor $1-(K_z/K)^2$ in Eq. (3.1.16) vanishes. This peculiarity can be used to establish the orientation of the magnetic moment in the ferromagnetic sample.

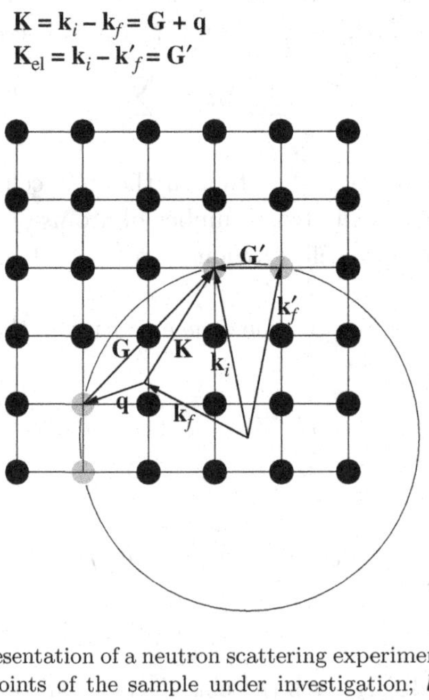


Fig. 3.1. Geometrical representation of a neutron scattering experiment: the grey and black circles are the reciprocal lattice points of the sample under investigation; $\boldsymbol{k}_i$ is the wavevector of the incident neutron, $\boldsymbol{k}_f$ and $\boldsymbol{k'}_f$ are the wavectors of two outgoing neutrons. The wavevctor $\boldsymbol{k'}_f$ lying on the circumference of radius $k_f' = k_i$ gives rise to an elastic Bragg peak since it falls on a reciprocal lattice point (grey circle). For $k_f < k_i$, the inelastic scattering corresponds to the creation of a magnon of wavevector $\boldsymbol{q} = \boldsymbol{K} - \boldsymbol{G}$ and frequency $\omega_{\boldsymbol{q}}$. $\boldsymbol{K}$ and $\boldsymbol{K}_{\text{el}} = \boldsymbol{G}'$ are the inelastic and elastic scattering wavevectors, respectively.

The inelastic cross-section (3.1.17) corresponds to a process in which *one* magnon is created or destroyed. In Fig. 3.1, the inelastic scattering corresponds to the creation of a magnon of wavevector $\boldsymbol{q} = \boldsymbol{K} - \boldsymbol{G}$ where $\boldsymbol{q}$ is restricted to the first BZ centered at the reciprocal lattice vector $\boldsymbol{G}$ and energy $\hbar\omega_{\boldsymbol{q}} = \frac{\hbar^2}{2M_n}(k_i^2 - k_f^2)$. As one can see, the choice of the vector $\boldsymbol{G}$ would not be unique as for the momentum conservation but it becomes unique when the energy conservation is satisfied. In general, a magnon of momentum $\hbar\boldsymbol{q}$ and energy $\hbar\omega_{\boldsymbol{q}}$ is created or destroyed at the expense of the incident neutron: the argument of the Dirac $\delta$-function appearing in the first term of (3.1.17) implies that $\hbar\omega > 0$ so that the incident neutron gives a fraction of its energy to the sample creating a magnon of energy $\hbar\omega_{\boldsymbol{q}}$: such an event can occur even at $T = 0$ explaining the factor $1 + n_{\boldsymbol{q}}$ in front of this term. On the contrary, the argument of the Dirac $\delta$-function appearing in the second term of (3.1.17) implies that $\hbar\omega < 0$ so that the neutron gets an amount of energy from the sample corresponding to the destruction of a magnon of energy $\hbar\omega_{\boldsymbol{q}}$: this event cannot occur at $T = 0$ since the magnons exist only at finite temperature, explaining the factor $n_{\boldsymbol{q}}$ in front of this term.

## 3.2. Boson Green Function

As shown in the previous section, the one-magnon inelastic neutron scattering (3.1.10) is an important probe to obtain the dispersion relation of the spin wave spectrum in a ferromagnet. Note that the cross-section (3.1.10) reduces to (3.1.17) in the harmonic approximation: this is a good approximation at very low temperatures where the magnon-magnon interaction can be ignored. As the temperature increases, however, the magnon-magnon interaction cannot be ignored and some systematic perturbative expansion has to be introduced in order to evaluate the correlation functions occurring in Eq. (3.1.10).

In this chapter, we illustrate the method of the Green function equation of motion introduced by Zubarev[24] in 1960. The correlation functions occurring in Eq. (3.1.10) may be written in terms of a boson Green function that satisfies a well defined equation of motion. Let us follow Zubarev[24] and define the *retarded* (R) and the *advanced* (A) Green functions as

$$G^{(\mathrm{R,A})}_{\boldsymbol{k},\boldsymbol{k}'}(t) = \mp i\theta(\pm t)\langle[a_{\boldsymbol{k}}(t), a^+_{\boldsymbol{k}'}]\rangle \equiv \langle\langle a_{\boldsymbol{k}}; a^+_{\boldsymbol{k}'}\rangle\rangle^{(\mathrm{R,A})} \tag{3.2.1}$$

where the upper (lower) sign refers to R(A), respectively; $\theta(t)$ is the step-function which is 1 if its argument is positive and 0 if its argument is negative. The equation of motion of the retarded or advanced Green function (3.2.1) is

$$i\hbar\frac{d}{dt}\langle\langle a_{\boldsymbol{k}}; a^+_{\boldsymbol{k}'}\rangle\rangle^{(\mathrm{R,A})} = \hbar\delta(t)\langle[a_{\boldsymbol{k}}, a^+_{\boldsymbol{k}'}]\rangle + \langle\langle[a_{\boldsymbol{k}}, \mathcal{H}]; a^+_{\boldsymbol{k}'}\rangle\rangle^{(\mathrm{R,A})}. \tag{3.2.2}$$

The first term of the right-hand side comes from the derivative of the step-function, the second term comes from the equation of motion of the operator $a_{\boldsymbol{k}}$ in the

Heisenberg representation

$$i\hbar\frac{da_{\boldsymbol{k}}}{dt} = [a_{\boldsymbol{k}}, \mathcal{H}] \tag{3.2.3}$$

where $\mathcal{H}$ is the boson Hamiltonian containing the dynamical interaction between the magnons. The correlation functions are given by

$$\begin{aligned}\langle a_{\boldsymbol{k}}(t)a^+_{\boldsymbol{k}'}\rangle &= \frac{1}{\mathcal{Q}}\sum_{\mu,\nu}\langle\mu|e^{i\mathcal{H}t/\hbar}a_{\boldsymbol{k}}e^{-i\mathcal{H}t/\hbar}|\nu\rangle\langle\nu|a^+_{\boldsymbol{k}'}|\mu\rangle e^{-\beta E_\mu}\\ &= \int_{-\infty}^{+\infty} d\omega' e^{-i\omega' t}\, e^{\beta\hbar\omega'} J(\omega')\end{aligned} \tag{3.2.4}$$

and

$$\langle a^+_{\boldsymbol{k}'}a_{\boldsymbol{k}}(t)\rangle = \int_{-\infty}^{+\infty} d\omega' e^{-i\omega' t}\, J(\omega') \tag{3.2.5}$$

where $|\mu\rangle$ and $|\nu\rangle$ are the eigenstates of the Hamiltonian $\mathcal{H}$ with eigenvalues $E_\mu$ and $E_\nu$, respectively; the time evolution of the boson operator $a_{\boldsymbol{k}}(t)$ in the Heisenberg representation has been used and the spectral intensity of the correlation function $J(\omega')$ is defined as

$$J(\omega') = \frac{1}{\mathcal{Q}}\sum_{\mu,\nu}\langle\mu|a_{\boldsymbol{k}}|\nu\rangle\langle\nu|a^+_{\boldsymbol{k}'}|\mu\rangle e^{-\beta E_\nu}\delta\left(\omega' - \frac{E_\nu - E_\mu}{\hbar}\right). \tag{3.2.6}$$

Then the commutator occurring in Eq. (3.2.1) becomes

$$\langle[a_{\boldsymbol{k}}(t), a^+_{\boldsymbol{k}'}]\rangle = \int_{-\infty}^{+\infty} d\omega' e^{-i\omega' t}(e^{\beta\hbar\omega'} - 1)J(\omega'). \tag{3.2.7}$$

Using the integral representation of the step-function[24]

$$\theta(t) = \frac{i}{2\pi}\lim_{\epsilon\to 0}\int_{-\infty}^{+\infty} dx\frac{e^{-ixt}}{x + i\epsilon} \tag{3.2.8}$$

and the $\delta$-function representation (3.1.12), the frequency Fourier transforms of the retarded and advanced Green functions (3.2.1) become

$$G^{(\mathrm{R,A})}_{\boldsymbol{k},\boldsymbol{k}'}(\omega) = \int_{-\infty}^{+\infty} dt e^{i\omega t}G^{(\mathrm{R,A})}_{\boldsymbol{k},\boldsymbol{k}'}(t) = \lim_{\epsilon\to 0}\int_{-\infty}^{+\infty} d\omega'(e^{\beta\hbar\omega'} - 1)\frac{J(\omega')}{\omega - \omega' \pm i\epsilon}. \tag{3.2.9}$$

If we define a generalized frequency Green function in the complex $z$-plane

$$G_{\boldsymbol{k},\boldsymbol{k}'}(z) = \int_{-\infty}^{+\infty} d\omega'(e^{\beta\hbar\omega'} - 1)\frac{J(\omega')}{z - \omega'}, \tag{3.2.10}$$

the retarded and advanced frequency Green functions are given by

$$G^{(\mathrm{R,A})}_{\boldsymbol{k},\boldsymbol{k}'}(\omega) = \lim_{\epsilon\to 0} G_{\boldsymbol{k},\boldsymbol{k}'}(\omega \pm i\epsilon). \tag{3.2.11}$$

Using the operator identity

$$\lim_{\epsilon\to 0}\frac{1}{\omega-\omega_0\pm i\epsilon}=\frac{\mathrm{P}}{\omega-\omega_0}\mp i\pi\delta(\omega-\omega_0) \tag{3.2.12}$$

where P means "principal value", from Eq. (3.2.11) one obtains

$$\lim_{\epsilon\to 0}[G_{\boldsymbol{k},\boldsymbol{k}'}(\omega+i\epsilon)-G_{\boldsymbol{k},\boldsymbol{k}'}(\omega-i\epsilon)]=-2\pi i(e^{\beta\hbar\omega}-1)J(\omega). \tag{3.2.13}$$

Equation (3.2.13) establishes the link between the spectral intensity $J(\omega)$ and the generalized frequency Green function. By means of Eq. (3.2.13), the correlation functions (3.2.4) and (3.2.5) become

$$\langle a_{\boldsymbol{k}}(t)a^+_{\boldsymbol{k}'}\rangle=\lim_{\epsilon\to 0}\frac{i}{2\pi}\int_{-\infty}^{+\infty}d\omega e^{-i\omega t}[1+n(\omega)][G_{\boldsymbol{k},\boldsymbol{k}'}(\omega+i\epsilon)-G_{\boldsymbol{k},\boldsymbol{k}'}(\omega-i\epsilon)] \tag{3.2.14}$$

and

$$\langle a^+_{\boldsymbol{k}'}a_{\boldsymbol{k}}(t)\rangle=\lim_{\epsilon\to 0}\frac{i}{2\pi}\int_{-\infty}^{+\infty}d\omega e^{-i\omega t}n(\omega)[G_{\boldsymbol{k},\boldsymbol{k}'}(\omega+i\epsilon)-G_{\boldsymbol{k},\boldsymbol{k}'}(\omega-i\epsilon)] \tag{3.2.15}$$

where $n(\omega)=(e^{\beta\hbar\omega}-1)^{-1}$. Due to the time-inedependence of the Hamiltonian $\mathcal{H}$, one has

$$\langle a_{\boldsymbol{q}}a^+_{\boldsymbol{q}}(t)\rangle=\langle a_{\boldsymbol{q}}(-t)a^+_{\boldsymbol{q}}\rangle \tag{3.2.16}$$

and the dynamical structure factor (3.1.10) becomes

$$\begin{aligned}S_\perp(\boldsymbol{K},\omega)&=i\frac{SN}{\hbar\pi}\sum_{\boldsymbol{q},\boldsymbol{G}}\delta_{\boldsymbol{K}+\boldsymbol{q},\boldsymbol{G}}\lim_{\epsilon\to 0}[1+n(\omega)][G_{\boldsymbol{q}}(\omega+i\epsilon)-G_{\boldsymbol{q}}(\omega-i\epsilon)]\\&\quad+i\frac{SN}{\hbar\pi}\sum_{\boldsymbol{q},\boldsymbol{G}}\delta_{\boldsymbol{K}-\boldsymbol{q},\boldsymbol{G}}n(-\omega)\lim_{\epsilon\to 0}[G_{\boldsymbol{q}}(-\omega+i\epsilon)-G_{\boldsymbol{q}}(-\omega-i\epsilon)]\end{aligned} \tag{3.2.17}$$

where $G_{\boldsymbol{q}}(\omega)\equiv G_{\boldsymbol{q},\boldsymbol{q}}(\omega)$. Finally, using the spectral representation of the retarded and advanced Green function

$$G^{(\mathrm{R,A})}_{\boldsymbol{k},\boldsymbol{k}'}(t)=\frac{1}{2\pi}\int_{-\infty}^{+\infty}d\omega e^{-i\omega t}G_{\boldsymbol{k},\boldsymbol{k}'}(\omega\pm i\epsilon) \tag{3.2.18}$$

from Eq. (3.2.2), one obtains the equation of motion for the generalized frequency Green function $G_{\boldsymbol{k},\boldsymbol{k}'}(\omega)\equiv\langle\langle a_{\boldsymbol{k}};a_{\boldsymbol{k}'}\rangle\rangle_\omega$

$$\hbar\omega\langle\langle a_{\boldsymbol{k}};a^+_{\boldsymbol{k}'}\rangle\rangle_\omega=\hbar\langle[a_{\boldsymbol{k}},a^+_{\boldsymbol{k}'}]\rangle+\langle\langle[a_{\boldsymbol{k}},\mathcal{H}];a^+_{\boldsymbol{k}'}\rangle\rangle_\omega. \tag{3.2.19}$$

In conclusion, if we are able to solve the equation of motion of the generalized Green function (3.2.19) to some degree of approximation, we can evaluate directly the dynamical structure factor (3.2.17) that enters the one-magnon inelastic neutron cross-section (3.1.4).

## 3.3. First-Order Approximation

In order to solve the equation of motion (3.2.19), we begin evaluating the commutator $[a_{\boldsymbol{k}}, \mathcal{H}]$: to do this, we split the boson Hamiltonian $\mathcal{H}$ in two parts

$$\mathcal{H} = \mathcal{H}_0 + \mathcal{H}_{\text{int}} \tag{3.3.1}$$

where the $\mathcal{H}_0$ is the *unperturbed* (harmonic) Hamiltonian given by

$$\mathcal{H}_0 = \sum_{\boldsymbol{q}} \hbar\omega_{\boldsymbol{q}} a_{\boldsymbol{q}}^+ a_{\boldsymbol{q}} \tag{3.3.2}$$

and $\mathcal{H}_{\text{int}}$ is the interaction Hamiltonian given by

$$\mathcal{H}_{\text{int}} = -\frac{zJ}{N} \sum_{\boldsymbol{q}_1,\boldsymbol{q}_2,\boldsymbol{q}_3,\boldsymbol{q}_4} \delta_{\boldsymbol{q}_1+\boldsymbol{q}_2,\boldsymbol{q}_3+\boldsymbol{q}_4} v_{\boldsymbol{q}_1,\boldsymbol{q}_2,\boldsymbol{q}_3,\boldsymbol{q}_4} a_{\boldsymbol{q}_1}^+ a_{\boldsymbol{q}_2}^+ a_{\boldsymbol{q}_3} a_{\boldsymbol{q}_4}. \tag{3.3.3}$$

The interaction potential $v$ occurring in Eq. (3.3.3) is

$$v^{DM}_{\boldsymbol{q}_1,\boldsymbol{q}_2,\boldsymbol{q}_3,\boldsymbol{q}_4} = \frac{1}{4}(\gamma_{\boldsymbol{q}_1-\boldsymbol{q}_3} + \gamma_{\boldsymbol{q}_1-\boldsymbol{q}_4} + \gamma_{\boldsymbol{q}_2-\boldsymbol{q}_3} + \gamma_{\boldsymbol{q}_2-\boldsymbol{q}_4}) - \frac{1}{2}(\gamma_{\boldsymbol{q}_1} + \gamma_{\boldsymbol{q}_2}) \tag{3.3.4}$$

or

$$\begin{aligned} v^{HP}_{\boldsymbol{q}_1,\boldsymbol{q}_2,\boldsymbol{q}_3,\boldsymbol{q}_4} &= \frac{1}{4}(\gamma_{\boldsymbol{q}_1-\boldsymbol{q}_3} + \gamma_{\boldsymbol{q}_1-\boldsymbol{q}_4} + \gamma_{\boldsymbol{q}_2-\boldsymbol{q}_3} + \gamma_{\boldsymbol{q}_2-\boldsymbol{q}_4}) - S\left(1 - \sqrt{1-\frac{1}{2S}}\right) \\ &\quad \times (\gamma_{\boldsymbol{q}_1} + \gamma_{\boldsymbol{q}_2} + \gamma_{\boldsymbol{q}_3} + \gamma_{\boldsymbol{q}_4}) = v^{DM}_{\boldsymbol{q}_1,\boldsymbol{q}_2,\boldsymbol{q}_3,\boldsymbol{q}_4} + \frac{1}{4}(\gamma_{\boldsymbol{q}_1} + \gamma_{\boldsymbol{q}_2} - \gamma_{\boldsymbol{q}_3} - \gamma_{\boldsymbol{q}_4}) \\ &\quad + \frac{1}{4}\left[1 - 4S\left(1 - \sqrt{1-\frac{1}{2S}}\right)\right](\gamma_{\boldsymbol{q}_1} + \gamma_{\boldsymbol{q}_2} + \gamma_{\boldsymbol{q}_3} + \gamma_{\boldsymbol{q}_4}) \end{aligned} \tag{3.3.5}$$

for the DM or HP spin-boson transformation, respectively. Note that both potentials (3.3.4) and (3.3.5) are invariant under the exchange $\boldsymbol{q}_1 \Leftrightarrow \boldsymbol{q}_2$ and/or $\boldsymbol{q}_3 \Leftrightarrow \boldsymbol{q}_4$. As for the HP spin-boson transformation, the NO expansion of Eq. (2.2.3) allows us to neglect all terms except $\mathcal{H}_4^{HP}$ of Eq. (2.2.18) with $v^{HP}$ given by Eq. (2.2.19) since all other terms give contributions of higher order in temperature. From Eqs. (3.3.1)–(3.3.3), the commutator in Eq. (3.2.19) gives

$$[a_{\boldsymbol{k}}, \mathcal{H}] = \hbar\omega_{\boldsymbol{k}} a_{\boldsymbol{k}} - \frac{2zJ}{N} \sum_{\boldsymbol{q}_1,\boldsymbol{q}_2} v_{\boldsymbol{k},\boldsymbol{q}_1,\boldsymbol{q}_2,\boldsymbol{k}+\boldsymbol{q}_1-\boldsymbol{q}_2} a_{\boldsymbol{q}_1}^+ a_{\boldsymbol{q}_2} a_{\boldsymbol{k}+\boldsymbol{q}_1-\boldsymbol{q}_2}. \tag{3.3.6}$$

Replacing Eq. (3.3.6) into the equation of motion (3.2.19), one has

$$(\omega - \omega_{\boldsymbol{k}})\langle\langle a_{\boldsymbol{k}}; a^+_{\boldsymbol{k}'}\rangle\rangle_\omega = \delta_{\boldsymbol{k},\boldsymbol{k}'} - \frac{2zJ}{\hbar N}\sum_{\boldsymbol{q}_1,\boldsymbol{q}_2} v_{\boldsymbol{k},\boldsymbol{q}_1,\boldsymbol{q}_2,\boldsymbol{k}+\boldsymbol{q}_1-\boldsymbol{q}_2}\langle\langle a^+_{\boldsymbol{q}_1} a_{\boldsymbol{q}_2} a_{\boldsymbol{k}+\boldsymbol{q}_1-\boldsymbol{q}_2}; a^+_{\boldsymbol{k}'}\rangle\rangle_\omega. \tag{3.3.7}$$

Equation (3.3.7), which is exact, shows that the evaluation of the two-operator (one particle) Green function needs the knowledge of the four-operator (two particle) Green function entered by the commutator of $a_{\boldsymbol{k}}$ with the interaction Hamiltonian $\mathcal{H}_{\text{int}}$. Analogously, the equation of motion of the four-operator Green function involves the six-operator (three-particle) Green function as we will see in the next section. In this way, an *infinite hierarchy* of equations of motion is generated so that an appropriate *decoupling scheme* at some order has to be assigned in order to truncate the infinite hierarchy. We will follow a *systematic* criterion of truncation that allows us to control the order of the perturbation at each step: in particular, we will consider a series expansion in the interaction potential $v$. To show the procedure, let us begin with a simple exercise: let us evaluate the zero-order ($v = 0$) perturbation theory corresponding to the harmonic approximation. For $v = 0$, the solution of Eq. (3.3.7) is

$$\langle\langle a_{\boldsymbol{k}}; a^+_{\boldsymbol{k}'}\rangle\rangle^{(0)}_\omega = \delta_{\boldsymbol{k},\boldsymbol{k}'} G^{(0)}_{\boldsymbol{k}}(\omega) \tag{3.3.8}$$

with

$$G^{(0)}_{\boldsymbol{k}}(\omega) = \frac{1}{\omega - \omega_{\boldsymbol{k}}}. \tag{3.3.9}$$

Using the operator identity (3.2.12), the zero-order Green function (3.3.9) becomes

$$G^{(0)}_{\boldsymbol{k}}(\omega \pm i\epsilon) = \frac{\mathrm{P}}{\omega - \omega_{\boldsymbol{k}}} \mp i\pi\delta(\omega - \omega_{\boldsymbol{k}}) \tag{3.3.10}$$

and

$$G^{(0)}_{\boldsymbol{k}}(-\omega \pm i\epsilon) = -\frac{\mathrm{P}}{\omega + \omega_{\boldsymbol{k}}} \mp i\pi\delta(\omega + \omega_{\boldsymbol{k}}). \tag{3.3.11}$$

Replacing Eqs. (3.3.10) and (3.3.11) into Eq. (3.2.17), one obtains

$$S_\perp(\boldsymbol{K},\omega) = 2\frac{NS}{\hbar}\sum_{\boldsymbol{q},\boldsymbol{G}}[(1 + n^{(0)}_{\boldsymbol{q}})\,\delta_{\boldsymbol{K}+\boldsymbol{q},\boldsymbol{G}}\,\delta(\omega - \omega_{\boldsymbol{q}}) + n^{(0)}_{\boldsymbol{q}}\,\delta_{\boldsymbol{K}-\boldsymbol{q},\boldsymbol{G}}\,\delta(\omega + \omega_{\boldsymbol{q}})] \tag{3.3.12}$$

where $n^{(0)}_{\boldsymbol{k}} = (e^{\beta\hbar\omega_{\boldsymbol{k}}} - 1)^{-1}$. The result (3.3.12) obviously coincides with the harmonic result (3.1.17).

To go beyond the zero-order approximation, we need to keep the second term in the right-hand side of Eq. (3.3.7) into account. Let us write explicitly the frequency Fourier transform of the four-operator frequency Green function appearing in Eq. (3.3.7)

$$\langle\langle a^+_{\boldsymbol{q}_1} a_{\boldsymbol{q}_2} a_{\boldsymbol{k}+\boldsymbol{q}_1-\boldsymbol{q}_2}; a^+_{\boldsymbol{k}'}\rangle\rangle_{\omega\pm i\epsilon} = \mp i \int_{-\infty}^{+\infty} dt\theta(\pm t) e^{i\omega t} \langle [a^+_{\boldsymbol{q}_1}(t) a_{\boldsymbol{q}_2}(t) a_{\boldsymbol{k}+\boldsymbol{q}_1-\boldsymbol{q}_2}(t), a^+_{\boldsymbol{k}'}]\rangle. \tag{3.3.13}$$

A systematic perturbation expansion in the interaction potential is established assuming that in the higher-order Green functions, the time evolution of the boson operators is ruled by the harmonic Hamiltonian. In this way, the time dependent boson operators appearing in Eq. (3.3.13) may be approximated as follows

$$a_{\boldsymbol{k}}(t) = e^{i\mathcal{H}t/\hbar} a_{\boldsymbol{k}} e^{-i\mathcal{H}t/\hbar} = e^{i\mathcal{H}_0 t/\hbar} a_{\boldsymbol{k}} e^{-i\mathcal{H}_0 t/\hbar} + O(\mathcal{H}_{\text{int}}) = a_{\boldsymbol{k}} e^{-i\omega_{\boldsymbol{k}} t} + O(v) \tag{3.3.14}$$

where $v$ is the magnon-magnon interaction potential. As one can see from Eq. (3.3.14), the neglected terms are at least of order $v$. Then the first-order calculation gives

$$\begin{aligned} \langle\langle a^+_{\boldsymbol{q}_1} a_{\boldsymbol{q}_2} a_{\boldsymbol{k}+\boldsymbol{q}_1-\boldsymbol{q}_2}; a^+_{\boldsymbol{k}'}\rangle\rangle_{\omega\pm i\epsilon} &= \mp i \int_{-\infty}^{+\infty} dt\theta(\pm t) e^{i(\omega+\omega_{\boldsymbol{q}_1}-\omega_{\boldsymbol{q}_2}-\omega_{\boldsymbol{k}+\boldsymbol{q}_1-\boldsymbol{q}_2})t} \\ &\quad \times \langle [a^+_{\boldsymbol{q}_1} a_{\boldsymbol{q}_2} a_{\boldsymbol{k}+\boldsymbol{q}_1-\boldsymbol{q}_2}, a^+_{\boldsymbol{k}'}]\rangle + O(v) \\ &= \frac{\langle [a^+_{\boldsymbol{q}_1} a_{\boldsymbol{q}_2} a_{\boldsymbol{k}+\boldsymbol{q}_1-\boldsymbol{q}_2}, a^+_{\boldsymbol{k}'}]\rangle}{\omega+\omega_{\boldsymbol{q}_1}-\omega_{\boldsymbol{q}_2}-\omega_{\boldsymbol{k}+\boldsymbol{q}_1-\boldsymbol{q}_2} \pm i\epsilon} + O(v) \end{aligned} \tag{3.3.15}$$

where the last step of Eq. (3.3.15) is obtained from the integral representation of the step-function (3.2.8). In Eq. (3.3.15), all the boson operators within the commutator are at $t = 0$ so that the commutator itself can be evaluated using the commutation rules (2.2.15) leading to

$$\langle [a^+_{\boldsymbol{q}_1} a_{\boldsymbol{q}_2} a_{\boldsymbol{k}+\boldsymbol{q}_1-\boldsymbol{q}_2}, a^+_{\boldsymbol{k}'}]\rangle = \langle a^+_{\boldsymbol{q}_1} a_{\boldsymbol{q}_1}\rangle \delta_{\boldsymbol{k},\boldsymbol{k}'} (\delta_{\boldsymbol{q}_1,\boldsymbol{q}_2} + \delta_{\boldsymbol{k},\boldsymbol{q}_2}). \tag{3.3.16}$$

Replacing Eq. (3.3.16) into Eq. (3.3.15), one has

$$\begin{aligned} &\langle\langle a^+_{\boldsymbol{q}_1} a_{\boldsymbol{q}_2} a_{\boldsymbol{k}+\boldsymbol{q}_1-\boldsymbol{q}_2}; a^+_{\boldsymbol{k}'}\rangle\rangle_{\omega\pm i\epsilon} \\ &\quad = \langle a^+_{\boldsymbol{q}_1} a_{\boldsymbol{q}_1}\rangle \delta_{\boldsymbol{k},\boldsymbol{k}'} (\delta_{\boldsymbol{q}_1,\boldsymbol{q}_2} + \delta_{\boldsymbol{k},\boldsymbol{q}_2}) \frac{1}{\omega - \omega_{\boldsymbol{k}} \pm i\epsilon} + O(v) \\ &\quad = n^{(0)}_{\boldsymbol{q}_1} \delta_{\boldsymbol{k},\boldsymbol{k}'} (\delta_{\boldsymbol{q}_1,\boldsymbol{q}_2} + \delta_{\boldsymbol{k},\boldsymbol{q}_2}) G^{(0)}_{\boldsymbol{k}}(\omega \pm i\epsilon) + O(v). \end{aligned} \tag{3.3.17}$$

Consistent with the first-order perturbation theory, the average $\langle a^+_{\boldsymbol{q}_1} a_{\boldsymbol{q}_1}\rangle = n^{(0)}_{\boldsymbol{k}}$ is obtained from Eq. (3.2.15) by putting $t = 0$ and replacing $G_{\boldsymbol{k}}(\omega \pm i\epsilon)$ by $G^{(0)}_{\boldsymbol{k}}(\omega \pm i\epsilon)$ given in Eq. (3.3.9). Using Eq. (3.3.17) and from Eq. (3.3.7), one obtains

$$(\omega - \omega_{\boldsymbol{k}}) \langle\langle a_{\boldsymbol{k}}; a^+_{\boldsymbol{k}'}\rangle\rangle^{(1)}_{\omega} = \delta_{\boldsymbol{k},\boldsymbol{k}'} \left[ 1 - \frac{4zJ}{\hbar N} \sum_{\boldsymbol{q}_1} v_{\boldsymbol{k},\boldsymbol{q}_1,\boldsymbol{k},\boldsymbol{q}_1} n^{(0)}_{\boldsymbol{q}_1} G^{(0)}_{\boldsymbol{k}}(\omega) \right]. \tag{3.3.18}$$

Equation (3.3.18) has been obtained using the invariance of the interaction potential $v$ under the exchange of the first two and/or the last two labels. The first-order two-operator Green function is then given by

$$\langle\langle a_{\boldsymbol{k}}; a^+_{\boldsymbol{k}'}\rangle\rangle^{(1)}_\omega \equiv \delta_{\boldsymbol{k},\boldsymbol{k}'} G^{(1)}_{\boldsymbol{k}}(\omega) \tag{3.3.19}$$

where

$$G^{(1)}_{\boldsymbol{k}}(\omega) = G^{(0)}_{\boldsymbol{k}}(\omega) + \Sigma^{(1)}(\boldsymbol{k})\big[G^{(0)}_{\boldsymbol{k}}(\omega)\big]^2 \tag{3.3.20}$$

where the first-order *self-energy* $\Sigma^{(1)}(\boldsymbol{k})$ is given by

$$\Sigma^{(1)}(\boldsymbol{k}) = -\frac{4zJ}{\hbar N}\sum_{\boldsymbol{q}} v_{\boldsymbol{k},\boldsymbol{q},\boldsymbol{k},\boldsymbol{q}} n^{(0)}_{\boldsymbol{q}}. \tag{3.3.21}$$

Using the DM potential (3.3.4), the self-energy (3.3.21) becomes

$$\Sigma^{(1)}_{DM}(\boldsymbol{k}) = -\frac{4zJ}{\hbar N}\sum_{\boldsymbol{q}} v^{DM}_{\boldsymbol{k},\boldsymbol{q},\boldsymbol{k},\boldsymbol{q}} n^{(0)}_{\boldsymbol{q}} = -\frac{2zJ}{\hbar N}\sum_{\boldsymbol{q}}(1+\gamma_{\boldsymbol{k}-\boldsymbol{q}}-\gamma_{\boldsymbol{k}}-\gamma_{\boldsymbol{q}}) n^{(0)}_{\boldsymbol{q}} \tag{3.3.22}$$

while the HP potential (3.3.5) gives

$$\begin{aligned}\Sigma^{(1)}_{HP}(\boldsymbol{k}) &= -\frac{4zJ}{\hbar N}\sum_{\boldsymbol{q}}\left\{v^{DM}_{\boldsymbol{k},\boldsymbol{q},\boldsymbol{k},\boldsymbol{q}} + \frac{1}{2}\left[1-4S\left(1-\sqrt{1-\frac{1}{2S}}\right)\right](\gamma_{\boldsymbol{k}}+\gamma_{\boldsymbol{q}})\right\} n^{(0)}_{\boldsymbol{q}}\\ &= \Sigma^{(1)}_{DM}(\boldsymbol{k}) - \frac{2zJ}{\hbar N}\sum_{\boldsymbol{q}}\left[1-4S\left(1-\sqrt{1-\frac{1}{2S}}\right)\right](\gamma_{\boldsymbol{k}}+\gamma_{\boldsymbol{q}}) n^{(0)}_{\boldsymbol{q}}\\ &= \Sigma^{(1)}_{DM}(\boldsymbol{k}) + \left[\frac{1}{8S}+O\left(\frac{1}{S^2}\right)\right]\frac{2zJ}{\hbar N}\sum_{\boldsymbol{q}}(\gamma_{\boldsymbol{k}}+\gamma_{\boldsymbol{q}})\, n^{(0)}_{\boldsymbol{q}}.\end{aligned} \tag{3.3.23}$$

As one can see, the DM and HP first-order self-energies differ at the order $\frac{1}{S}$. Note that the unperturbed spectrum $\omega_{\boldsymbol{k}}$ is of order $O(S)$ and $\Sigma^{(1)}_{DM}(\boldsymbol{k})$ is of order $O(1)$. Moreover, the HP self-energy contains an infinite number of terms when expanded in powers of $\frac{1}{S}$ and it violates the Goldstone theorem since Eq. (3.3.23) does not vanish for $k = 0$, giving a spurious contribution to the Goldstone mode of the order $T^{\frac{3}{2}}$. This puzzle will be solved in the next section. Finally, note that both the DM and HP first-order self-energies are real functions independent of the frequency.

## 3.4. Second-Order Approximation

In this section, we perform the second-order perturbation expansion and we show that the second-order self-energy is no longer a real function. Instead of treating the four-operator Green function appearing in Eq. (3.3.7) in a perturbative way, we write the exact equation of the motion of the four-operator Green function and we apply the approximation (3.1.14) to the six-operator Green function. The equation of motion of the four-operator Green function is

$$\hbar\omega\langle\langle a^+_{\boldsymbol{q}_1}a_{\boldsymbol{q}_2}a_{\boldsymbol{k}+\boldsymbol{q}_1-\boldsymbol{q}_2};a^+_{\boldsymbol{k}'}\rangle\rangle_\omega = \hbar\langle[a^+_{\boldsymbol{q}_1}a_{\boldsymbol{q}_2}a_{\boldsymbol{k}+\boldsymbol{q}_1-\boldsymbol{q}_2},a^+_{\boldsymbol{k}'}]\rangle + \langle\langle[a^+_{\boldsymbol{q}_1}a_{\boldsymbol{q}_2}a_{\boldsymbol{k}+\boldsymbol{q}_1-\boldsymbol{q}_2},\mathcal{H}_0];a^+_{\boldsymbol{k}'}\rangle\rangle_\omega + \langle\langle[a^+_{\boldsymbol{q}_1}a_{\boldsymbol{q}_2}a_{\boldsymbol{k}+\boldsymbol{q}_1-\boldsymbol{q}_2},\mathcal{H}_{\rm int}];a^+_{\boldsymbol{k}'}\rangle\rangle_\omega. \quad (3.4.1)$$

The evaluation of the three terms on the right-hand side of Eq. (3.4.1) gives

$$\langle[a^+_{\boldsymbol{q}_1}a_{\boldsymbol{q}_2}a_{\boldsymbol{k}+\boldsymbol{q}_1-\boldsymbol{q}_2},a^+_{\boldsymbol{k}'}]\rangle = \langle a^+_{\boldsymbol{q}_1}a_{\boldsymbol{q}_1}\rangle\delta_{\boldsymbol{k},\boldsymbol{k}'}(\delta_{\boldsymbol{q}_1,\boldsymbol{q}_2}+\delta_{\boldsymbol{k},\boldsymbol{q}_2}), \quad (3.4.2)$$

$$\langle\langle[a^+_{\boldsymbol{q}_1}a_{\boldsymbol{q}_2}a_{\boldsymbol{k}+\boldsymbol{q}_1-\boldsymbol{q}_2},\mathcal{H}_0];a^+_{\boldsymbol{k}'}\rangle\rangle_\omega = \hbar(\omega_{\boldsymbol{k}+\boldsymbol{q}_1-\boldsymbol{q}_2}+\omega_{\boldsymbol{q}_2}-\omega_{\boldsymbol{q}_1})\langle\langle a^+_{\boldsymbol{q}_1}a_{\boldsymbol{q}_2}a_{\boldsymbol{k}+\boldsymbol{q}_1-\boldsymbol{q}_2};a^+_{\boldsymbol{k}'}\rangle\rangle_\omega \quad (3.4.3)$$

and

$$\begin{aligned}\langle\langle[a^+_{\boldsymbol{q}_1}a_{\boldsymbol{q}_2}a_{\boldsymbol{k}+\boldsymbol{q}_1-\boldsymbol{q}_2},\mathcal{H}_{\rm int}];a^+_{\boldsymbol{k}'}\rangle\rangle_\omega &= -\frac{2zJ}{N}\Bigg\{\sum_{\boldsymbol{p}}v_{\boldsymbol{k}+\boldsymbol{q}_1-\boldsymbol{q}_2,\boldsymbol{q}_2,\boldsymbol{p},\boldsymbol{k}+\boldsymbol{q}_1-\boldsymbol{p}}\langle\langle a^+_{\boldsymbol{q}_1}a_{\boldsymbol{p}}a_{\boldsymbol{k}+\boldsymbol{q}_1-\boldsymbol{p}};a^+_{\boldsymbol{k}'}\rangle\rangle_\omega \\ &+\sum_{\boldsymbol{p},\boldsymbol{q}}v_{\boldsymbol{k}+\boldsymbol{q}_1-\boldsymbol{q}_2,\boldsymbol{p},\boldsymbol{q},\boldsymbol{k}+\boldsymbol{q}_1-\boldsymbol{q}_2+\boldsymbol{p}-\boldsymbol{q}}\langle\langle a^+_{\boldsymbol{q}_1}a^+_{\boldsymbol{p}}a_{\boldsymbol{q}_2}a_{\boldsymbol{q}}a_{\boldsymbol{k}+\boldsymbol{q}_1-\boldsymbol{q}_2+\boldsymbol{p}-\boldsymbol{q}};a^+_{\boldsymbol{k}'}\rangle\rangle_\omega \\ &+\sum_{\boldsymbol{p},\boldsymbol{q}}v_{\boldsymbol{q}_2,\boldsymbol{p},\boldsymbol{q},\boldsymbol{q}_2+\boldsymbol{p}-\boldsymbol{q}}\langle\langle a^+_{\boldsymbol{q}_1}a^+_{\boldsymbol{p}}a_{\boldsymbol{k}+\boldsymbol{q}_1-\boldsymbol{q}_2}a_{\boldsymbol{q}}a_{\boldsymbol{q}_2+\boldsymbol{p}-\boldsymbol{q}};a^+_{\boldsymbol{k}'}\rangle\rangle_\omega \\ &-\sum_{\boldsymbol{p},\boldsymbol{q}}v_{\boldsymbol{p},\boldsymbol{q},\boldsymbol{q}_1,\boldsymbol{p}+\boldsymbol{q}-\boldsymbol{q}_1}\langle\langle a^+_{\boldsymbol{p}}a^+_{\boldsymbol{q}}a_{\boldsymbol{p}+\boldsymbol{q}-\boldsymbol{q}_1}a_{\boldsymbol{q}_2}a_{\boldsymbol{k}+\boldsymbol{q}_1-\boldsymbol{q}_2};a^+_{\boldsymbol{k}'}\rangle\rangle_\omega\Bigg\}. \quad (3.4.4)\end{aligned}$$

By means of Eqs. (3.4.2)–(3.4.4), the equation of motion of the four-operator Green function (3.4.1) becomes

$$\begin{aligned}&(\omega-\omega_{\boldsymbol{k}+\boldsymbol{q}_1-\boldsymbol{q}_2}-\omega_{\boldsymbol{q}_2}+\omega_{\boldsymbol{q}_1})\langle\langle a^+_{\boldsymbol{q}_1}a_{\boldsymbol{q}_2}a_{\boldsymbol{k}+\boldsymbol{q}_1-\boldsymbol{q}_2};a^+_{\boldsymbol{k}'}\rangle\rangle_\omega \\ &= \langle a^+_{\boldsymbol{q}_1}a_{\boldsymbol{q}_1}\rangle\delta_{\boldsymbol{k},\boldsymbol{k}'}(\delta_{\boldsymbol{q}_1,\boldsymbol{q}_2}+\delta_{\boldsymbol{k},\boldsymbol{q}_2}) \\ &-\frac{2zJ}{\hbar N}\sum_{\boldsymbol{p}}v_{\boldsymbol{k}+\boldsymbol{q}_1-\boldsymbol{q}_2,\boldsymbol{q}_2,\boldsymbol{p},\boldsymbol{k}+\boldsymbol{q}_1-\boldsymbol{p}}\langle\langle a^+_{\boldsymbol{q}_1}a_{\boldsymbol{p}}a_{\boldsymbol{k}+\boldsymbol{q}_1-\boldsymbol{p}};a^+_{\boldsymbol{k}'}\rangle\rangle_\omega \\ &-\frac{2zJ}{\hbar N}\sum_{\boldsymbol{p},\boldsymbol{q}}v_{\boldsymbol{k}+\boldsymbol{q}_1-\boldsymbol{q}_2,\boldsymbol{p},\boldsymbol{q},\boldsymbol{k}+\boldsymbol{q}_1-\boldsymbol{q}_2+\boldsymbol{p}-\boldsymbol{q}}\langle\langle a^+_{\boldsymbol{q}_1}a^+_{\boldsymbol{p}}a_{\boldsymbol{q}_2}a_{\boldsymbol{q}}a_{\boldsymbol{k}+\boldsymbol{q}_1-\boldsymbol{q}_2+\boldsymbol{p}-\boldsymbol{q}};a^+_{\boldsymbol{k}'}\rangle\rangle_\omega \\ &-\frac{2zJ}{\hbar N}\sum_{\boldsymbol{p},\boldsymbol{q}}v_{\boldsymbol{q}_2,\boldsymbol{p},\boldsymbol{q},\boldsymbol{q}_2+\boldsymbol{p}-\boldsymbol{q}}\langle\langle a^+_{\boldsymbol{q}_1}a^+_{\boldsymbol{p}}a_{\boldsymbol{k}+\boldsymbol{q}_1-\boldsymbol{q}_2}a_{\boldsymbol{q}}a_{\boldsymbol{q}_2+\boldsymbol{p}-\boldsymbol{q}};a^+_{\boldsymbol{k}'}\rangle\rangle_\omega \\ &+\frac{2zJ}{\hbar N}\sum_{\boldsymbol{p},\boldsymbol{q}}v_{\boldsymbol{p},\boldsymbol{q},\boldsymbol{q}_1,\boldsymbol{p}+\boldsymbol{q}-\boldsymbol{q}_1}\langle\langle a^+_{\boldsymbol{p}}a^+_{\boldsymbol{q}}a_{\boldsymbol{p}+\boldsymbol{q}-\boldsymbol{q}_1}a_{\boldsymbol{q}_2}a_{\boldsymbol{k}+\boldsymbol{q}_1-\boldsymbol{q}_2};a^+_{\boldsymbol{k}'}\rangle\rangle_\omega. \quad (3.4.5)\end{aligned}$$

Note that Eq. (3.4.5) is exact but it is of no use since it involves the six-operator Green functions. Following the systematic approach introduced in the previous section, we replace the time evolution of the boson operators in the Green functions of the right-hand side of Eq. (3.4.5) by the time evolution given by Eq. (3.3.14). In so doing, we neglect terms of order $O(v^2)$ in Eq. (3.4.5) and terms of order $O(v^3)$ in Eq. (3.3.7). Within the second-order perturbation expansion, the terms occurring in Eq. (3.4.5) are given by

$$\langle\langle a^+_{\boldsymbol{q}_1} a_{\boldsymbol{p}} a_{\boldsymbol{k}+\boldsymbol{q}_1-\boldsymbol{p}}; a^+_{\boldsymbol{k}'}\rangle\rangle_\omega = \delta_{\boldsymbol{k},\boldsymbol{k}'} G^{(0)}_{\boldsymbol{k}}(\omega) n^{(0)}_{\boldsymbol{q}_1}(\delta_{\boldsymbol{q}_1,\boldsymbol{p}} + \delta_{\boldsymbol{k},\boldsymbol{p}}) + O(v), \tag{3.4.6}$$

$$\begin{aligned}\langle\langle a^+_{\boldsymbol{q}_1} a^+_{\boldsymbol{p}} a_{\boldsymbol{q}_2} a_{\boldsymbol{q}} a_{\boldsymbol{k}+\boldsymbol{q}_1-\boldsymbol{q}_2+\boldsymbol{p}-\boldsymbol{q}}; a^+_{\boldsymbol{k}'}\rangle\rangle_\omega &\\ = \delta_{\boldsymbol{k},\boldsymbol{k}'} G^{(0)}_{\boldsymbol{k}}(\omega) n^{(0)}_{\boldsymbol{q}_1} n^{(0)}_{\boldsymbol{p}} (\delta_{\boldsymbol{q}_1,\boldsymbol{q}_2}\delta_{\boldsymbol{p},\boldsymbol{q}} + \delta_{\boldsymbol{q},\boldsymbol{q}_1}\delta_{\boldsymbol{p},\boldsymbol{q}_2} + \delta_{\boldsymbol{q}_1,\boldsymbol{q}_2}\delta_{\boldsymbol{k},\boldsymbol{q}} &\\ + \delta_{\boldsymbol{p},\boldsymbol{q}_2}\delta_{\boldsymbol{k},\boldsymbol{q}} + \delta_{\boldsymbol{q},\boldsymbol{q}_1}\delta_{\boldsymbol{k},\boldsymbol{q}_2} + \delta_{\boldsymbol{p},\boldsymbol{q}}\delta_{\boldsymbol{k},\boldsymbol{q}_2}) + O(v),\end{aligned} \tag{3.4.7}$$

$$\begin{aligned}\langle\langle a^+_{\boldsymbol{q}_1} a^+_{\boldsymbol{p}} a_{\boldsymbol{k}+\boldsymbol{q}_1-\boldsymbol{q}_2} a_{\boldsymbol{q}} a_{\boldsymbol{q}_2+\boldsymbol{p}-\boldsymbol{q}}; a^+_{\boldsymbol{k}'}\rangle\rangle_\omega = \delta_{\boldsymbol{k},\boldsymbol{k}'} G^{(0)}_{\boldsymbol{k}}(\omega) n^{(0)}_{\boldsymbol{q}_1} n^{(0)}_{\boldsymbol{p}} (\delta_{\boldsymbol{k},\boldsymbol{q}_2}\delta_{\boldsymbol{p},\boldsymbol{q}} &\\ + \delta_{\boldsymbol{q},\boldsymbol{q}_1}\delta_{\boldsymbol{p},\boldsymbol{k}+\boldsymbol{q}_1-\boldsymbol{q}_2} + \delta_{\boldsymbol{k},\boldsymbol{q}_2}\delta_{\boldsymbol{k},\boldsymbol{q}} + \delta_{\boldsymbol{k},\boldsymbol{q}}\delta_{\boldsymbol{p},\boldsymbol{k}+\boldsymbol{q}_1-\boldsymbol{q}_2} &\\ + \delta_{\boldsymbol{q},\boldsymbol{q}_1}\delta_{\boldsymbol{q}_1,\boldsymbol{q}_2} + \delta_{\boldsymbol{p},\boldsymbol{q}}\delta_{\boldsymbol{q}_1,\boldsymbol{q}_2}) + O(v)\end{aligned} \tag{3.4.8}$$

and

$$\begin{aligned}\langle\langle a^+_{\boldsymbol{p}} a^+_{\boldsymbol{q}} a_{\boldsymbol{p}+\boldsymbol{q}-\boldsymbol{q}_1} a_{\boldsymbol{q}_2} a_{\boldsymbol{k}+\boldsymbol{q}_1-\boldsymbol{q}_2}; a^+_{\boldsymbol{k}'}\rangle\rangle_\omega &\\ = \delta_{\boldsymbol{k},\boldsymbol{k}'} G^{(0)}_{\boldsymbol{k}}(\omega) n^{(0)}_{\boldsymbol{p}} n^{(0)}_{\boldsymbol{q}} (\delta_{\boldsymbol{q},\boldsymbol{q}_1}\delta_{\boldsymbol{q}_1,\boldsymbol{q}_2} + \delta_{\boldsymbol{p},\boldsymbol{q}_2}\delta_{\boldsymbol{q}_1,\boldsymbol{q}_2} + \delta_{\boldsymbol{q},\boldsymbol{q}_1}\delta_{\boldsymbol{k},\boldsymbol{q}_2} &\\ + \delta_{\boldsymbol{p},\boldsymbol{q}_1}\delta_{\boldsymbol{k},\boldsymbol{q}_2} + \delta_{\boldsymbol{p},\boldsymbol{q}_2}\delta_{\boldsymbol{q},\boldsymbol{k}+\boldsymbol{q}_1-\boldsymbol{q}_2} + \delta_{\boldsymbol{q},\boldsymbol{q}_2}\delta_{\boldsymbol{q},\boldsymbol{k}+\boldsymbol{q}_1-\boldsymbol{q}_2}) + O(v).\end{aligned} \tag{3.4.9}$$

By means of Eqs. (3.4.6)–(3.4.9), the sums in Eq. (3.4.5) become

$$\begin{aligned}&-\frac{2zJ}{\hbar N}\sum_{\boldsymbol{p}} v_{\boldsymbol{k}+\boldsymbol{q}_1-\boldsymbol{q}_2,\boldsymbol{q}_2,\boldsymbol{p},\boldsymbol{k}+\boldsymbol{q}_1-\boldsymbol{p}}\langle\langle a^+_{\boldsymbol{q}_1} a_{\boldsymbol{p}} a_{\boldsymbol{k}+\boldsymbol{q}_1-\boldsymbol{p}}; a^+_{\boldsymbol{k}'}\rangle\rangle_\omega \\ &= -\frac{4zJ}{\hbar N}\delta_{\boldsymbol{k},\boldsymbol{k}'} G^{(0)}_{\boldsymbol{k}}(\omega) v_{\boldsymbol{k}+\boldsymbol{q}_1-\boldsymbol{q}_2,\boldsymbol{q}_2,\boldsymbol{q}_1,\boldsymbol{k}} n^{(0)}_{\boldsymbol{q}_1} + O(v^2),\end{aligned} \tag{3.4.10}$$

$$\begin{aligned}&-\frac{2zJ}{\hbar N}\sum_{\boldsymbol{p},\boldsymbol{q}} v_{\boldsymbol{k}+\boldsymbol{q}_1-\boldsymbol{q}_2,\boldsymbol{p},\boldsymbol{q},\boldsymbol{k}+\boldsymbol{q}_1-\boldsymbol{q}_2+\boldsymbol{p}-\boldsymbol{q}}\langle\langle a^+_{\boldsymbol{q}_1} a^+_{\boldsymbol{p}} a_{\boldsymbol{q}_2} a_{\boldsymbol{q}} a_{\boldsymbol{k}+\boldsymbol{q}_1-\boldsymbol{q}_2+\boldsymbol{p}-\boldsymbol{q}}; a^+_{\boldsymbol{k}'}\rangle\rangle_\omega \\ &= \delta_{\boldsymbol{k},\boldsymbol{k}'} G^{(0)}_{\boldsymbol{k}}(\omega) \\ &\quad\times\left\{-\frac{4zJ}{\hbar N} v_{\boldsymbol{k}+\boldsymbol{q}_1-\boldsymbol{q}_2,\boldsymbol{q}_2,\boldsymbol{k},\boldsymbol{q}_1} n^{(0)}_{\boldsymbol{q}_1} n^{(0)}_{\boldsymbol{q}_2} + [\Sigma^{(1)}(\boldsymbol{k})\delta_{\boldsymbol{q}_1,\boldsymbol{q}_2} + \Sigma^{(1)}(\boldsymbol{q}_1)\delta_{\boldsymbol{k},\boldsymbol{q}_2}] n^{(0)}_{\boldsymbol{q}_1}\right\} \\ &\quad + O(v^2),\end{aligned} \tag{3.4.11}$$

$$\begin{aligned}&-\frac{2zJ}{\hbar N}\sum_{\boldsymbol{p},\boldsymbol{q}} v_{\boldsymbol{q}_2,\boldsymbol{p},\boldsymbol{q},\boldsymbol{q}_2+\boldsymbol{p}-\boldsymbol{q}}\langle\langle a^+_{\boldsymbol{q}_1} a^+_{\boldsymbol{p}} a_{\boldsymbol{k}+\boldsymbol{q}_1-\boldsymbol{q}_2} a_{\boldsymbol{q}} a_{\boldsymbol{q}_2+\boldsymbol{p}-\boldsymbol{q}}; a^+_{\boldsymbol{k}'}\rangle\rangle_\omega = \delta_{\boldsymbol{k},\boldsymbol{k}'} G^{(0)}_{\boldsymbol{k}}(\omega) \\ &\quad\times\left\{-\frac{4zJ}{\hbar N} v_{\boldsymbol{k}+\boldsymbol{q}_1-\boldsymbol{q}_2,\boldsymbol{q}_2,\boldsymbol{k},\boldsymbol{q}_1} n^{(0)}_{\boldsymbol{q}_1} n^{(0)}_{\boldsymbol{k}+\boldsymbol{q}_1-\boldsymbol{q}_2} + [\Sigma^{(1)}(\boldsymbol{k})\delta_{\boldsymbol{k},\boldsymbol{q}_2} + \Sigma^{(1)}(\boldsymbol{q}_1)\delta_{\boldsymbol{q}_1,\boldsymbol{q}_2}] n^{(0)}_{\boldsymbol{q}_1}\right\} \\ &\quad + O(v^2)\end{aligned} \tag{3.4.12}$$

and

$$\frac{2zJ}{\hbar N}\sum_{\boldsymbol{p},\boldsymbol{q}} v_{\boldsymbol{p},\boldsymbol{q},\boldsymbol{q}_1,\boldsymbol{p}+\boldsymbol{q}-\boldsymbol{q}_1}\langle\langle a^+_{\boldsymbol{p}} a^+_{\boldsymbol{q}} a_{\boldsymbol{p}+\boldsymbol{q}-\boldsymbol{q}_1} a_{\boldsymbol{q}_2} a_{\boldsymbol{k}+\boldsymbol{q}_1-\boldsymbol{q}_2}; a^+_{\boldsymbol{k}'}\rangle\rangle_\omega$$

$$= \delta_{\boldsymbol{k},\boldsymbol{k}'} G^{(0)}_{\boldsymbol{k}}(\omega)\left[\frac{4zJ}{\hbar N} v_{\boldsymbol{k}+\boldsymbol{q}_1-\boldsymbol{q}_2,\boldsymbol{q}_2,\boldsymbol{q}_1,\boldsymbol{k}} n^{(0)}_{\boldsymbol{q}_2} n^{(0)}_{\boldsymbol{k}+\boldsymbol{q}_1-\boldsymbol{q}_2}\right.$$

$$\left. - \Sigma^{(1)}(\boldsymbol{q}_1) n^{(0)}_{\boldsymbol{q}_1}(\delta_{\boldsymbol{q}_1,\boldsymbol{q}_2} + \delta_{\boldsymbol{k},\boldsymbol{q}_2})\right] + O(v^2) \tag{3.4.13}$$

where $\Sigma^{(1)}(\boldsymbol{q})$ is the first-order self-energy (3.3.21). Replacing Eqs. (3.4.10)–(3.4.13) into the equation of motion (3.4.5), one obtains

$$(\omega - \omega_{\boldsymbol{k}+\boldsymbol{q}_1-\boldsymbol{q}_2} - \omega_{\boldsymbol{q}_2} + \omega_{\boldsymbol{q}_1})\langle\langle a^+_{\boldsymbol{q}_1} a_{\boldsymbol{q}_2} a_{\boldsymbol{k}+\boldsymbol{q}_1-\boldsymbol{q}_2}; a^+_{\boldsymbol{k}'}\rangle\rangle_\omega$$

$$= \langle a^+_{\boldsymbol{q}_1} a_{\boldsymbol{q}_1}\rangle \delta_{\boldsymbol{k},\boldsymbol{k}'}(\delta_{\boldsymbol{q}_1,\boldsymbol{q}_2} + \delta_{\boldsymbol{k},\boldsymbol{q}_2})$$

$$+ \delta_{\boldsymbol{k},\boldsymbol{k}'} G^{(0)}_{\boldsymbol{k}}(\omega)\left\{(\delta_{\boldsymbol{q}_1,\boldsymbol{q}_2} + \delta_{\boldsymbol{k},\boldsymbol{q}_2})\Sigma^{(1)}(\boldsymbol{k}) n^{(0)}_{\boldsymbol{q}_1} - \frac{4zJ}{\hbar N} v_{\boldsymbol{k}+\boldsymbol{q}_1-\boldsymbol{q}_2,\boldsymbol{q}_2,\boldsymbol{q}_1,\boldsymbol{k}}\right.$$

$$\left.\times\left[n^{(0)}_{\boldsymbol{q}_1}(1 + n^{(0)}_{\boldsymbol{q}_2} + n^{(0)}_{\boldsymbol{k}+\boldsymbol{q}_1-\boldsymbol{q}_2}) - n^{(0)}_{\boldsymbol{q}_2} n^{(0)}_{\boldsymbol{k}+\boldsymbol{q}_1-\boldsymbol{q}_2}\right]\right\} + O(v^2). \tag{3.4.14}$$

It remains to evaluate the average $\langle a^+_{\boldsymbol{q}_1} a_{\boldsymbol{q}_1}\rangle$ at the consistent order in $v$. To do this, we replace in the correlation function (3.2.15) $G_{\boldsymbol{k}}(\omega)$ by $G^{(1)}_{\boldsymbol{k}}(\omega)$ given by Eq. (3.3.20), thus obtaining

$$\langle a^+_{\boldsymbol{q}_1} a_{\boldsymbol{q}_1}\rangle = \lim_{\epsilon\to 0}\frac{i}{2\pi}\int_{-\infty}^{+\infty} d\omega n(\omega)[G^{(1)}_{\boldsymbol{q}_1}(\omega + i\epsilon) - G^{(1)}_{\boldsymbol{q}_1}(\omega - i\epsilon)]$$

$$= \lim_{\epsilon\to 0}\frac{i}{2\pi}\int_{-\infty}^{+\infty} d\omega n(\omega)[G^{(0)}_{\boldsymbol{q}_1}(\omega + i\epsilon) - G^{(0)}_{\boldsymbol{q}_1}(\omega - i\epsilon)]$$

$$+ \Sigma^{(1)}(\boldsymbol{q}_1)\lim_{\epsilon\to 0}\frac{i}{2\pi}\int_{-\infty}^{+\infty} d\omega n(\omega)\{[G^{(0)}_{\boldsymbol{q}_1}(\omega + i\epsilon)]^2 - [G^{(0)}_{\boldsymbol{q}_1}(\omega - i\epsilon)]^2\}$$

$$= n^{(0)}_{\boldsymbol{q}_1} + \Sigma^{(1)}(\boldsymbol{q}_1)\lim_{\epsilon\to 0}\frac{1}{\pi}\int_{-\infty}^{+\infty} d\omega n(\omega)\frac{2\epsilon(\omega - \omega_{\boldsymbol{q}_1})}{[(\omega - \omega_{\boldsymbol{q}_1})^2 + \epsilon^2]^2} + O(v^2). \tag{3.4.15}$$

The integral occurring in Eq. (3.4.15) may be evaluated by parts leading to

$$\int_{-\infty}^{+\infty} d\omega n(\omega)\frac{2\epsilon(\omega - \omega_{\boldsymbol{q}_1})}{[(\omega - \omega_{\boldsymbol{q}_1})^2 + \epsilon^2]^2} = \int_{-\infty}^{+\infty} d\omega\left[\frac{dn(\omega)}{d\omega}\right]\frac{\epsilon}{(\omega - \omega_{\boldsymbol{q}_1})^2 + \epsilon^2} \tag{3.4.16}$$

since the first term vanishes when evaluated for $\omega = \pm\infty$ and

$$\frac{dn(\omega)}{d\omega} = -\beta\hbar\frac{e^{\beta\hbar\omega}}{(e^{\beta\hbar\omega} - 1)^2} = -\beta\hbar n(\omega)[1 + n(\omega)].$$

Using the definition of the $\delta$-function

$$\delta(\omega - \omega_0) = \lim_{\epsilon\to 0}\frac{1}{\pi}\frac{\epsilon}{(\omega - \omega_0)^2 + \epsilon^2},$$

one obtains

$$\lim_{\epsilon\to 0}\frac{1}{\pi}\int_{-\infty}^{+\infty} d\omega n(\omega)\frac{2\epsilon(\omega-\omega_{\boldsymbol{q}_1})}{[(\omega-\omega_{\boldsymbol{q}_1})^2+\epsilon^2]^2} = -\beta\hbar n_{\boldsymbol{q}_1}^{(0)}(1+n_{\boldsymbol{q}_1}^{(0)}) \tag{3.4.17}$$

so that

$$\langle a_{\boldsymbol{q}_1}^+ a_{\boldsymbol{q}_1}\rangle \equiv n_{\boldsymbol{q}_1}^{(1)} = n_{\boldsymbol{q}_1}^{(0)}[1-\beta\hbar(1+n_{\boldsymbol{q}_1}^{(0)})\Sigma^{(1)}(\boldsymbol{q}_1)]. \tag{3.4.18}$$

Replacing Eq. (3.4.18) into Eq. (3.4.14), one obtains the two-particle Green function

$$\begin{aligned}
&\langle\langle a_{\boldsymbol{q}_1}^+ a_{\boldsymbol{q}_2} a_{\boldsymbol{k}+\boldsymbol{q}_1-\boldsymbol{q}_2}; a_{\boldsymbol{k}'}^+\rangle\rangle_\omega \\
&= \delta_{\boldsymbol{k},\boldsymbol{k}'} G_{\boldsymbol{k}}^{(0)}(\omega)\Big\{ n_{\boldsymbol{q}_1}^{(0)}(\delta_{\boldsymbol{q}_1,\boldsymbol{q}_2}+\delta_{\boldsymbol{k},\boldsymbol{q}_2})[1-\beta\hbar(1+n_{\boldsymbol{q}_1}^{(0)})\Sigma^{(1)}(\boldsymbol{q}_1)+G_{\boldsymbol{k}}^{(0)}(\omega)\Sigma^{(1)}(\boldsymbol{k})] \\
&\quad - \frac{4zJ}{\hbar N} v_{\boldsymbol{k}+\boldsymbol{q}_1-\boldsymbol{q}_2,\boldsymbol{q}_2,\boldsymbol{q}_1,\boldsymbol{k}} \frac{n_{\boldsymbol{q}_1}^{(0)}(1+n_{\boldsymbol{q}_2}^{(0)}+n_{\boldsymbol{k}+\boldsymbol{q}_1-\boldsymbol{q}_2}^{(0)}) - n_{\boldsymbol{q}_2}^{(0)} n_{\boldsymbol{k}+\boldsymbol{q}_1-\boldsymbol{q}_2}^{(0)}}{\omega-\omega_{\boldsymbol{k}+\boldsymbol{q}_1-\boldsymbol{q}_2}-\omega_{\boldsymbol{q}_2}+\omega_{\boldsymbol{q}_1}}\Big\} \\
&\quad + O(v^2).
\end{aligned} \tag{3.4.19}$$

Substituting Eq. (3.4.19) into Eq. (3.3.7), the second-order one-particle Green function becomes

$$\begin{aligned}
(\omega-\omega_{\boldsymbol{k}})\langle\langle a_{\boldsymbol{k}}; a_{\boldsymbol{k}'}^+\rangle\rangle_\omega^{(2)} = \delta_{\boldsymbol{k},\boldsymbol{k}'}\{1&+\Sigma^{(1)}(\boldsymbol{k})G_{\boldsymbol{k}}^{(0)}(\omega)+[\Sigma_a^{(2)}(\boldsymbol{k})+\Sigma_b^{(2)}(\boldsymbol{k},\omega) \\
&+\Sigma_c^{(2)}(\boldsymbol{k},\omega)]G_{\boldsymbol{k}}^{(0)}(\omega)\}
\end{aligned} \tag{3.4.20}$$

where

$$\Sigma_a^{(2)}(\boldsymbol{k},\omega) = G_{\boldsymbol{k}}^{(0)}(\omega)[\Sigma^{(1)}(\boldsymbol{k})]^2, \tag{3.4.21}$$

$$\Sigma_b^{(2)}(\boldsymbol{k}) = 4\beta\hbar\left(\frac{zJ}{\hbar N}\right)\sum_{\boldsymbol{q}} v_{\boldsymbol{k},\boldsymbol{q},\boldsymbol{k},\boldsymbol{q}} n_{\boldsymbol{q}}^{(0)}(1+n_{\boldsymbol{q}}^{(0)})\Sigma^{(1)}(\boldsymbol{q}) \tag{3.4.22}$$

and

$$\Sigma_c^{(2)}(\boldsymbol{k},\omega) = 2\left(\frac{2zJ}{\hbar N}\right)^2 \sum_{\boldsymbol{q}_1,\boldsymbol{q}_2} A_{\boldsymbol{k}}(\boldsymbol{q}_1,\boldsymbol{q}_2)\frac{n_{\boldsymbol{q}_1}^{(0)}(1+n_{\boldsymbol{q}_2}^{(0)}+n_{\boldsymbol{k}+\boldsymbol{q}_1-\boldsymbol{q}_2}^{(0)}) - n_{\boldsymbol{q}_2}^{(0)} n_{\boldsymbol{k}+\boldsymbol{q}_1-\boldsymbol{q}_2}^{(0)}}{\omega-\omega_{\boldsymbol{k}+\boldsymbol{q}_1-\boldsymbol{q}_2}-\omega_{\boldsymbol{q}_2}+\omega_{\boldsymbol{q}_1}} \tag{3.4.23}$$

with

$$A_{\boldsymbol{k}}(\boldsymbol{q}_1,\boldsymbol{q}_2) = v_{\boldsymbol{k},\boldsymbol{q}_1,\boldsymbol{q}_2,\boldsymbol{k}+\boldsymbol{q}_1-\boldsymbol{q}_2} v_{\boldsymbol{k}+\boldsymbol{q}_1-\boldsymbol{q}_2,\boldsymbol{q}_2,\boldsymbol{q}_1,\boldsymbol{k}}. \tag{3.4.23a}$$

Within the perturbation theory, in Eq. (3.4.23) $\omega$ has to be replaced by $\omega_{\boldsymbol{k}}\pm i\epsilon$ where the infinitesimal imaginary part is reminiscent of the choice of the argument of the generalized Green function $G_{\boldsymbol{k}}(\omega \pm i\epsilon)$ corresponding to the retarded or advanced Green function, respectively. Using Eq. (3.2.12), the self-energy (3.4.23) becomes

$$\Sigma_c^{(2)}(\boldsymbol{k}, \omega_{\boldsymbol{k}} \pm i\epsilon) = \Sigma_c^{(2)'}(\boldsymbol{k}, \omega_{\boldsymbol{k}}) \mp i\Sigma_c^{(2)''}(\boldsymbol{k}, \omega_{\boldsymbol{k}}) \tag{3.4.24}$$

where the real part (renormalization) and the imaginary part (damping) of the self-energy (3.4.24) are given by

$$\Sigma_c^{(2)'}(\boldsymbol{k}, \omega_{\boldsymbol{k}}) = 2\left(\frac{2zJ}{\hbar N}\right)^2 \mathrm{P} \sum_{\boldsymbol{q}_1, \boldsymbol{q}_2} A_{\boldsymbol{k}}(\boldsymbol{q}_1, \boldsymbol{q}_2) \frac{n_{\boldsymbol{q}_1}^{(0)}(1 + n_{\boldsymbol{q}_2}^{(0)} + n_{\boldsymbol{k}+\boldsymbol{q}_1-\boldsymbol{q}_2}^{(0)}) - n_{\boldsymbol{q}_2}^{(0)} n_{\boldsymbol{k}+\boldsymbol{q}_1-\boldsymbol{q}_2}^{(0)}}{\omega_{\boldsymbol{k}} - \omega_{\boldsymbol{k}+\boldsymbol{q}_1-\boldsymbol{q}_2} - \omega_{\boldsymbol{q}_2} + \omega_{\boldsymbol{q}_1}} \tag{3.4.25}$$

and

$$\Sigma_c^{(2)''}(\boldsymbol{k}, \omega_{\boldsymbol{k}}) = 2\pi \left(\frac{2zJ}{\hbar N}\right)^2 \sum_{\boldsymbol{q}_1, \boldsymbol{q}_2} [n_{\boldsymbol{q}_1}^{(0)}(1 + n_{\boldsymbol{q}_2}^{(0)} + n_{\boldsymbol{k}+\boldsymbol{q}_1-\boldsymbol{q}_2}^{(0)}) - n_{\boldsymbol{q}_2}^{(0)} n_{\boldsymbol{k}+\boldsymbol{q}_1-\boldsymbol{q}_2}^{(0)}]$$
$$\times A_{\boldsymbol{k}}(\boldsymbol{q}_1, \boldsymbol{q}_2)\, \delta(\omega_{\boldsymbol{k}} - \omega_{\boldsymbol{k}+\boldsymbol{q}_1-\boldsymbol{q}_2} - \omega_{\boldsymbol{q}_2} + \omega_{\boldsymbol{q}_1}), \tag{3.4.26}$$

respectively. Defining

$$\langle\langle a_{\boldsymbol{k}}; a_{\boldsymbol{k}'}^+ \rangle\rangle_\omega^{(2)} = \delta_{\boldsymbol{k},\boldsymbol{k}'} G_{\boldsymbol{k}}^{(2)}(\omega) \tag{3.4.27}$$

from Eq. (3.4.20), we obtain the second-order one-particle Green function

$$G_{\boldsymbol{k}}^{(2)}(\omega \pm i\epsilon) = G_{\boldsymbol{k}}^{(0)}(\omega \pm i\epsilon)\{1 + [\Sigma^{(1)}(\boldsymbol{k}) + \Sigma_b^{(2)}(\boldsymbol{k}) + \Sigma_c^{(2)}(\boldsymbol{k}, \omega_{\boldsymbol{k}} \pm i\epsilon)] G_{\boldsymbol{k}}^{(0)}(\omega \pm i\epsilon)$$
$$+ [\Sigma^{(1)}(\boldsymbol{k}) G_{\boldsymbol{k}}^{(0)}(\omega \pm i\epsilon)]^2\} \tag{3.4.28}$$

where the self-energy $\Sigma_a^{(2)}$ has been replaced by Eq. (3.4.21). Note that the second-order self-energy $\Sigma_a^{(2)}$ defined in Eq. (3.4.21) is a *reducible* self-energy since it is the square of the first-order self-energy $\Sigma^{(1)}$ times an unperturbed one-particle Green function. The self-energy $\Sigma_b^{(2)}$ defined in Eq. (3.4.22) is a real function of the wavevector $\boldsymbol{k}$ and of the temperature $T$ and contributes only to the renormalization but not to the damping while the self-energy $\Sigma_c^{(2)}$ defined in Eq. (3.4.24) is characterized by both a real contribution (renormalization) given by Eq. (3.4.25) and an imaginary contribution (damping) given by Eq. (3.4.26).

Before evaluating explicitly the second-order self-energies $\Sigma_b^{(2)}$ and $\Sigma_c^{(2)}$, we wish to solve the puzzle about the serious discrepancy between the DM and HP first-order self-energies pointed out at the end of Section 3.3. First of all, we note that the main contribution in temperature to the second-order self-energy renormalization comes from $\Sigma_c^{(2)'}$ given by Eq. (3.4.25) where a term with a single Bose factor $n_{\boldsymbol{q}_1}^{(0)}$ occurs. All other contributions contain products of two Bose factors leading to higher powers in temperature. Using the DM potential (3.3.4) in Eq. (3.4.23a), one has

$$A_{\boldsymbol{k}}^{DM}(\boldsymbol{q}_1,\boldsymbol{q}_2) = \frac{1}{4}(\gamma_{\boldsymbol{k}-\boldsymbol{q}_2} + \gamma_{\boldsymbol{q}_1-\boldsymbol{q}_2} - \gamma_{\boldsymbol{k}} - \gamma_{\boldsymbol{q}_1})(\gamma_{\boldsymbol{k}-\boldsymbol{q}_2} + \gamma_{\boldsymbol{q}_1-\boldsymbol{q}_2} - \gamma_{\boldsymbol{k}+\boldsymbol{q}_1-\boldsymbol{q}_2} - \gamma_{\boldsymbol{q}_2}) \tag{3.4.29}$$

while taking the HP potential (3.3.5), one has

$$\begin{aligned} A_{\boldsymbol{k}}^{HP}(\boldsymbol{q}_1,\boldsymbol{q}_2) &= A_{\boldsymbol{k}}^{DM}(\boldsymbol{q}_1,\boldsymbol{q}_2) + \left\{\frac{1}{16}(\gamma_{\boldsymbol{k}} + \gamma_{\boldsymbol{q}_1} - \gamma_{\boldsymbol{k}+\boldsymbol{q}_1-\boldsymbol{q}_2} - \gamma_{\boldsymbol{q}_2})^2\right. \\ &\quad + \frac{1}{4}\left[1 - 4S\left(1 - \sqrt{1-\frac{1}{2S}}\right)\right](\gamma_{\boldsymbol{k}-\boldsymbol{q}_2} + \gamma_{\boldsymbol{q}_1-\boldsymbol{q}_2})(\gamma_{\boldsymbol{k}} + \gamma_{\boldsymbol{q}_1} + \gamma_{\boldsymbol{q}_2} + \gamma_{\boldsymbol{k}+\boldsymbol{q}_1-\boldsymbol{q}_2}) \\ &\quad \left. - \frac{1}{16}\left[1 - 16S^2\left(1 - \sqrt{1-\frac{1}{2S}}\right)^2\right](\gamma_{\boldsymbol{k}} + \gamma_{\boldsymbol{q}_1} + \gamma_{\boldsymbol{q}_2} + \gamma_{\boldsymbol{k}+\boldsymbol{q}_1-\boldsymbol{q}_2})^2\right\}. \end{aligned} \tag{3.4.30}$$

Neglecting terms with more than one Bose factor (this is consistent with the hypothesis of neglecting the potential $w^{HP}$ with respect to $v^{HP}$ as discussed in Section 3.3) and expanding the terms in the square brackets in powers of $\frac{1}{S}$, the real part of the second-order HP self-energy becomes

$$\begin{aligned} \Sigma_{HP}^{(2)}(\boldsymbol{k},\omega_{\boldsymbol{k}}) &= \Sigma_{DM}^{(2)}(\boldsymbol{k},\omega_{\boldsymbol{k}}) + \left(\frac{2zJ}{\hbar N}\right)^2 \\ &\quad \times \sum_{\boldsymbol{q}_1,\boldsymbol{q}_2}\left\{\frac{1}{8}(\gamma_{\boldsymbol{k}} + \gamma_{\boldsymbol{q}_1} - \gamma_{\boldsymbol{k}+\boldsymbol{q}_1-\boldsymbol{q}_2} - \gamma_{\boldsymbol{q}_2})^2\right. \\ &\quad - \frac{1}{16S}\left(1 + \frac{1}{4S} + \cdots\right)(\gamma_{\boldsymbol{k}-\boldsymbol{q}_2} + \gamma_{\boldsymbol{q}_1-\boldsymbol{q}_2})(\gamma_{\boldsymbol{k}} + \gamma_{\boldsymbol{q}_1} + \gamma_{\boldsymbol{q}_2} + \gamma_{\boldsymbol{k}+\boldsymbol{q}_1-\boldsymbol{q}_2}) \\ &\quad \left. + \frac{1}{32S}\left(1 + \frac{5}{16S} + \cdots\right)(\gamma_{\boldsymbol{k}} + \gamma_{\boldsymbol{q}_1} + \gamma_{\boldsymbol{q}_2} + \gamma_{\boldsymbol{k}+\boldsymbol{q}_1-\boldsymbol{q}_2})^2\right\} \\ &\quad \times \frac{n_{\boldsymbol{q}_1}^{(0)}}{\omega_{\boldsymbol{k}} - \omega_{\boldsymbol{k}+\boldsymbol{q}_1-\boldsymbol{q}_2} - \omega_{\boldsymbol{q}_2} + \omega_{\boldsymbol{q}_1}}. \end{aligned} \tag{3.4.31}$$

Replacing $\omega_{\boldsymbol{k}}$ with $\frac{2zJS}{\hbar}(1-\gamma_{\boldsymbol{k}})$ in the denominator of Eq. (3.4.31), one obtains

$$\Sigma_{HP}^{(2)}(\boldsymbol{k},\omega_{\boldsymbol{k}}) = \Sigma_{DM}^{(2)}(\boldsymbol{k},\omega_{\boldsymbol{k}}) - \left[\frac{1}{8S} + O\left(\frac{1}{S^2}\right)\right]\left(\frac{2zJ}{\hbar}\right)\frac{1}{N}\sum_{\boldsymbol{q}}(\gamma_{\boldsymbol{k}} + \gamma_{\boldsymbol{q}})n_{\boldsymbol{q}}^{(0)}. \tag{3.4.32}$$

As well as the first-order self-energies (3.3.22) and (3.3.23), the second-order HP and DM self-energies also differ from each other. However, summing the first-order result (3.3.23) to the second-order result (3.4.32), we obtain

$$\Sigma_{HP}^{(1)}(\boldsymbol{k}) + \Sigma_{HP}^{(2)}(\boldsymbol{k}, \omega_{\boldsymbol{k}}) = \Sigma_{DM}^{(1)}(\boldsymbol{k}) + \Sigma_{DM}^{(2)}(\boldsymbol{k}, \omega_{\boldsymbol{k}}) + O\left(\frac{1}{S^2}\right). \tag{3.4.33}$$

Even though the DM and HP spin-boson transformations lead to different results in a perturbation expansion in powers of the interaction potential $v$, they agree when all terms of the same order $\frac{1}{S}$ and temperature are taken into account. The perturbation expansion in $v^{DM}$ coincides with the expansion in $\frac{1}{S}$ since $v^{DM}$ does not depend on $S$. On the contrary, a further expansion in $\frac{1}{S}$ is required when the perturbation expansion in $v^{HP}$ is used since $v^{HP}$ does depend on $S$. Incidentally, an agreement between DM and HP perturbation expansion was also proven for terms of order $\frac{1}{S^2}$ when the third-order terms of the DM and HP self-energies are taken into account.[18] This result confirms that both DM and HP spin-boson transformations lead to the same low temperature self-energy when the perturbation parameter is assumed to be $\frac{1}{S}$ instead of the interaction potential $v$. Moreover, since at low temperature one finds $\Sigma_{HP} = \Sigma_{DM}$, the conclusion is that the effect of the unphysical states (kinematical interaction) is negligible for both the HP and the DM transformation even though a rigorous proof exists only for the DM transformation. Obviously, the perturbation calculation is greatly simplified by the use of the DM spin-boson transformation.

## 3.5. Dyson's Equation

Equation (3.4.28) is a good starting point to infer the Dyson's equation of the many-body theory. Indeed, pushing the perturbation expansion to the infinity, one expects to obtain

$$\begin{aligned} G_{\boldsymbol{k}}(\omega \pm i\epsilon) &= G_{\boldsymbol{k}}^{(0)}(\omega \pm i\epsilon) \sum_{n=0}^{\infty} \{[\Sigma^{(1)}(\boldsymbol{k}) + \Sigma_b^{(2)}(\boldsymbol{k}) \\ &\quad + \Sigma_c^{(2)}(\boldsymbol{k}, \omega_{\boldsymbol{k}} \pm i\epsilon) + \cdots] G_{\boldsymbol{k}}^{(0)}(\omega \pm i\epsilon)\}^n \\ &= \frac{G_{\boldsymbol{k}}^{(0)}(\omega \pm i\epsilon)}{1 - [\Sigma^{(1)}(\boldsymbol{k}) + \Sigma_b^{(2)}(\boldsymbol{k}) + \Sigma_c^{(2)}(\boldsymbol{k}, \omega_{\boldsymbol{k}} \pm i\epsilon) + \cdots] G_{\boldsymbol{k}}^{(0)}(\omega \pm i\epsilon)}. \end{aligned} \tag{3.5.1}$$

In the geometric series (3.5.1), the common ratio is given by the sum of all *irreducible* contributions to the *proper* self-energy[13] $\Sigma^*(\boldsymbol{k}, \omega_{\boldsymbol{k}} \pm i\epsilon)$ times the unperturbed generalized Green function $G_{\boldsymbol{k}}^{(0)}(\omega \pm i\epsilon)$ while all the powers in the geometric series with $n > 1$ come from the *reducible* contributions to the self-energy. Using Eq. (3.3.9), the generalized Green function of Eq. (3.5.1) becomes

$$G_{\boldsymbol{k}}(\omega \pm i\epsilon) = \frac{1}{\omega - \omega_{\boldsymbol{k}} - \Sigma^*(\boldsymbol{k}, \omega_{\boldsymbol{k}} \pm i\epsilon)} \tag{3.5.2}$$

where

$$\Sigma^*(\boldsymbol{k}, \omega \pm i\epsilon) = \Sigma^{*'}(\boldsymbol{k}, \omega_{\boldsymbol{k}}) \mp i\Sigma^{*''}(\boldsymbol{k}, \omega_{\boldsymbol{k}}). \tag{3.5.3}$$

As one can see, the generalized Green function (3.5.2) has a simple pole at a complex frequency

$$\omega = \omega_{\boldsymbol{k}} + \Sigma^{*'}(\boldsymbol{k}, \omega_{\boldsymbol{k}}) \mp i\Sigma^{*''}(\boldsymbol{k}, \omega_{\boldsymbol{k}}). \tag{3.5.4}$$

Note that only after the sum over the infinite (reducible) terms of the geometric series (3.5.1), the pole of the one-particle Green function is shifted from its (real) unperturbed value $\omega = \omega_{\boldsymbol{k}}$. Equation (3.5.1) may also be written as

$$G_{\boldsymbol{k}}(\omega) = G_{\boldsymbol{k}}^{(0)}(\omega) + G_{\boldsymbol{k}}^{(0)}(\omega)\Sigma^*(\boldsymbol{k}, \omega)G_{\boldsymbol{k}}(\omega), \tag{3.5.5}$$

that is the celebrated Dyson's equation of the many-body theory.[13]

Since the first-order self-energy $\Sigma^{(1)}(\boldsymbol{k})$ is frequency-independent, the first-order perturbation theory leads to a simple renormalization of the magnon frequency

$$\Delta\omega_{\boldsymbol{k}}^{(1)} = \Sigma^{(1)}(\boldsymbol{k}). \tag{3.5.6}$$

On the contrary, the second-order self-energy leads to either a renormalization given by

$$\Delta\omega_{\boldsymbol{k}}^{(2)} = \Sigma_b^{(2)}(\boldsymbol{k}) + \Sigma_c^{(2)'}(\boldsymbol{k}, \omega_{\boldsymbol{k}}) \tag{3.5.7}$$

and a damping given by

$$\Gamma_{\boldsymbol{k}}^{(2)} = |\Sigma_c^{(2)''}(\boldsymbol{k}, \omega_{\boldsymbol{k}})|. \tag{3.5.8}$$

Both renormalization and damping of the magnon frequency are due to the interaction with all other magnons.

Using Eq. (3.5.2), the dynamical structure factor (3.2.17) becomes

$$\begin{aligned} S_\perp(\boldsymbol{K}, \omega) = {} & 2\frac{SN}{\hbar\pi}\sum_{\boldsymbol{q},\boldsymbol{G}} \delta_{\boldsymbol{K}+\boldsymbol{q},\boldsymbol{G}}[1 + n(\omega)]\frac{\Sigma^{*''}(\boldsymbol{q}, \omega_{\boldsymbol{q}})}{[\omega - \omega_{\boldsymbol{q}} - \Sigma^{*'}(\boldsymbol{q}, \omega_{\boldsymbol{q}})]^2 + [\Sigma^{*''}(\boldsymbol{q}, \omega_{\boldsymbol{q}})]^2} \\ & + 2\frac{SN}{\hbar\pi}\sum_{\boldsymbol{q},\boldsymbol{G}} \delta_{\boldsymbol{K}-\boldsymbol{q},\boldsymbol{G}}\, n(-\omega)\frac{\Sigma^{*''}(\boldsymbol{q}, \omega_{\boldsymbol{q}})}{[\omega + \omega_{\boldsymbol{q}} + \Sigma^{*'}(\boldsymbol{q}, \omega_{\boldsymbol{q}})]^2 + [\Sigma^{*''}(\boldsymbol{q}, \omega_{\boldsymbol{q}})]^2}. \end{aligned} \tag{3.5.9}$$

As one can see by comparing Eqs. (3.5.9) and (3.3.12), the existence of a damping $\Sigma^{*''}$ changes the $\delta$-like peaks of the unperturbed system into Lorentzian peaks located at the renormalized magnon frequency

$$\omega = \pm[\omega_{\boldsymbol{q}} + \Sigma^{*'}(\boldsymbol{q}, \omega_{\boldsymbol{q}})]. \tag{3.5.10}$$

The peak width is proportional to the magnon damping $\Sigma^{*''}$ and the height of the peak is proportional to $[\Sigma^{*''}]^{-1}$. Only for $T \to 0$, the poles of the Green function lead to $\delta$-like peaks located at the unperturbed magnon frequencies wich are the lowest eigenvalues of the Heisenberg Hamiltonian. For $T \neq 0$, the poles of the Green function do not correspond to eigenvalues of the Heisenberg Hamiltonian but their shift and broadening correspond to the effect on the correlation function of the interaction between the spin waves. Note that for $\Sigma^{*''} \to 0$, the Lorentian peaks reduce to $\delta$-peaks located at the renormalized frequencies (3.5.10).

Several truncations of the Green function hierarchy were proposed in literature. Some of them[26,27] give specific rules similar to (3.3.14) in order to write the many-particle Green functions in terms of the one-particle Green function. However, many other uncontrolled decouplings have been proposed to write the two-particle Green function of Eq. (3.3.7) in terms of the one-particle Green function. Among them, the most well known is perhaps that suggested by Tahir-Kheli and Ter Haar[28]

$$\langle\langle a^+_{\boldsymbol{q}_1} a_{\boldsymbol{q}_2} a_{\boldsymbol{k}+\boldsymbol{q}_1-\boldsymbol{q}_2}; a^+_{\boldsymbol{k}'}\rangle\rangle_\omega \simeq \langle a^+_{\boldsymbol{q}_1} a_{\boldsymbol{q}_1}\rangle(\delta_{\boldsymbol{q}_1,\boldsymbol{q}_2} + \delta_{\boldsymbol{k},\boldsymbol{q}_2})\langle\langle a_{\boldsymbol{k}}; a^+_{\boldsymbol{k}'}\rangle\rangle_\omega, \tag{3.5.11}$$

leading to the one-particle Green function

$$\langle\langle a_{\boldsymbol{k}}; a^+_{\boldsymbol{k}'}\rangle\rangle_{\omega\pm i\epsilon} = \frac{\delta_{\boldsymbol{k},\boldsymbol{k}'}}{\omega - \omega_{\boldsymbol{k}} - \widetilde{\Sigma}(\boldsymbol{k}) \pm i\epsilon} \tag{3.5.12}$$

where the self-energy $\widetilde{\Sigma}(\boldsymbol{k})$ is given by the first-order self-energy (3.3.22) in which the unperturbed Bose factor $n^{(0)}_{\boldsymbol{k}}$ is replaced by a Bose factor with a renormalized frequency $\omega_{\boldsymbol{k}} + \widetilde{\Sigma}(\boldsymbol{k})$ that is

$$\widetilde{\Sigma}(\boldsymbol{k}) = -\omega_{\boldsymbol{k}}\frac{\widetilde{\alpha}}{S} \tag{3.5.13}$$

with

$$\widetilde{\alpha} = \frac{1}{N}\sum_{\boldsymbol{p}}(1-\gamma_{\boldsymbol{p}})\frac{1}{e^{\beta\hbar\omega_{\boldsymbol{p}}(1-\widetilde{\alpha}/S)} - 1}. \tag{3.5.14}$$

Equation (3.5.14) is a self-consistent equation for $\widetilde{\alpha}$. The self-energy (3.5.13) reduces to the first-order result (3.3.22) putting $\widetilde{\alpha} = 0$ in the right-hand side of Eq. (3.5.14). The decoupling (3.5.11) allows us to get some (uncontrolled) informations about the behaviour of a ferromagnet at higher temperatures. Following a procedure very similar to that illustrated in Section 2.4, Eq. (3.5.14) for a SC lattice becomes

$$\widetilde{\alpha} = \sum_{n=1}^{\infty} e^{-12\beta JSn(1-\widetilde{\alpha}/S)}\left\{I_0\left[4\beta JSn\left(1-\frac{\widetilde{\alpha}}{S}\right)\right]\right\}^2 \times\left\{I_0\left[4\beta JSn\left(1-\frac{\widetilde{\alpha}}{S}\right)\right] - I_1\left[4\beta JSn\left(1-\frac{\widetilde{\alpha}}{S}\right)\right]\right\} \tag{3.5.15}$$

where $I_0$ and $I_1$ are the modified Bessel functions of order 0 and 1, respectively.[4] For $T \to 0$, Equation (3.5.15) has two solutions: $\widetilde{\alpha}_1 = 0$ and $\widetilde{\alpha}_2 = S$. In the low temperature limit, $\widetilde{\alpha}_1$ can be obtained using the asymptotic expansions of $I_0$ given by Eq. (2.4.16) and of $I_1$ given by

$$I_1(z) = \frac{e^z}{\sqrt{2\pi z}}\left(1 - \frac{3}{8z} - \frac{15}{128z^2} + \cdots\right), \tag{3.5.16}$$

Table 3.1. Physical ($\tilde{\alpha}_1$) and unphysical ($\tilde{\alpha}_2$) solutions of Eq. (3.5.15) with the corresponding magnetizations $m_1$ and $m_2$ obtained from Eq. (3.5.19) for a SC lattice with $S = 1$ for several temperatures. In the numerical calculation, the sums occurring in Eqs. (3.5.15) and (3.5.19) are truncated at $n = 30$, consistent with the precision quoted in the Table.

| $\frac{k_BT}{2JS}$ | $\tilde{\alpha}_1$ | $\tilde{\alpha}_2$ | $m_1$ | $m_2$ |
|---|---|---|---|---|
| 0 | 0 | 1 | 1 | −0.5286 |
| 0.1 | $2.439 \times 10^{-5}$ | 0.9883 | 0.9984 | −0.5130 |
| 0.5 | $1.574 \times 10^{-3}$ | 0.9394 | 0.9809 | −0.4479 |
| 1 | $1.144 \times 10^{-2}$ | 0.8721 | 0.9397 | −0.3578 |
| 1.5 | $3.784 \times 10^{-2}$ | 0.7951 | 0.8738 | −0.2541 |
| 2 | $8.851 \times 10^{-2}$ | 0.7025 | 0.7761 | −0.1282 |
| 2.5 | 0.1814 | 0.5753 | 0.6215 | 0.0472 |
| 2.7 | 0.2496 | 0.4952 | 0.5166 | 0.1591 |
| 2.8 | 0.3151 | 0.4241 | 0.4182 | 0.2601 |
| 2.82426 | 0.3689 | 0.3689 | 0.3395 | 0.3395 |

leading to

$$\tilde{\alpha}_1 = \pi\zeta\left(\frac{5}{2}\right)\left(\frac{k_BT}{8\pi JS}\right)^{\frac{5}{2}}. \tag{3.5.17}$$

In the same limit $\tilde{\alpha}_2$ is given by

$$\tilde{\alpha}_2 = S - A\left(\frac{k_BT}{8\pi JS}\right) \tag{3.5.18}$$

where $A$ cannot be obtained by the asymptotic expansion since the argument of the Bessel functions is finite for $T \to 0$ and $\tilde{\alpha}_2 \to S$. For $S = 1$, the numerical calculation gives $A = 1.456$. The two solutions of Eq. (3.5.15) are given in Table 3.1 for $S = 1$ and several temperatures. The smaller solution $\tilde{\alpha}_1$ corresponds to a minimum of the free energy and represents the *physical* solution of Eq. (3.5.15) while the solution $\tilde{\alpha}_2$ is unphysical. As shown in Table 3.1, Eq. (3.5.15) has two solutions for $T < T_{\text{max}} = 2.8243(2JS/k_B)$, one solution for $T = T_{\text{max}}$ and no solution for $T > T_{\text{max}}$. From Eq. (2.4.21), one obtains the spontaneous magnetization corresponding to the solutions of Eq. (3.5.15):

$$m_i = \langle S_i^z \rangle = S - \sum_{n=1}^{\infty} e^{-12\beta JSn(1-\tilde{\alpha}_i/S)} \left\{ I_0 \left[ 4\beta JSn \left( 1 - \frac{\tilde{\alpha}_i}{S} \right) \right] \right\}^3 \tag{3.5.19}$$

where $i = 1, 2$ corresponds to the physical and unphysical solution of Eq. (3.5.15), respectively. In the last two columns of Table 3.1, the values of $m_1$ and $m_2$ are given for a SC lattice with $S = 1$ at various temperatures. As one can see, the magnetization $m_1$ decreses from 1 to 0.3395 as the temperature increase from 0 to $T_{\text{max}}$. The unphysical solution $m_2$ decreases in a re-entrant way from $T_{\text{max}}$ to 0, crossing the temperature axis at $T \simeq 2.5(2JS/k_B)$. The self-consistent Eq. (3.5.14) was obtained independently using a variational theorem[29] by Micheline Bloch in 1962. The author pointed out the surprising good agreement between $T_{\text{max}}$ and the critical temperature of the SC Heisenberg model obtained by a high temperature series expansion[30]

$$\frac{k_B T_c}{2J} = \frac{5}{2}[0.579\ S(S+1) - 0.072] \tag{3.5.20}$$

that gives $T_c \simeq 2.715(2J/k_B)$ for $S = 1$. This agreement was not to be taken too seriously since at $T_{\max}$, a discontinuity in the magnetization ($\Delta m = 0.3395$) is found in the present calculation pointing out the occurrence of a wrong first-order phase transition. On the other hand, it is well known that the critical phenomena are driven by excitations that cannot be described in terms of spin waves only. However, in the low temperature limit, where the spin waves are the only excitations of the system, the physical solution (3.5.17) coincides with the first term of the low temperature series expansion. Moreover, using the asymptotic expansion (2.4.16), the low temperature expansion of the magnetization $m_1$ becomes

$$\begin{aligned} m_1 &= S - \zeta\left(\frac{3}{2}\right)\left(\frac{k_B T}{8\pi J S}\right)^{\frac{3}{2}} \frac{1}{(1-\tilde{\alpha}_1/S)} \\ &= S - \zeta\left(\frac{3}{2}\right)\left(\frac{k_B T}{8\pi J S}\right)^{\frac{3}{2}} - \frac{3\pi}{2S}\zeta\left(\frac{3}{2}\right)\zeta\left(\frac{5}{2}\right)\left(\frac{k_B T}{8\pi J S}\right)^4, \end{aligned} \tag{3.5.21}$$

leading to the correct $T^{3/2}$ and $T^4$ contributions obtained by Bloch[21] and by Dyson,[15] respectively.

## 3.6. Renormalization and Damping

In Section 3.5, we have seen how the spin wave spectrum is modified by the interaction between the spin waves. Within the second-order perturbation theory, the real part of the renormalized spectrum reads

$$\omega_{\boldsymbol{k}}^{(2)} = \omega_{\boldsymbol{k}} + \Sigma^{(1)}(\boldsymbol{k}) + \Sigma_b^{(2)}(\boldsymbol{k}) + \Sigma_c^{(2)'}(\boldsymbol{k}, \omega_{\boldsymbol{k}}) \tag{3.6.1}$$

where $\Sigma^{(1)}(\boldsymbol{k})$, $\Sigma_b^{(2)}(\boldsymbol{k})$ and $\Sigma_c^{(2)'}(\boldsymbol{k}, \omega_{\boldsymbol{k}})$ are given by Eqs. (3.3.21), (3.4.22) and (3.4.31), respectively. The magnon damping is given by Eq. (3.4.32): note that the inverse of the damping is the lifetime of the magnon. In this section, we evaluate the low temperature behaviour of such contributions assuming that the interaction potential is the DM potential $v^{DM}$ given by Eq. (3.3.4). The first-order self-energy becomes

$$\Sigma^{(1)}(\boldsymbol{k}) = -\frac{2zJ}{\hbar N}\sum_{\boldsymbol{q}}(1 + \gamma_{\boldsymbol{k}-\boldsymbol{q}} - \gamma_{\boldsymbol{k}} - \gamma_{\boldsymbol{q}})n_{\boldsymbol{q}}^{(0)} = -\alpha\frac{\omega_{\boldsymbol{k}}}{S} \tag{3.6.2}$$

where

$$\alpha = \frac{1}{N}\sum_{\boldsymbol{q}}(1 - \gamma_{\boldsymbol{q}})n_{\boldsymbol{q}}^{(0)}. \tag{3.6.3}$$

To obtain the last step of Eq. (3.6.2), the relationships $\sum_{\boldsymbol{q}} \gamma_{\boldsymbol{k}-\boldsymbol{q}} n_{\boldsymbol{q}}^{(0)} = \sum_{\boldsymbol{q}} \gamma_{\boldsymbol{k}} \gamma_{\boldsymbol{q}} n_{\boldsymbol{q}}^{(0)}$ and $\hbar\omega_{\boldsymbol{k}} = 2zJS(1-\gamma_{\boldsymbol{k}})$ have to be used. The low temperature limit of $\alpha$ for a SC lattice was obtained in Eq. (3.5.17), making use of the asymptotic expansions of the Bessel functions occurring in Eq. (3.5.15). A more direct way to arrive at the same result consists of taking the long wavelength limit in the factor $(1-\gamma_{\boldsymbol{q}})$ as well as in the magnon frequency occurring in the Bose factor and replacing the cubic cell by a sphere of infinite radius. In doing so, Eq. (3.6.3) becomes

$$\begin{aligned}\alpha &\simeq \frac{1}{(2\pi)^3} \int_0^\infty dq q^2 \frac{1}{6} q^2 \frac{1}{e^{2\beta JSq^2}-1} \int_0^\pi d\theta \sin\theta \int_0^{2\pi} d\phi \\ &= \frac{1}{24\pi^2} \left(\frac{k_B T}{2JS}\right)^{\frac{5}{2}} \int_0^\infty dx \frac{x^{\frac{3}{2}}}{e^x - 1}. \end{aligned} \tag{3.6.4}$$

Using the relationship[3]

$$\int_0^\infty dx \frac{x^{\frac{3}{2}}}{e^x-1} = \frac{3}{4}\sqrt{\pi}\zeta\left(\frac{5}{2}\right), \tag{3.6.5}$$

Eq. (3.6.4) becomes

$$\alpha = \pi\zeta\left(\frac{5}{2}\right)\left(\frac{k_B T}{8\pi JS}\right)^{\frac{5}{2}} = 7.53 \times 10^{-3} \left(\frac{k_B T}{2JS}\right)^{\frac{5}{2}} \tag{3.6.6}$$

and the first-order renormalized spin wave spectrum reduces to

$$\omega_{\boldsymbol{k}}^{(1)} = \omega_{\boldsymbol{k}}\left(1 - \frac{\alpha}{S}\right) \simeq \omega_{\boldsymbol{k}}\left[1 - \frac{\pi}{S}\zeta\left(\frac{5}{2}\right)\left(\frac{k_B T}{8\pi JS}\right)^{\frac{5}{2}}\right]. \tag{3.6.7}$$

Equation (3.6.7) shows that the renormalized magnon frequency is lower than the unperturbed one leading to the conclusion that the effect of the magnon-magnon interaction is to lower the spin wave energy. The fact that the renormalization is the same for every wavevector $\boldsymbol{k}$ is peculiar of the first-order self-energy. The first-order spectrum (3.6.7) leads to the inelastic cross-section (3.1.17) with $\omega_{\boldsymbol{q}}$ replaced by $\omega_{\boldsymbol{q}}^{(1)}$. We stress that no damping occurs in the first-order perturbation theory. The spontaneous magnetization is obtained from Eq. (2.4.10) replacing $\omega_{\boldsymbol{q}}$ by $\omega_{\boldsymbol{q}}^{(1)}$. In the long wavelength limit, Eq. (3.5.21) is recovered.

From Eq. (3.4.22), one can see that the second-order self-energy $\Sigma_b^{(2)}$ is a real function independent of the frequency that can be written

$$\Sigma_b^{(2)}(\boldsymbol{k}) = -\frac{\omega_{\boldsymbol{k}}}{S}\left(\frac{2zJ}{k_B T}\right)\alpha\rho \tag{3.6.8}$$

where $\alpha$ is given by Eq. (3.6.3) and $\rho$ is given by

$$\rho = \frac{1}{N}\sum_{\boldsymbol{p}}(1-\gamma_{\boldsymbol{p}})^2 n_{\boldsymbol{p}}^{(0)}(1+n_{\boldsymbol{p}}^{(0)}). \tag{3.6.8a}$$

At low temperature for a SC lattice, $\alpha$ is given by Eq. (3.6.6) and $\rho$ becomes

$$\rho \simeq \frac{1}{2\pi^2}\int_0^\infty p^2dp\left(\frac{1}{6}p^2\right)^2\frac{e^{2JS\beta p^2}}{(e^{2JS\beta p^2}-1)^2}$$
$$=\frac{1}{144\pi^2}\left(\frac{k_BT}{2JS}\right)^{\frac{7}{2}}\int_0^\infty dx\frac{x^{5/2}e^x}{(e^x-1)^2}. \tag{3.6.8b}$$

Integrating by part and using Eq. (3.6.5), one obtains

$$\rho=\frac{5}{3}\pi^2\zeta\left(\frac{5}{2}\right)\left(\frac{k_BT}{8\pi JS}\right)^{\frac{7}{2}}=3.14\times10^{-3}\left(\frac{k_BT}{2JS}\right)^{\frac{7}{2}} \tag{3.6.8c}$$

so that Eq. (3.6.8) becomes

$$\Sigma_b^{(2)}=-\frac{5}{2}\frac{\omega_{\boldsymbol{k}}}{S^2}\left[\pi\zeta\left(\frac{5}{2}\right)\right]^2\left(\frac{k_BT}{8\pi JS}\right)^5. \tag{3.6.9}$$

From Eq. (3.6.9), one can see that $\Sigma_b^{(2)}$ is of the order $T^5$, therefore it is negligible with respect to the first-order result given by Eq. (3.6.7). On the contrary, the main term coming from the temperature expansion of $\Sigma_c^{(2)'}$ is of the same order as $\Sigma^{(1)}$ ($T^{5/2}$). Indeed, replacing $v^{DM}$ given by Eq. (3.3.4) in Eq. (3.4.25), the real part (renormalization) of the self-energy $\Sigma_c^{(2)}$ becomes

$$\Sigma_c^{(2)'}(\boldsymbol{k},\omega_{\boldsymbol{k}})=\frac{1}{2}\left(\frac{2zJ}{\hbar N}\right)^2\sum_{\boldsymbol{q}_1,\boldsymbol{q}_2}[n_{\boldsymbol{q}_1}^{(0)}(1+n_{\boldsymbol{q}_2}^{(0)}+n_{\boldsymbol{k}+\boldsymbol{q}_1-\boldsymbol{q}_2}^{(0)})-n_{\boldsymbol{q}_2}^{(0)}n_{\boldsymbol{k}+\boldsymbol{q}_1-\boldsymbol{q}_2}^{(0)}]$$
$$\times\frac{(\gamma_{\boldsymbol{k}-\boldsymbol{q}_2}+\gamma_{\boldsymbol{q}_1-\boldsymbol{q}_2}-\gamma_{\boldsymbol{k}}-\gamma_{\boldsymbol{q}_1})(\gamma_{\boldsymbol{k}-\boldsymbol{q}_2}+\gamma_{\boldsymbol{q}_1-\boldsymbol{q}_2}-\gamma_{\boldsymbol{k}+\boldsymbol{q}_1-\boldsymbol{q}_2}-\gamma_{\boldsymbol{q}_2})}{\omega_{\boldsymbol{k}}-\omega_{\boldsymbol{k}+\boldsymbol{q}_1-\boldsymbol{q}_2}-\omega_{\boldsymbol{q}_2}+\omega_{\boldsymbol{q}_1}}. \tag{3.6.10}$$

The main temperature contribution of Eq. (3.6.10) is obtained, neglecting all terms containing products of two Bose factors: this is correct when the Bose factors refer to "independent" frequencies like in this case. However, this is no longer correct when the presence of a $\delta$-function of the frequencies, like in Eq. (3.4.26), makes them no more independent.[31] Using the relationship $\hbar\omega_{\boldsymbol{k}}=2zJS(1-\gamma_{\boldsymbol{k}})$, adding and subtracting the term $\gamma_{\boldsymbol{k}+\boldsymbol{q}_1-\boldsymbol{q}_2}+\gamma_{\boldsymbol{q}_2}$ in the first factor of the numerator, the function within the sum of Eq. (3.6.10) becomes

$$\left[\frac{(\gamma_{\boldsymbol{k}-\boldsymbol{q}_2}+\gamma_{\boldsymbol{q}_1-\boldsymbol{q}_2}-\gamma_{\boldsymbol{k}+\boldsymbol{q}_1-\boldsymbol{q}_2}-\gamma_{\boldsymbol{q}_2})^2}{\gamma_{\boldsymbol{k}+\boldsymbol{q}_1-\boldsymbol{q}_2}+\gamma_{\boldsymbol{q}_2}-\gamma_{\boldsymbol{k}}-\gamma_{\boldsymbol{q}_1}}+\gamma_{\boldsymbol{k}-\boldsymbol{q}_2}+\gamma_{\boldsymbol{q}_1-\boldsymbol{q}_2}-\gamma_{\boldsymbol{k}+\boldsymbol{q}_1-\boldsymbol{q}_2}-\gamma_{\boldsymbol{q}_2}\right]n_{\boldsymbol{q}_1}^{(0)}. \tag{3.6.10a}$$

Summing over $\boldsymbol{q}_1$ and $\boldsymbol{q}_2$, the last term of Eq. (3.6.10a) becomes

$$\sum_{q_1,q_2}(\gamma_{k-q_2}+\gamma_{q_1-q_2}-\gamma_{k+q_1-q_2}-\gamma_{q_2})n^{(0)}_{q_1}$$
$$=\sum_{q_1,q_2}(\gamma_k\gamma_{q_2}+\gamma_{q_1}\gamma_{q_2}-\gamma_{k+q_1}\gamma_{q_2}-\gamma_{q_2})n^{(0)}_{q_1}$$
$$=\sum_{q_1}(\gamma_k+\gamma_{q_1}-\gamma_{k+q_1}-1)n^{(0)}_{q_1}\sum_{q_2}\gamma_{q_2}=0 \qquad (3.6.10b)$$

since $\sum_{q_2}\gamma_{q_2}=0$. Then the main contribution of the real part of the second-order self-energy $\Sigma_c^{(2)}$ becomes

$$\Sigma_c^{(2)\prime}(\boldsymbol{k},\omega_{\boldsymbol{k}})=\frac{1}{2S^2}\left(\frac{2zJS}{\hbar}\right)\frac{1}{N^2}\sum_{q_1,q_2}\frac{(\gamma_{k-q_2}+\gamma_{q_1-q_2}-\gamma_{k+q_1-q_2}-\gamma_{q_2})^2}{\gamma_{k+q_1-q_2}+\gamma_{q_2}-\gamma_k-\gamma_{q_1}}n^{(0)}_{q_1}. \qquad (3.6.11)$$

The explicit calculation of (3.6.11) will be performed for a SC lattice for which $\gamma_{\boldsymbol{k}}=\frac{1}{3}\sum_\alpha\cos k_\alpha$ where $\alpha=x,y,z$ and the lattice constant is assumed to be $a=1$. Replacing $\boldsymbol{q}_1$ by $\boldsymbol{p}$ and $\boldsymbol{q}_2$ by $\frac{1}{2}(\boldsymbol{k}+\boldsymbol{p})-\boldsymbol{\rho}$, Equation (3.6.11) becomes

$$\Sigma_c^{(2)\prime}(\boldsymbol{k},\omega_{\boldsymbol{k}})=\frac{16J}{\hbar S}\frac{1}{N^2}\sum_{\boldsymbol{p},\boldsymbol{\rho}}\frac{\sum_{\alpha,\beta}\sin\frac{k_\alpha}{2}\sin\frac{k_\beta}{2}\sin\frac{p_\alpha}{2}\sin\frac{p_\beta}{2}\cos\rho_\alpha\cos\rho_\beta}{\sum_\gamma\cos\frac{1}{2}(k_\gamma+p_\gamma)[\cos\rho_\gamma-\cos\frac{1}{2}(k_\gamma-p_\gamma)]} \qquad (3.6.12)$$

where $\alpha,\beta,\gamma=x,y,z$. The main tepmerature contribution to the self-energy (3.6.12) is obtained by expanding the function under the sum in powers of $\boldsymbol{p}$ and retaining the first term of such expansion. In view of the sum over $\boldsymbol{p}$, only the terms with $\alpha=\beta$ in Eq. (3.6.12) give non zero contributions so that the self-energy (3.6.12) becomes

$$\Sigma_c^{(2)\prime}(\boldsymbol{k},\omega_{\boldsymbol{k}})=\frac{4J}{\hbar S}\sum_\alpha\sin^2\frac{k_\alpha}{2}\frac{1}{N}\sum_{\boldsymbol{p}}p_\alpha^2n^{(0)}_{\boldsymbol{p}}\frac{1}{N}\sum_{\boldsymbol{\rho}}\frac{\cos^2\rho_\alpha}{\sum_\gamma\cos\frac{k_\gamma}{2}[\cos\rho_\gamma-\cos\frac{k_\gamma}{2}]}. \qquad (3.6.13)$$

For symmetry reasons, the sum over $\boldsymbol{p}$ in Eq. (3.6.13) is the same for $\alpha=x,y$ and $z$ so that one has

$$\frac{1}{N}\sum_{\boldsymbol{p}}p_\alpha^2n^{(0)}_{\boldsymbol{p}}=\frac{1}{3}\frac{1}{N}\sum_{\boldsymbol{p}}p^2n^{(0)}_{\boldsymbol{p}}=2\pi\,\zeta\left(\frac{5}{2}\right)\left(\frac{k_BT}{8\pi JS}\right)^{\frac{5}{2}} \qquad (3.6.14)$$

where we have used Eqs. (3.6.3) and (3.6.6). The sum over $\boldsymbol{\rho}$ in Eq. (3.6.13) may be written as

$$I_\alpha(\boldsymbol{k})=-\frac{1}{\pi^3}\int_0^\pi\int_0^\pi\int_0^\pi d\rho_x\,d\rho_y\,d\rho_z\frac{\cos^2\rho_\alpha}{\sum_\gamma\cos\frac{k_\gamma}{2}[\cos\frac{k_\gamma}{2}-\cos\rho_\gamma]} \qquad (3.6.15)$$

so that the second-order renormalization (3.6.13) becomes

$$\Sigma_c^{(2)'}(\boldsymbol{k},\omega_{\boldsymbol{k}}) = \frac{8J}{\hbar S}\,\pi\,\zeta\left(\frac{5}{2}\right)\left(\frac{k_BT}{8\pi JS}\right)^{\frac{5}{2}}\sum_{\alpha} I_\alpha(\boldsymbol{k})\sin^2\frac{k_\alpha}{2}. \tag{3.6.16}$$

As announced, the main temperature contribution of the real part of the second-order self-energy $\Sigma_c^{(2)'}$ given by Eq. (3.6.16) is of the same order of the main temperature contribution to the first-order self-energy $\Sigma^{(1)}$ given by Eqs. (3.6.2)–(3.6.6). In the long wavelength limit ($\boldsymbol{k}\to 0$), Eq. (3.6.16) reduces to

$$\Sigma_c^{(2)'}(\boldsymbol{k}\to 0,\omega_{\boldsymbol{k}}) = -\frac{2J}{3\hbar S}I_c^{(2)}|\boldsymbol{k}|^2\pi\zeta\left(\frac{5}{2}\right)\left(\frac{k_BT}{8\pi JS}\right)^{\frac{5}{2}} \tag{3.6.17}$$

where

$$I_c^{(2)} = \frac{1}{\pi^3}\int_0^\pi\int_0^\pi\int_0^\pi dx\,dy\,dz\,\frac{\cos^2 x}{1-\frac{1}{3}(\cos x+\cos y+\cos z)} = 0.88686. \tag{3.6.17a}$$

and the second-order magnon spectrum reads

$$\omega_{\boldsymbol{k}}^{(2)} \simeq \frac{2JS}{\hbar}k^2\left[1-\frac{\pi}{S}\,\zeta\left(\frac{5}{2}\right)\left(\frac{k_BT}{8\pi JS}\right)^{\frac{5}{2}}\left(1+\frac{I_c^{(2)}}{3S}\right)\right]. \tag{3.6.18}$$

The evaluation of the main temperature contribution of the damping coming from the second-order self-energy $\Sigma_c^{(2)''}$ is more subtle. As mentioned previously[31] in the calculation of the imaginary part of the self-energy, we cannot neglect terms containing two Bose factors since the energies occurring in the Bose factors are not independent due to the $\delta$-function condition on the frequencies appearing in Eq. (3.4.32). Indeed, using the identities

$$n(\omega_{\boldsymbol{q}}+\omega_{\boldsymbol{p}}) = \frac{n_{\boldsymbol{q}}^{(0)}n_{\boldsymbol{p}}^{(0)}}{1+n_{\boldsymbol{q}}^{(0)}+n_{\boldsymbol{p}}^{(0)}}, \tag{3.6.19}$$

$$\begin{aligned}1-\frac{n_{\boldsymbol{p}}^{(0)}}{1+n_{\boldsymbol{q}}^{(0)}+n_{\boldsymbol{p}}^{(0)}} &= \frac{1+n_{\boldsymbol{q}}^{(0)}}{1+n_{\boldsymbol{q}}^{(0)}+n_{\boldsymbol{p}}^{(0)}} = [1+n(\omega_{\boldsymbol{q}}+\omega_{\boldsymbol{p}})]\frac{1}{1+n_{\boldsymbol{p}}^{(0)}}\\ &= [1+n(\omega_{\boldsymbol{q}}+\omega_{\boldsymbol{p}})](1-e^{-\beta\hbar\omega_{\boldsymbol{p}}})\end{aligned} \tag{3.6.20}$$

and the relationship between the frequencies occurring in the argument of the $\delta$-function ($\omega_{\boldsymbol{q}_2}+\omega_{\boldsymbol{k}+\boldsymbol{q}_1-\boldsymbol{q}_2}=\omega_{\boldsymbol{k}}+\omega_{\boldsymbol{q}_1}$), one has

$$\begin{aligned}&n_{\boldsymbol{q}_1}^{(0)}\left(1+n_{\boldsymbol{q}_2}^{(0)}+n_{\boldsymbol{k}+\boldsymbol{q}_1-\boldsymbol{q}_2}^{(0)}\right)-n_{\boldsymbol{q}_2}^{(0)}n_{\boldsymbol{k}+\boldsymbol{q}_1-\boldsymbol{q}_2}^{(0)}\\ &\quad=\left(1+n_{\boldsymbol{q}_2}^{(0)}+n_{\boldsymbol{k}+\boldsymbol{q}_1-\boldsymbol{q}_2}^{(0)}\right)\left[n_{\boldsymbol{q}_1}^{(0)}-n(\omega_{\boldsymbol{k}}+\omega_{\boldsymbol{q}_1})\right]\\ &\quad=n_{\boldsymbol{q}_1}^{(0)}\left(1+n_{\boldsymbol{q}_2}^{(0)}+n_{\boldsymbol{k}+\boldsymbol{q}_1-\boldsymbol{q}_2}^{(0)}\right)[1+n(\omega_{\boldsymbol{k}}+\omega_{\boldsymbol{q}_1})](1-e^{-\beta\hbar\omega_{\boldsymbol{k}}}).\end{aligned} \tag{3.6.21}$$

Replacing Eq. (3.6.21) into Eq. (3.4.26), the imaginary part (damping) of the self-energy $\Sigma_c^{(2)}$ becomes

$$\Sigma_c^{(2)''}(\boldsymbol{k},\omega_{\boldsymbol{k}})$$
$$= (1-e^{-\beta\hbar\omega_{\boldsymbol{k}}})\frac{\pi}{2S^2}\left(\frac{2zJS}{\hbar N^2}\right)\sum_{\boldsymbol{q}_1,\boldsymbol{q}_2}(\gamma_{\boldsymbol{k}-\boldsymbol{q}_2}+\gamma_{\boldsymbol{q}_1-\boldsymbol{q}_2}-\gamma_{\boldsymbol{k}}-\gamma_{\boldsymbol{q}_1})^2$$
$$\times\,\delta(\gamma_{\boldsymbol{k}+\boldsymbol{q}_1-\boldsymbol{q}_2}+\gamma_{\boldsymbol{q}_2}-\gamma_{\boldsymbol{k}}-\gamma_{\boldsymbol{q}_1})n_{\boldsymbol{q}_1}^{(0)}(1+n_{\boldsymbol{q}_2}^{(0)}+n_{\boldsymbol{k}+\boldsymbol{q}_1-\boldsymbol{q}_2}^{(0)})[1+n(\omega_{\boldsymbol{k}}+\omega_{\boldsymbol{q}_1})]. \tag{3.6.22}$$

For magnons of small wavevectors such that $\beta\hbar\omega_{\boldsymbol{k}} \ll 1$, the presence of the Bose factor $n_{\boldsymbol{q}_1}^{(0)}$ implies that $\boldsymbol{q}_2$ is also small and all the $\gamma$'s in the argument of the $\delta$-function can be expanded in powers of their wavevectors so that

$$\delta(\gamma_{\boldsymbol{k}+\boldsymbol{q}_1-\boldsymbol{q}_2}+\gamma_{\boldsymbol{q}_2}-\gamma_{\boldsymbol{k}}-\gamma_{\boldsymbol{q}_1}) \simeq \frac{z}{2}\delta[\boldsymbol{k}\cdot\boldsymbol{q}_1-\boldsymbol{q}_2\cdot(\boldsymbol{k}+\boldsymbol{q}_1-\boldsymbol{q}_2)] \tag{3.6.23}$$

where the lattice constant $a = 1$ has been assumed. In the same way, the interaction potential becomes

$$(\gamma_{\boldsymbol{k}-\boldsymbol{q}_2}+\gamma_{\boldsymbol{q}_1-\boldsymbol{q}_2}-\gamma_{\boldsymbol{k}}-\gamma_{\boldsymbol{q}_1})^2 = \frac{4}{z^2}[\boldsymbol{q}_2\cdot(\boldsymbol{k}+\boldsymbol{q}_1-\boldsymbol{q}_2)]^2 \tag{3.6.24}$$

so that Eq. (3.6.22) reduces to

$$\Sigma_c^{(2)''}(\boldsymbol{k}\to 0,\omega_{\boldsymbol{k}}) = \pi\frac{2J}{\hbar S}\left(\frac{2JS}{k_BT}\right)k^2\frac{1}{N^2}\sum_{\boldsymbol{q}_1,\boldsymbol{q}_2}(\boldsymbol{k}\cdot\boldsymbol{q}_1)^2$$
$$\times\,\delta[\boldsymbol{k}\cdot\boldsymbol{q}_1-\boldsymbol{q}_2\cdot(\boldsymbol{k}+\boldsymbol{q}_1-\boldsymbol{q}_2)]n_{\boldsymbol{q}_1}^{(0)}(1+2n_{\boldsymbol{q}_2}^{(0)})[1+n(\omega_{\boldsymbol{k}}+\omega_{\boldsymbol{q}_1})] \tag{3.6.25}$$

where the Bose factor $n_{\boldsymbol{k}+\boldsymbol{q}_1-\boldsymbol{q}_2}^{(0)}$ in Eq. (3.6.22) has been replaced by $n_{\boldsymbol{q}_2}^{(0)}$ due to the symmetry of the interaction potential and of the argument of the $\delta$-function under the exchange of $\boldsymbol{k}+\boldsymbol{q}_1-\boldsymbol{q}_2$ with $\boldsymbol{q}_2$. Moreover, in the low temperature limit, the spin wave spectra occurring in the Bose factors of Eq. (3.6.25) may be replaced by their expressions in the long wavelength limit. In order to evaluate the sums over $\boldsymbol{q}_1$ and $\boldsymbol{q}_2$ in Eq. (3.6.25), we choose spherical coordinates in such a way that $\theta_1$ is the polar angle between $\boldsymbol{k}$ and $\boldsymbol{q}_1$ and $\theta_2$ is the polar angle between $\boldsymbol{q}_2$ and $\boldsymbol{k}+\boldsymbol{q}_1$. This choice implies that

$$(\boldsymbol{k}+\boldsymbol{q}_1)^2 = k^2+q_1^2+2kq_1\cos\theta_1 \tag{3.6.26a}$$

and

$$\boldsymbol{q}_2\cdot(\boldsymbol{k}+\boldsymbol{q}_1) = q_2\cos\theta_2\sqrt{k^2+q_1^2+2kq_1\cos\theta_1} \tag{3.6.26b}$$

so that

$$(\boldsymbol{k}\cdot\boldsymbol{q}_1)^2\delta[\boldsymbol{k}\cdot\boldsymbol{q}_1-\boldsymbol{q}_2\cdot(\boldsymbol{k}+\boldsymbol{q}_1-\boldsymbol{q}_2)]$$
$$= k^2q_1^2\cos^2\theta_1\delta(kq_1\cos\theta_1-q_2\cos\theta_2\sqrt{k^2+q_1^2+2kq_1\cos\theta_1}+q_2^2). \tag{3.6.26c}$$

From Eq. (3.6.26c), one can see that all the functions occurring in Eq. (3.6.25) do not depend explicitly on the azimuthal angles $\phi_1$ and $\phi_2$ so that a direct integration can be performed over these two angles leading to a factor $(2\pi)^2$. Putting $\boldsymbol{q}_1 = \boldsymbol{p}$, $\boldsymbol{q}_2 = \boldsymbol{q}$, $\cos\theta_1 = x$ and $\cos\theta_2 = y$, Eq. (3.6.25) becomes

$$\Sigma_c^{(2)''}(\boldsymbol{k} \to 0, \omega_{\boldsymbol{k}}) = \frac{J}{\hbar S}\left(\frac{2JS}{k_BT}\right) k^4 \frac{1}{(2\pi)^3}\int_0^\infty dp\, p^4 n_p^{(0)}[1 + n(\omega_k + \omega_p)]$$
$$\times \int_{-1}^1 dx \frac{x^2}{\sqrt{k^2+p^2+2kpx}} \int_0^\infty dq\; q(1 + 2n_q^{(0)})$$
$$\times \int_{-1}^1 dy\, \delta\left(\frac{q^2 + kpx}{q\sqrt{k^2+p^2+2kpx}} - y\right). \tag{3.6.27}$$

Due to the argument of the $\delta$-function, the integral over $y$ is 1 if

$$-1 < \frac{q^2 + kpx}{q\sqrt{k^2+p^2+2kpx}} < 1 \tag{3.6.28}$$

and zero otherwise. The condition (3.6.28) implies that

$$-q^{(-)}(x) < q < q^{(+)}(x) \tag{3.6.29a}$$

for $x < 0$ ($\frac{\pi}{2} < \theta_1 < \pi$) and

$$q^{(-)}(x) < q < q^{(+)}(x) \tag{3.6.29b}$$

for $x > 0$ ($0 < \theta_1 < \frac{\pi}{2}$) where

$$q^{(\pm)}(x) = \frac{1}{2}\left[\sqrt{k^2+p^2+2kpx} \pm \sqrt{k^2+p^2-2kpx}\right]. \tag{3.6.29c}$$

Noticing that $q^{(-)}(-x) = -q^{(-)}(x)$ and $q^{(+)}(-x) = q^{(+)}(x)$, the last three integrals of Eq. (3.6.27) give

$$\int_{-1}^1 dx \frac{x^2}{\sqrt{k^2+p^2+2kpx}} \int_0^\infty dq\, q(1 + 2n_q^{(0)}) \int_{-1}^1 dy\, \delta\left(\frac{q^2 + kpx}{q\sqrt{k^2+p^2+2kpx}} - y\right)$$
$$= \int_0^1 dx\; x^2\left(\frac{1}{\sqrt{k^2+p^2+2kpx}} + \frac{1}{\sqrt{k^2+p^2-2kpx}}\right)\int_{q^{(-)}(x)}^{q^{(+)}(x)} dq\, q(1 + 2n_q^{(0)}). \tag{3.6.30}$$

The last integration of Eq. (3.6.30) can be easily performed if we make the variable change $q^2 = \rho$. Indeed,

$$\int_{q^{(-)}(x)}^{q^{(+)}(x)} dq\, q(1 + 2n_q^{(0)}) = \int_{\rho^{(-)}(x)}^{\rho^{(+)}(x)} d\rho \left(\frac{1}{2} + \frac{1}{e^{2\beta JS\rho} - 1}\right)$$
$$= \frac{1}{2}[\rho^{(+)}(x) - [\rho^{(-)}(x)] + \left(\frac{k_BT}{2JS}\right)\ln\frac{1 - e^{-2\beta JS\rho^{(+)}(x)}}{1 - e^{-2\beta JS\rho^{(-)}(x)}} \tag{3.6.31}$$

where

$$\rho^{(\pm)}(x) = \frac{1}{2}[k^2 + p^2 \pm \sqrt{(k^2+p^2)^2 - 4k^2p^2x^2}]. \tag{3.6.31a}$$

By means of Eq. (3.6.31), the damping (3.6.27) becomes

$$\Sigma_c^{(2)''}(\boldsymbol{k} \to 0, \omega_{\boldsymbol{k}}) = \frac{J}{\hbar S}\left(\frac{2JS}{k_BT}\right) k^4 \frac{1}{8\pi^3}[I_1(k) + I_2(k)] \tag{3.6.32}$$

where

$$I_1(k) = \frac{1}{2}\int_0^\infty dp\; p^4 n_p^{(0)}[1 + n(\omega_k + \omega_p)] \int_{-1}^{1} dx\; x^2 \sqrt{k^2 + p^2 - 2kpx} \tag{3.6.33}$$

and

$$I_2(k) = \left(\frac{k_BT}{2JS}\right)\int_0^\infty dp\; p^4 n_p^{(0)}[1 + n(\omega_k + \omega_p)]$$
$$\times \int_{-1}^{1} dx \frac{x^2}{\sqrt{k^2 + p^2 + 2kpx}} \ln \frac{1 - e^{-2\beta JS\rho^{(+)}(x)}}{1 - e^{-2\beta JS\rho^{(-)}(x)}}. \tag{3.6.34}$$

In the long wavelength limit the integral in Eq. (3.6.33) may be evaluated at $k = 0$ since it remains finite in that limit. One has[3]

$$I_1(0) = \frac{1}{3}\int_0^\infty dp\; p^5 n_p^{(0)}(1 + n_p^{(0)}) = \frac{1}{6}\left(\frac{k_BT}{2JS}\right)^3 \int_0^\infty dx\, x^2 \frac{e^x}{(e^x - 1)^2}$$
$$= \frac{1}{6}\left(\frac{k_BT}{2JS}\right)^3 \left[\left|-\frac{x^2}{e^x - 1}\right|_0^\infty + 2\int_0^\infty dx\; \frac{x}{e^x - 1}\right] = \frac{\pi^2}{18}\left(\frac{k_BT}{2JS}\right)^3. \tag{3.6.35}$$

The evaluation of the integral (3.6.34) is more difficult since it diverges for $k \to 0$. In order to evaluate this singular contribution,[31] the integral $I_2(k)$ may be broken in two parts:

$$I_2(k) = I_2^{(1)} + I_2^{(2)} \tag{3.6.36}$$

where

$$I_2^{(1)} = \left(\frac{k_BT}{2JS}\right)\int_0^{\lambda k} dp\, p^4 n_p^{(0)}[1 + n(\omega_k + \omega_p)]$$
$$\times \int_{-1}^{1} dx \frac{x^2}{\sqrt{k^2 + p^2 + 2kpx}} \ln \frac{1 - e^{-2\beta JS\rho^{(+)}(x)}}{1 - e^{-2\beta JS\rho^{(-)}(x)}} \tag{3.6.37}$$

and

$$I_2^{(2)} = \left(\frac{k_BT}{2JS}\right)\int_{\lambda k}^\infty dp\, p^4 n_p^{(0)}[1 + n(\omega_k + \omega_p)]$$
$$\times \int_{-1}^{1} dx \frac{x^2}{\sqrt{k^2 + p^2 + 2kpx}} \ln \frac{1 - e^{-2\beta JS\rho^{(+)}(x)}}{1 - e^{-2\beta JS\rho^{(-)}(x)}} \tag{3.6.38}$$

with the assumption that $\lambda \gg 1$ even though $\lambda k \ll 1$. For instance, this can be achieved by choosing[31] $\lambda = (\frac{k_BT}{2JSk^2})^{1/4}$. Since $\lambda k \ll 1$, the Bose factors in Eq. (3.6.37) may by replaced by

$$n_p^{(0)}[1+n(\omega_k+\omega_p)] \simeq \left(\frac{k_BT}{2JS}\right)^2 \frac{1}{p^2(k^2+p^2)} \tag{3.6.39}$$

and the exponential functions may be expanded in powers of $2\beta JS\rho^{(\pm)}$. By means of the variable change $p = \xi k$, Eq. (3.6.37) becomes

$$I_2^{(1)} = \left(\frac{k_BT}{2JS}\right)^3 \int_0^\lambda d\xi \frac{\xi^2}{1+\xi^2} \int_{-1}^1 dx \frac{x^2}{\sqrt{1+\xi^2+2x\xi}} \times \ln \frac{1+\xi^2+\sqrt{(1+\xi^2)^2-4\xi^2x^2}}{1+\xi^2-\sqrt{(1+\xi^2)^2-4\xi^2x^2}}. \tag{3.6.40}$$

Since $\lambda \gg 1$, we can split the integral over $\xi$ in Eq. (3.6.40) in two integrals: the former going from 0 to 1 and the latter going from 1 to $\lambda$. In the second integral, a variable change from $\xi$ to $\frac{1}{\xi}$ may be performed so that Eq. (3.6.40) becomes

$$I_2^{(1)} = \left(\frac{k_BT}{2JS}\right)^3 \left[\int_0^1 d\xi \frac{\xi^2}{1+\xi^2} S(\xi) + \int_{1/\lambda}^1 d\xi \frac{1}{\xi(1+\xi^2)} S(\xi)\right] \tag{3.6.41}$$

where

$$S(\xi) = \int_{-1}^1 dx \frac{x^2}{\sqrt{1+\xi^2+2x\xi}} \ln \frac{1+\xi^2+\sqrt{(1+\xi^2)^2-4x^2\xi^2}}{1+\xi^2-\sqrt{(1+\xi^2)^2-4x^2\xi^2}}. \tag{3.6.42}$$

The first integral of Eq. (3.6.41) may be split in two parts: one going from 0 to $1/\lambda$ and the other from $1/\lambda$ to 1. Then Eq. (3.6.41) becomes

$$I_2^{(1)} = \left(\frac{k_BT}{2JS}\right)^3 \left[I_2^{(1a)} + I_2^{(1b)}\right] \tag{3.6.43}$$

where

$$I_2^{(1a)} = \int_{1/\lambda}^1 d\xi \frac{1+\xi^3}{\xi(1+\xi^2)} S(\xi) \tag{3.6.44}$$

and

$$I_2^{(1b)} = \int_0^{1/\lambda} d\xi \frac{\xi^2}{1+\xi^2} S(\xi). \tag{3.6.45}$$

Making the variable change $x = \frac{1+\xi^2}{2\xi} y$ in the integral of Eq. (3.6.42), one obtains[3]

$$S(\xi) = \frac{(1+\xi^2)^{5/2}}{8\xi^3} \int_{-\frac{2\xi}{1+\xi^2}}^{\frac{2\xi}{1+\xi^2}} dy \frac{y^2}{\sqrt{1+y}} \ln \frac{1+\sqrt{1-y^2}}{1-\sqrt{1-y^2}} = -\frac{4}{15}(5+2\xi^2)\ln\xi - \frac{4}{15}\frac{(2+\xi^2)(1+\xi^2)}{\xi^2} + \frac{8}{15}\frac{(1+\xi^2)^{\frac{5}{2}}}{\xi^3}\ln(\sqrt{1+\xi^2}+\xi). \tag{3.6.46}$$

Replacing Eq. (3.6.46) into Eqs. (3.6.44) and (3.6.45) and expanding for large values of $\lambda$, one obtains

$$I_2^{(1a)} = \frac{2}{3}\ln^2\lambda + \frac{4}{9}\ln\lambda + C + O\left(\frac{1}{\lambda}\ln\lambda\right) \tag{3.6.47}$$

and

$$I_2^{(1b)} = \frac{2}{3\lambda^2} + O\left(\frac{1}{\lambda^3}\ln\lambda\right) \tag{3.6.48}$$

with

$$C = \frac{56}{135} + \frac{7}{60}\pi^2 - \frac{4}{15}\ln^2(\sqrt{2}+1) - \frac{4}{5}G$$
$$- \frac{8}{15}[\mathrm{Li}_2(\sqrt{2}-1) - \mathrm{Li}_2(-\sqrt{2}+1)] = 0.17552256 \tag{3.6.49}$$

where $G = 0.91596559$ is the Catalan's constant and $\mathrm{Li}_2(x)$ the polylogarithm function given by

$$\mathrm{Li}_s(x) = \sum_{n=1}^{\infty}\frac{x^n}{n^s}. \tag{3.6.50}$$

The numerical value of the constant $C$ in Eq. (3.6.49) differs from that given by Harris[31] in Eq. (B8) where the loss of a factor $\frac{4}{15}$ in Eq. (B7) leads to a negative value of $C$. Replacing Eqs. (3.6.47) and (3.6.48) into Eq. (3.6.43), one obtains

$$I_2^{(1)} = \left(\frac{k_BT}{2JS}\right)^3\left[\frac{2}{3}\ln^2\lambda + \frac{4}{9}\ln\lambda + C + O\left(\frac{1}{\lambda}\ln\lambda\right)\right]. \tag{3.6.51}$$

The integration range over $p$ in Eq. (3.6.38) allows us to neglect $k$ in comparison to $p$ in the square root and $\omega_k$ with respect to $\omega_p$ in the Bose factor. Recalling that $\rho^{(\pm)}(x)$ are even functions of $x$ and making the variable change $y = x^2$, Eq. (3.6.38) becomes

$$I_2^{(2)} = \left(\frac{k_BT}{2JS}\right)\int_{\lambda k}^{\infty} dp\, p^3 n_p^{(0)}(1+n_p^{(0)})\int_0^1 dy\,\sqrt{y}\,\ln\frac{1-e^{-2\beta JS\rho^{(+)}(y)}}{1-e^{-2\beta JS\rho^{(-)}(y)}} \tag{3.6.52}$$

where

$$\rho^{(\pm)}(y) = \frac{1}{2}[k^2 + p^2 \pm \sqrt{(k^2+p^2)^2 - 4k^2p^2y}]. \tag{3.6.52a}$$

Integrating by parts the last integral of Eq. (3.6.52), one obtains

$$\int_0^1 dy\,\sqrt{y}\,\ln\frac{1-e^{-2\beta JS\rho^{(+)}(y)}}{1-e^{-2\beta JS\rho^{(-)}(y)}}$$
$$= \frac{2}{3}\ln\frac{1-e^{-2\beta JS\rho^{(+)}(1)}}{1-e^{-2\beta JS\rho^{(-)}(1)}}$$
$$- \frac{2}{3}\left(\frac{2JS}{k_BT}\right)\int_0^1 dy\, y^{3/2}\left[\frac{1}{e^{2\beta JS\rho^{(+)}(y)}-1}\frac{d\rho^{(+)}}{dy} - \frac{1}{e^{2\beta JS\rho^{(-)}(y)}-1}\frac{d\rho^{(-)}}{dy}\right]$$

$$= \frac{2}{3}\ln\frac{1-e^{-2\beta JS\rho^{(+)}(1)}}{1-e^{-2\beta JS\rho^{(-)}(1)}} - \frac{2}{3}\left(\frac{2JS}{k_BT}\right)\int_{\rho^{(+)}(0)}^{\rho^{(+)}(1)} d\rho^{(+)}\frac{y[\rho^{(+)}]^{\frac{3}{2}}}{e^{2\beta JS\rho^{(+)}}-1}$$

$$+ \frac{2}{3}\left(\frac{2JS}{k_BT}\right)\int_{\rho^{(-)}(0)}^{\rho^{(-)}(1)} d\rho^{(-)}\frac{y[\rho^{(-)}]^{\frac{3}{2}}}{e^{2\beta JS\rho^{(-)}}-1} \tag{3.6.53}$$

where

$$y(\rho) = \frac{\rho\left(k^2+p^2-\rho\right)}{(kp)^2} \tag{3.6.54}$$

is the inverse function of Eq. (3.6.52a). Using the relationships $\rho^{(+)}(1) = p^2$, $\rho^{(+)}(0) = k^2+p^2$, $\rho^{(-)}(1) = k^2$ and $\rho^{(-)}(0) = 0$, obtained from Eq. (3.6.52a), Eq. (3.6.53) becomes

$$I_2^{(2)} = I_2^{(2a)} + I_2^{(2b)} + I_2^{(2c)} \tag{3.6.55}$$

where

$$I_2^{(2a)} = \frac{2}{3k^3}\int_{\lambda k}^{\infty} dp\, n_p^{(0)}(1+n_p^{(0)})\int_0^{k^2} d\rho\,\frac{[\rho(k^2+p^2-\rho)]^{\frac{3}{2}}}{e^{2\beta JS\rho}-1}, \tag{3.6.56}$$

$$I_2^{(2b)} = \frac{2}{3k^3}\int_{\lambda k}^{\infty} dp\, n_p^{(0)}(1+n_p^{(0)})\int_{p^2}^{p^2+k^2} d\rho\,\frac{[\rho(k^2+p^2-\rho)]^{\frac{3}{2}}}{e^{2\beta JS\rho}-1} \tag{3.6.57}$$

and

$$I_2^{(2c)} = \frac{2}{3}\left(\frac{k_BT}{2JS}\right)\int_{\lambda k}^{\infty} dp\; p^3\; n_p^{(0)}(1+n_p^{(0)})\ln\frac{1-e^{-2\beta JSp^2}}{1-e^{-2\beta JSk^2}}. \tag{3.6.58}$$

Because of the chain of inequalities

$$0 < \rho < k^2 \ll (\lambda k)^2 < p^2, \tag{3.6.59}$$

the integral of Eq. (3.6.56) may be approximated by

$$\begin{aligned} I_2^{(2a)} &\simeq \frac{2}{3k^3}\left(\frac{k_BT}{2JS}\right)\int_{\lambda k}^{\infty} dp\; p^3\; n_p^{(0)}(1+n_p^{(0)})\int_0^{k^2} d\rho\sqrt{\rho} \\ &= \frac{2}{9}\left(\frac{k_BT}{2JS}\right)^3\int_{2\beta JS(\lambda k)^2}^{\infty} dx\,\frac{xe^x}{(e^x-1)^2} \\ &= \frac{2}{9}\left(\frac{k_BT}{2JS}\right)^3\left[\frac{2\beta JS(\lambda k)^2}{e^{2\beta JS(\lambda k)^2}-1} - \ln(1-e^{-2\beta JS\lambda^2k^2})\right]. \end{aligned} \tag{3.6.60}$$

The second row of Eq. (3.6.60) is obtained by the replacement $x = 2\beta JSp^2$ and the third row is obtained by an integration by parts. Expanding Eq. (3.6.60) for

$2\beta JS(\lambda k)^2 \ll 1$, one obtains

$$I_2^{(2a)} = \frac{2}{9}\left(\frac{k_BT}{2JS}\right)^3 \left\{ \ln\frac{k_BT}{2JSk^2} - 2\ln\lambda + 1 + O\left[\left(\frac{2JS\lambda^2k^2}{k_BT}\right)^2\right]\right\}. \qquad (3.6.61)$$

From the chain of inequalities given in Eq. (3.6.59), the integral over $p$ of Eq. (3.6.57) may be evaluated by means of the mean value theorem leading to

$$\int_{p^2}^{p^2+k^2} d\rho \frac{[\rho(k^2+p^2-\rho)]^{\frac{3}{2}}}{e^{2\beta JS\rho}-1} \simeq (1-\xi)^{\frac{3}{2}} k^5 p^3 \frac{1}{e^{2\beta JSp^2}-1} \qquad (3.6.62)$$

where $0 < \xi < 1$ is due to the mean value theorem. Then Eq. (3.6.57) reduces to

$$\begin{aligned} I_2^{(2b)} &= \frac{2}{3}(1-\xi)^{\frac{3}{2}} k^2 \int_{\lambda k}^{\infty} dp\, p^3\, n_p^{(0)2}(1+n_p^{(0)}) \\ &= \frac{(1-\xi)^{3/2}}{3}\left(\frac{k_BT}{2JS}\right)^2 k^2 \int_{2\beta JS(\lambda k)^2}^{\infty} dx \frac{xe^x}{(e^x-1)^3}. \end{aligned} \qquad (3.6.63)$$

After some integrations by parts, Eq. (3.6.63) becomes

$$\begin{aligned} I_2^{(2b)} &= \frac{(1-\xi)^{\frac{3}{2}}}{6}\left(\frac{k_BT}{2JS}\right)^2 k^2 \left[\frac{2\beta JS(\lambda k)^2}{(e^{2\beta JS\lambda^2k^2}-1)^2}\right. \\ &\quad \left. + \frac{1}{e^{2\beta JS(\lambda k)^2}-1} + 2\ln(1-e^{-2\beta JS\lambda^2k^2})\right]. \end{aligned} \qquad (3.6.64)$$

Expanding Eq. (3.6.64) in powers of $2\beta JS(\lambda k)^2$, one obtains

$$I_2^{(2b)} = \frac{(1-\xi)^{\frac{3}{2}}}{3}\left(\frac{k_BT}{2JS}\right)^3 \left[\frac{1}{\lambda^2} + \frac{2JSk^2}{k_BT}\left(\frac{1}{2}\ln\frac{2JSk^2}{k_BT} + \ln\lambda - \frac{3}{4} + \cdots\right)\right] \qquad (3.6.65)$$

By the variable change $x = 2\beta JSp^2$, the integral of Eq. (3.6.58) becomes

$$\begin{aligned} I_2^{(2c)} &= -\frac{1}{3}\left(\frac{k_BT}{2JS}\right)^3 \ln(1-e^{-2\beta JS\lambda^2k^2}) \int_{2\beta JS(\lambda k)^2}^{\infty} dx \frac{xe^x}{(e^x-1)^2} \\ &\quad + \frac{1}{3}\left(\frac{k_BT}{2JS}\right)^3 \int_{2\beta JS(\lambda k)^2}^{\infty} dx \frac{xe^x}{(e^x-1)^2}\ln(1-e^x). \end{aligned} \qquad (3.6.66)$$

After some integrations by parts, one obtains

$$\begin{aligned} I_2^{(2c)} &= -\frac{1}{3}\left(\frac{k_BT}{2JS}\right)^3 \ln(1-e^{-2\beta JSk^2}) \left[\frac{2\beta JS(\lambda k)^2}{e^{2\beta JS(\lambda k)^2}-1} - \ln(1-e^{-2\beta JS\lambda^2k^2})\right] \\ &\quad + \frac{1}{3}\left(\frac{k_BT}{2JS}\right)^3 \left[\left(\frac{2\beta JS\lambda^2k^2}{e^{2\beta JS\lambda^2k^2}-1} - 1 + 2\beta JS\lambda^2k^2\right)\ln(1-e^{-2\beta JS\lambda^2k^2})\right. \\ &\quad \left. + \frac{2\beta JS(\lambda k)^2}{e^{2\beta JS(\lambda k)^2}-1} - \frac{1}{2}\ln^2(1-e^{-2\beta JS\lambda^2k^2}) - \mathrm{Li}_2(e^{-2\beta JS\lambda^2k^2})\right] \end{aligned} \qquad (3.6.67)$$

where the polylogarithm function $\mathrm{Li}_2(x)$ is given by Eq. (3.6.50). Expanding Eq. (3.6.67) in powers of $2\beta JS(\lambda k)^2$, one obtains

$$I_2^{(2c)} = \frac{1}{3}\left(\frac{k_BT}{2JS}\right)^3 \left[\frac{1}{2}\ln^2\frac{k_BT}{2JSk^2} + \ln\frac{k_BT}{2JSk^2} - 2\ln^2\lambda + 1 - \frac{\pi^2}{6} + O\left(2\beta JS\lambda^2k^2\right)\right]. \tag{3.6.68}$$

Replacing Eqs. (3.6.61), (3.6.65) and (3.6.68) into Eq. (3.6.55), one obtains

$$I_2^{(2)} = \left(\frac{k_BT}{2JS}\right)^3 \left[\frac{1}{6}\ln^2\frac{k_BT}{2JSk^2} + \frac{5}{9}\ln\frac{k_BT}{2JSk^2} - \frac{2}{3}\ln^2\lambda - \frac{4}{9}\ln\lambda + \frac{5}{9} - \frac{\pi^2}{18} + O\left(\frac{2JS\lambda^2k^2}{k_BT}\right)\right]. \tag{3.6.69}$$

From Eqs. (3.6.36), (3.6.51) and (3.6.69), one obtains

$$I_2(k\to 0) = \left(\frac{k_BT}{2JS}\right)^3 \left[\frac{1}{6}\ln^2\frac{k_BT}{2JSk^2} + \frac{5}{9}\ln\frac{k_BT}{2JSk^2} + C + \frac{5}{9} - \frac{\pi^2}{18}\right] \tag{3.6.70}$$

where $C$ is given by Eq. (3.6.49). Finally, from Eqs. (3.6.32), (3.6.35), and (3.6.70), one obtains

$$\Sigma_c^{(2)''}(\boldsymbol{k}\to 0, \omega_{\boldsymbol{k}}) = \frac{Jk^4}{8\pi^3\hbar S}\left(\frac{k_BT}{2JS}\right)^2 \left[\frac{1}{6}\ln^2\left(\frac{k_BT}{2JSk^2}\right) + \frac{5}{9}\ln\left(\frac{k_BT}{2JSk^2}\right) + K\right] \tag{3.6.71}$$

where

$$K = \frac{1}{3}C + \frac{5}{9} = 0.73107812. \tag{3.6.72}$$

The damping given by Eq. (3.6.71) was obtained by A. B. Harris.[31] As one can see from Eq. (3.6.18) and (3.6.71), in the long wavelength and low temperature limit, the renormalization and damping of a spin wave of momentum $\boldsymbol{k}$ are of order $k^2T^{5/2}$ and $k^4T^2$, respectively. Moreover, singular terms appear in the damping.

In the short wavelength limit, that is for magnons with $\beta\hbar\omega_{\boldsymbol{k}} \gg 1$, in Eq. (3.6.22) one has $1 - e^{\beta\hbar\omega_{\boldsymbol{k}}} \simeq 1$, $n(\omega_{\boldsymbol{k}} + \omega_{\boldsymbol{q}_1}) \simeq n_{\boldsymbol{k}}^{(0)} \simeq 0$, $n_{\boldsymbol{q}_2}^{(0)} \simeq 0$ since $\boldsymbol{q}_2 \simeq \boldsymbol{k}$ due to the $\delta$-function argument so that the imaginary part of the second-order self-energy becomes

$$\Sigma_c^{(2)''}(\boldsymbol{k}, \omega_{\boldsymbol{k}}) = \frac{\pi}{2S^2}\left(\frac{2zJS}{\hbar N^2}\right) \sum_{\boldsymbol{q}_1,\boldsymbol{q}_2} (\gamma_{\boldsymbol{k}-\boldsymbol{q}_2} + \gamma_{\boldsymbol{q}_1-\boldsymbol{q}_2} - \gamma_{\boldsymbol{k}} - \gamma_{\boldsymbol{q}_1})^2 \times \delta(\gamma_{\boldsymbol{k}+\boldsymbol{q}_1-\boldsymbol{q}_2} + \gamma_{\boldsymbol{q}_2} - \gamma_{\boldsymbol{k}} - \gamma_{\boldsymbol{q}_1}) n_{\boldsymbol{q}_1}^{(0)}. \tag{3.6.73}$$

Putting $\boldsymbol{q}_1 = \boldsymbol{p}$ and $\boldsymbol{q}_2 = \frac{1}{2}(\boldsymbol{k}+\boldsymbol{p}) - \boldsymbol{\rho}$, for a SC lattice the main temperature contribution of Eq. (3.6.73) becomes

$$\Sigma_c^{(2)''}(\boldsymbol{k}, \omega_{\boldsymbol{k}}) = \frac{4\pi J}{\hbar S} \sum_\alpha \sin^2 \frac{k_\alpha}{2} \frac{1}{N} \sum_{\boldsymbol{p}} p_\alpha^2 n_{\boldsymbol{p}}^{(0)} \frac{1}{N} \sum_{\boldsymbol{\rho}}$$

$$\times \cos^2 \rho_\alpha \delta \left[ \sum_\gamma \cos \frac{k_\gamma}{2} \left( \cos \rho_\gamma - \cos \frac{k_\gamma}{2} \right) \right]$$

$$= \frac{8J}{\hbar S} \pi^2 \zeta\left(\frac{5}{2}\right) \left(\frac{k_B T}{8\pi J S}\right)^{\frac{5}{2}} \sum_\alpha \sin^2 \frac{k_\alpha}{2} L_\alpha(\boldsymbol{k}) \tag{3.6.74}$$

where $\alpha, \gamma = x, y, z$ and Eq. (3.6.14) has been used. Moreover,

$$L_\alpha(\boldsymbol{k}) = \frac{1}{(2\pi)^3} \int_{-\pi}^{\pi} \int_{-\pi}^{\pi} \int_{-\pi}^{\pi} d\rho_x d\rho_y d\rho_z \cos^2 \rho_\alpha\, \delta \left[ \sum_\gamma \cos \frac{k_\gamma}{2} \left( \cos \rho_\gamma - \cos \frac{k_\gamma}{2} \right) \right]. \tag{3.6.75}$$

As one can see, the damping is proportional to $T^{5/2}$ for $\hbar\omega_{\boldsymbol{k}} \gg k_B T$. For simplicity, we evaluate the integral in Eq. (3.6.75) for $\boldsymbol{k} = (k, k, k)$. One has

$$L_\alpha(\boldsymbol{k}) = L(k) = \frac{1}{4\pi^3} \frac{1}{\cos \frac{k}{2}} \iint_D d\rho_x d\rho_y \frac{(3\cos\frac{k}{2} - \cos\rho_x - \cos\rho_y)^2}{\sqrt{1 - (3\cos\frac{k}{2} - \cos\rho_x - \cos\rho_y)^2}} \tag{3.6.76}$$

where the domain $D$ is given by

$$\cos\rho_x + \cos\rho_y > 3\cos\frac{k}{2} - 1 \tag{3.6.77}$$

for $k < k_0 = \arccos\frac{1}{3} = 0.78365\pi$ and

$$3\cos\frac{k}{2} + 1 > \cos\rho_x + \cos\rho_y > 3\cos\frac{k}{2} - 1 \tag{3.6.78}$$

for $k > k_0$. For $k \lesssim \frac{\pi}{2}$, the domain (3.6.77) reduces to

$$\rho_x^2 + \rho_y^2 < 6\left(1 - \cos\frac{k}{2}\right) \tag{3.6.79}$$

and the integral of Eq. (3.6.76) becomes

$$L(k) \simeq \frac{1}{2\pi^2} \frac{1}{\cos\frac{k}{2}} \int_0^{\sqrt{6(1-\cos k/2)}} \rho\, d\rho \frac{\left(3\cos\frac{k}{2} - 2 + \frac{1}{2}\rho^2\right)^2}{\sqrt{1 - \left(3\cos\frac{k}{2} - 2 + \frac{1}{2}\rho^2\right)^2}}$$

$$= \frac{1}{2\pi^2} \frac{1}{\cos\frac{k}{2}} \int_{3\cos(k/2)-2}^{1} d\xi \frac{\xi^2}{\sqrt{1-\xi^2}}$$

$$= \frac{1}{2\pi^2} \frac{1}{\cos\frac{k}{2}} \left[ \frac{\pi}{4} + \frac{1}{2}\left(3\cos\frac{k}{2} - 2\right)\sqrt{1 - \left(3\cos\frac{k}{2} - 2\right)^2} \right.$$

$$\left. - \frac{1}{2}\arcsin\left(3\cos\frac{k}{2} - 2\right)\right]. \tag{3.6.80}$$

Expanding Eq. (3.6.80) in powers of $k$, one has

$$L(k) \simeq \frac{1}{4\pi^2}\sqrt{3}k\left(1 - \frac{5}{48}k^2 + \cdots\right). \tag{3.6.81}$$

Replacing Eq. (3.6.81) into Eq. (3.6.74), the imaginary part of the second order self-energy for $\hbar\omega_{\boldsymbol{k}} \gg k_BT$ and $k \lesssim \frac{\pi}{2}$ becomes

$$\Sigma_c^{(2)''}(\boldsymbol{k}, \omega_{\boldsymbol{k}}) = \frac{J}{2\hbar S}|\boldsymbol{k}|^3\zeta\left(\frac{5}{2}\right)\left(\frac{k_BT}{8\pi JS}\right)^{\frac{5}{2}} \tag{3.6.82}$$

where $|\boldsymbol{k}| = \sqrt{3}k$. The damping (3.6.82) was obtained by Lovesey.[23] Note that in the Lovesey's damping formula, a factor 2 in the denominator is missing.

In conclusion, for $\hbar\omega_{\boldsymbol{k}} \ll k_BT$, the damping is given by Eq. (3.6.71) while for $\hbar\omega_{\boldsymbol{k}} \gg k_BT$, it is given by Eq. (3.6.74) that reduces to Eq. (3.6.82) for $\boldsymbol{k}$ neither too small nor too near the ZB.

# Chapter 4

# FEYNMAN DIAGRAMS EXPANSION IN FERROMAGNETS

## 4.1. Temperature Green Function and Perturbation Expansion

We have seen in Eq. (3.2.17) that the neutron scattering cross-section may be written in terms of the generalized boson Green function $G_{\boldsymbol{k}}(\omega \pm i\epsilon)$ solution of the equation of motion given by Eq. (3.2.19). In the diagram expansion, it is more convenient to define the *temperature Green function*[13]

$$\mathcal{G}_{\boldsymbol{k}}(\tau) = -\langle T[a^{\mathrm{H}}_{\boldsymbol{k}}(\tau)a^{+}_{\boldsymbol{k}}]\rangle \tag{4.1.1}$$

where $\tau = it$ is an imaginary time and $T$ is the ordering time operator that puts the operators within the square brakets in such a way that the time decreases going from the first to the last. In particular, for $\tau > 0$, the temperature Green function (4.1.1) reduces to $\mathcal{G}_{\boldsymbol{k}}(\tau) = -\langle a^{\mathrm{H}}_{\boldsymbol{k}}(\tau)a^{+}_{\boldsymbol{k}}\rangle$. The "time evolution" of the boson operator in the Heisenberg picture is given by

$$a^{\mathrm{H}}_{\boldsymbol{k}}(\tau) = e^{\mathcal{H}\tau/\hbar}a_{\boldsymbol{k}}e^{-\mathcal{H}\tau/\hbar} \tag{4.1.2}$$

where

$$\mathcal{H} = \mathcal{H}_0 + \mathcal{H}_{\mathrm{int}} \tag{4.1.3}$$

with

$$\mathcal{H}_0 = \sum_{\boldsymbol{q}} \hbar\omega_{\boldsymbol{q}} a^{+}_{\boldsymbol{q}} a_{\boldsymbol{q}}, \quad \hbar\omega_{\boldsymbol{q}} = 2zJS(1-\gamma_{\boldsymbol{q}}), \tag{4.1.4}$$

$$\mathcal{H}_{\mathrm{int}} = \sum_{\boldsymbol{q}_1,\boldsymbol{q}_2,\boldsymbol{q}_3,\boldsymbol{q}_4} V_{\boldsymbol{q}_1,\boldsymbol{q}_2,\boldsymbol{q}_3,\boldsymbol{q}_4} a^{+}_{\boldsymbol{q}_1} a^{+}_{\boldsymbol{q}_2} a_{\boldsymbol{q}_3} a_{\boldsymbol{q}_4} \tag{4.1.5}$$

and

$$V_{\boldsymbol{q}_1,\boldsymbol{q}_2,\boldsymbol{q}_3,\boldsymbol{q}_4} = -\frac{zJ}{N}\delta_{\boldsymbol{q}_1+\boldsymbol{q}_2,\boldsymbol{q}_3+\boldsymbol{q}_4} v_{\boldsymbol{q}_1,\boldsymbol{q}_2,\boldsymbol{q}_3,\boldsymbol{q}_4} \tag{4.1.6a}$$

where

$$v_{q_1,q_2,q_3,q_4} = \frac{1}{4}\left(\gamma_{q_1-q_3} + \gamma_{q_1-q_4} + \gamma_{q_2-q_3} + \gamma_{q_2-q_4}\right) - \frac{1}{2}\left(\gamma_{q_1} + \gamma_{q_2}\right) \quad (4.1.6b)$$

is the DM interaction potential. Now, we intend to illustrate the main steps to arrive at a perturbation expansion for the temperature Green function. Let us begin by introducing the "interaction picture" in which the time dependent operators evolve according to the non-interacting Hamiltonian $\mathcal{H}_0$ that is

$$a_{\boldsymbol{k}}(\tau) = e^{\mathcal{H}_0\tau/\hbar} a_{\boldsymbol{k}} e^{-\mathcal{H}_0\tau/\hbar}. \quad (4.1.7)$$

The interaction picture is particularly convenient in view of a perturbative expansion since the unperturbed Hamiltonian is involved. However, since in the temperature Green function of Eq. (4.1.1), the time evolution is ruled by the full Hamiltonian (Heisenberg picture), the relationship between the Heisenberg and the interaction picture becomes of fundamental importance. From Eqs. (4.1.2) and (4.1.7), one obtains

$$a_{\boldsymbol{k}}^{\mathrm{H}}(\tau) = e^{\mathcal{H}\tau/\hbar} e^{-\mathcal{H}_0\tau/\hbar} a_{\boldsymbol{k}}(\tau) e^{\mathcal{H}_0\tau/\hbar} e^{-\mathcal{H}\tau/\hbar} = \mathcal{U}(0,\tau) a_{\boldsymbol{k}}(\tau) \mathcal{U}(\tau,0) \quad (4.1.8)$$

where

$$\mathcal{U}(\tau,\tau') = e^{\mathcal{H}_0\tau/\hbar} e^{-\mathcal{H}(\tau-\tau')/\hbar} e^{-\mathcal{H}_0\tau'/\hbar}. \quad (4.1.9)$$

From Eq. (4.1.9), one has

$$\mathcal{U}(\tau,\tau) = 1 \quad (4.1.9a)$$

and

$$\mathcal{U}(\beta\hbar, 0) = e^{\beta\mathcal{H}_0} e^{-\beta\mathcal{H}}. \quad (4.1.9b)$$

The time derivative of $\mathcal{U}(\tau,\tau')$ with respect to $\tau$ leads to the integral-differential equation

$$\mathcal{U}(\tau,\tau') = 1 - \frac{1}{\hbar}\int_{\tau'}^{\tau} \mathcal{H}_{\mathrm{int}}(\tau'')\mathcal{U}(\tau'',\tau')d\tau'' \quad (4.1.10)$$

where

$$\mathcal{H}_{\mathrm{int}}(\tau'') = e^{\mathcal{H}_0\tau''/\hbar} \mathcal{H}_{\mathrm{int}} e^{-\mathcal{H}_0\tau''/\hbar}. \quad (4.1.11)$$

The iterative solution of Eq. (4.1.11) is given by[13]

$$\mathcal{U}(\tau,\tau') = \sum_{n=0}^{\infty}\left(-\frac{1}{\hbar}\right)^n \frac{1}{n!}\int_{\tau'}^{\tau} d\tau_1 \cdots \int_{\tau'}^{\tau} d\tau_n T[\mathcal{H}_{\mathrm{int}}(\tau_1)\cdots\mathcal{H}_{\mathrm{int}}(\tau_n)] \quad (4.1.12)$$

where the time ordering operator is due to the fact that the interaction Hamiltonians $\mathcal{H}_{\mathrm{int}}$ do not commute at different times. By means of Eq. (4.1.9b), one may write the granpartition funtion as

$$\mathcal{Q} = \mathrm{Tr}(e^{-\beta\mathcal{H}}) = \mathrm{Tr}[e^{-\beta\mathcal{H}_0}\mathcal{U}(\beta\hbar, 0)] \tag{4.1.13}$$

and using Eq. (4.1.12), one obtains

$$\mathcal{Q} = \mathrm{Tr}(e^{-\beta\mathcal{H}_0})\sum_{n=0}^{\infty}\left(-\frac{1}{\hbar}\right)^n \frac{1}{n!}\int_0^{\beta\hbar} d\tau_1 \cdots \int_0^{\beta\hbar} d\tau_n \, \langle T[\mathcal{H}_{\mathrm{int}}(\tau_1)\cdots\mathcal{H}_{\mathrm{int}}(\tau_n)]\rangle_0 \tag{4.1.14}$$

where

$$\langle \mathcal{O}\rangle_0 = \frac{\mathrm{Tr}(\mathcal{O}e^{-\beta\mathcal{H}_0})}{\mathrm{Tr}(e^{-\beta\mathcal{H}_0})}. \tag{4.1.15}$$

By means of the relationship[13]

$$\mathcal{U}(\beta\hbar, \tau)a_{\boldsymbol{k}}(\tau)\mathcal{U}(\tau, 0) = \sum_{n=0}^{\infty}\left(-\frac{1}{\hbar}\right)^n \frac{1}{n!}\int_0^{\beta\hbar} d\tau_1 \cdots \times \int_0^{\beta\hbar} d\tau_n \langle T[\mathcal{H}_{\mathrm{int}}(\tau_1)\cdots\mathcal{H}_{\mathrm{int}}(\tau_n)a_{\boldsymbol{k}}(\tau)]\rangle_0 \tag{4.1.16}$$

and of Eqs. (4.1.8), (4.1.14) and (4.1.16), the temperature Green function (4.1.1) becomes

$$\begin{aligned}\mathcal{G}_{\boldsymbol{k}}(\tau) &= -\frac{1}{\mathcal{Q}}\mathrm{Tr}[e^{-\beta\mathcal{H}_0}\mathcal{U}(\beta\hbar, \tau)a_{\boldsymbol{k}}(\tau)\mathcal{U}(\tau, 0)a_{\boldsymbol{k}}^+] \\ &= -\frac{\sum_{n=0}^{\infty}\left(-\frac{1}{\hbar}\right)^n \frac{1}{n!}\int_0^{\beta\hbar} d\tau_1 \cdots \int_0^{\beta\hbar} d\tau_n \langle T[\mathcal{H}_{\mathrm{int}}(\tau_1)\cdots\mathcal{H}_{\mathrm{int}}(\tau_n)a_{\boldsymbol{k}}(\tau)a_{\boldsymbol{k}}^+]\rangle_0}{\sum_{n=0}^{\infty}\left(-\frac{1}{\hbar}\right)^n \frac{1}{n!}\int_0^{\beta\hbar} d\tau_1 \cdots \int_0^{\beta\hbar} d\tau_n \langle T[\mathcal{H}_{\mathrm{int}}(\tau_1)\cdots\mathcal{H}_{\mathrm{int}}(\tau_n)]\rangle_0}.\end{aligned} \tag{4.1.17}$$

From Eq. (4.1.5), one realizes that the $n$-th term in the numerator of Eq. (4.1.17) consists of a time ordered product containing $(2n+1)$ creation and $(2n+1)$ destruction Bose operators while the $n$-th term in the denominator consists of a time ordered product containing $2n$ creation and $2n$ destruction Bose operators.

Two very important theorems of the many-body theory are used to greatly simplify the series expansions of Eq. (4.1.17). The first theorem is the "generalized Wick's theorem"[13,32] proved by Matsubara:[33] such a theorem, applied to the numerator of Eq. (4.1.17), establishes that the average of a time ordered product containing $(2n + 1)$ creation and $(2n + 1)$ destruction operators is equal to the sum of $(2n+1)!$ permutations of the fully contracted terms consisting of the product of $(2n + 1)$ averages of time ordered products containing one creation and one destruction operator. The second theorem is the "linked cluster theorem"[13,34] that establishes that the series occurring in the numerator of Eq. (4.1.17) can be written as a product of a series containing only "connected diagrams" times a series containing only "disconnected diagrams" which is coincident with the series occurring in the denominator of Eq. (4.1.17). By means of the linked cluster theorem,

Eq. (4.1.17) reduces to

$$\mathcal{G}_{\boldsymbol{k}}(\tau) = -\sum_{n=0}^{\infty}\left(-\frac{1}{\hbar}\right)^n \frac{1}{n!}\int_0^{\beta\hbar} d\tau_1 \cdots \int_0^{\beta\hbar} d\tau_n \langle T[\mathcal{H}_{\text{int}}(\tau_1)\cdots\mathcal{H}_{\text{int}}(\tau_n)a_{\boldsymbol{k}}(\tau)a_{\boldsymbol{k}}^{+}]\rangle_0^{\text{c}} \tag{4.1.18}$$

where the superscript "c" means that only contributions corresponding to connected diagrams have to be retained in the sum. We will show how to handle Eq. (4.1.18) in the next sections by evaluating explicitly the first few terms of the series expansion. In particular, the zero-order term ($n = 0$) corresponds to the choice $\mathcal{H}_{\text{int}} = 0$. Then $\mathcal{H} = \mathcal{H}_0$ and the Heisenberg and interaction pictures coincide so that

$$a_{\boldsymbol{k}}^{\text{H}}(\tau) = a_{\boldsymbol{k}}(\tau) = e^{\mathcal{H}_0\tau/\hbar}a_{\boldsymbol{k}}e^{-\mathcal{H}_0\tau/\hbar} = a_{\boldsymbol{k}}e^{-\omega_{\boldsymbol{k}}\tau}. \tag{4.1.19}$$

The zero-order or "unperturbed" temperature Green function becomes

$$\mathcal{G}_{\boldsymbol{k}}^{(0)}(\tau) = -\langle T[a_{\boldsymbol{k}}(\tau)a_{\boldsymbol{k}}^{+}]\rangle_0 = \begin{cases} -(1+n_{\boldsymbol{k}}^{(0)})e^{-\omega_{\boldsymbol{k}}\tau} & \text{for } \tau > 0 \\ -n_{\boldsymbol{k}}^{(0)}e^{-\omega_{\boldsymbol{k}}\tau} & \text{for } \tau < 0 \end{cases} \tag{4.1.20}$$

where $n_{\boldsymbol{k}}^{(0)} = \langle a_{\boldsymbol{k}}^{+}a_{\boldsymbol{k}}\rangle_0 = (e^{\beta\hbar\omega_{\boldsymbol{k}}} - 1)^{-1}$.

Starting from Eqs. (4.1.1) and (4.1.2), one can prove that the boson temperature Green function $\mathcal{G}_{\boldsymbol{k}}(\tau)$ is a periodic function of $\tau$ with period $\beta\hbar$. Indeed, assuming $0 < \tau < \beta\hbar$ and using the cyclic property of the trace, one has

$$\begin{aligned}\mathcal{G}_{\boldsymbol{k}}(\tau) &= -\frac{1}{\mathcal{Q}}\text{Tr}\left[e^{-\beta\mathcal{H}}e^{\mathcal{H}\tau/\hbar}a_{\boldsymbol{k}}e^{-\mathcal{H}\tau/\hbar}a_{\boldsymbol{k}}^{+}\right] \\ &= -\frac{1}{\mathcal{Q}}\text{Tr}\left[a_{\boldsymbol{k}}^{+}e^{(\tau-\beta\hbar)\mathcal{H}/\hbar}a_{\boldsymbol{k}}e^{-(\tau-\beta\hbar)\mathcal{H}/\hbar}e^{-\beta\mathcal{H}}\right] \\ &= -\frac{1}{\mathcal{Q}}\text{Tr}\left[e^{-\beta\mathcal{H}}a_{\boldsymbol{k}}^{+}a_{\boldsymbol{k}}^{\text{H}}(\tau-\beta\hbar)\right] = -\left\langle T\left[a_{\boldsymbol{k}}^{\text{H}}(\tau-\beta\hbar)a_{\boldsymbol{k}}^{+}\right]\right\rangle = \mathcal{G}_{\boldsymbol{k}}(\tau-\beta\hbar).\end{aligned} \tag{4.1.21}$$

Since the temperature Green function is a periodic function of $\tau$ with a period $\beta\hbar$, it may be expanded in a Fourier series

$$\mathcal{G}_{\boldsymbol{k}}(\tau) = \frac{1}{\beta\hbar}\sum_n e^{-i\omega_n\tau}\mathcal{G}_{\boldsymbol{k}}(i\omega_n) \tag{4.1.22}$$

with

$$\omega_n = \frac{2\pi n}{\beta\hbar} \tag{4.1.23}$$

where the associated Fourier coefficients are given by

$$\mathcal{G}_{\boldsymbol{k}}(i\omega_n) = \int_0^{\beta\hbar} d\tau e^{i\omega_n\tau}\mathcal{G}_{\boldsymbol{k}}(\tau). \tag{4.1.24}$$

For the zero-order temperature Green function (4.1.20), the associated Fourier coefficients given by Eq. (4.1.24) become

$$\mathcal{G}_{\boldsymbol{k}}^{(0)}(i\omega_n) = \int_0^{\beta\hbar} d\tau e^{i\omega_n\tau}\mathcal{G}_{\boldsymbol{k}}^{(0)}(\tau) = -(1+n_{\boldsymbol{k}}^{(0)})\int_0^{\beta\hbar} d\tau e^{(i\omega_n-\omega_{\boldsymbol{k}})\tau}$$
$$= -(1+n_{\boldsymbol{k}}^{(0)})\frac{e^{(i\omega_n-\omega_{\boldsymbol{k}})\beta\hbar}-1}{i\omega_n-\omega_{\boldsymbol{k}}} = \frac{1}{i\omega_n-\omega_{\boldsymbol{k}}}. \tag{4.1.25}$$

The relationship between the temperature Green function (4.1.24) and the generalized Green functions (3.2.10) is easily obtained using Eqs.(4.1.1), (4.1.2), (3.2.6) and (4.1.23). Indeed,

$$\mathcal{G}_{\boldsymbol{k}}(i\omega_n) = -\int_0^{\beta\hbar} d\tau e^{i\omega_n\tau}\frac{1}{\mathcal{Q}}\sum_{\mu,\nu}\langle\mu|a_{\boldsymbol{k}}|\nu\rangle\langle\nu|a_{\boldsymbol{k}}^{+}|\mu\rangle e^{-\beta E_\mu}e^{(E_\mu-E_\nu)\tau/\hbar}$$
$$= -\int_0^{\beta\hbar} d\tau e^{i\omega_n\tau}\int_{-\infty}^{+\infty} d\omega' e^{-\omega'\tau}e^{\beta\hbar\omega'}J(\omega')$$
$$= \int_{-\infty}^{+\infty} d\omega'\left(e^{\beta\hbar\omega'}-1\right)J(\omega')\frac{1}{i\omega_n-\omega'} = G_{\boldsymbol{k}}(z=i\omega_n). \tag{4.1.26}$$

As one can see from Eq. (4.1.26), the temperature Green function coincides with the generalized Green function (3.2.10) evaluated at $z = i\omega_n$. Therefore, a suitable[13] analytic continuation of the temperature Green function to the complex $z$-plane recovers the generalized Green function that enters the neutron scattering cross-section.

## 4.2. First-Order Perturbation Theory

The first-order temperature Green function corresponds to the term with $n = 1$ of the expansion (4.1.18). From now on, we will omit the subscript "0" since all the averages occurring in the series expansion (4.1.18) are performed over the eigenstates of the unperturbed Hamiltonian $\mathcal{H}_0$ (interaction picture) while we keep the label "c" because only connected diagrams have to be retained in the expansion. Then the first-order term of the expansion (4.1.18) reads

$$\mathcal{G}_{\boldsymbol{k}}^{(1)}(\tau) = \frac{1}{\hbar}\int_0^{\beta\hbar} d\tau_1\langle T[\mathcal{H}_{\rm int}(\tau_1)a_{\boldsymbol{k}}(\tau)a_{\boldsymbol{k}}^{+}]\rangle_{\rm c}$$
$$= \frac{1}{\hbar}\sum_{q_1,q_2,q_3,q_4} V_{q_1,q_2,q_3,q_4}\int_0^{\beta\hbar} d\tau_1\langle T[a_{q_1}^{+}(\tau_1)a_{q_2}^{+}(\tau_1)$$
$$\times a_{q_3}(\tau_1)a_{q_4}(\tau_1)a_{\boldsymbol{k}}(\tau)a_{\boldsymbol{k}}^{+}]\rangle_{\rm c}. \tag{4.2.1}$$

The generalized Wick's theorem[13] applied to the time ordered product of the 3 creation and 3 destruction operators of Eq. (4.2.1) generates $3! = 6$ fully contracted terms. Each term is made up of the product of 3 time ordered products of one creation and one destruction operator. Four out of six terms are given in Eq. (4.2.2) (connected terms); the remaining two (disconnected) terms are given in Eq. (4.2.5). The linked cluster theorem[13] allows us to disregard the disconnected terms so that Eq. (4.2.1) gives

$$\begin{aligned}
&\langle T[a^+_{q_1}(\tau_1)a^+_{q_2}(\tau_1)a_{q_3}(\tau_1)a_{q_4}(\tau_1)a_{k}(\tau)a^+_{k}]\rangle_c \\
&= \langle T[a^+_{q_1}(\tau_1)a_{k}(\tau)]\rangle\langle T[a^+_{q_2}(\tau_1)a_{q_3}(\tau_1)]\rangle\langle T[a_{q_4}(\tau_1)a^+_{k}]\rangle \\
&+\langle T[a^+_{q_1}(\tau_1)a_{k}(\tau)]\rangle\langle T[a^+_{q_2}(\tau_1)a_{q_4}(\tau_1)]\rangle\langle T[a_{q_3}(\tau_1)a^+_{k}]\rangle \\
&+\langle T[a^+_{q_1}(\tau_1)a_{q_3}(\tau_1)]\rangle\langle T[a^+_{q_2}(\tau_1)a_{k}(\tau)]\rangle\langle T[a_{q_4}(\tau_1)a^+_{k}]\rangle \\
&+\langle T[a^+_{q_1}(\tau_1)a_{q_4}(\tau_1)]\rangle\langle T[a^+_{q_2}(\tau_1)a_{k}(\tau)]\rangle\langle T[a_{q_3}(\tau_1)a^+_{k}]\rangle. \qquad (4.2.2)
\end{aligned}$$

Using Eq. (4.1.20), we replace the time ordered products in Eq. (4.2.2) by the zero-order temperature Green function. For instance,

$$\langle T[a^+_{q_1}(\tau_1)a_{k}(\tau)]\rangle = -\delta_{q_1,k}\mathcal{G}^{(0)}_{k}(\tau-\tau_1). \qquad (4.2.3a)$$

The time difference appearing in the argument of the unperturbed termperature Green function is given by the time argument of the destruction operator (the time at which the boson is destroyed) minus the time argument of the creation operator (the time at which the boson was created). A problem arises when both the destruction and creation operators have the *same* time argument. This occurs when both operators belong to the same interaction Hamiltonian, for instance the Hamiltonian $\mathcal{H}_{\text{int}}(\tau_1)$ in Eq. (4.2.1). Since the creation operator always appears on the left of the destruction operator in the interaction Hamiltonian [see Eq. (4.1.5)], the time ordering operator must operate in such a way to assure this ordering. In particular, one has to assume that if the time of the destruction operator is $\tau_1$, the time of the creation operator belonging to the same interaction Hamiltonian has to be chosen as $\tau_1+\eta$ where $\eta$ is an infinitesimal *positive* time. For instance, the equal time ordered product occurring in the first term of the right-hand side of Eq. (4.2.2) has to be intended as

$$\lim_{\eta\to 0^+}\langle T[a^+_{q_2}(\tau_1+\eta)a_{q_3}(\tau_1)]\rangle = -\delta_{q_2,q_3}\mathcal{G}^{(0)}_{q_2}(0^-) \qquad (4.2.3b)$$

where $0^-$ is an infinitesimal *negative* time. By means of Eqs. (4.2.2), (4.2.3a), (4.2.3b) and from the invariance of the interaction potential (4.1.6b) under the permutation of the first two or the last two labels, Eq. (4.2.1) becomes

$$\mathcal{G}^{(1)}_{k}(\tau) = -\frac{4}{\hbar}\sum_{p} V_{k,p,k,p}\int_0^{\beta\hbar} d\tau_1\mathcal{G}^{(0)}_{k}(\tau-\tau_1)\mathcal{G}^{(0)}_{p}(0^-)\mathcal{G}^{(0)}_{k}(\tau_1). \qquad (4.2.4)$$

It is convenient to represent the right-hand side of Eq. (4.2.4) by a Feynman diagram as shown in Fig.4.1(a). Indeed, we have represented the potential interaction $V_{k,p,k,p}$ by a small full circle with two outgoing lines representing unperturbed temperature Green functions with momenta given by the first two labels of the interaction potential and two incoming lines representing unperturbed temperature Green functions with momenta given by the last two labels of the interaction potential. The invariance of the interaction potential leads to a multiplicity factor of 4. In Fig. 4.1(a),

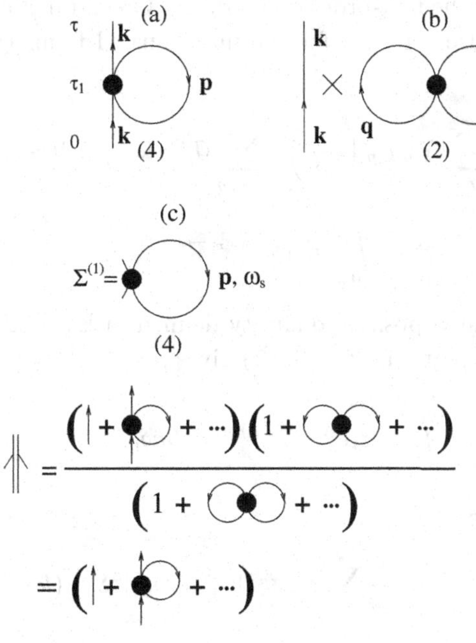


(4.1.17)

(4.1.18)

Fig. 4.1. First-order (a) connected and (b) disconnected Feynman diagrams and (c) first-order self-enrgy. The numbers in brackets below the diagrams give the multiplicity of the diagram. The lower part of the figure shows the first-order diagrammatic representation of Eqs. (4.1.17) and (4.1.18), respectively.

the lines going upwards (↑) and downwards (↓) are often called "particle"and "hole" lines, respectively. Note that the lines outoing from and entering the same interaction potential are *always* hole lines due to the choice (4.2.3b). The diagram shown in Fig. 4.1(a) is the only first-order "connected" diagram corresponding to Eq. (4.2.4) while the diagram shown in Fig. 4.1(b) is the only first-order "disconnected diagram" occurring in the numerator of Eq. (4.1.17) but not in Eq. (4.1.18). Indeed, the two disconnected terms occurring in the first-order expansion of the numerator of Eq. (4.1.17) are given by

$$\begin{aligned}&\langle T[a^+_{q_1}(\tau_1)a_{q_3}(\tau_1)]\rangle\langle T[a^+_{q_2}(\tau_1)a_{q_4}(\tau_1)]\rangle\langle T[a_{\boldsymbol{k}}(\tau)a^+_{\boldsymbol{k}}]\rangle\\&\quad+\langle T[a^+_{q_1}(\tau_1)a_{q_4}(\tau_1)]\rangle\langle T[a^+_{q_2}(\tau_1)a_{q_3}(\tau_1)]\rangle\langle T[a_{\boldsymbol{k}}(\tau)a^+_{\boldsymbol{k}}]\rangle,\end{aligned} \tag{4.2.5}$$

leading to the contribution

$$-\frac{2}{\hbar}\sum_{p,q} V_{q_1,q_2,q_1,q_2}\ \beta\hbar\mathcal{G}^{(0)}_{\boldsymbol{k}}(\tau)\mathcal{G}^{(0)}_{q_1}(0^-)\mathcal{G}^{(0)}_{q_2}(0^-). \tag{4.2.6}$$

The same contribution without $\mathcal{G}^{(0)}_{k}(\tau)$ comes from the first-order term of the denominator of Eq. (4.1.17). The first terms of the series expansion of the numerator and denominator of Eq. (4.1.17) are shown in the lower part of Fig. 4.1. One can infer the great importance of the linked cluster theorem by comparing the diagrammatic expansion of Eq. (4.1.17) and (4.1.18) in Fig. 4.1.

The Fourier series of the first-order temperature Green function can be obtained from Eq. (4.2.4) using Eq. (4.1.22) for the unperturbed temperature Green functions. One has

$$\mathcal{G}_{\boldsymbol{k}}^{(1)}(\tau) = -\frac{4}{\hbar}\sum_{\boldsymbol{p}} V_{\boldsymbol{k},\boldsymbol{p},\boldsymbol{k},\boldsymbol{p}} \left(\frac{1}{\beta\hbar}\right)^3 \sum_{n,s,m} \mathcal{G}_{\boldsymbol{k}}^{(0)}(i\omega_n)\mathcal{G}_{\boldsymbol{p}}^{(0)}(i\omega_s)\mathcal{G}_{\boldsymbol{k}}^{(0)}(i\omega_m)$$
$$\times e^{-i\omega_n\tau} e^{i\omega_s\eta} \int_0^{\beta\hbar} d\tau_1 e^{-i(\omega_m-\omega_n)\tau_1} \tag{4.2.7}$$

where $\eta$ is the infinitesimal positive quantity defined in Eq. (4.2.3b). By means of Eq. (4.1.23), the time integral in Eq. (4.2.7) gives

$$\int_0^{\beta\hbar} d\tau_1 e^{i(\omega_n-\omega_m)\tau_1} = \beta\hbar\delta_{m,n} \tag{4.2.8}$$

and Eq. (4.2.7) becomes

$$\mathcal{G}_{\boldsymbol{k}}^{(1)}(\tau) = \frac{1}{\beta\hbar}\sum_n e^{-i\omega_n\tau}[\mathcal{G}_{\boldsymbol{k}}^{(0)}(i\omega_n)]^2\Sigma^{(1)}(\boldsymbol{k}) \tag{4.2.9}$$

where the first-order self-energy is given by

$$\Sigma^{(1)}(\boldsymbol{k}) = -\frac{4}{\hbar}\sum_{\boldsymbol{p}} V_{\boldsymbol{k},\boldsymbol{p},\boldsymbol{k},\boldsymbol{p}} \frac{1}{\beta\hbar}\sum_s \frac{e^{i\omega_s\eta}}{i\omega_s - \omega_{\boldsymbol{p}}}. \tag{4.2.10}$$

Comparing Eq. (4.2.9) with Eq. (4.1.22), one has

$$\mathcal{G}_{\boldsymbol{k}}^{(1)}(i\omega_n) = [\mathcal{G}_{\boldsymbol{k}}^{(0)}(i\omega_n)]^2\Sigma^{(1)}(\boldsymbol{k}). \tag{4.2.11}$$

The diagram corresponding to the first-order self-energy (4.2.10) is shown in Fig. 4.1(c). Let us give some general rules that make it possible to pass from Fyenman diagrams to analytic forms and vice versa:

i) Associate a full circle with two outgoing and two incoming lines to each interaction potential. In Figs. 4.1(a) and (c), the *multiplicity* factor 4 is due to the number of equivalent diagrams that can be obtained after exchanging the outgoing lines (2) and the incoming lines (2).
ii) Associate an unperturbed temperature Green function $\mathcal{G}_{\boldsymbol{p}}^{(0)}(\tau-\tau')$ (time diagrams) or

$$\mathcal{G}_{\boldsymbol{p}}^{(0)}(i\omega_s) = \frac{1}{i\omega_s - \omega_{\boldsymbol{p}}} \tag{4.2.12}$$

in the self-energy (frequency) diagrams to each line of the diagram. If the unperturbed temperature Green function makes a closed loop, the time argument is $\tau = 0^-$ and a factor $e^{i\omega_s\eta}$ with $\eta \to 0^+$ has to be associated to the frequency Green function (4.2.12), in agreement with Eqs. (4.2.3b) and (4.1.22). Note that the two short "external" lines in the diagram (c) of Fig. 4.1 are reminiscent of the term $[\mathcal{G}_{\boldsymbol{k}}^{(0)}(i\omega_n)]^2$ of Eq. (4.2.11).

iii) The $\delta$-function occurring in the interaction potential (4.1.6a) implies the momentum conservation at each interaction: the sum of the momenta of the outgoing lines must be equal to the sum of the momenta of the incoming lines.

iv) Equation (4.2.8) implies the "frequency" conservation at each interaction. Note that the arrows that in the time space correspond to "particle" or "hole" lines, in the frequency space (self-energy diagrams) they correspond to a sort of "current" that has to be conserved at each "node".

v) A sum over the "internal" momenta in the time diagrams and a sum over both the internal momenta and frequencies times a factor $\frac{1}{\beta\hbar}$ for the self-energy diagrams have to be performed. Obviously, no sum has to be performed over the "external" momentum $\boldsymbol{k}$ and the external frequency $\omega_n$. In the self-energy diagram of Fig. 4.1(c), a sum over $\boldsymbol{p}$ and $\omega_s = \frac{2\pi s}{\beta\hbar}$ (or simply a sum over $s$) has to be carried out.

vi) A factor $(-1/\hbar)^n$ has to be added where $n$ is the order of the perturbation.

Using the rules i)–vi), the translation of Eq. (4.2.10) into the diagram of Fig. 4.1(c) and vice versa is direct. Let us now evaluate explicitly the frequency sum occurring in Eq. (4.2.10). To do this,[13] we introduce the meromorphic function on the complex plane $z$

$$f_1(z) = \frac{e^{\eta z}}{e^{\beta\hbar z} - 1}\,\frac{1}{z - \omega_{\boldsymbol{p}}}. \tag{4.2.13}$$

A meromorphic[35] function is analytic everywhere in the complex plane except at a discrete number of poles. As one can see, the function $f_1(z)$ defined in Eq. (4.2.13) has isolated simple poles on the imaginary axis at the points $z = i\omega_s = i\frac{2\pi s}{\beta\hbar}$ where $s$ is an integer and on the real axis at $z = \omega_{\boldsymbol{p}}$. Moreover, the function $f_1(z)$ vanishes on the circle $\Gamma$ of radius $|z| \to \infty$. Indeed, for $\Re z > 0$, one has $f_1(z) \to \frac{1}{z}e^{-\beta\hbar z}$ and for $\Re z < 0$, one has $f_1(z) \to -\frac{1}{z}e^{\eta z}$. If we integrate the function $f_1(z)$ along the circle $\Gamma$ and take the limit $|z| \to \infty$, we obtain

$$\lim_{|z|\to\infty} \oint_\Gamma f_1(z)dz = 0 \tag{4.2.14}$$

since the function $f_1(z)$ goes to zero faster than $1/z$ for $|z| \to \infty$. On the other hand, if we apply the residue theorem[35] to the meromorphic function $f_1(z)$ along the contour $\Gamma$, we obtain

$$\oint_\Gamma f_1(z)dz = 2\pi i\left(\frac{1}{\beta\hbar}\sum_s \frac{e^{i\omega_s\eta}}{i\omega_s - \omega_{\boldsymbol{p}}} + \frac{e^{\eta\omega_{\boldsymbol{p}}}}{e^{\beta\hbar\omega_{\boldsymbol{p}}} - 1}\right) \tag{4.2.15}$$

since inside the contour $\Gamma$, the function $f_1(z)$ has simple poles on the imaginary axis at $z = i\omega_s$ each with residue

$$\frac{1}{\beta\hbar}\,\frac{e^{i\omega_s\eta}}{i\omega_s - \omega_{\boldsymbol{p}}}$$

and on the real axis at $z = \omega_p$ with residue

$$\frac{e^{\eta\omega_p}}{e^{\beta\hbar\omega_p} - 1}.$$

In the limit $\eta \to 0$, Eqs. (4.2.14) and (4.2.15) lead to the relationship

$$\lim_{\eta\to 0}\frac{1}{\beta\hbar}\sum_s \frac{e^{i\omega_s\eta}}{i\omega_s - \omega_p} = -n_p^{(0)}. \tag{4.2.16}$$

Using Eqs. (4.2.16), (4.1.6a) and (4.1.6b), the first-order self-energy (4.2.10) becomes

$$\Sigma^{(1)}(\boldsymbol{k}) = \frac{4}{\hbar}\sum_{\boldsymbol{p}} V_{\boldsymbol{k},\boldsymbol{p},\boldsymbol{k},\boldsymbol{p}} n_{\boldsymbol{p}}^{(0)} = -\alpha\frac{\omega_{\boldsymbol{k}}}{S} \tag{4.2.17}$$

where $\alpha$ is given by Eq. (3.6.3). Equation (4.2.17) coincides with the first-order self-energy (3.6.2) obtained by means of the method of the Green function equation of motion.

## 4.3. Second-Order Perturbation Theory

The second-order temperature Green function, given by the term with $n = 2$ of Eq. (4.1.18), reads

$$\begin{aligned}
\mathcal{G}_{\boldsymbol{k}}^{(2)}(\tau) &= -\frac{1}{2!\,\hbar^2}\int_0^{\beta\hbar} d\tau_1 \int_0^{\beta\hbar} d\tau_2 \langle T[\mathcal{H}_{\text{int}}(\tau_1)\mathcal{H}_{\text{int}}(\tau_2)a_{\boldsymbol{k}}(\tau)a_{\boldsymbol{k}}^+]\rangle_c \\
&= -\frac{1}{2!\,\hbar^2}\sum_{\boldsymbol{q}_1,\boldsymbol{q}_2,\boldsymbol{q}_3,\boldsymbol{q}_4} V_{\boldsymbol{q}_1,\boldsymbol{q}_2,\boldsymbol{q}_3,\boldsymbol{q}_4} \sum_{\boldsymbol{q}_5,\boldsymbol{q}_6,\boldsymbol{q}_7,\boldsymbol{q}_8} V_{\boldsymbol{q}_5,\boldsymbol{q}_6,\boldsymbol{q}_7,\boldsymbol{q}_8}\int_0^{\beta\hbar} d\tau_1 \int_0^{\beta\hbar} d\tau_2 \\
&\quad \times \langle T[a_{\boldsymbol{q}_1}^+(\tau_1)a_{\boldsymbol{q}_2}^+(\tau_1)a_{\boldsymbol{q}_3}(\tau_1)a_{\boldsymbol{q}_4}(\tau_1)a_{\boldsymbol{q}_5}^+(\tau_2) \\
&\quad \times a_{\boldsymbol{q}_6}^+(\tau_2)a_{\boldsymbol{q}_7}(\tau_2)a_{\boldsymbol{q}_8}(\tau_2)a_{\boldsymbol{k}}(\tau)a_{\boldsymbol{k}}^+]\rangle_c.
\end{aligned} \tag{4.3.1}$$

The generalized Wick's theorem applied to Eq. (4.3.1) generates $5! = 120$ fully contracted terms: 80 terms lead to connected diagrams, 40 to disconnected diagrams. The invariance property of the interaction potential allows us to group the connected diagrams into 6 distinct terms:

$$\begin{aligned}
\mathcal{G}_{\boldsymbol{k}}^{(2)}(\tau) &= \frac{16}{2!\,\hbar^2}\sum_{\boldsymbol{p},\boldsymbol{q}} V_{\boldsymbol{k},\boldsymbol{p},\boldsymbol{k},\boldsymbol{p}}V_{\boldsymbol{k},\boldsymbol{q},\boldsymbol{k},\boldsymbol{q}}\int_0^{\beta\hbar} d\tau_1 \int_0^{\beta\hbar} d\tau_2 [\mathcal{G}_{\boldsymbol{k}}^{(0)}(\tau-\tau_2) \\
&\quad \times \mathcal{G}_{\boldsymbol{p}}^{(0)}(0^-)\mathcal{G}_{\boldsymbol{k}}^{(0)}(\tau_2-\tau_1)\mathcal{G}_{\boldsymbol{q}}^{(0)}(0^-)\mathcal{G}_{\boldsymbol{k}}^{(0)}(\tau_1) + \mathcal{G}_{\boldsymbol{k}}^{(0)}(\tau-\tau_1) \\
&\quad \times \mathcal{G}_{\boldsymbol{p}}^{(0)}(0^-)\mathcal{G}_{\boldsymbol{k}}^{(0)}(\tau_1-\tau_2)\mathcal{G}_{\boldsymbol{q}}^{(0)}(0^-)\mathcal{G}_{\boldsymbol{k}}^{(0)}(\tau_2)] \\
&\quad + \frac{16}{2!\,\hbar^2}\sum_{\boldsymbol{p},\boldsymbol{q}} V_{\boldsymbol{k},\boldsymbol{q},\boldsymbol{k},\boldsymbol{q}}V_{\boldsymbol{q},\boldsymbol{p},\boldsymbol{q},\boldsymbol{p}}\int_0^{\beta\hbar} d\tau_1 \int_0^{\beta\hbar} d\tau_2 [\mathcal{G}_{\boldsymbol{k}}^{(0)}(\tau-\tau_2)
\end{aligned}$$

$$\times \mathcal{G}_{q}^{(0)}(\tau_2-\tau_1)\mathcal{G}_{q}^{(0)}(\tau_1-\tau_2)\mathcal{G}_{p}^{(0)}(0^-)\mathcal{G}_{k}^{(0)}(\tau_2)+\mathcal{G}_{k}^{(0)}(\tau-\tau_1)$$

$$\times \mathcal{G}_{q}^{(0)}(\tau_1-\tau_2)\mathcal{G}_{q}^{(0)}(\tau_2-\tau_1)\mathcal{G}_{p}^{(0)}(0^-)\mathcal{G}_{k}^{(0)}(\tau_1)]$$

$$+\frac{8}{2!\,\hbar^2}\sum_{q_1,q_2,p} V_{k,p,q_1,q_2}V_{q_1,q_2,k,p}\int_0^{\beta\hbar} d\tau_1 \int_0^{\beta\hbar} d\tau_2 \big[\mathcal{G}_{k}^{(0)}(\tau-\tau_2)$$

$$\times \mathcal{G}_{q_1}^{(0)}(\tau_2-\tau_1)\mathcal{G}_{q_2}^{(0)}(\tau_2-\tau_1)\mathcal{G}_{p}^{(0)}(\tau_1-\tau_2)\mathcal{G}_{k}^{(0)}(\tau_1)+G_{k}^{(0)}(\tau-\tau_1)$$

$$\times \mathcal{G}_{q_1}^{(0)}(\tau_1-\tau_2)\mathcal{G}_{q_2}^{(0)}(\tau_1-\tau_2)\mathcal{G}_{p}^{(0)}(\tau_2-\tau_1)\mathcal{G}_{k}^{(0)}(\tau_2)\big]. \qquad (4.3.2)$$

The first and second term in the right-hand side of Eq. (4.3.2) can be identified with the Feynman diagrams ($a_1$) and ($a_2$) of Fig. 4.2, respectively; the third and fourth term with the diagrams ($b_1$) and ($b_2$); the fifth and sixth term with diagrams ($c_1$) and ($c_2$). Note that the second, fourth and sixth integral of Eq. (4.3.2) reduce to the first, third and fifth integral, respectively, by the change of the integration variables. From a topological point of view, this means that the diagrams ($a_2$), ($b_2$) and ($c_2$)

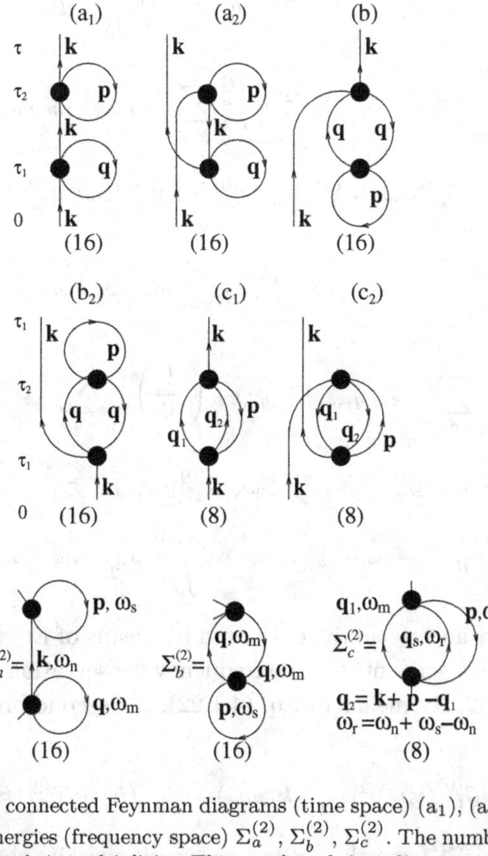


Fig. 4.2. Second-order connected Feynman diagrams (time space) ($a_1$), ($a_2$), ($b_1$), ($b_2$), ($c_1$), ($c_2$) and second-order self-energies (frequency space) $\Sigma_a^{(2)}$, $\Sigma_b^{(2)}$, $\Sigma_c^{(2)}$. The numbers whitin the brackets under the diagrams give their multiplicity. The couples of time diagrams ($a_1$) and ($a_2$), ($b_1$) and ($b_2$), ($c_1$) and ($c_2$) are topologically equivalent.

reduce to ($a_1$), ($b_1$) and ($c_1$), respectively, pulling both ends of the external lines. Such a symmetry suggests a more general rule:

vii) Each of the $n!$ permutations of the time variables $\tau_1, \tau_2, \cdots, \tau_n$ in the $n$-th term of series in Eq. (4.1.18) leads to the same contribution such that the corresponding diagrams are defined as "topologically equivalent". Then, we may suppress the factor $\frac{1}{n!}$ in the $n$-th term of the series expansion, keeping only *topologically distinct* diagrams.

For example, in Fig. 4.2, associating a multiplicity $16 \times 2!$ to diagrams ($a_1$) and ($b_1$) and a multiplicity $8 \times 2!$ to diagram ($c_1$), these only three diagrams represent the 80 connected terms coming from the second-order perturbation expansion of Eq. (4.3.1). Using Eq. (4.1.22), the second-order temperature Green function (4.3.2) becomes

$$
\begin{aligned}
\mathcal{G}_{\boldsymbol{k}}^{(2)}(\tau) = {} & \frac{16}{\hbar^2} \sum_{\boldsymbol{p},\boldsymbol{q}} V_{\boldsymbol{k},\boldsymbol{p},\boldsymbol{k},\boldsymbol{p}} V_{\boldsymbol{k},\boldsymbol{q},\boldsymbol{k},\boldsymbol{q}} \left(\frac{1}{\beta\hbar}\right)^5 \sum_{n,r,s,t,m} \mathcal{G}_{\boldsymbol{k}}^{(0)}(i\omega_n)\mathcal{G}_{\boldsymbol{p}}^{(0)}(i\omega_s)\mathcal{G}_{\boldsymbol{k}}^{(0)}(i\omega_t) \\
& \times \mathcal{G}_{\boldsymbol{q}}^{(0)}(i\omega_r)\mathcal{G}_{\boldsymbol{k}}^{(0)}(i\omega_m) e^{-i\omega_n\tau} e^{i\omega_s\eta} e^{i\omega_r\eta} \int_0^{\beta\hbar} d\tau_1 \, e^{-i(\omega_m-\omega_t)\tau_1} \\
& \times \int_0^{\beta\hbar} d\tau_2 \, e^{-i(\omega_t-\omega_m)\tau_2} + \frac{16}{\hbar^2} \sum_{\boldsymbol{p},\boldsymbol{q}} V_{\boldsymbol{k},\boldsymbol{q},\boldsymbol{k},\boldsymbol{q}} V_{\boldsymbol{q},\boldsymbol{p},\boldsymbol{q},\boldsymbol{p}} \left(\frac{1}{\beta\hbar}\right)^5 \\
& \sum_{n,r,s,t,m} \mathcal{G}_{\boldsymbol{k}}^{(0)}(i\omega_n)\mathcal{G}_{\boldsymbol{q}}^{(0)}(i\omega_r)\mathcal{G}_{\boldsymbol{q}}^{(0)}(i\omega_t)\mathcal{G}_{\boldsymbol{p}}^{(0)}(i\omega_s)\mathcal{G}_{\boldsymbol{k}}^{(0)}(i\omega_m) e^{-i\omega_n\tau} e^{i\omega_s\eta} \\
& \times \int_0^{\beta\hbar} d\tau_1 e^{-i(\omega_t-\omega_r)\tau_1} \int_0^{\beta\hbar} d\tau_2 e^{-i(\omega_r-\omega_m-\omega_t+\omega_m)\tau_2} \\
& + \frac{8}{\hbar^2} \sum_{\boldsymbol{q}_1,\boldsymbol{q}_2,\boldsymbol{p}} V_{\boldsymbol{k},\boldsymbol{p},\boldsymbol{q}_1,\boldsymbol{q}_2} V_{\boldsymbol{q}_1,\boldsymbol{q}_2,\boldsymbol{k},\boldsymbol{p}} \left(\frac{1}{\beta\hbar}\right)^5 \sum_{n,r,s,t,m} \mathcal{G}_{\boldsymbol{k}}^{(0)}(i\omega_n) \\
& \times \mathcal{G}_{\boldsymbol{q}_1}^{(0)}(i\omega_r)\mathcal{G}_{\boldsymbol{q}_2}^{(0)}(i\omega_t)\mathcal{G}_{\boldsymbol{p}}^{(0)}(i\omega_s)\mathcal{G}_{\boldsymbol{k}}^{(0)}(i\omega_m) e^{-i\omega_n\tau} \\
& \times \int_0^{\beta\hbar} d\tau_1 \, e^{-i(\omega_r+\omega_t-\omega_s-\omega_m)\tau_1} \int_0^{\beta\hbar} d\tau_2 e^{-i(\omega_r-\omega_n+\omega_t-\omega_s)\tau_2}. 
\end{aligned}
\tag{4.3.3}
$$

The integrals over $\tau_1$ and $\tau_2$ can be evaluated by means of Eq. (4.2.8): each integral gives a $\delta$-function that accounts for the frequency conservation as established in the iv) rule of Section 4.2. By means of Eq. (4.1.22), the second-order frequency Green function becomes

$$
\mathcal{G}_{\boldsymbol{k}}^{(2)}(i\omega_n) = [\mathcal{G}_{\boldsymbol{k}}^{(0)}(i\omega_n)]^2[\Sigma_a^{(2)}(\boldsymbol{k}, i\omega_n) + \Sigma_b^{(2)}(\boldsymbol{k}) + \Sigma_c^{(2)}(\boldsymbol{k}, i\omega_n)] \tag{4.3.4}
$$

where $\Sigma_a^{(2)}$, $\Sigma_b^{(2)}$ and $\Sigma_c^{(2)}$ are the second-order self-energies shown in Fig. 4.2. The analytic form of such self-energies can be obtained directly from Eq. (4.3.3) or using the rules i)–vi) of Section 4.2. Indeed, the self-enegy $\Sigma_a^{(2)}$ is given by

$$\Sigma_a^{(2)}(\boldsymbol{k}, i\omega_n) = \mathcal{G}_{\boldsymbol{k}}^{(0)}(i\omega_n)\frac{16}{\hbar^2}\sum_{\boldsymbol{p},\boldsymbol{q}} V_{\boldsymbol{k},\boldsymbol{p},\boldsymbol{k},\boldsymbol{p}}V_{\boldsymbol{k},\boldsymbol{q},\boldsymbol{k},\boldsymbol{q}}\ \frac{1}{\beta\hbar}\sum_s \frac{e^{i\omega_s\eta}}{i\omega_s - \omega_{\boldsymbol{p}}}$$
$$\times\frac{1}{\beta\hbar}\sum_m \frac{e^{i\omega_m\eta}}{i\omega_m - \omega_{\boldsymbol{q}}} = \mathcal{G}_{\boldsymbol{k}}^{(0)}(i\omega_n)\left[\Sigma^{(1)}(\boldsymbol{k})\right]^2 \tag{4.3.5}$$

where the first-order self-energy $\Sigma^{(1)}$ is given by Eq. (4.2.10) or (4.2.17). The self-energy $\Sigma_b^{(2)}$ is given by

$$\Sigma_b^{(2)}(\boldsymbol{k}) = \frac{16}{\hbar^2}\sum_{\boldsymbol{p},\boldsymbol{q}} V_{\boldsymbol{k},\boldsymbol{q},\boldsymbol{k},\boldsymbol{q}}V_{\boldsymbol{q},\boldsymbol{p},\boldsymbol{q},\boldsymbol{p}}\ \frac{1}{\beta\hbar}\sum_m \frac{1}{(i\omega_m - \omega_{\boldsymbol{q}})^2}\frac{1}{\beta\hbar}\sum_s \frac{e^{i\omega_s\eta}}{i\omega_s - \omega_{\boldsymbol{p}}}$$
$$= -4\beta\sum_{\boldsymbol{q}} V_{\boldsymbol{k},\boldsymbol{q},\boldsymbol{k},\boldsymbol{q}}\ n_{\boldsymbol{q}}^{(0)}(1+n_{\boldsymbol{q}}^{(0)})\ \Sigma^{(1)}(\boldsymbol{q}) \tag{4.3.6}$$

where $\Sigma^{(1)}$ is given by Eq. (4.2.10) and the relationship

$$\frac{1}{\beta\hbar}\sum_m \frac{1}{(i\omega_m - \omega_{\boldsymbol{q}})^2} = \beta\hbar\ n_{\boldsymbol{q}}^{(0)}(1+n_{\boldsymbol{q}}^{(0)}) \tag{4.3.7}$$

has been used. To prove Eq. (4.3.7), we introduce the meromorphic function

$$f_2(z) = \frac{1}{e^{\beta\hbar z}-1}\ \frac{1}{(z-\omega_{\boldsymbol{q}})^2} \tag{4.3.8}$$

which has a discrete set of simple poles along the imaginary axis located at $z = i\omega_m = \frac{2i\pi r}{\beta\hbar}$ with residue

$$\frac{1}{\beta\hbar}\ \frac{1}{(i\omega_m - \omega_{\boldsymbol{q}})^2}$$

and a pole of order two on the real axis at $z = \omega_{\boldsymbol{q}}$ with residue

$$-\beta\hbar\frac{e^{\beta\hbar\omega_{\boldsymbol{q}}}}{(e^{\beta\hbar\omega_{\boldsymbol{q}}}-1)^2} = -\beta\hbar\ n_{\boldsymbol{q}}^{(0)}(1+n_{\boldsymbol{q}}^{(0)}).$$

Then the residue theorem applied to the function $f_2(z)$ around the the circle $\Gamma$ with radius $|z| \to \infty$ gives

$$\oint_\Gamma f_2(z)dz = 2\pi i\left[\frac{1}{\beta\hbar}\sum_m \frac{1}{(i\omega_m - \omega_{\boldsymbol{q}})^2} - \beta\hbar\ n_{\boldsymbol{q}}^{(0)}(1+n_{\boldsymbol{q}}^{(0)})\right]. \tag{4.3.9}$$

Since the function $f_2(z)$ on the circle $\Gamma$ goes to zero at least as $\frac{1}{z^2}$ for $|z| \to \infty$, the integral on the left-hand side of Eq. (4.3.9) vanishes and Eq. (4.3.7) is proven. The self-energy $\Sigma_c^{(2)}$ is given by

$$\Sigma_c^{(2)}(\boldsymbol{k}, i\omega_n) = \frac{8}{\hbar^2} \sum_{\boldsymbol{q}_1,\boldsymbol{q}_2,\boldsymbol{p}} V_{\boldsymbol{k},\boldsymbol{p},\boldsymbol{q}_1,\boldsymbol{q}_2} V_{\boldsymbol{q}_1,\boldsymbol{q}_2,\boldsymbol{k},\boldsymbol{p}} \frac{1}{\beta\hbar} \sum_s \frac{1}{i\omega_s - \omega_{\boldsymbol{p}}}$$
$$\times \frac{1}{\beta\hbar} \sum_m \frac{1}{i\omega_m - \omega_{\boldsymbol{q}_1}} \frac{1}{i\omega_n + i\omega_s - i\omega_m - \omega_{\boldsymbol{q}_2}}. \qquad (4.3.10)$$

The sum over $m$ in Eq. (4.3.10) can be evaluated as follows

$$\frac{1}{\beta\hbar} \sum_m \frac{1}{i\omega_m - \omega_{\boldsymbol{q}_1}} \frac{1}{i\omega_n + i\omega_s - i\omega_m - \omega_{\boldsymbol{q}_2}}$$
$$= \frac{1}{i\omega_n + i\omega_s - \omega_{\boldsymbol{q}_1} - \omega_{\boldsymbol{q}_2}} \frac{1}{\beta\hbar} \sum_m \left( \frac{1}{i\omega_m - \omega_{\boldsymbol{q}_1}} - \frac{1}{i\omega_m - i\omega_n - i\omega_s + \omega_{\boldsymbol{q}_2}} \right)$$
$$= \frac{-n_{\boldsymbol{q}_1}^{(0)} + n(i\omega_n + i\omega_s - \omega_{\boldsymbol{q}_2})}{i\omega_n + i\omega_s - \omega_{\boldsymbol{q}_1} - \omega_{\boldsymbol{q}_2}} = -\frac{1 + n_{\boldsymbol{q}_1}^{(0)} + n_{\boldsymbol{q}_2}^{(0)}}{i\omega_n + i\omega_s - \omega_{\boldsymbol{q}_1} - \omega_{\boldsymbol{q}_2}} \qquad (4.3.11)$$

where the last step is obtained using Eqs. (4.2.16) and (4.1.23) that gives

$$n(i\omega_n + i\omega_s - \omega_{\boldsymbol{q}_2}) = n(-\omega_{\boldsymbol{q}_2}) = -(1 + n_{\boldsymbol{q}_2}^{(0)}). \qquad (4.3.12)$$

Replacing Eq. (4.3.11) into Eq. (4.3.10), one obtains

$$\Sigma_c^{(2)}(\boldsymbol{k}, i\omega_n) = -\frac{8}{\hbar^2} \sum_{\boldsymbol{q}_1,\boldsymbol{q}_2,\boldsymbol{p}} V_{\boldsymbol{k},\boldsymbol{p},\boldsymbol{q}_1,\boldsymbol{q}_2} V_{\boldsymbol{q}_1,\boldsymbol{q}_2,\boldsymbol{k},\boldsymbol{p}} (1 + n_{\boldsymbol{q}_1}^{(0)} + n_{\boldsymbol{q}_2}^{(0)})$$
$$\times \frac{1}{\beta\hbar} \sum_s \frac{1}{i\omega_s - \omega_{\boldsymbol{p}}} \frac{1}{i\omega_s + i\omega_n - \omega_{\boldsymbol{q}_1} - \omega_{\boldsymbol{q}_2}}. \qquad (4.3.13)$$

The sum over $s$ in Eq. (4.3.13) can be evaluated by repeating the same steps used in Eq. (4.3.11). One obtains

$$\frac{1}{\beta\hbar} \sum_s \frac{1}{i\omega_s - \omega_{\boldsymbol{p}}} \frac{1}{i\omega_s + i\omega_n - \omega_{\boldsymbol{q}_1} - \omega_{\boldsymbol{q}_2}}$$
$$= \frac{-n_{\boldsymbol{p}}^{(0)} + n(-i\omega_n + \omega_{\boldsymbol{q}_1} + \omega_{\boldsymbol{q}_2})}{i\omega_n - \omega_{\boldsymbol{q}_1} - \omega_{\boldsymbol{q}_2} + \omega_{\boldsymbol{p}}}$$
$$= -\frac{1}{i\omega_n - \omega_{\boldsymbol{q}_1} - \omega_{\boldsymbol{q}_2} + \omega_{\boldsymbol{p}}} \frac{n_{\boldsymbol{p}}^{(0)}(1 + n_{\boldsymbol{q}_1}^{(0)} + n_{\boldsymbol{q}_2}^{(0)}) - n_{\boldsymbol{q}_1}^{(0)} n_{\boldsymbol{q}_2}^{(0)}}{1 + n_{\boldsymbol{q}_1}^{(0)} + n_{\boldsymbol{q}_2}^{(0)}} \qquad (4.3.14)$$

where the relationship

$$n(-i\omega_n + \omega_{\boldsymbol{q}_1} + \omega_{\boldsymbol{q}_2}) = n(\omega_{\boldsymbol{q}_1} + \omega_{\boldsymbol{q}_2}) = \frac{n_{\boldsymbol{q}_1}^{(0)} n_{\boldsymbol{q}_2}^{(0)}}{1 + n_{\boldsymbol{q}_1}^{(0)} + n_{\boldsymbol{q}_2}^{(0)}} \qquad (4.3.15)$$

has been used. Replacing Eq. (4.3.14) into Eq. (4.3.13), one obtains

$$\Sigma_c^{(2)}(\boldsymbol{k}, i\omega_n) = \frac{8}{\hbar^2} \sum_{q_1,q_2,p} V_{k,p,q_1,q_2} V_{q_1,q_2,k,p} \frac{n_p^{(0)}(1+n_{q_1}^{(0)}+n_{q_2}^{(0)}) - n_{q_1}^{(0)} n_{q_2}^{(0)}}{i\omega_n - \omega_{q_1} - \omega_{q_2} + \omega_p}. \tag{4.3.16}$$

By means of Eq. (4.1.6a), one can see that $\Sigma_a^{(2)}$ given by Eq. (4.3.5), $\Sigma_b^{(2)}$ given by Eq. (4.3.6) and $\Sigma_c^{(2)}$ given by Eq. (4.3.16) coincide with the corresponding second-order self-energies (3.4.21), (3.4.22) and (3.4.23) obtained by the method of the Green function equation of motion. This result supports the equivalence of the two perturbation approaches. The diagrams correspoding to $\Sigma_a^{(2)}$, $\Sigma_b^{(2)}$ and $\Sigma_c^{(2)}$ are shown in the lower part of Fig. 4.2. As explained in Section 3.5, the self-energy $\Sigma_a^{(2)}$ is "reducible" so that it does not appear in the denominator of Eq. (3.5.1). In the frame of the Feynman diagram expansion, a *reducible* self-energy diagram is a diagram that becomes disconnected "cutting" one line of the diagram so that the next general rule can be established:

viii) Only *irreducible* diagrams have to be retained in the *proper* self-energy $\Sigma^*(\boldsymbol{k}, i\omega_n)$. As one can see from Fig. 4.2, the diagram corresponding to the self-energy $\Sigma_a^{(2)}$ is *reducible* because it becomes disconnected cutting the line corresponding to $\mathcal{G}_{\boldsymbol{k}}^{(0)}(i\omega_n)$.

## 4.4. Third-order Perturbation Theory

As remarked in Section 4.3, both the diagram expansion and the method of the Green function equation of motion lead to the same second-order self-energy. However, to illustrate the power of the diagram expansion, we evaluate the third-order proper self-energy in this section, an almost hopeless attempt using the method of the equation of motion. The third-order self-energy, that never appears in literature, can be an useful exercise to become familiarised with the diagram technique in magnetism. The third-order temperature Green function is given by the term with $n = 3$ of the perturbation expansion (4.1.18) that is

$$\begin{aligned}
\mathcal{G}_{\boldsymbol{k}}^{(3)}(\tau) &= \frac{1}{3!\,\hbar^3} \int_0^{\beta\hbar} d\tau_1 \int_0^{\beta\hbar} d\tau_2 \int_0^{\beta\hbar} d\tau_3 \langle T[\mathcal{H}_{\rm int}(\tau_1)\mathcal{H}_{\rm int}(\tau_2)\mathcal{H}_{\rm int}(\tau_3) a_{\boldsymbol{k}}(\tau) a_{\boldsymbol{k}}^+]\rangle_{\rm c} \\
&= \frac{1}{3!\,\hbar^3} \sum_{q_1,q_2,q_3,q_4} V_{q_1,q_2,q_3,q_4} \sum_{q_5,q_6,q_7,q_8} V_{q_5,q_6,q_7,q_8} \\
&\quad \times \sum_{q_9,q_{10},q_{11},q_{12}} V_{q_9,q_{10},q_{11},q_{12}} \int_0^{\beta\hbar} d\tau_1 \int_0^{\beta\hbar} d\tau_2 \int_0^{\beta\hbar} d\tau_2 \langle T[a_{q_1}^+(\tau_1) \\
&\quad \times a_{q_2}^+(\tau_1) a_{q_3}(\tau_1) a_{q_4}(\tau_1) a_{q_5}^+(\tau_2) a_{q_6}^+(\tau_2) a_{q_7}(\tau_2) a_{q_8}(\tau_2) \\
&\quad \times a_{q_9}^+(\tau_3) a_{q_{10}}^+(\tau_3) a_{q_{11}}(\tau_3) a_{q_{12}}(\tau_3) a_{\boldsymbol{k}}(\tau) a_{\boldsymbol{k}}^+]\rangle_{\rm c}.
\end{aligned} \tag{4.4.1}$$

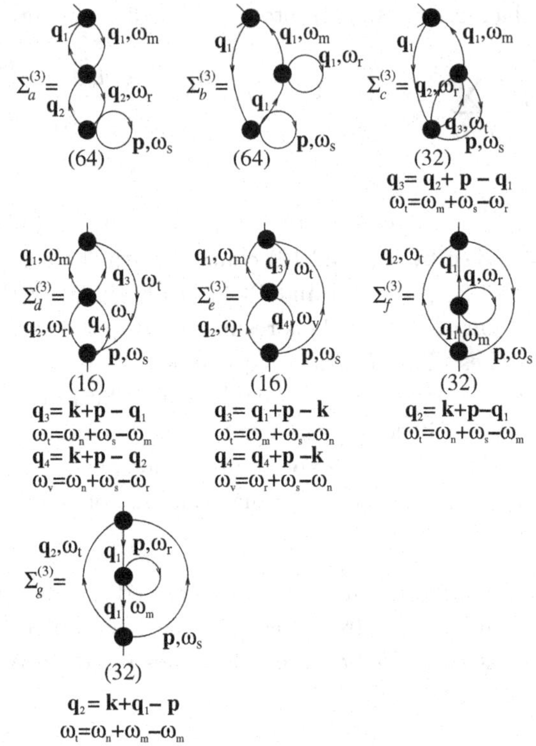


Fig. 4.3. Third-order irreducible diagrams in the frequency space. The multiplicity of each diagram is given by the number in brackets below the corresponding diagram.

The generalized Wick's theorem applied to Eq. (4.4.1) gives $7! = 5040$ fully contracted terms: of these $3!\times 328 = 1968$ lead to disconnected diagrams, $3!\times 256 = 1536$ lead to reducible connected diagrams and $3! \times 256 = 1536$ lead to irreducible connected diagrams. According to the vii) rule, the number of topologically non-equivalent diagrams is reduced by a factor 3!. Finally, using the invariance property of the interaction potential (4.1.6b), the 256 irreducible topologically non-equivalent, connected diagrams may be grouped into 7 classes of diagrams (in the frequency space) shown in Fig. 4.3. The third-order frequency Green function obtained from Eq. (4.4.1) becomes

$$\mathcal{G}_{\boldsymbol{k}}^{(3)}(i\omega_n) = [\mathcal{G}_{\boldsymbol{k}}^{(0)}(i\omega_n)]^2\,[\Sigma_{\rm red}^{(3)}(\boldsymbol{k}, i\omega_n) + \Sigma^{(3)*}(\boldsymbol{k}, i\omega_n)] \tag{4.4.2}$$

where $\Sigma_{\rm red}^{(3)}$ is the third-order reducible self-energy and $\Sigma^{(3)*}$ is the third-order proper self-energy given by

$$\Sigma^{(3)*}(\boldsymbol{k}, i\omega_n) = \Sigma_a^{(3)} + \Sigma_b^{(3)} + \Sigma_c^{(3)} + \Sigma_d^{(3)} + \Sigma_e^{(3)} + \Sigma_f^{(3)} + \Sigma_g^{(3)} \tag{4.4.3}$$

where the self-energies occurring in Eq. (4.4.3) are shown in Fig. 4.3. According to the i)–viii) rules, we obtain the analytic form of each self-energy diagram shown in Fig. 4.3.

The self-energy $\Sigma_a^{(3)}$ is given by

$$\Sigma_a^{(3)}(\boldsymbol{k}, i\omega_n) = -\frac{64}{\hbar^3}\sum_{\boldsymbol{q}_1,\boldsymbol{q}_2,\boldsymbol{p}} V_{\boldsymbol{k},\boldsymbol{q}_1,\boldsymbol{k},\boldsymbol{q}_1} V_{\boldsymbol{q}_1,\boldsymbol{q}_2,\boldsymbol{q}_1,\boldsymbol{q}_2} V_{\boldsymbol{q}_2,\boldsymbol{p},\boldsymbol{q}_2,\boldsymbol{p}}$$
$$\times \frac{1}{\beta\hbar}\sum_m \frac{1}{(i\omega_m - \omega_{\boldsymbol{q}_1})^2}\,\frac{1}{\beta\hbar}\sum_r \frac{1}{(i\omega_r - \omega_{\boldsymbol{q}_2})^2}\,\frac{1}{\beta\hbar}\sum_s \frac{e^{i\omega_s\eta}}{i\omega_s - \omega_{\boldsymbol{p}}}. \tag{4.4.4}$$

By means of Eqs. (4.3.6) and (4.3.7), the self-energy (4.4.4) becomes

$$\Sigma_a^{(3)}(\boldsymbol{k}) = -4\beta\sum_{\boldsymbol{q}_1} V_{\boldsymbol{k},\boldsymbol{q}_1,\boldsymbol{k},\boldsymbol{q}_1} n_{\boldsymbol{q}_1}^{(0)}(1 + n_{\boldsymbol{q}_1}^{(0)})\, \Sigma_b^{(2)}(\boldsymbol{q}_1). \tag{4.4.5}$$

In order to obtain the main temperature contribution to $\Sigma_a^{(3)}(\boldsymbol{k})$ in Eq. (4.4.5), we replace the temperature dependence of $\Sigma_b^{(2)}$ given by Eq. (3.6.9) and the interaction potential (4.1.6b). For a SC lattice, one has

$$\Sigma_a^{(3)}(\boldsymbol{k}) = -\frac{15}{4}\pi\left[\zeta\left(\frac{5}{2}\right)\right]^2 \frac{\omega_{\boldsymbol{k}}}{S^3}\left(\frac{k_BT}{8\pi JS}\right)^4 \frac{1}{N}\sum_{\boldsymbol{q}_1}(1-\gamma_{\boldsymbol{q}_1})^2 n_{\boldsymbol{q}_1}^{(0)}(1+n_{\boldsymbol{q}_1}^{(0)})$$
$$= -\frac{25}{4}\left[\pi\,\zeta\left(\frac{5}{2}\right)\right]^3 \frac{\omega_{\boldsymbol{k}}}{S^3}\left(\frac{k_BT}{8\pi JS}\right)^{\frac{15}{2}} \tag{4.4.6}$$

where we have used the relationship[3]

$$\frac{1}{N}\sum_{\boldsymbol{q}_1}(1-\gamma_{\boldsymbol{q}_1})^2 n_{\boldsymbol{q}_1}^{(0)}(1+n_{\boldsymbol{q}_1}^{(0)}) = \frac{1}{72\pi^2}\int_0^\infty dq_1 q_1^6 \frac{e^{2JSq_1^2}}{\left(e^{2JSq_1^2}-1\right)^2}$$
$$= \left(\frac{k_BT}{2JS}\right)^{\frac{7}{2}} \frac{1}{144\pi^2}\int_0^\infty dx \frac{x^{5/2}\,e^x}{(e^x-1)^2}$$
$$= \frac{5}{3}\pi^2\zeta\left(\frac{5}{2}\right)\left(\frac{k_BT}{8\pi JS}\right)^{\frac{7}{2}}.$$

As one can see from Eq. (4.4.6), $\Sigma_a^{(3)}$ is a real function of the wavevector independent of the frequency and proportional to $T^{15/2}$. Therefore, the self-energy $\Sigma_a^{(3)}$ is negligible with respect to the first-order contribution $\Sigma^{(1)}$ propotional to $T^{5/2}$.

The self-energy $\Sigma_b^{(3)}$ is given by

$$\Sigma_b^{(3)}(\boldsymbol{k}) = -\frac{64}{\hbar^3}\sum_{\boldsymbol{q}_1,\boldsymbol{q},\boldsymbol{p}} V_{\boldsymbol{k},\boldsymbol{q}_1,\boldsymbol{k},\boldsymbol{q}_1} V_{\boldsymbol{q}_1,\boldsymbol{q},\boldsymbol{q}_1,\boldsymbol{q}} V_{\boldsymbol{q}_1,\boldsymbol{p},\boldsymbol{q}_1,\boldsymbol{p}}$$
$$\times \frac{1}{\beta\hbar}\sum_m \frac{1}{(i\omega_m - \omega_{\boldsymbol{q}_1})^3}\,\frac{1}{\beta\hbar}\sum_r \frac{e^{i\omega_r\eta}}{i\omega_r - \omega_{\boldsymbol{q}}}\,\frac{1}{\beta\hbar}\sum_s \frac{e^{i\omega_s\eta}}{i\omega_s - \omega_{\boldsymbol{p}}}. \tag{4.4.7}$$

To evaluate the sum over $m$, we introduce the meromorphic function

$$f_3(z) = \frac{1}{e^{\beta\hbar z} - 1} \frac{1}{(z - \omega_{q_1})^3} \tag{4.4.8}$$

that vanishes on the circle $\Gamma$ of radius $|z| \to \infty$ at least as $\frac{1}{z^3}$. Making use of the residue theorem around the circle $\Gamma$, one obtains

$$\frac{1}{\beta\hbar} \sum_r \frac{1}{(i\omega_r - \omega_{q_1})^3} = -\frac{1}{2} (\beta\hbar)^2 n^{(0)}_{q_1} (1 + n^{(0)}_{q_1})(1 + 2n^{(0)}_{q_1}). \tag{4.4.9}$$

From Eqs. (4.4.9) and (4.2.16), the self-energy $\Sigma^{(3)}_b$ becomes

$$\Sigma^{(3)}_b(\boldsymbol{k}) = 2\beta^2\hbar \sum_{q_1} V_{\boldsymbol{k},q_1,\boldsymbol{k},q_1}\; n^{(0)}_{q_1} (1 + n^{(0)}_{q_1})(1 + 2\, n^{(0)}_{q_1}) \left[\Sigma^{(1)}(\boldsymbol{q}_1)\right]^2. \tag{4.4.10}$$

By means of Eqs. (3.6.2), (3.6.6), (4.1.6a) and (4.1.6b), one obtains

$$\begin{aligned} \Sigma^{(3)}_b(\boldsymbol{k}) &= -\frac{9}{8} \frac{\omega_{\boldsymbol{k}}}{S^3} \left[\zeta\left(\frac{5}{2}\right)\right]^2 \left(\frac{k_B T}{8\pi JS}\right)^3 \frac{1}{N} \\ &\quad \times \sum_{q_1} (1 - \gamma_{q_1})^3\; n^{(0)}_{q_1} (1 + n^{(0)}_{q_1})(1 + 2\, n^{(0)}_{q_1}) \\ &= -\frac{2}{3} \frac{\omega_{\boldsymbol{k}}}{S^3} \pi^{5/2} \left[\zeta\left(\frac{5}{2}\right)\right]^2 \left(\frac{k_B T}{8\pi JS}\right)^{\frac{15}{2}} \int_0^\infty dx \frac{x^{7/2}\, e^x (e^x + 1)}{(e^x - 1)^3} \\ &= -\frac{35}{8} \left[\pi\zeta\left(\frac{5}{2}\right)\right]^3 \frac{\omega_{\boldsymbol{k}}}{S^3} \left(\frac{k_B T}{8\pi JS}\right)^{\frac{15}{2}} \end{aligned} \tag{4.4.11}$$

where we have used the relationship[3]

$$\int_0^\infty dx \frac{x^{7/2}\, e^x (e^x + 1)}{(e^x - 1)^3} = \frac{105}{16} \sqrt{\pi}\, \zeta\left(\frac{5}{2}\right).$$

From Eq. (4.4.11), one can see that $\Sigma^{(3)}_b$ is real and proportional to $T^{15/2}$ like $\Sigma^{(3)}_a$, negligible with respect to the first-order self-energy $\Sigma^{(1)}$.

The self-energy $\Sigma^{(3)}_c$ is given by

$$\begin{aligned} \Sigma^{(3)}_c(\boldsymbol{k}) &= -\frac{32}{\hbar^3} \sum_{q_1,q_2,q_3,p} V_{\boldsymbol{k},q_1,\boldsymbol{k},q_1} V_{q_1,p,q_2,q_3} V_{q_2,q_3,q_1,p} \frac{1}{\beta\hbar} \sum_m \frac{1}{(i\omega_m - \omega_{q_1})^2} \\ &\quad \times \frac{1}{\beta\hbar} \sum_r \frac{1}{i\omega_r - \omega_{q_2}} \frac{1}{\beta\hbar} \sum_s \frac{1}{i\omega_s - \omega_p} \frac{1}{i\omega_m + i\omega_s - i\omega_r - \omega_{q_3}} \\ &= \frac{4}{\hbar} \sum_{q_1} V_{\boldsymbol{k},q_1,\boldsymbol{k},q_1} \frac{1}{\beta\hbar} \sum_m \frac{1}{(i\omega_m - \omega_{q_1})^2} \Sigma^{(2)}_c(\boldsymbol{q}_1, i\omega_m) \end{aligned} \tag{4.4.12}$$

where we have used Eq. (4.3.10). From Eq. (4.3.16), we obtain

$$\Sigma_c^{(3)}(\boldsymbol{k}) = -\frac{32}{\hbar^3} \sum_{\boldsymbol{q}_1,\boldsymbol{q}_2,\boldsymbol{q}_3,\boldsymbol{p}} V_{\boldsymbol{k},\boldsymbol{q}_1,\boldsymbol{k},\boldsymbol{q}_1} V_{\boldsymbol{q}_1,\boldsymbol{p},\boldsymbol{q}_2,\boldsymbol{q}_3} V_{\boldsymbol{q}_2,\boldsymbol{q}_3,\boldsymbol{q}_1,\boldsymbol{p}}$$
$$\times \left[ n_{\boldsymbol{p}}^{(0)} (1 + n_{\boldsymbol{q}_2}^{(0)} + n_{\boldsymbol{q}_3}^{(0)}) - n_{\boldsymbol{q}_2}^{(0)} n_{\boldsymbol{q}_3}^{(0)} \right]$$
$$\times \frac{1}{\beta\hbar} \sum_m \frac{1}{(i\omega_m - \omega_{\boldsymbol{q}_1})^2} \frac{1}{i\omega_m - \omega_{\boldsymbol{q}_2} - \omega_{\boldsymbol{q}_3} + \omega_{\boldsymbol{p}}}. \tag{4.4.13}$$

To evaluate the sum over $m$ in Eq. (4.4.13), we may proceed in a way similar to that used in Eq. (4.3.12) or introducing a meromorphic function $f_4(z)$ given by

$$f_4(z) = \frac{1}{e^{\beta\hbar z} - 1} \frac{1}{(z - \omega_{\boldsymbol{q}_1})^2} \frac{1}{z - \omega_{\boldsymbol{q}_2} - \omega_{\boldsymbol{q}_3} + \omega_{\boldsymbol{q}_4}} \tag{4.4.14}$$

and using the residue theorem around a circle $\Gamma$ with $|z| \to \infty$ on which the function $f_4(z)$ vanishes at least as $\frac{1}{z^3}$. Both procedures lead to the relationship

$$\frac{1}{\beta\hbar} \sum_m \frac{1}{(i\omega_m - \omega_{\boldsymbol{q}_1})^2} \frac{1}{i\omega_m - \omega_{\boldsymbol{q}_2} - \omega_{\boldsymbol{q}_3} + \omega_{\boldsymbol{p}}}$$
$$= \frac{\beta\hbar\, n_{\boldsymbol{q}_1}^{(0)} (1 + n_{\boldsymbol{q}_1}^{(0)})}{\omega_{\boldsymbol{q}_1} - \omega_{\boldsymbol{q}_2} - \omega_{\boldsymbol{q}_3} + \omega_{\boldsymbol{p}}} + \frac{1}{(\omega_{\boldsymbol{q}_1} - \omega_{\boldsymbol{q}_2} - \omega_{\boldsymbol{q}_3} + \omega_{\boldsymbol{p}})^2}$$
$$\times \frac{n_{\boldsymbol{q}_1}^{(0)} n_{\boldsymbol{p}}^{(0)} (1 + n_{\boldsymbol{q}_2}^{(0)} + n_{\boldsymbol{q}_3}^{(0)}) - n_{\boldsymbol{q}_2}^{(0)} n_{\boldsymbol{q}_3}^{(0)} (1 + n_{\boldsymbol{q}_1}^{(0)} + n_{\boldsymbol{p}}^{(0)})}{n_{\boldsymbol{p}}^{(0)} (1 + n_{\boldsymbol{q}_2}^{(0)} + n_{\boldsymbol{q}_3}^{(0)}) - n_{\boldsymbol{q}_2}^{(0)} n_{\boldsymbol{q}_3}^{(0)}}. \tag{4.4.15}$$

Replacing Eq. (4.4.15) into Eq. (4.4.13), one obtains

$$\Sigma_c^{(3)}(\boldsymbol{k}) = -\frac{32}{\hbar^3} \sum_{\boldsymbol{q}_1,\boldsymbol{q}_2,\boldsymbol{q}_3,\boldsymbol{p}} V_{\boldsymbol{q}_1,\boldsymbol{p},\boldsymbol{q}_2,\boldsymbol{q}_3} V_{\boldsymbol{q}_2,\boldsymbol{q}_3,\boldsymbol{q}_1,\boldsymbol{p}} \frac{1}{\omega_{\boldsymbol{q}_1} - \omega_{\boldsymbol{q}_2} - \omega_{\boldsymbol{q}_3} + \omega_{\boldsymbol{p}}}$$
$$\times \left\{ \beta\hbar\, V_{\boldsymbol{k},\boldsymbol{q}_1,\boldsymbol{k},\boldsymbol{q}_1}\, n_{\boldsymbol{q}_1}^{(0)} (1 + n_{\boldsymbol{q}_1}^{(0)}) \left[ n_{\boldsymbol{p}}^{(0)} (1 + n_{\boldsymbol{q}_2}^{(0)} + n_{\boldsymbol{q}_3}^{(0)}) - n_{\boldsymbol{q}_2}^{(0)} n_{\boldsymbol{q}_3}^{(0)} \right] \right.$$
$$\left. + (V_{\boldsymbol{k},\boldsymbol{q}_1,\boldsymbol{k},\boldsymbol{q}_1} - V_{\boldsymbol{k},\boldsymbol{q}_2,\boldsymbol{k},\boldsymbol{q}_2}) \frac{n_{\boldsymbol{q}_1}^{(0)} n_{\boldsymbol{p}}^{(0)} (1 + n_{\boldsymbol{q}_2}^{(0)} + n_{\boldsymbol{q}_3}^{(0)})}{\omega_{\boldsymbol{q}_1} - \omega_{\boldsymbol{q}_2} - \omega_{\boldsymbol{q}_3} + \omega_{\boldsymbol{p}}} \right\}.. \tag{4.4.16}$$

The self-energy $\Sigma_d^{(3)}$ is given by

$$\Sigma_d^{(3)}(\boldsymbol{k}, i\omega_n) = -\frac{16}{\hbar^3} \sum_{\boldsymbol{q}_1,\boldsymbol{q}_2,\boldsymbol{q}_3,\boldsymbol{q}_4,\boldsymbol{p}} V_{\boldsymbol{k},\boldsymbol{p},\boldsymbol{q}_1,\boldsymbol{q}_3} V_{\boldsymbol{q}_1,\boldsymbol{q}_3,\boldsymbol{q}_2,\boldsymbol{q}_4} V_{\boldsymbol{q}_2,\boldsymbol{q}_4,\boldsymbol{k},\boldsymbol{p}} \frac{1}{\beta\hbar} \sum_s \frac{1}{i\omega_s - \omega_{\boldsymbol{p}}}$$
$$\times \frac{1}{\beta\hbar} \sum_m \frac{1}{i\omega_m - \omega_{\boldsymbol{q}_1}} \frac{1}{i\omega_n + i\omega_s - i\omega_m - \omega_{\boldsymbol{q}_3}}$$

$$\times \frac{1}{\beta\hbar}\sum_r \frac{1}{i\omega_r - \omega_{q_2}} \frac{1}{i\omega_n + i\omega_s - i\omega_r - \omega_{q_4}}$$

$$= -\frac{16}{\hbar^3} \sum_{q_1,q_2,q_3,q_4,p} V_{k,p,q_1,q_3} V_{q_1,q_2,q_2,q_4} V_{q_2,q_4,k,p} \frac{1}{\beta\hbar}\sum_s \frac{1}{i\omega_s - \omega_p}$$

$$\times \frac{1 + n_{q_1}^{(0)} + n_{q_3}^{(0)}}{i\omega_n + i\omega_s - \omega_{q_1} - \omega_{q_3}} \frac{1 + n_{q_2}^{(0)} + n_{q_4}^{(0)}}{i\omega_n + i\omega_s - \omega_{q_2} - \omega_{q_4}} \tag{4.4.17}$$

where the sums over $m$ and $r$ have been performed using Eq. (4.3.11). The sum over $s$ can be performed in a similar way leading to

$$\Sigma_d^{(3)}(k, i\omega_n) = \frac{16}{\hbar^3} \sum_{q_1,q_2,q_3,q_4,p} V_{k,p,q_1,q_3} V_{q_1,q_3,q_2,q_4} V_{q_2,q_4,k,p} \frac{1}{\omega_{q_1} + \omega_{q_3} - \omega_{q_2} - \omega_{q_4}}$$

$$\times \left[ (1 + n_{q_2}^{(0)} + n_{q_4}^{(0)}) \frac{n_p^{(0)}(1 + n_{q_1}^{(0)} + n_{q_3}^{(0)}) - n_{q_1}^{(0)} n_{q_3}^{(0)}}{i\omega_n - \omega_{q_1} - \omega_{q_3} + \omega_p} \right.$$

$$\left. -(1 + n_{q_1}^{(0)} + n_{q_3}^{(0)}) \frac{n_p^{(0)}(1 + n_{q_2}^{(0)} + n_{q_4}^{(0)}) - n_{q_2}^{(0)} n_{q_4}^{(0)}}{i\omega_n - \omega_{q_2} - \omega_{q_4} + \omega_p} \right]. \tag{4.4.18}$$

The self-energy $\Sigma_e^{(3)}$ is given by

$$\Sigma_e^{(3)}(k, i\omega_n) = -\frac{16}{\hbar^3} \sum_{q_1,q_2,q_3,q_4,p} V_{k,q_3,q_1,p} V_{q_1,q_4,q_2,q_3} V_{q_2,p,k,q_4}$$

$$\times \frac{1}{\beta\hbar}\sum_s \frac{1}{i\omega_s - \omega_p} \frac{1}{\beta\hbar}\sum_m \frac{1}{i\omega_m - \omega_{q_1}} \frac{1}{i\omega_m + i\omega_s - i\omega_n - \omega_{q_3}}$$

$$\times \frac{1}{\beta\hbar}\sum_r \frac{1}{i\omega_r - \omega_{q_2}} \frac{1}{i\omega_r + i\omega_s - i\omega_n - \omega_{q_4}}$$

$$= -\frac{16}{\hbar^3} \sum_{q_1,q_2,q_3,q_4,p} V_{k,q_3,q_1,p} V_{q_1,q_4,q_2,q_3} V_{q_2,p,k,q_4}$$

$$\times (n_{q_1}^{(0)} - n_{q_3}^{(0)})(n_{q_2}^{(0)} - n_{q_4}^{(0)}) \frac{1}{\beta\hbar}\sum_s \frac{1}{i\omega_s - \omega_p}$$

$$\times \frac{1}{i\omega_s - i\omega_n + \omega_{q_1} - \omega_{q_3}} \frac{1}{i\omega_s - i\omega_n + \omega_{q_2} - \omega_{q_4}}. \tag{4.4.19}$$

The sums over $m$ and $r$ in Eq. (4.4.19) have been performed using the relationship

$$\frac{1}{\beta\hbar}\sum_m \frac{1}{i\omega_m - \omega_{q_1}} \frac{1}{i\omega_m + i\omega_s - i\omega_n - \omega_{q_3}} = \frac{n_{q_3}^{(0)} - n_{q_1}^{(0)}}{i\omega_s - i\omega_n + \omega_{q_1} - \omega_{q_3}} \tag{4.4.20}$$

obtained in a way similar to that used to write Eq. (4.3.11). The remaining sum over $s$ in Eq. (4.4.19) may be performed using an analogous procedure. One obtains

$$\Sigma_e^{(3)}(\boldsymbol{k}, i\omega_n) = \frac{16}{\hbar^3} \sum_{\boldsymbol{q}_1,\boldsymbol{q}_2,\boldsymbol{q}_3,\boldsymbol{q}_4,\boldsymbol{p}} \Big( V_{\boldsymbol{k},\boldsymbol{q}_3,\boldsymbol{q}_1,\boldsymbol{p}} V_{\boldsymbol{q}_1,\boldsymbol{q}_4,\boldsymbol{q}_2,\boldsymbol{q}_3} V_{\boldsymbol{q}_2,\boldsymbol{p},\boldsymbol{k},\boldsymbol{q}_4}$$
$$+ V_{\boldsymbol{q}_4,\boldsymbol{k},\boldsymbol{p},\boldsymbol{q}_2} V_{\boldsymbol{q}_3,\boldsymbol{q}_2,\boldsymbol{q}_4,\boldsymbol{q}_1} V_{\boldsymbol{p},\boldsymbol{q}_1,\boldsymbol{q}_3,\boldsymbol{k}} \Big) \frac{n^{(0)}_{\boldsymbol{q}_1} - n^{(0)}_{\boldsymbol{q}_3}}{\omega_{\boldsymbol{q}_2} + \omega_{\boldsymbol{q}_3} - \omega_{\boldsymbol{q}_1} - \omega_{\boldsymbol{q}_4}}$$
$$\times \frac{n^{(0)}_{\boldsymbol{p}}(n^{(0)}_{\boldsymbol{q}_2} - n^{(0)}_{\boldsymbol{q}_4}) - n^{(0)}_{\boldsymbol{q}_4}(1 + n^{(0)}_{\boldsymbol{q}_2})}{i\omega_n + \omega_{\boldsymbol{q}_4} - \omega_{\boldsymbol{q}_2} - \omega_{\boldsymbol{p}}}. \tag{4.4.21}$$

The self-energy $\Sigma_f^{(3)}$ is given by

$$\Sigma_f^{(3)}(\boldsymbol{k}, i\omega_n) = -\frac{32}{\hbar^3} \sum_{\boldsymbol{q}_1,\boldsymbol{q}_2,\boldsymbol{q}_3,\boldsymbol{p}} V_{\boldsymbol{k},\boldsymbol{p},\boldsymbol{q}_1,\boldsymbol{q}_2} V_{\boldsymbol{q}_1,\boldsymbol{q},\boldsymbol{q}_1,\boldsymbol{q}} V_{\boldsymbol{q}_1,\boldsymbol{q}_2,\boldsymbol{k},\boldsymbol{p}} \frac{1}{\beta\hbar} \sum_r \frac{e^{i\omega_r \eta}}{i\omega_r - \omega_{\boldsymbol{q}}}$$
$$\times \frac{1}{\beta\hbar} \sum_m \frac{1}{(i\omega_m - \omega_{\boldsymbol{q}_1})^2} \frac{1}{\beta\hbar} \sum_s \frac{1}{i\omega_s - \omega_{\boldsymbol{p}}} \frac{1}{i\omega_s + i\omega_n - i\omega_m - \omega_{\boldsymbol{q}_2}}$$
$$= \frac{8}{\hbar^2} \sum_{\boldsymbol{q}_1,\boldsymbol{q}_2,\boldsymbol{p}} V_{\boldsymbol{k},\boldsymbol{p},\boldsymbol{q}_1,\boldsymbol{q}_2} V_{\boldsymbol{q}_1,\boldsymbol{q}_2,\boldsymbol{k},\boldsymbol{p}} \, \Sigma^{(1)}(\boldsymbol{q}_1) \, (n^{(0)}_{\boldsymbol{q}_2} - n^{(0)}_{\boldsymbol{p}})$$
$$\times \frac{1}{\beta\hbar} \sum_m \frac{1}{(i\omega_m - \omega_{\boldsymbol{q}_1})^2} \frac{1}{i\omega_n - i\omega_m - \omega_{\boldsymbol{q}_2} + \omega_{\boldsymbol{p}}} \tag{4.4.22}$$

where Eq. (4.2.10) has been used and the sum over $s$ is taken from Eq. (4.4.20). Moreover,

$$\frac{1}{\beta\hbar} \sum_m \frac{1}{(i\omega_m - \omega_{\boldsymbol{q}_1})^2} \frac{1}{i\omega_n - i\omega_m - \omega_{\boldsymbol{q}_2} + \omega_{\boldsymbol{p}}}$$
$$= \frac{\beta\hbar \, n^{(0)}_{\boldsymbol{q}_1}(1 + n^{(0)}_{\boldsymbol{q}_1})}{i\omega_n + \omega_{\boldsymbol{p}} - \omega_{\boldsymbol{q}_1} - \omega_{\boldsymbol{q}_2}} + \frac{1}{(i\omega_n + \omega_{\boldsymbol{p}} - \omega_{\boldsymbol{q}_1} - \omega_{\boldsymbol{q}_2})^2}$$
$$\times \frac{n^{(0)}_{\boldsymbol{p}}(1 + n^{(0)}_{\boldsymbol{q}_2}) - n^{(0)}_{\boldsymbol{q}_1}(n^{(0)}_{\boldsymbol{q}_2} - n^{(0)}_{\boldsymbol{p}})}{n^{(0)}_{\boldsymbol{q}_2} - n^{(0)}_{\boldsymbol{p}}}. \tag{4.4.23}$$

Replacing Eq. (4.4.23) into Eq. (4.4.22), one obtains

$$\Sigma_f^{(3)}(\boldsymbol{k}, i\omega_n) = \frac{8}{\hbar^2} \sum_{\boldsymbol{q}_1,\boldsymbol{q}_2,\boldsymbol{p}} V_{\boldsymbol{k},\boldsymbol{p},\boldsymbol{q}_1,\boldsymbol{q}_2} V_{\boldsymbol{q}_1,\boldsymbol{q}_2,\boldsymbol{k},\boldsymbol{p}} \, \Sigma^{(1)}(\boldsymbol{q}_1)$$
$$\times \left[ \beta\hbar \, \frac{n^{(0)}_{\boldsymbol{q}_1}(1 + n^{(0)}_{\boldsymbol{q}_1})(n^{(0)}_{\boldsymbol{q}_2} - n^{(0)}_{\boldsymbol{p}})}{i\omega_n + \omega_{\boldsymbol{p}} - \omega_{\boldsymbol{q}_1} - \omega_{\boldsymbol{q}_2}} \right.$$
$$\left. + \frac{n^{(0)}_{\boldsymbol{p}}(1 + n^{(0)}_{\boldsymbol{q}_2}) - n_{\boldsymbol{q}_1}(n^{(0)}_{\boldsymbol{q}_2} - n^{(0)}_{\boldsymbol{p}})}{(i\omega_n + \omega_{\boldsymbol{p}} - \omega_{\boldsymbol{q}_1} - \omega_{\boldsymbol{q}_2})^2} \right]. \tag{4.4.24}$$

The self-energy $\Sigma_g^{(3)}$ is given by

$$\Sigma_g^{(3)}(\boldsymbol{k}, i\omega_n) = -\frac{32}{\hbar^3} \sum_{\boldsymbol{q}_1,\boldsymbol{q}_2,\boldsymbol{q},\boldsymbol{p}} V_{\boldsymbol{k},\boldsymbol{q}_1,\boldsymbol{q}_2,\boldsymbol{p}} V_{\boldsymbol{q}_1,\boldsymbol{q},\boldsymbol{q}_1,\boldsymbol{q}} V_{\boldsymbol{q}_2,\boldsymbol{p},\boldsymbol{k},\boldsymbol{q}_1} \frac{1}{\beta\hbar} \sum_r \frac{e^{i\omega_r \eta}}{i\omega_r - \omega_{\boldsymbol{q}}}$$

$$\times \frac{1}{\beta\hbar} \sum_m \frac{1}{(i\omega_m - \omega_{\boldsymbol{q}_1})^2} \frac{1}{\beta\hbar} \sum_s \frac{1}{i\omega_s - \omega_{\boldsymbol{p}}} \frac{1}{i\omega_m + i\omega_n - i\omega_s - \omega_{\boldsymbol{q}_2}}$$

$$= -\frac{8}{\hbar^2} \sum_{\boldsymbol{q}_1,\boldsymbol{q}_2,\boldsymbol{p}} V_{\boldsymbol{k},\boldsymbol{q}_1,\boldsymbol{q}_2,\boldsymbol{p}} V_{\boldsymbol{q}_2,\boldsymbol{p},\boldsymbol{k},\boldsymbol{q}_1} \Sigma^{(1)}(\boldsymbol{q}_1) \left(1 + n_{\boldsymbol{p}}^{(0)} + n_{\boldsymbol{q}_2}^{(0)}\right)$$

$$\times \frac{1}{\beta\hbar} \sum_m \frac{1}{(i\omega_m - \omega_{\boldsymbol{q}_1})^2} \frac{1}{i\omega_m + i\omega_n - \omega_{\boldsymbol{q}_2} - \omega_{\boldsymbol{p}}} \tag{4.4.25}$$

where Eq. (4.2.10) has been used and the sum over $s$ is taken from Eq. (4.3.11). Using the relationship

$$\frac{1}{\beta\hbar} \sum_m \frac{1}{(i\omega_m - \omega_{\boldsymbol{q}_1})^2} \frac{1}{i\omega_m + i\omega_n - \omega_{\boldsymbol{q}_2} - \omega_{\boldsymbol{p}}}$$

$$= \frac{\beta\hbar\, n_{\boldsymbol{q}_1}^{(0)}(1 + n_{\boldsymbol{q}_1}^{(0)})}{i\omega_n - \omega_{\boldsymbol{q}_2} - \omega_{\boldsymbol{p}} + \omega_{\boldsymbol{q}_1}} + \frac{1}{(i\omega_n - \omega_{\boldsymbol{q}_2} - \omega_{\boldsymbol{p}} + \omega_{\boldsymbol{q}_1})^2}$$

$$\times \frac{n_{\boldsymbol{q}_1}^{(0)}(1 + n_{\boldsymbol{q}_2}^{(0)} + n_{\boldsymbol{p}}^{(0)}) - n_{\boldsymbol{q}_2}^{(0)} n_{\boldsymbol{p}}^{(0)}}{1 + n_{\boldsymbol{q}_2}^{(0)} + n_{\boldsymbol{p}}^{(0)}} \tag{4.4.26}$$

the self-energy $\Sigma_g^{(3)}$ becomes

$$\Sigma_g^{(3)}(\boldsymbol{k}, i\omega_n) = -\frac{8}{\hbar^2} \sum_{\boldsymbol{q}_1,\boldsymbol{q}_2,\boldsymbol{p}} V_{\boldsymbol{k},\boldsymbol{q}_1,\boldsymbol{q}_2,\boldsymbol{p}} V_{\boldsymbol{q}_2,\boldsymbol{p},\boldsymbol{k},\boldsymbol{q}_1}\, \Sigma^{(1)}(\boldsymbol{q}_1)$$

$$\times \left[ \frac{\beta\hbar\, n_{\boldsymbol{q}_1}^{(0)}(1 + n_{\boldsymbol{q}_1}^{(0)})(1 + n_{\boldsymbol{q}_2}^{(0)} + n_{\boldsymbol{p}}^{(0)})}{i\omega_n - \omega_{\boldsymbol{q}_2} - \omega_{\boldsymbol{p}} + \omega_{\boldsymbol{q}_1}} \right.$$

$$\left. + \frac{n_{\boldsymbol{q}_1}^{(0)}(1 + n_{\boldsymbol{q}_2}^{(0)} + n_{\boldsymbol{p}}^{(0)}) - n_{\boldsymbol{q}_2}^{(0)} n_{\boldsymbol{p}}^{(0)}}{(i\omega_n - \omega_{\boldsymbol{q}_2} - \omega_{\boldsymbol{p}} + \omega_{\boldsymbol{q}_1})^2} \right]. \tag{4.4.27}$$

## 4.5. T-matrix Approximation

As shown in the previous sections, the main temperature contributions to the proper self-energy come from the diagrams containing a single downward line like $\Sigma^{(1)}$ [see Fig. 4.1 and Eq. (4.2.10)], $\Sigma_c^{(2)}$ [see Fig. 4.2 and Eq. (4.3.13) or (4.3.16] and $\Sigma_d^{(3)}$ [see Fig. 4.3 and Eq. (4.4.17) or (4.4.18]. The peculiar shape of these diagrams suggests how to select *all* the infinite contributions to the lowest order in temperature. Indeed, in order to build a temperature series expansion, particularly useful to fit the experimental data at low temperature, one has to keep all diagrams containing only one downward line. Such diagrams look like a "ladder", the rungs of which

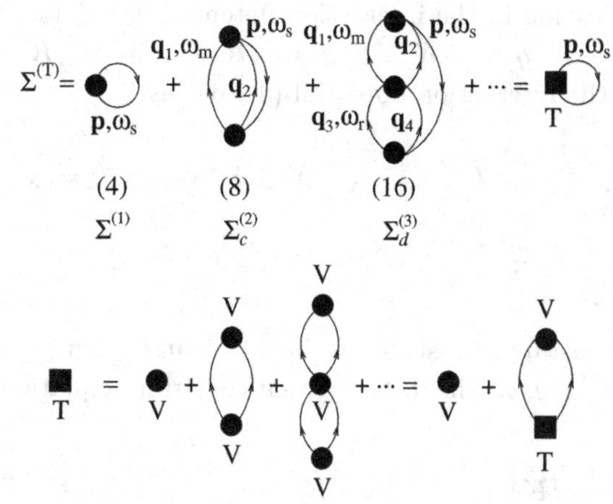


Fig. 4.4. T-matrix contribution to the self-energy (upper) and self-consistent equation for the T-matrix (lower). $V$ is the interaction potential.

are the full circles representing the interaction potential. Each rung is connected to the next by a couple of upward lines except the highest and lowest rung that are connected by a downward line. The self-energy corresponding to the sum of such diagrams is called *T-matrix* self-energy and it is shown in Fig. 4.4 along with the T-matrix self-consistent equation. Analytically, the first few terms of the T-matrix self-energy are given by Eqs. (4.2.10), (4.3.13) and (4.4.17) that is

$$\Sigma^{(\mathrm{T})}(\boldsymbol{k}, i\omega_n) = \Sigma^{(1)}(\boldsymbol{k}) + \Sigma_c^{(2)}(\boldsymbol{k}, i\omega_n) + \Sigma_d^{(3)}(\boldsymbol{k}, i\omega_n) + \cdots = -2\sum_{\boldsymbol{p}} \frac{1}{\beta\hbar}$$

$$\times \sum_s \frac{1}{i\omega_s - \omega_{\boldsymbol{p}}} \left[ \frac{2}{\hbar} V_{\boldsymbol{k},\boldsymbol{p},\boldsymbol{k},\boldsymbol{p}} + \sum_{\boldsymbol{q}_1,\boldsymbol{q}_2} \left(\frac{2}{\hbar}\right)^2 V_{\boldsymbol{k},\boldsymbol{p},\boldsymbol{q}_1,\boldsymbol{q}_2} V_{\boldsymbol{q}_1,\boldsymbol{q}_2,\boldsymbol{k},\boldsymbol{p}} \right.$$

$$\times \frac{1 + n^{(0)}_{\boldsymbol{q}_1} + n(0)_{\boldsymbol{q}_2}}{i\omega_s + i\omega_n - \omega_{\boldsymbol{q}_1} - \omega_{\boldsymbol{q}_2}} + \sum_{\boldsymbol{q}_1,\boldsymbol{q}_2,\boldsymbol{q}_3,\boldsymbol{q}_4} \left(\frac{2}{\hbar}\right)^3 V_{\boldsymbol{k},\boldsymbol{p},\boldsymbol{q}_1,\boldsymbol{q}_2} V_{\boldsymbol{q}_1,\boldsymbol{q}_2,\boldsymbol{q}_3,\boldsymbol{q}_4}$$

$$\left. \times V_{\boldsymbol{q}_3,\boldsymbol{q}_4,\boldsymbol{k},\boldsymbol{p}} \frac{1 + n^{(0)}_{\boldsymbol{q}_1} + n^{(0)}_{\boldsymbol{q}_2}}{i\omega_s + i\omega_n - \omega_{\boldsymbol{q}_1} - \omega_{\boldsymbol{q}_2}} \, \frac{1 + n^{(0)}_{\boldsymbol{q}_3} + n^{(0)}_{\boldsymbol{q}_4}}{i\omega_s + i\omega_n - \omega_{\boldsymbol{q}_3} - \omega_{\boldsymbol{q}_4}} + \cdots \right]$$

$$= -2\sum_{\boldsymbol{p}} \frac{1}{\beta\hbar} \sum_s \frac{1}{i\omega_s - \omega_{\boldsymbol{p}}} \frac{2}{\hbar} T_{\boldsymbol{k},\boldsymbol{p},\boldsymbol{k},\boldsymbol{p}}(i\omega_n + i\omega_s). \tag{4.5.1}$$

From Eq. (4.5.1), one can obtain the self-consistent equation for the T-matrix

$$T_{\boldsymbol{q}_1,\boldsymbol{q}_2,\boldsymbol{q}_3,\boldsymbol{q}_4}(\omega) = V_{\boldsymbol{q}_1,\boldsymbol{q}_2,\boldsymbol{q}_3,\boldsymbol{q}_4} + \frac{2}{\hbar} \sum_{\boldsymbol{q}_5,\boldsymbol{q}_6} V_{\boldsymbol{q}_1,\boldsymbol{q}_2,\boldsymbol{q}_5,\boldsymbol{q}_6} \frac{1 + n^{(0)}_{\boldsymbol{q}_5} + n^{(0)}_{\boldsymbol{q}_6}}{\omega - \omega_{\boldsymbol{q}_5} - \omega_{\boldsymbol{q}_6}} T_{\boldsymbol{q}_5,\boldsymbol{q}_6,\boldsymbol{q}_3,\boldsymbol{q}_4}(\omega). \tag{4.5.2}$$

The $\delta$-function occurring in the interaction potential of Eq. (4.1.6a) is satisfied by choosing $\boldsymbol{q}_1 = \frac{1}{2}\boldsymbol{K} + \boldsymbol{\lambda}$, $\boldsymbol{q}_2 = \frac{1}{2}\boldsymbol{K} - \boldsymbol{\lambda}$, $\boldsymbol{q}_3 = \frac{1}{2}\boldsymbol{K} + \boldsymbol{\mu}$, $\boldsymbol{q}_4 = \frac{1}{2}\boldsymbol{K} - \boldsymbol{\mu}$. Such a choice allows us to write the interaction potential (4.1.6a) as

$$\begin{aligned} V_{q_1,q_2,q_3,q_4} = V_{\boldsymbol{\lambda},\boldsymbol{\mu}}(\boldsymbol{K}) &= -\frac{zJ}{2N}(\gamma_{\boldsymbol{\lambda}-\boldsymbol{\mu}} + \gamma_{\boldsymbol{\lambda}+\boldsymbol{\mu}} - \gamma_{\frac{1}{2}\boldsymbol{K}+\boldsymbol{\lambda}} - \gamma_{\frac{1}{2}\boldsymbol{K}-\boldsymbol{\lambda}}) \\ &= -\frac{2J}{N}\sum_{\alpha} \cos\lambda_\alpha \left(\cos\mu_\alpha - \cos\frac{K_\alpha}{2}\right) \end{aligned} \tag{4.5.3}$$

where the lattice constant is assigned to be $a = 1$ and $\alpha$ runs over half of the NN: in a SC lattice, $\alpha = x, y, z$. The T-matrix self-consistent equation (4.5.2) becomes

$$T_{\boldsymbol{\lambda},\boldsymbol{\mu}}(\boldsymbol{K},\omega) = V_{\boldsymbol{\lambda},\boldsymbol{\mu}}(\boldsymbol{K}) + \frac{2}{\hbar}\sum_{\boldsymbol{\rho}} V_{\boldsymbol{\lambda},\boldsymbol{\rho}}(\boldsymbol{K}) \, \frac{1 + n^{(0)}_{\frac{1}{2}\boldsymbol{K}+\boldsymbol{\rho}} + n^{(0)}_{\frac{1}{2}\boldsymbol{K}-\boldsymbol{\rho}}}{\omega - \omega_{\frac{1}{2}\boldsymbol{K}+\boldsymbol{\rho}} - \omega_{\frac{1}{2}\boldsymbol{K}-\boldsymbol{\rho}}} \, T_{\boldsymbol{\rho},\boldsymbol{\mu}}(\boldsymbol{K},\omega). \tag{4.5.4}$$

The solution of Eq. (4.5.4) is given by

$$T_{\boldsymbol{\lambda},\boldsymbol{\mu}}(\boldsymbol{K},\omega) = -\frac{2J}{N}\sum_{\alpha,\beta} \cos\lambda_\alpha \left[\mathbf{1} - \frac{1}{2S}\mathbf{B}(\boldsymbol{K},\omega)\right]^{-1}_{\alpha\beta} \left(\cos\mu_\beta - \cos\frac{K_\beta}{2}\right) \tag{4.5.5}$$

where $\mathbf{1}$ is the unit matrix and

$$\mathrm{B}_{\alpha\beta}(\boldsymbol{K},\omega) = \frac{1}{N}\sum_{\boldsymbol{\rho}} \frac{\left(\cos\frac{K_\alpha}{2} - \cos\rho_\alpha\right)\cos\rho_\beta}{x + \sum_\sigma \cos\frac{K_\sigma}{2}\cos\rho_\sigma} \left(1 + n^{(0)}_{\frac{1}{2}\boldsymbol{K}+\boldsymbol{\rho}} + n^{(0)}_{\frac{1}{2}\boldsymbol{K}-\boldsymbol{\rho}}\right) \tag{4.5.6}$$

with

$$x = \frac{\hbar\omega}{8JS} - \frac{z}{2}. \tag{4.5.7}$$

For a SC lattice, both $\mathbf{1}$ and $\mathbf{B}$ are $3\times3$ matrices and the 9 elements of the $\mathbf{B}$-matrix are given by Eq. (4.5.6) with $\alpha, \beta = x, y, z$. Note that Eq. (4.5.5) involves the elements of the *inverse* matrix of $(\mathbf{1}\text{-}\frac{1}{2S}\mathbf{B})$. The T-matrix self-energy of Eq. (4.5.1) becomes

$$\Sigma^{(\mathrm{T})}(\boldsymbol{k}, i\omega_n) = -\frac{4}{\hbar}\sum_{\boldsymbol{p}} \frac{1}{\beta\hbar}\sum_s \frac{1}{i\omega_s - \omega_{\boldsymbol{p}}} T_{\boldsymbol{\lambda},\boldsymbol{\lambda}}(\boldsymbol{K}, i\omega_n + i\omega_s) \tag{4.5.8}$$

where $\boldsymbol{\lambda} = \frac{1}{2}(\boldsymbol{k} - \boldsymbol{p})$ and $\boldsymbol{K} = \boldsymbol{k} + \boldsymbol{p}$. The dispersion relation[13] for the T-matrix reads

$$T'_{\boldsymbol{\lambda},\boldsymbol{\mu}}(\boldsymbol{K},\omega) = V_{\boldsymbol{\lambda},\boldsymbol{\mu}}(\boldsymbol{K}) + \frac{\mathrm{P}}{\pi}\int_{-\infty}^{+\infty} d\omega' \frac{T''_{\boldsymbol{\lambda},\boldsymbol{\mu}}(\boldsymbol{K},\omega')}{\omega - \omega'} \tag{4.5.9a}$$

where

$$T_{\boldsymbol{\lambda},\boldsymbol{\mu}}(\boldsymbol{K},\omega \pm i\epsilon) = T'_{\boldsymbol{\lambda},\boldsymbol{\mu}}(\boldsymbol{K},\omega) \mp iT''_{\boldsymbol{\lambda},\boldsymbol{\mu}}(\boldsymbol{K},\omega). \tag{4.5.9b}$$

Equations (4.5.9a) and (4.5.9b) may be grouped in the relationship[36]

$$T_{\boldsymbol{\lambda},\boldsymbol{\mu}}(\boldsymbol{K},z) = V_{\boldsymbol{\lambda},\boldsymbol{\mu}}(\boldsymbol{K}) + \frac{1}{\pi}\int_{-\infty}^{+\infty} d\omega' \frac{T''_{\boldsymbol{\lambda},\boldsymbol{\mu}}(\boldsymbol{K},\omega')}{z-\omega'} \tag{4.5.9c}$$

where $z$ is the complex variable $\omega \pm i\epsilon$. By means of Eq. (4.5.9c), the self-energy (4.5.8) becomes

$$\begin{aligned}\Sigma^{(\mathrm{T})}(\boldsymbol{k},i\omega_n) &= \frac{4}{\hbar}\sum_{\boldsymbol{p}} V_{\boldsymbol{\lambda},\boldsymbol{\lambda}}(\boldsymbol{K})\, n_{\boldsymbol{p}}^{(0)} - \frac{4}{\hbar}\sum_{\boldsymbol{p}}\frac{1}{\pi}\int_{-\infty}^{+\infty} d\omega' \frac{T''_{\boldsymbol{\lambda},\boldsymbol{\lambda}}(\boldsymbol{K},\omega')}{i\omega_n-\omega'+\omega_{\boldsymbol{p}}} \\ &\quad \times \frac{1}{\beta\hbar}\sum_{s}\left(\frac{1}{i\omega_s-\omega_{\boldsymbol{p}}} - \frac{1}{i\omega_s+i\omega_n-\omega'}\right) \\ &= \frac{4}{\hbar}\sum_{\boldsymbol{p}} n_{\boldsymbol{p}}^{(0)}\left\{V_{\boldsymbol{\lambda},\boldsymbol{\lambda}}(\boldsymbol{K}) + \frac{1}{\pi}\int_{-\infty}^{+\infty} d\omega \frac{T''_{\boldsymbol{\lambda},\boldsymbol{\lambda}}(\boldsymbol{K},\omega')}{i\omega_n-\omega'+\omega_{\boldsymbol{p}}}\right\} \\ &\quad -\frac{4}{\hbar}\sum_{\boldsymbol{p}}\frac{1}{\pi}\int_{-\infty}^{+\infty} d\omega' \frac{T''_{\boldsymbol{\lambda},\boldsymbol{\lambda}}(\boldsymbol{K},\omega')}{i\omega_n-\omega'+\omega_{\boldsymbol{p}}} n(\omega') \end{aligned} \tag{4.5.10}$$

where Eq. (4.2.16) has been used. By means of Eq. (4.5.9c), the self-energy (4.5.1) becomes

$$\Sigma^{(\mathrm{T})}(\boldsymbol{k},\omega_{\boldsymbol{k}} \pm i\epsilon) = \Sigma^{(\mathrm{T})'}(\boldsymbol{k},\omega_{\boldsymbol{k}}) \pm i\Sigma^{(\mathrm{T})''}(\boldsymbol{k},\omega_{\boldsymbol{k}}) \tag{4.5.11}$$

where

$$\begin{aligned}\Sigma^{(\mathrm{T})'}(\boldsymbol{k},\omega_{\boldsymbol{k}}) &= \frac{4}{\hbar}\sum_{\boldsymbol{p}} T'_{\boldsymbol{\lambda},\boldsymbol{\lambda}}(\boldsymbol{K},\omega_{\boldsymbol{k}}+\omega_{\boldsymbol{p}})\, n_{\boldsymbol{p}}^{(0)} \\ &\quad -\frac{4}{\hbar}\sum_{\boldsymbol{p}}\frac{\mathrm{P}}{\pi}\int_{\omega_{\mathrm{bottom}}}^{\omega_{\mathrm{top}}} d\omega' \frac{T''_{\boldsymbol{\lambda},\boldsymbol{\lambda}}(\boldsymbol{K},\omega')}{\omega_{\boldsymbol{k}}+\omega_{\boldsymbol{p}}-\omega'}\, n(\omega') \end{aligned} \tag{4.5.11a}$$

and

$$\Sigma^{(\mathrm{T})''}(\boldsymbol{k},\omega_{\boldsymbol{k}}) = \frac{4}{\hbar}\sum_{\boldsymbol{p}} T''_{\boldsymbol{\lambda},\boldsymbol{\lambda}}(\boldsymbol{K},\omega_{\boldsymbol{k}}+\omega_{\boldsymbol{p}})[n_{\boldsymbol{p}}^{(0)} - n(\omega_{\boldsymbol{k}}+\omega_{\boldsymbol{p}})]. \tag{4.5.11b}$$

In the second term of Eq. (4.5.11a), the restricted integration range is due to the nature of the imaginary part of the T-matrix that is non zero only within the two-magnon band. Moreover, the presence of the Bose factor $n(\omega')$ implies that the temperature dependence of such a term is exponentially small compared to the first one which is proportional to $k^2T^{5/2}$ so that such a term can be neglected. The damping given by Eq. (4.5.11b) leads to two different results according to $\hbar\omega_{\boldsymbol{k}} \ll k_BT$ (long wavelength limit) or $\hbar\omega_{\boldsymbol{k}} \gg k_BT$ (short wavelength): in the first case, the evaluation of the damping[36] reduces to the second order result given by

Eq. (3.6.71). In the second case, the second Bose factor occurring in Eq. (4.5.11b) can be neglected and one recovers the damping given by Eq. (3.6.74). In the low temperature, short wavelength limit, the T-matrix self-energy (4.5.11) becomes

$$\Sigma^{(\mathrm{T})}(\boldsymbol{k}, \omega_{\boldsymbol{k}} \pm i\epsilon) = \frac{4}{\hbar} \sum_{\boldsymbol{p}} T_{\boldsymbol{\lambda},\boldsymbol{\lambda}}(\boldsymbol{K}, \omega_{\boldsymbol{k}} + \omega_{\boldsymbol{p}} \pm i\epsilon) n_{\boldsymbol{p}}^{(0)}. \tag{4.5.12a}$$

Making use of Eqs. (4.5.5) and (4.5.6), the self-energy (4.5.12a) becomes

$$\begin{aligned}\Sigma^{(\mathrm{T})}(\boldsymbol{k}, \omega_{\boldsymbol{k}} \pm i\epsilon) &= -\frac{16J}{\hbar N} \sum_{\boldsymbol{p}} n_{\boldsymbol{p}}^{(0)} \sum_{\alpha\beta} \cos\frac{k_\alpha - p_\alpha}{2} \sin\frac{p_\beta}{2} \sin\frac{k_\beta}{2} \\ &\quad \times \left[\mathbf{1} - \frac{1}{2S}\mathbf{B}(\boldsymbol{k} + \boldsymbol{p}, \omega_{\boldsymbol{k}} + \omega_{\boldsymbol{p}} \pm i\epsilon)\right]^{-1}_{\alpha\beta},\end{aligned} \tag{4.5.12b}$$

where the Bose factors appearing in Eq. (4.5.6) have been neglected. We will perform the explicit calculation of the T-matrix at zero temperature in 1D and 3D.

In 1D, an analytic calculation can be performed for any wavevector $k$. Indeed, the T-matrix of Eq. (4.5.5) reduces to

$$T_{\lambda,\mu}(K, \omega \pm i\epsilon) = -\frac{2J}{N} \frac{\cos\lambda\left(\cos\mu - \cos\frac{K}{2}\right)}{1 - \frac{1}{2S}\mathrm{B}(K, \omega \pm i\epsilon)}. \tag{4.5.13}$$

Using Eq. (3.2.12), one may write Eq. (4.5.6) in the form

$$\mathrm{B}(K, \omega \pm i\epsilon) = \mathrm{B}'(K, x) \pm i\mathrm{B}''(K, x) \tag{4.5.14}$$

where

$$\mathrm{B}'(K, x) = \frac{\mathrm{P}}{\pi} \int_0^\pi d\rho \, \frac{\left(\cos\frac{K}{2} - \cos\rho\right)\cos\rho}{x + \cos\frac{K}{2}\cos\rho} = \left(1 + \frac{x}{\cos^2\frac{K}{2}}\right)[1 - xI_0(K, x)] \tag{4.5.15}$$

and $x$ is given by Eq. (4.5.7) with $z = 2$. The integral $I_0(K, x)$ appearing in Eq. (4.5.15) is given by[3]

$$\begin{aligned}I_0(K, x) &= \frac{\mathrm{P}}{\pi} \int_0^\pi \frac{d\rho}{x + \cos\frac{K}{2}\cos\rho} \\ &= \begin{cases} -\dfrac{1}{\sqrt{x^2 - \cos^2\frac{K}{2}}} & \text{for } x < -\cos\frac{K}{2} \\ 0 & \text{for } -\cos\frac{K}{2} < x < \cos\frac{K}{2} \\ \dfrac{1}{\sqrt{x^2 - \cos^2\frac{K}{2}}} & \text{for } x > \cos\frac{K}{2} \end{cases}\end{aligned} \tag{4.5.15a}$$

The imaginary part of $\mathrm{B}(K,x)$ in Eq. (4.5.14) is given by

$$\mathrm{B}''(K,x) = -\int_0^\pi d\rho \cos\rho \left(\cos\frac{K}{2} - \cos\rho\right)\,\delta\left(x + \cos\frac{K}{2}\cos\rho\right)$$
$$= \frac{x}{\sqrt{\cos^2\frac{K}{2} - x^2}}\left(1 + \frac{x}{\cos^2\frac{K}{2}}\right) \tag{4.5.16}$$

for $-\cos\frac{K}{2} < x < \cos\frac{K}{2}$ and zero otherwise. The last row of Eq. (4.5.16) is obtained accounting for the property of the Dirac's $\delta$-function: indeed, its argument vanishes for $\rho = \rho_0 = \arccos(\frac{-x}{\cos K/2})$ only if $|-x/\cos\frac{K}{2}| < 1$. By means of Eqs. (4.5.15) and (4.5.16), Eq. (4.5.14) becomes

$$\mathrm{B}(K,\omega\pm i\epsilon) = \begin{cases} \left(1 + \dfrac{x}{\cos^2\frac{K}{2}}\right)\left(1 + \dfrac{x}{\sqrt{x^2 - \cos^2\frac{K}{2}}}\right) & \text{for } x < -\cos\frac{K}{2} \\ \left(1 + \dfrac{x}{\cos^2\frac{K}{2}}\right)\left(1 \pm i\dfrac{x}{\sqrt{\cos^2\frac{K}{2} - x^2}}\right) & \text{for } -\cos\frac{K}{2} < x < \cos\frac{K}{2} \\ \left(1 + \dfrac{x}{\cos^2\frac{K}{2}}\right)\left(1 - \dfrac{x}{\sqrt{x^2 - \cos^2\frac{K}{2}}}\right) & \text{for } x > \cos\frac{K}{2} \end{cases} \tag{4.5.17}$$

Equations (4.5.13) and (4.5.17) provide the T-matrix for a 1D ferromagnet at zero temperature. Using Eq. (4.5.7), one can see that the first row of Eq. (4.5.17) corresponds to the value of $\mathrm{B}(K,\omega)$ for $\hbar\omega < \hbar\omega_{\rm bottom} = 8JS(1 - \cos\frac{K}{2})$ that is for frequencies below the bottom of the two-spin wave band

$$\hbar\omega_{\rm 2sw} = 4JS\,(1 - \cos q_1) + 4JS\,(1 - \cos q_2) = 8JS\left(1 - \cos\frac{K}{2}\cos q\right)$$

with $q_1 = \frac{1}{2}K + q$ and $q_2 = \frac{1}{2}K - q$. Indeed, $\hbar\omega_{\rm 2sw}$ is the sum of the energies of two non-interacting spin waves in which $K$ is the wavevector of the complex formed by the two spin waves and $q$ is the relative wavevector of the the pair. The two-spin wave band is the energy region delimited between $\hbar\omega_{\rm bottom} = 8JS(1 - \cos\frac{K}{2})$ and $\hbar\omega_{\rm top} = 8JS(1 + \cos\frac{K}{2})$. The second row of Eq. (4.5.17) corresponds to the value of $\mathrm{B}(K,\omega)$ inside the two-spin wave band $\omega_{\rm bottom} < \omega < \omega_{\rm top}$ and the third row corresponds to the value of $\mathrm{B}(K,\omega)$ for $\omega > \omega_{\rm top}$ that is for frequencies above the band. Note that $\mathrm{B}(K,\omega \pm i\epsilon)$ is a real function outside and a complex function inside the two-spin wave band. Moreover, since $\omega_{\rm bottom} < \omega_k + \omega_p < \omega_{\rm top}$, in the self-energy (4.5.12b), one has to select the B-function given by the second row of Eq. (4.5.17) evaluated at $x = -\frac{1}{2}(\cos k + \cos p) = -\cos\frac{k+p}{2}\cos\frac{k-p}{2}$ so that

$$\Sigma^{(\mathrm{T})}(k,\omega_k \pm i\epsilon) = -\frac{16J}{\hbar N}\sum_p n_p^{(0)} \frac{\cos\frac{k-p}{2}\sin\frac{k}{2}\sin\frac{p}{2}}{1 + \frac{1}{S}\frac{\sin\frac{k}{2}\sin\frac{p}{2}}{\cos\frac{k+p}{2}}\left(1 \mp i\frac{\cos\frac{k-p}{2}}{\left|\sin\frac{k-p}{2}\right|}\right)}. \tag{4.5.18}$$

Expanding Eq. (4.5.18) in powers of $p$ and retaining only the main contribution, one has

$$\Sigma^{(\mathrm{T})'}(k,\omega_k) = -\frac{8J}{\hbar N}\sin\frac{k}{2}\sum_p p\, n_p^{(0)}\frac{\cos\frac{k}{2}+\frac{p}{2}\left(1+\frac{1}{S}\right)\sin\frac{k}{2}+O(p^2)}{1+\frac{p}{S}\tan\frac{k}{2}+O(p^2)}$$

$$= -\frac{4J}{\hbar N}\left(1-\frac{1}{S}\right)\sin^2\frac{k}{2}\sum_p p^2\, n_p^{(0)}$$

$$= -\frac{\omega_k}{S}\left(1-\frac{1}{S}\right)\pi\,\zeta\left(\frac{3}{2}\right)\left(\frac{k_BT}{8\pi JS}\right)^{\frac{3}{2}} \tag{4.5.19}$$

for the real part and

$$\Sigma^{(\mathrm{T})''}(k,\omega_k) = \frac{2J}{\hbar SN}\sum_p p^2\, n_p^{(0)}[\sin k + O(p)] = \frac{4J}{\hbar S}\pi\,\zeta\left(\frac{3}{2}\right)\left(\frac{k_BT}{8\pi JS}\right)^{\frac{3}{2}}\sin k \tag{4.5.20}$$

for the imaginary part, where the relationship[3]

$$\frac{1}{N}\sum_p p^2 n_p^{(0)} = \frac{1}{2\pi}\left(\frac{k_BT}{2JS}\right)^{\frac{3}{2}}\int_0^\infty dx\frac{\sqrt{x}}{e^x-1} = 2\pi\,\zeta\left(\frac{3}{2}\right)\left(\frac{k_BT}{8\pi JS}\right)^{\frac{3}{2}}. \tag{4.5.21}$$

has been used. The renormalization (4.5.19) and the damping (4.5.20) of the spin wave frequency for the 1D ferromagnet agree with the renormalization and damping obtained using the method of the equation of motion of the Green function.[37] An interesting feature of the renormalization (4.5.19) is that it is positive (upword renormalization) for $S=\frac{1}{2}$, 0 (at the order $T^{3/2}$) for $S=1$ and negative (downward renormalization) for $S>1$. Moreover, Eq. (4.5.19) points out that only the first two terms of the perturbation expansion contribute to the renormalization at the order $T^{3/2}$ since only terms proportional to 1 and $\frac{1}{S}$ appear in Eq. (4.5.19). Similarly, Eq. (4.5.20) points out that only the term coming from the second-order perturbation expansion contributes to the damping at the order $T^{3/2}$. In particular, only $\Sigma^{(1)}(k)$ and $\Sigma_c^{(2)}(k,\omega_k\pm i\epsilon)$ contribute to the T-matrix of a 1D ferromagnet at the lowest order in temperature. To check this statement, we evaluate directly the first-order term (4.2.16) and the second-order one (4.3.16) for the 1D ferromagnet obtaining

$$\Sigma^{(1)}(\boldsymbol{k}) = -\frac{\omega_k}{2S}\frac{1}{N}\sum_p p^2\, n_p^{(0)} = -\pi\,\zeta\left(\frac{3}{2}\right)\frac{\omega_k}{S}\left(\frac{k_BT}{8\pi JS}\right)^{\frac{3}{2}} \tag{4.5.22}$$

and

$$\Sigma_c^{(2)'}(k,\omega_k) = \frac{\omega_k}{2S^2}\frac{1}{N}\sum_p p^2\, n_p^{(0)} = \pi\,\zeta\left(\frac{3}{2}\right)\frac{\omega_k}{S^2}\left(\frac{k_BT}{8\pi JS}\right)^{\frac{3}{2}}, \tag{4.5.23}$$

respectively. Summing Eqs. (4.5.22) and (4.5.23), one recovers the T-matrix result given in Eq. (4.5.19). The third-order term that contributes to the T-matrix is

$\Sigma_d^{(3)'}(k,\omega_k)$ given by Eq. (4.4.17). In 1D, we make the replacements $K = \frac{1}{2}(k+p)$, $\lambda = \frac{1}{2}(k-p)$, $q_1 = \frac{1}{2}K + \mu$, $q_3 = \frac{1}{2}K - \mu$, $q_2 = \frac{1}{2}K + \rho$ and $q_4 = \frac{1}{2}K - \rho$, obtaining

$$\Sigma_d^{(3)'}(k,\omega_k) = -\frac{2J}{\hbar S^2}\frac{\mathrm{P}}{N^3}\sum_{\lambda,\mu,\rho}\frac{\cos\lambda(\cos\lambda - \cos\frac{K}{2})}{\cos^2\frac{K}{2}}n^{(0)}_{\frac{1}{2}K-\lambda}$$
$$\times\left\{\left[\frac{1}{N}\sum_\rho\frac{\cos\rho(\cos\rho-\cos\frac{K}{2})}{\cos\rho-\cos\lambda}\right]^2\right.$$
$$\left.-\pi^2\left[\frac{1}{N}\sum_\rho\cos\rho\left(\cos\rho-\cos\frac{K}{2}\right)\ \delta(\cos\rho-\cos\lambda)\right]^2\right\}. \tag{4.5.24}$$

Using the relationship[3]

$$\frac{\mathrm{P}}{\pi}\int_0^\pi d\rho\frac{\cos\rho(\cos\rho-\cos\frac{K}{2})}{\cos\rho-\cos\lambda} = \cos\lambda - \cos\frac{K}{2} \tag{4.5.25a}$$

and

$$\frac{1}{\pi}\int_0^\pi d\rho\ \cos\rho\left(\cos\rho-\cos\frac{K}{2}\right)\ \delta(\cos\rho-\cos\lambda) = \frac{1}{\pi}\frac{\cos\lambda\left(\cos\lambda-\cos\frac{K}{2}\right)}{|\sin\lambda|}, \tag{4.5.25b}$$

Equation (4.5.24) becomes

$$\Sigma_d^{(3)'}(k,\omega_k) = -\frac{2J}{\hbar S^2}\frac{1}{N}\sum_\lambda\frac{\cos\lambda(\cos\lambda-\cos\frac{K}{2})^3}{\cos^2\frac{K}{2}}\left(1-\frac{\cos^2\lambda}{\sin^2\lambda}\right)n^{(0)}_{\frac{1}{2}K-\lambda}. \tag{4.5.26}$$

Replacing $K$ by $k+p$, $\lambda$ by $\frac{1}{2}(k-p)$ and expanding in powers of $p$ due to the presence of the Bose factor, the main temperature contribution of Eq. (4.5.26) is given by

$$\Sigma_d^{(3)'}(k,\omega_k) = \frac{2J}{\hbar S^2}\frac{(2-\cos k)(1+3\cos k)}{1+\cos k}\frac{1}{N}\sum_p p^4\ n_p^{(0)}$$
$$= \frac{2J}{\hbar S^2}\frac{(2-\cos k)(1+3\cos k)}{1+\cos k}3\pi^2\ \zeta\left(\frac{5}{2}\right)\left(\frac{k_BT}{8\pi JS}\right)^{\frac{5}{2}} \tag{4.5.27}$$

which is of order $T^{5/2}$ and so it is negligible compared to the terms coming from the first and second-order self-energies of order $T^{3/2}$. The conclusion is that in 1D, the first two terms of the T-matrix expansion give contributions proportional to $\sum_p p^2 n_p^{(0)}$ while from the third term onwards, only contributions of order $\sum_p p^4 n_p^{(0)}$ or higher occur even though all these terms have a single Bose factor. The different powers in $p$ are entered by the function of $p$ that multiplies the Bose factor. In any case, we may check that in the 1D case, the temperature contributions to the self-energy that come from diagrams containing more than one downward line, are of higher order in temperature: to do this, we evaluate explicitly some diagrams that do not belong to the T-matrix expansion. For instance, the second-order term (4.2.6) corresponding to the diagram $\Sigma_b^{(2)}$ of Fig. 4.2, in 1D reads

$$\Sigma_b^{(2)}(k) = -\frac{3\omega_k}{2S^2}\left[\pi\,\zeta\left(\frac{3}{2}\right)\right]^2\left(\frac{k_BT}{8\pi JS}\right)^3. \qquad (4.5.28)$$

The presence of two downward lines in the diagram leads to a contribution proportional to $(T^{3/2})^2 = T^3$ which is the square of the contribution coming from the term with one downward line proportional to $T^{3/2}$. The third-order term (4.4.5) corresponding to the diagram $\Sigma_a^{(3)}$ of Fig. 4.3, in 1D reads

$$\Sigma_a^{(3)}(k) = -\frac{9\omega_k}{4S^3}\left[\pi\,\zeta\left(\frac{3}{2}\right)\right]^3\left(\frac{k_BT}{8\pi JS}\right)^{\frac{9}{2}}. \qquad (4.5.29)$$

The three downward lines in the diagram lead to a contribution proportional to $(T^{3/2})^3 = T^{9/2}$ which is the cube of the term containing one downward line. The third-order term (4.4.10) corresponding to the diagram $\Sigma_b^{(3)}$ of Fig. 4.3, in 1D reads

$$\Sigma_b^{(3)}(k) = -\frac{15\omega_k}{8S^3}\left[\pi\,\zeta\left(\frac{3}{2}\right)\right]^3\left(\frac{k_BT}{8\pi JS}\right)^{\frac{9}{2}}. \qquad (4.5.30)$$

Again, the three downward lines of the diagram lead to a contribution proportional to $(T^{3/2})^3 = T^{9/2}$. The third-order term (4.4.16) corresponding to the diagram $\Sigma_c^{(3)}$ of Fig. 4.3, in 1D reads

$$\Sigma_c^{(3)}(k) = \frac{2\omega_k}{S^3}\left[\pi\,\zeta\left(\frac{3}{2}\right)\right]^2\left(\frac{k_BT}{8\pi JS}\right)^3. \qquad (4.5.31)$$

The presence of two downward lines leads to a contribution proportional to $(T^{3/2})^2 = T^3$. The real part of the third-order term (4.4.24) corresponding to the diagram $\Sigma_f^{(3)}$ of Fig. 4.3, in 1D reads

$$\Sigma_f^{(3)'}(k) = \frac{\omega_k}{4S^3}\left[\pi\,\zeta\left(\frac{3}{2}\right)\right]^2\frac{5\cos k - 1}{1+\cos k}\left(\frac{k_BT}{8\pi JS}\right)^3. \qquad (4.5.32)$$

The real part of the third-order term (4.4.27) corresponding to the diagram $\Sigma_g^{(3)}$ of Fig. 4.3, in 1D reads

$$\Sigma_g^{(3)'}(k) = -\frac{3\omega_k}{S^3}\left[\pi\,\zeta\left(\frac{3}{2}\right)\right]^2\left(\frac{k_BT}{8\pi JS}\right)^3. \qquad (4.5.33)$$

The expectation that any downward line making a contribution proportional to $T^{3/2}$ in 1D appears to be confirmed. As for the $\frac{1}{S}$-expansion, one can see that any second-order self-energy contribution is proportional to $\frac{1}{S^2}$ as shown by Eqs. (4.5.23) and (4.5.28); any third-order self-energy contribution is proportional to $\frac{1}{S^3}$ as shown by Eqs. (4.5.27) and (4.5.29)–(4.5.33). However, *different* powers in temperature come from terms of the *same* order in $\frac{1}{S}$. This means that $\frac{1}{S}$ is a good perturbation parameter, not the temperature which is, however, the most important expansion parameter to fit experimental data. Moreover, the presence of a single Bose factor, corresponding to diagrams with only one downward line, does not imply (in

1D) that the corresponding temperature contribution is the same at each order of perturbation in $\frac{1}{S}$ since the function that multiplies the Bose factor when expanded in $p$ gives different powers as illustrated when comparing Eqs. (4.5.22) and (4.5.23) with Eq. (4.5.27): all diagrams corresponding to these terms have one downward line as shown in Fig. 4.1 (see $\Sigma^{(1)}$), Fig. 4.2 (see $\Sigma_c^{(2)}$) and Fig. 4.3 (see $\Sigma_d^{(3)}$) but the first two diagrams lead to a contribution proportional to $T^{3/2}$ whereas the third diagram leads to a contribution proportional to $T^{5/2}$. This peculiarity, however, does not apply to 3D systems where any diagram containing one downward line leads to a contribution proportional to $T^{5/2}$. To show this, we evaluate the T-matrix self-energy (4.5.12b) for a 3D lattice. Due to the Bose factor in Eq. (4.5.12b), we can expand the function that multiplies $n_{\boldsymbol{p}}^{(0)}$ in powers of $\boldsymbol{p}$. Then the sum over $\boldsymbol{p}$ in Eq. (4.5.12b) gives non zero contributions only for the even powers of the wavevector $\boldsymbol{p}$, leading to

$$\Sigma^{(\mathrm{T})}(\boldsymbol{k},\omega) = -\frac{16J}{\hbar N}\sum_{\boldsymbol{p}} n_{\boldsymbol{p}}^{(0)}\left\{\sum_{\alpha}\frac{1}{4}\,p_\alpha^2\sin^2\frac{k_\alpha}{2}\left[\mathbf{1}-\frac{1}{2S}\mathbf{B}(\boldsymbol{k},\omega)\right]^{-1}_{\alpha\alpha}\right.$$
$$\left. +\sum_{\alpha,\beta}\frac{1}{2}\,p_\beta^2\cos\frac{k_\alpha}{2}\sin\frac{k_\beta}{2}\left|\frac{\partial}{\partial p_\beta}\left[\mathbf{1}-\frac{1}{2S}\mathbf{B}(\boldsymbol{k}+\boldsymbol{p},\omega)\right]^{-1}_{\alpha\beta}\right|_{\boldsymbol{p}=0}\right\}. \tag{4.5.34}$$

By use of the relationship

$$\left|\frac{\partial}{\partial p_\beta}\left[\mathbf{1}-\frac{1}{2S}\mathbf{B}(\boldsymbol{k}+\boldsymbol{p},\omega)\right]^{-1}_{\alpha\beta}\right|_{\boldsymbol{p}=0} = \frac{\partial}{\partial k_\beta}\left[\mathbf{1}-\frac{1}{2S}\mathbf{B}(\boldsymbol{k},\omega)\right]^{-1}_{\alpha\beta}, \tag{4.5.35}$$

Equation (4.5.34) becomes

$$\Sigma^{(\mathrm{T})}(\boldsymbol{k},\omega_{\boldsymbol{k}}\pm i\epsilon) = -\frac{8J}{\hbar}\,\pi\zeta\left(\frac{5}{2}\right)\left(\frac{k_BT}{8\pi JS}\right)^{\frac{5}{2}}\left\{\sum_\alpha \sin^2\frac{k_\alpha}{2}\right.$$
$$\times\left[\mathbf{1}-\frac{1}{2S}\mathbf{B}(\boldsymbol{k},\omega_{\boldsymbol{k}}\pm i\epsilon)\right]^{-1}_{\alpha\alpha} + 2\sum_{\alpha,\beta}\cos\frac{k_\alpha}{2}\sin\frac{k_\beta}{2}$$
$$\left.\times\left|\frac{\partial}{\partial k_\beta}\left[\mathbf{1}-\frac{1}{2S}\mathbf{B}(\boldsymbol{k},\omega\pm i\epsilon)\right]^{-1}_{\alpha\beta}\right|_{\omega=\omega_{\boldsymbol{k}}}\right\} \tag{4.5.36}$$

where

$$\mathrm{B}_{\alpha\beta}(\boldsymbol{k},\omega_{\boldsymbol{k}}\pm i\epsilon) = \cos\frac{k_\alpha}{2}D_\beta - D_{\alpha\beta} \tag{4.5.37}$$

with

$$D_{\alpha\beta} = \frac{1}{N}\sum_{\boldsymbol{\rho}}\frac{\cos\rho_\alpha\cos\rho_\beta}{\sum_\sigma\cos\frac{k_\sigma}{2}\left(\cos\rho_\sigma-\cos\frac{k_\sigma}{2}\right)\pm i\epsilon} \tag{4.5.37a}$$

and

$$D_\alpha = \frac{1}{N}\sum_{\rho} \frac{\cos\rho_\alpha}{\sum_\sigma \cos\frac{k_\sigma}{2}\left(\cos\rho_\sigma - \cos\frac{k_\sigma}{2}\right) \pm i\epsilon}, \tag{4.5.37b}$$

where $\sigma = x, y, z$. For a SC lattice with $\boldsymbol{k} = (k, k, k)$, one has $D_\alpha = D_1$ for $\alpha = x, y, z$; $D_{\alpha\beta} = D_{11}$ for $\alpha = \beta = x, y, z$; $D_{\alpha\beta} = D_{12}$ for $\alpha \neq \beta$ and the self-energy (4.5.36) becomes

$$\Sigma^{(\mathrm{T})}(k, \omega_k \pm i\epsilon) = -\frac{8J}{\hbar}\,\pi\,\zeta\left(\frac{5}{2}\right)\left(\frac{k_B T}{8\pi JS}\right)^{\frac{5}{2}} \left\{ \sin^2\frac{k}{2} \sum_\alpha \left[\mathbf{1} - \frac{1}{2S}\mathbf{B}(k, \omega_k \pm i\epsilon)\right]^{-1}_{\alpha\alpha} \right.$$
$$\left. + 2\cos\frac{k}{2}\sin\frac{k}{2}\sum_{\alpha,\beta}\left|\frac{\partial}{\partial k_\beta}\left[\mathbf{1} - \frac{1}{2S}\mathbf{B}(\boldsymbol{k}, \omega \pm i\epsilon)\right]^{-1}_{\alpha\beta}\right|_{\omega=\omega_k} \right\} \tag{4.5.38}$$

where $\hbar\omega_k = 12JS(1 - \cos k)$. The evaluation of the trace of the inverse matrix of $(\mathbf{1} - \frac{1}{2S}\mathbf{B})$ occurring in the first term of Eq. (4.5.38) is direct: one has

$$\sum_\alpha \left[\mathbf{1} - \frac{1}{2S}\mathbf{B}(k, \omega_k \pm i\epsilon)\right]^{-1}_{\alpha\alpha} = 3\,\frac{1 + \frac{1}{2S}\mathrm{B}_{12}}{1 + \frac{3}{2S}\mathrm{B}_{12}} \tag{4.5.39}$$

where

$$\mathrm{B}_{12} = D_1 \cos\frac{k}{2} - D_{12} \tag{4.5.39a}$$

with

$$D_1 = -\frac{1}{3\cos\frac{k}{2}}\alpha(k), \tag{4.5.39b}$$

$$D_{12} = -\frac{1}{3\cos\frac{k}{2}}\frac{1}{N}\sum_\rho \frac{\cos\rho_x \cos\rho_y}{\cos\frac{k}{2} - \frac{1}{3}(\cos\rho_x + \cos\rho_y + \cos\rho_z) \mp i\epsilon} \tag{4.5.39c}$$

and

$$\alpha(k) = \frac{1}{N}\sum_\rho \frac{\cos\rho_x}{\cos\frac{k}{2} - \frac{1}{3}(\cos\rho_x + \cos\rho_y + \cos\rho_z) \mp i\epsilon}. \tag{4.5.40}$$

As for the second term of Eq. (4.5.38), where the partial derivative with respect to $k_\beta$ appears, one can proceed as follows

$$\sum_{\alpha,\beta}\left|\frac{\partial}{\partial k_\beta}\left[\mathbf{1} - \frac{1}{2S}\mathbf{B}(\boldsymbol{k}, \omega \pm i\epsilon)\right]^{-1}_{\alpha\beta}\right|_{\omega=\omega_k} = \sum_{\alpha,\beta}\left|\frac{\partial}{\partial k_\beta}\frac{A_{\alpha\beta}}{A}\right|_{k,\omega_k}$$
$$= \left|\frac{1}{A}\sum_{\alpha,\beta}\frac{\partial A_{\alpha\beta}}{\partial k_\beta} - \frac{1}{A^2}\sum_{\alpha,\beta} A_{\alpha\beta}\frac{\partial A}{\partial k_\beta}\right|_{k,\omega_k} \tag{4.5.41}$$

where $A_{\alpha\beta}/A$ and $A$ are the elements and the determinant of the matrix $(\mathbf{1}\text{-}\frac{1}{2S}\mathbf{B})^{-1}$, respectively. The explicit evaluation of the terms appearing in Eq. (4.5.41) gives

$$\left|\frac{1}{A}\sum_{\alpha,\beta}\frac{\partial A_{\alpha\beta}}{\partial k_\beta}\right|_{k,\omega_k} = -\frac{1}{4S}\sin\frac{k}{2}\cos\frac{k}{2}\frac{1}{1+\frac{3}{2S}\mathrm{B}_{12}}\left(\sum_{\alpha,\beta}G_{\alpha\beta}-3\sum_{\alpha}G_{\alpha\alpha}\right) \tag{4.5.42}$$

and

$$\begin{aligned}\left|\frac{1}{A^2}\sum_{\alpha,\beta}A_{\alpha\beta}\frac{\partial A}{\partial k_\beta}\right|_{k,\omega_k} &= \frac{1}{4S}\sin\frac{k}{2}\frac{1}{1+\frac{3}{2S}\mathrm{B}_{12}}\left[\left(1+\frac{3}{2S}\mathrm{B}_{12}\right)\right.\\ &\quad\times\left(3\,D_1-\cos\frac{k}{2}\sum_{\alpha,\beta}G_{\alpha\beta}\right)\\ &\quad\left.+\sum_{\alpha,\beta}G_{\alpha\alpha\beta}+\frac{1}{2S}\mathrm{B}_{12}\sum_{\alpha,\beta,\gamma}G_{\alpha\beta\gamma}\right].\end{aligned} \tag{4.5.43}$$

$D_1$ is given by Eq. (4.5.39b),

$$G_{\alpha\beta} = \frac{1}{N}\sum_\rho\frac{\cos\rho_\alpha\cos\rho_\beta}{\left[\cos\frac{k}{2}\sum_\sigma\left(\cos\frac{k}{2}-\cos\rho\right)\right]^2}, \tag{4.5.44}$$

$$G_{\alpha\beta\gamma} = \frac{1}{N}\sum_\rho\frac{\cos\rho_\alpha\cos\rho_\beta\cos\rho_\gamma}{\left[\cos\frac{k}{2}\sum_\sigma\left(\cos\frac{k}{2}-\cos\rho\right)\right]^2} \tag{4.5.45}$$

and the sum rule $\mathrm{B}_{11}+2\mathrm{B}_{12} = 0$ has been used. By means of Eqs. (4.5.39) and (4.5.41)–(4.5.43), Equation (4.5.38) becomes

$$\begin{aligned}\Sigma^{(\mathrm{T})}(k,\omega_k\pm i\epsilon) &= -\frac{8J}{\hbar}\pi\zeta\left(\frac{5}{2}\right)\left(\frac{k_BT}{8\pi JS}\right)^{\frac{5}{2}}\sin^2\frac{k}{2}\left\{3\frac{1+\frac{1}{2S}\mathrm{B}_{12}}{1+\frac{3}{2S}\mathrm{B}_{12}}\right.\\ &\quad-\frac{1}{2S}\frac{\cos\frac{k}{2}}{1+\frac{3}{2S}\mathrm{B}_{12}}\left[3\,D_1\left(1+\frac{3}{2S}\mathrm{B}_{12}\right)-\frac{3}{2S}\mathrm{B}_{12}\cos\frac{k}{2}\sum_{\alpha,\beta}G_{\alpha\beta}\right.\\ &\quad\left.\left.-3\cos\frac{k}{2}\sum_\alpha G_{\alpha\alpha}+\sum_{\alpha,\beta}G_{\alpha\alpha\beta}+\frac{1}{2S}\mathrm{B}_{12}\sum_{\alpha,\beta,\gamma}G_{\alpha\beta\gamma}\right]\right\}.\end{aligned} \tag{4.5.46}$$

Finally, using the sum rules

$$\sum_\gamma G_{\alpha\beta\gamma} = \frac{D_{\alpha\beta}}{\cos\frac{k}{2}}+3\,G_{\alpha\beta}\cos\frac{k}{2} \tag{4.5.47}$$

and

$$\sum_\beta D_{\alpha\beta} = 3\,D_1\cos\frac{k}{2}, \tag{4.5.48}$$

the T-matrix self-energy (4.5.46) becomes

$$\Sigma^{(T)}(k, \omega_k \pm i\epsilon) = \Sigma^{(1)}(k) Q(k) \tag{4.5.49}$$

where

$$\Sigma^{(1)}(k) = -\frac{\omega_k}{S}\,\pi\,\zeta\left(\frac{5}{2}\right)\left(\frac{k_B T}{8\pi J S}\right)^{\frac{5}{2}} \tag{4.5.49a}$$

is the first-order self-energy,

$$Q(k) = \frac{1 - \frac{1}{2S}B_{12}}{1 + \frac{3}{2S}B_{12}} - \frac{1}{S}D_1 \cos\frac{k}{2} = \frac{1 + \frac{1}{6S}\Gamma(k)}{1 - \frac{1}{2S}\Gamma(k)} + \frac{1}{3S}\alpha(k), \tag{4.5.49b}$$

where $\alpha(k)$ is given by Eq. (4.5.40), and

$$\Gamma(k) = \frac{1}{\cos\frac{k}{2}}\frac{1}{N}\sum_{\rho}\frac{\cos\rho_x\left(\cos\frac{k}{2} - \cos\rho_y\right)}{\cos\frac{k}{2} - \frac{1}{3}\left(\cos\rho_x + \cos\rho_y + \cos\rho_z\right) \mp i\epsilon}. \tag{4.5.50}$$

The self-energy (4.5.49) was obtained by Silberglitt and Harris.[36] Notice that unlike the T-matrix self-energy of the LC (1D) given by Eqs. (4.5.19) and (4.5.20), the self-energy of the SC (3D) lattice (4.5.49) consists of an infinite power series in $\frac{1}{S}$. Indeed, Eq. (4.5.49b) may be written as

$$Q(k) = 1 + \frac{\alpha(k)}{3S} + \frac{4}{3}\sum_{n=1}^{\infty}\left[\frac{\Gamma(k)}{2S}\right]^n \tag{4.5.51}$$

showing that all the *infinite* diagrams containing one downward line shown in Fig. 4.4 give the same contribution $T^{5/2}$ in temperature.

Before concluding this section, we evaluate explicitly the real and imaginary part of the self-energy (4.5.49). Let us begin evaluating the real and imaginary part of $\alpha(k)$ and $\Gamma(k)$ given by Eqs. (4.5.40) and (4.5.50). From Eq. (3.2.12), one has

$$\alpha(k) = \alpha'(k) \pm i\alpha''(k) \tag{4.5.52}$$

with

$$\alpha'(k) = \frac{\mathrm{P}}{\pi^3}\int_0^{\pi}\int_0^{\pi}\int_0^{\pi} dx\, dy\, dz \frac{\cos x}{\cos\frac{k}{2} - \frac{1}{3}(\cos x + \cos y + \cos z)}, \tag{4.5.53}$$

$$\alpha''(k) = \frac{1}{\pi^2}\int_0^{\pi}\int_0^{\pi}\int_0^{\pi} dx\, dy\, dz \cos x \quad \delta\left[\cos\frac{k}{2} - \frac{1}{3}\left(\cos x + \cos y + \cos z\right)\right] \tag{4.5.54}$$

and

$$\Gamma(k) = \Gamma'(k) \pm i\Gamma''(k) \tag{4.5.55}$$

with

$$\Gamma'(k) = \alpha'(k) - \frac{\mathrm{P}}{\cos\frac{k}{2}}\frac{1}{\pi^3}\int_0^\pi\int_0^\pi\int_0^\pi dx\, dy\, dz\, \frac{\cos x\, \cos y}{\cos\frac{k}{2} - \frac{1}{3}(\cos x + \cos y + \cos z)}, \tag{4.5.56}$$

$$\Gamma''(k) = \alpha''(k) - \frac{1}{\cos\frac{k}{2}}\frac{1}{\pi^2}\int_0^\pi\int_0^\pi\int_0^\pi dx\, dy\, dz \cos x \cos y \times \delta\left[\cos\frac{k}{2} - \frac{1}{3}(\cos x + \cos y + \cos z)\right]. \tag{4.5.57}$$

The integration over $z$ is direct: indeed, using Eq. (4.5.15a) one has

$$\frac{1}{\pi}\int_0^\pi dz\, \frac{1}{\cos\frac{k}{2} - \frac{1}{3}(\cos x + \cos y + \cos z)} = \begin{cases} -\dfrac{1}{\sqrt{\left[\cos\frac{k}{2} - \frac{1}{3}(\cos x + \cos y)\right]^2 - \frac{1}{9}}} & \text{for } x, y \in D_1 \\ 0 & \text{for } x, y \in D_2 \\ \dfrac{1}{\sqrt{\left[\cos\frac{k}{2} - \frac{1}{3}(\cos x + \cos y)\right]^2 - \frac{1}{9}}} & \text{for } x, y \in D_3 \end{cases} \tag{4.5.58}$$

and

$$\int_0^\pi dz\, \delta\left[\cos\frac{k}{2} - \frac{1}{3}(\cos x + \cos y + \cos z)\right] = \begin{cases} 0 & \text{for } x, y \in D_1 \text{ or } D_3 \\ \dfrac{1}{\sqrt{\frac{1}{9} - \left[\cos\frac{k}{2} - \frac{1}{3}(\cos x + \cos y)\right]^2}} & \text{for } x, y \in D_2 \end{cases} \tag{4.5.59}$$

where the domains $D_1, D_2$ and $D_3$ are given by

$$\mathrm{D}_1 : \left\{x, y > 0 \middle| \cos x + \cos y > 1 + 3\cos\frac{k}{2}\right\}, \tag{4.5.60}$$

$$\mathrm{D}_2 : \left\{0 < x, y < \pi \middle| -1 + 3\cos\frac{k}{2} < \cos x + \cos y < 1 + 3\cos\frac{k}{2}\right\} \tag{4.5.61}$$

and

$$\mathrm{D}_3 : \left\{0 < x, y < \pi \middle| \cos x + \cos y < -1 + 3\cos\frac{k}{2}\right\}, \tag{4.5.62}$$

respectively. The domain $\mathrm{D}_1$ exists only if $\cos\frac{k}{2} < \frac{1}{3}$, that is for $k > k_0 = 2\arccos\frac{1}{3} = 0.78365\pi$. The domain $\mathrm{D}_2$ vanishes for $k = 0$ where $\mathrm{D}_3$ extends to the square $0 < x, y < \pi$. In Fig. 4.5a, the domains $\mathrm{D}_2$ and $\mathrm{D}_3$ are shown for $k = \frac{\pi}{2}$. The scenario is qualitatively the same for any $k$ in the range $0 < k < k_0$, where $\mathrm{D}_1 = 0$. On the other hand, the scenario for $k_0 < k < \pi$ is illustrated in Fig. 4.5b

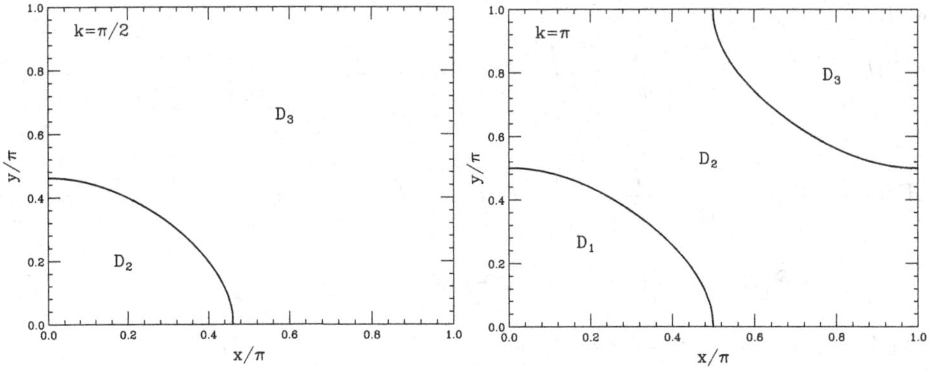


Fig. 4.5. Domains of integration (a) for $k = \frac{\pi}{2}$ ($D_2$ and $D_3$) and (b) for $k = \pi$ ($D_1$, $D_2$ and $D_3$).

where the domains $D_1$, $D_2$ and $D_3$ are shown for $k = \pi$. By means of Eqs. (4.5.58) and (4.5.59), Equations (4.5.53), (4.5.54), (4.5.56) and (4.5.57) reduce to

$$\alpha'(k) = \frac{1}{\pi^2}\left\{\iint_{D_3} dx\, dy \frac{\cos x}{\sqrt{\left[\cos\frac{k}{2} - \frac{1}{3}(\cos x + \cos y)\right]^2 - \frac{1}{9}}} \right.$$

$$\left. -\theta(k-k_0)\iint_{D_1} dx\, dy \frac{\cos x}{\sqrt{\left[\cos\frac{k}{2} - \frac{1}{3}(\cos x + \cos y)\right]^2 - \frac{1}{9}}}\right\}, \quad (4.5.63)$$

$$\alpha''(k) = \frac{1}{\pi^2}\iint_{D_2} dx\, dy \frac{\cos x}{\sqrt{\frac{1}{9} - \left[\cos\frac{k}{2} - \frac{1}{3}(\cos x + \cos y)\right]^2}}, \quad (4.5.64)$$

$$\Gamma'(k) = \alpha'(k) - \frac{1}{\cos\frac{k}{2}}\frac{1}{\pi^2}\left\{\iint_{D_3} dx\, dy \frac{\cos x\ \cos y}{\sqrt{\left[\cos\frac{k}{2} - \frac{1}{3}(\cos x + \cos y)\right]^2 - \frac{1}{9}}}\right.$$

$$\left. -\theta(k-k_0)\iint_{D_1} dx\, dy \frac{\cos x\ \cos y}{\sqrt{\left[\cos\frac{k}{2} - \frac{1}{3}(\cos x + \cos y)\right]^2 - \frac{1}{9}}}\right\} \quad (4.5.65)$$

and

$$\Gamma''(k) = \alpha''(k) - \frac{1}{\cos\frac{k}{2}}\frac{1}{\pi^2}\iint_{D_2} dx\, dy \frac{\cos x\ \cos y}{\sqrt{\frac{1}{9} - \left[\cos\frac{k}{2} - \frac{1}{3}(\cos x + \cos y)\right]^2}}, \quad (4.5.66)$$

respectively. The step-function $\theta(k - k_0)$ is 0 for $k < k_0$ and 1 for $k > k_0$. The 2D integrals occurring in Eqs. (4.5.63)–(4.5.66) have to be evaluated numerically. In Fig. 4.6, we show $\alpha'(k)$ (full circles), $\alpha''(k)$ (full squares), $\Gamma'(k)$ (open circles) and $\Gamma''(k)$ (open squares) versus $k$. Note that $\alpha'(k)$ and $\Gamma'(k)$ are finite at $k = 0$:

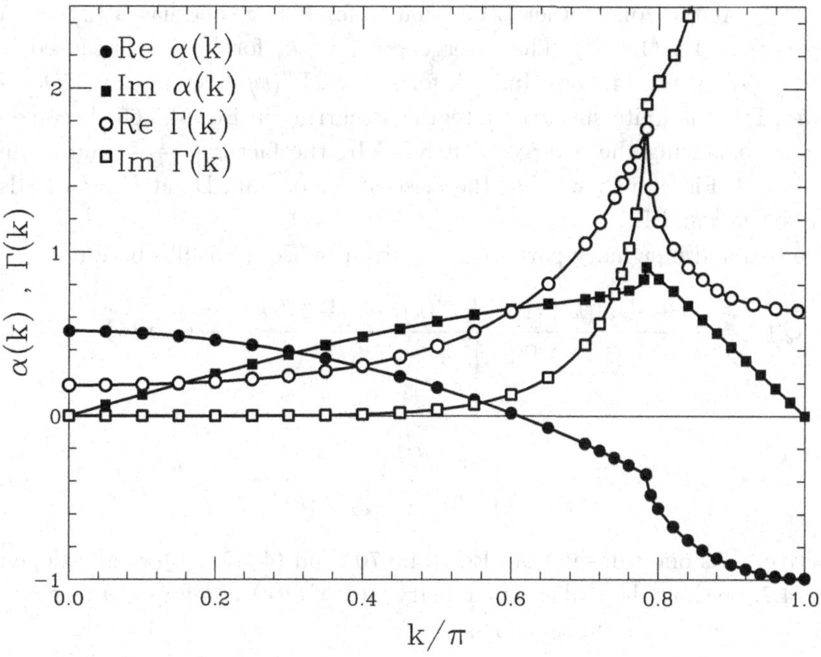


Fig. 4.6. Real part (full circles) and imaginary part (full squares) of $\alpha(k)$; real part (open circles) and imaginary part (open squares) of $\Gamma(k)$ for a SC lattice with $\boldsymbol{k} = (k, k, k)$.

$$\alpha'(0) = \frac{1}{\pi^2}\int_0^\pi \int_0^\pi dx\, dy \frac{\cos x}{\sqrt{\left[1 - \frac{1}{3}(\cos x + \cos y)\right]^2 - \frac{1}{9}}} = 0.51639 \qquad (4.5.67)$$

and

$$\Gamma'(0) = \alpha'(0) - \frac{1}{\pi^2}\int_0^\pi \int_0^\pi dxdy \frac{\cos x\; cosy}{\sqrt{\left[1 - \frac{1}{3}(\cos x + \cos y)\right]^2 - \frac{1}{9}}}$$
$$= 0.18524. \qquad (4.5.68)$$

On the contrary, $\alpha''(k)$ and $\Gamma''(k)$ vanish at $k = 0$. In particular, in the limit $k \to 0$, the domain $D_2$ reduces to the quarter of the circle $x^2 + y^2 < \frac{3}{4}k^2$ with $x, y > 0$, so that

$$\alpha''(k \to 0) \simeq \frac{3}{\pi^2}\int_0^{\frac{\pi}{2}} d\theta \int_0^{\frac{\sqrt{3}}{2}k} qdq \frac{1}{\sqrt{\frac{3}{4}k^2 - q^2}} = \frac{3\sqrt{3}}{4\pi}k \qquad (4.5.69)$$

and $\Gamma''(k \to 0)$ vanishes with a power higher than $k^3$. Both the linear dependence on $k$ of $\alpha''(k)$ and the flatness of the function $\Gamma''(k)$ for $k \to 0$ are clearly seen in Fig. 4.6. At the zone corner (ZC), that is for $k = \pi$, one has $\alpha'(\pi) = -1$ and $\alpha''(k \to \pi) \simeq 4.21(1 - \frac{k}{\pi})$. The divergence of $\Gamma''(k)$ for $k \to \pi$ is caused by the factor $\frac{1}{\cos(k/2)}$ in Eq. (4.5.66). Indeed, for $k \to \pi$, $\Gamma''(k)$ diverses as $0.37(1 - \frac{k}{\pi})^{-1}$. However, $\Gamma'(\pi)$ is finite since the integrals occurring in Eq. (4.5.65) become equal at $k = \pi$, balancing the divergence entered by the factor $\frac{1}{\cos(k/2)}$ and leading to $\Gamma'(\pi) \simeq 0.64$. Fig. 4.6 shows that the arise of the domain $D_1$ at $k = k_0$ leads to a cusp in both $\alpha$ and $\Gamma$.

The real and imaginary parts of $Q(k)$ given by Eq. (4.5.49b) become

$$Q'(k) = \frac{\left[1 + \frac{1}{6S}\Gamma'(k)\right]\left[1 - \frac{1}{2S}\Gamma'(k)\right] - \frac{1}{12S^2}\Gamma''(k)^2}{\left[1 - \frac{1}{2S}\Gamma'(k)\right]^2 + \frac{1}{4S^2}\Gamma''(k)^2} + \frac{1}{3S}\alpha'(k) \tag{4.5.70}$$

and

$$Q''(k) = \frac{\frac{2}{3S}\Gamma''(k)}{\left[1 - \frac{1}{2S}\Gamma'(k)\right]^2 + \frac{1}{4S^2}\Gamma''(k)^2} + \frac{1}{3S}\alpha''(k), \tag{4.5.71}$$

respectively. As one can see from Eqs. (4.5.70) and (4.5.71), $Q(k)$ is $S$-dependent. In Fig. 4.7, we show the real and imaginary part of $Q(k)$ vs $k$ for several spin values. In the long wavelength limit one has

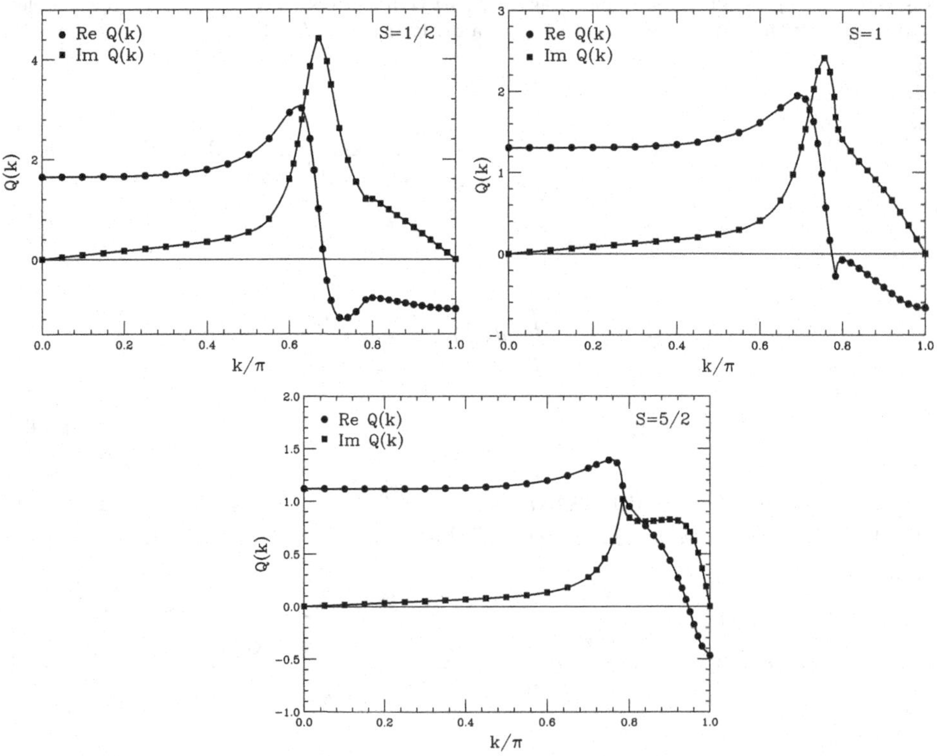


Fig. 4.7. Real (full circles) and imaginary (full squares) parts of $Q(k)$ for $S = 1/2$, 1 and 5/2.

$$Q'(0) = \frac{1+\frac{1}{6S}\Gamma'(0)}{1-\frac{1}{2S}\Gamma'(0)} + \frac{\alpha'(0)}{3S} = \frac{1+0.03087/S}{1-0.09262/S} + \frac{0.17213}{S} \tag{4.5.72a}$$

and

$$Q''(k \to 0) = \frac{\alpha''(k \to 0)}{3S} = \frac{|\boldsymbol{k}|}{4\pi S} + O(k^3). \tag{4.5.72b}$$

For $k \to \pi$, one obtains

$$Q'(k \to \pi) = -\frac{S+1}{3S} \tag{4.5.73a}$$

and

$$Q''(k \to \pi) = 0. \tag{4.5.73b}$$

Note that $Q'(k)$ changes its sign going from $k = 0$ to $k = \pi$ for any $S$ as shown in Fig. 4.7. For $S = 1/2$ and $S = 1$, $Q''(k)$ shows a maximum located very close to the value of $k$ at which $Q'(k)$ changes its sign: $k = 0.683\pi$ and $0.775\pi$ for $S = 1/2$ and 1, respectively. The real part (renormalization) and the imaginary part (damping) of the T-matrix self-energy are given by

$$\Sigma^{(\mathrm{T})'}(k, \omega_k) = \Sigma^{(1)}(k)\ Q'(k) \tag{4.5.74}$$

and

$$\Sigma^{(\mathrm{T})''}(k, \omega_k) = -\Sigma^{(1)}(k)\ Q''(k) \tag{4.5.75}$$

with

$$\Sigma^{(1)}(k) = -0.18064\,\frac{J}{\hbar}\left(\frac{k_B T}{2JS}\right)^{\frac{5}{2}} \sin^2\frac{k}{2}. \tag{4.5.76}$$

In Fig. 4.8, the renormalization (full circles) and damping (full squares) of the spin wave with $S = 1/2$ are shown together with the first-order result (full stars). Fig. 4.8 coincides with Fig. 2 of Silberglitt and Harris[36] except for a factor 2 due to a different choice of the scale. Note that the T-matrix renormalization seems to agree with the first-order result for $k \to 0$. However, for small $k$, one has

$$\Sigma^{(\mathrm{T})'}(k \to 0) = -\frac{2J}{\hbar}\,\pi\,\zeta\left(\frac{5}{2}\right)\left(\frac{k_B T}{8\pi JS}\right)^{\frac{5}{2}} |\boldsymbol{k}|^2 \left(\frac{1+0.03087/S}{1-0.09262/S} + \frac{0.17213}{S}\right) \tag{4.5.77}$$

so that Eq. (4.5.77) is in accord with the first-order result as regards the powers $k^2 T^{5/2}$ but the coefficient (1.6474, for $S = 1/2$) recovers the first-order result (1, for any $S$) only for $S \to \infty$. As $k$ increases, the difference between the first-order result (full stars) and the T-matrix renormalization (full circles) appears more and more marked up to become dramatic close to the ZC where the first-order and the T-matrix renormalization have opposite signs. For $k$ small enough, while respecting

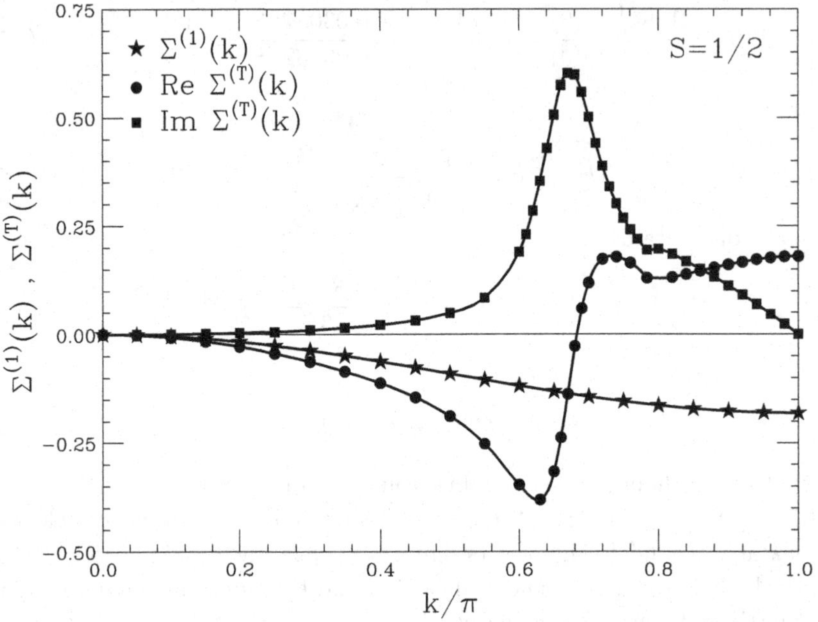


Fig. 4.8. First-order self energy $\Sigma^{(1)}$ (stars), real (full circles) and imaginary (full squares) part of the T-matrix self-energy $\Sigma^{(T)}$ for a SC lattice with $\boldsymbol{k} = (k, k, k)$ and $S = 1/2$ in units of $\frac{J}{\hbar}\left(\frac{k_B T}{2JS}\right)^{5/2}$.

the inequality $\hbar\omega_k \gg k_B T$, as the basis of the calculations of this section, the damping given by Eq. (4.5.75) reduces to

$$\Sigma^{(T)''} \simeq \frac{J}{2\hbar S} |\boldsymbol{k}|^3 \zeta\left(\frac{5}{2}\right)\left(\frac{k_B T}{8\pi JS}\right)^{\frac{5}{2}} \tag{4.5.78}$$

that coincides with the second-order result given by Eq. (3.6.82). Figure 4.8 shows a resonant-like behaviour at $k = 0.683\pi$ that can be traced back to the cross between the renormalized and damped spin wave with the two-magnon (damped) bound state inside the two-spin wave band.[36] For $k > 0.683\pi$, the renormalization becomes positive. This fact is understood as a repulsion between the single particle excitation and the bound states.[36]

# Chapter 5

# TWO-MAGNON BOUND STATES IN FERROMAGNETS

## 5.1. Two-Spin Deviation Eigenstates

In Chapter 1, we have seen how to build the excited states of a ferromagnet by adding one or two spin deviations. We have shown that a state with one spin deviation delocalized over the whole lattice (spin wave) is an eigenstate of the Heisenberg Hamiltonian. On the contrary, the state with two spin deviations delocalized over the lattice (two "free" spin waves) is not an eigenstate of the Heisenberg Hamiltonian: Only in the thermodynamic limit, the two "interacting" spin waves become asymptotically free if their energy is inside the two-magnon band. Moreover, in Chapter 4, we have shown that the poles of the T-matrix correspond to the bound states of the system. In this chapter, we will obtain the two-magnon bound states starting directly from the boson operators corresponding to the creation and destruction operators of magnons. This procedure[38] is the most direct way to arrive at the two-magnon bound state of a ferromagnet.

Consider the state made up of a linear combination of states with two magnons of momenta $\boldsymbol{k}_1 = \frac{1}{2}\boldsymbol{K} + \boldsymbol{q}$ and $\boldsymbol{k}_2 = \frac{1}{2}\boldsymbol{K} - \boldsymbol{q}$

$$|\boldsymbol{K}\rangle = \sum_{\boldsymbol{q}} f_{\boldsymbol{K}}(\boldsymbol{q})\, a^{+}_{\frac{1}{2}\boldsymbol{K}+\boldsymbol{q}} a^{+}_{\frac{1}{2}\boldsymbol{K}-\boldsymbol{q}} |0\rangle, \tag{5.1.1}$$

where $f_{\boldsymbol{K}}(\boldsymbol{q})$ are coefficients to be determined in such a way that the state (5.1.1) is eigenstate of the Heisenberg Hamiltonian realized in terms of Bose operators using the DM transformation as given by Eqs. (4.1.3)–(4.1.5a). Imposing that $\mathcal{H}|\boldsymbol{K}\rangle = E|\boldsymbol{K}\rangle$ and defining $\hbar\omega = E - E_{\mathrm{GS}}$ where $E_{\mathrm{GS}}$ is the ground-state energy of the ferromagnet, one obtains

$$\begin{aligned}(\hbar\omega - \hbar\omega_{\frac{1}{2}\boldsymbol{K}+\boldsymbol{q}} - \hbar\omega_{\frac{1}{2}\boldsymbol{K}-\boldsymbol{q}}) f_{\boldsymbol{K}}(\boldsymbol{q})\\ = \frac{zJ}{N}\sum_{\boldsymbol{q}'}(\gamma_{\frac{1}{2}\boldsymbol{K}+\boldsymbol{q}} + \gamma_{\frac{1}{2}\boldsymbol{K}-\boldsymbol{q}} - \gamma_{\boldsymbol{q}+\boldsymbol{q}'} - \gamma_{\boldsymbol{q}-\boldsymbol{q}'}) f_{\boldsymbol{K}}(\boldsymbol{q}').\end{aligned} \tag{5.1.2}$$

Equation (5.1.2) illustrates clearly that the state with two free spin waves (left-hand side) is never the eigenstate of the boson Hamiltonian because of the presence of the right-hand side coming from the interaction potential between the spin waves. Using the definition $\gamma_{\boldsymbol{q}} = \frac{1}{z}\sum_{\boldsymbol{\delta}} e^{i\boldsymbol{q}\cdot\boldsymbol{\delta}}$, Eq. (5.1.2) becomes

$$\left[\hbar\omega - 4JS\sum_{\boldsymbol{\delta}}\left(1 - \cos\frac{\boldsymbol{K}}{2}\cdot\boldsymbol{\delta}\,\cos\boldsymbol{q}\cdot\boldsymbol{\delta}\right)\right] f_{\boldsymbol{K}}(\boldsymbol{q})$$

$$= \frac{2J}{N}\sum_{\boldsymbol{\delta}}\cos\boldsymbol{q}\cdot\boldsymbol{\delta}\sum_{\boldsymbol{q}'}\left(\cos\frac{\boldsymbol{K}}{2}\cdot\boldsymbol{\delta} - \cos\boldsymbol{q}'\cdot\boldsymbol{\delta}\right) f_{\boldsymbol{K}}(\boldsymbol{q}') \qquad (5.1.3)$$

which coincides with Eq. (1.4.13). As we have shown in Chapter 1, the two-spin wave bound states are determined as the solutions of the Eq. (1.4.30) that is

$$\det\left[\mathbf{1} - \frac{1}{2S}\mathbf{B}(\boldsymbol{K},\omega)\right] = 0. \qquad (5.1.4)$$

For cubic lattices (LC, SQ, SC), defining $\alpha, \beta = x, y, z$ one has

$$B_{\alpha\beta}(\boldsymbol{K},\omega) = \frac{8JS}{N}\sum_{\boldsymbol{q}}\frac{\cos q_\alpha\left(\cos K_\beta/2 - \cos q_\beta\right)}{\hbar\omega - \hbar\omega_{\boldsymbol{K}}(\boldsymbol{q})} \qquad (5.1.5)$$

and

$$f_{\boldsymbol{K}}(\boldsymbol{q}) = \frac{4J}{N}\sum_{\alpha}\frac{\cos q_\alpha}{\hbar\omega - \hbar\omega_{\boldsymbol{K}}(\boldsymbol{q})}G^{\alpha}_{\boldsymbol{K}}, \qquad (5.1.6)$$

where $G^{\alpha}_{\boldsymbol{K}}$ are obtained from the system

$$\begin{pmatrix} 1 - \frac{1}{2S}B_{xx} & -\frac{1}{2S}B_{xy} & -\frac{1}{2S}B_{xz} \\ -\frac{1}{2S}B_{yx} & 1 - \frac{1}{2S}B_{yy} & -\frac{1}{2S}B_{yz} \\ -\frac{1}{2S}B_{zx} & -\frac{1}{2S}B_{zy} & 1 - \frac{1}{2S}B_{zz} \end{pmatrix}\begin{pmatrix} G^{x}_{\boldsymbol{K}} \\ G^{y}_{\boldsymbol{K}} \\ G^{z}_{\boldsymbol{K}} \end{pmatrix} = 0. \qquad (5.1.7)$$

In 2D (SQ lattice) and in 1D (LC), Eq. (5.1.7) reduces to a $2 \times 2$ matrix equation and a scalar equation, respectively.

## 5.2. Bound States in 1D

Let us begin considering the LC for which Eq. (5.1.7) reduces to the scalar equation

$$\left[1 - \frac{1}{2S\pi}\int_0^{\pi} dq\,\frac{\cos q\left(\cos\frac{K}{2} - \cos q\right)}{x + \cos\frac{K}{2}\cos q}\right] G_K = 0 \qquad (5.2.1)$$

where $x = \frac{\hbar\omega}{8JS} - 1$. The amplitude of the wave function is given by

$$f_K(q) = \frac{\cos q}{2S\left(x + \cos\frac{K}{2}\cos q\right)}G_K, \qquad (5.2.2)$$

where $G_K$ is an arbitrary constant to be fixed by the normalization of the wave function. The integral appearing in Eq. (5.2.1) is evaluated[3] as "principal part" and it gives different contributions accordingly if $x$ is lesser than $-\cos\frac{K}{2}$ or greater than $\cos\frac{K}{2}$. Consequently, two scalar equations are obtained for the bound states in 1D: for $-1 < x < -\cos\frac{K}{2}$, corresponding to an energy below the bottom of the two-spin wave band, that is $0 < \hbar\omega < \hbar\omega_{\text{bottom}} = 8JS(1-\cos\frac{K}{2})$, one has

$$1 - \frac{1}{2S}\left(1+\frac{x}{\cos^2\frac{K}{2}}\right)\left(1+\frac{x}{\sqrt{x^2-\cos^2\frac{K}{2}}}\right) = 0, \tag{5.2.3}$$

while for $x > \cos\frac{K}{2}$, corresponding to an energy above the top of the two-spin wave band, that is $\hbar\omega > \hbar\omega_{\text{top}} = 8JS(1+\cos\frac{K}{2})$, one has

$$1 - \frac{1}{2S}\left(1+\frac{x}{\cos^2\frac{K}{2}}\right)\left(1-\frac{x}{\sqrt{x^2-\cos^2\frac{K}{2}}}\right) = 0. \tag{5.2.4}$$

Equation (5.2.3) has always at least one root since its left-hand side is $[2S+(2S-1)\sin\frac{K}{2}]/[2S(1+\sin\frac{K}{2})] > 0$ for $x=-1$ and goes to $-\infty$ for $x \to -\cos\frac{K}{2}$ crossing the $x$-axis in one point $x_{\text{BS}}$ such that $-1 < x_{\text{BS}} < -\cos\frac{K}{2}$. On the contrary, Eq. (5.2.4) has no solution since its left-hand side is a positive monotonic decreasing function from $+\infty$ for $x \to \cos\frac{K}{2}$ to 1 for $x \to \infty$. Indeed, it is direct to prove that the first derivative with respect to $x$ of the left-hand side of Eq. (5.2.4) is negative for any $x > \cos\frac{K}{2}$. Then in 1D, only one two-magnon bound state exists for any wavevector $K$ below the two-magnon band. To obtain explicitly the energy of this bound state, we look for the solution of Eq. (5.2.3) for a generic $K$ and $S$. The most direct way is to square Eq. (5.2.3) that leads to the cubic equation

$$4Sx^3 + \left[1-4S(S-1)\cos^2\frac{K}{2}\right]x^2 - 2(2S-1)x\cos^2\frac{K}{2} + (2S-1)^2\cos^4\frac{K}{2} = 0. \tag{5.2.5}$$

Equation (5.2.5) has only one root that satisfies the condition $-1 < x < -\cos\frac{K}{2}$ as required by the original Eq. (5.2.3). Indeed, for $S < 3$, Eq. (5.2.5) has one real root and a pair of complex conjugate roots. For $S \geq 3$, Eq. (5.2.5) has three real roots for $K < K_c(S)$ and one real root and a pair of complex conjugate roots for $K > K_c(S)$ where $K_c(S)$ is given by $0.19781\pi$, $0.46624\pi, 0.58283\pi, 0.65488\pi, 0.70483\pi, \ldots$ for $S = 3, \frac{7}{2}, 4, \frac{9}{2}, 5, \ldots$, respectively. In any case, these real roots are always positive and do not satisfy the condition $-1 < x < -\cos\frac{K}{2}$, so that they have to be rejected. Note that at the ZB ($K=\pi$), one obtains $x_{\text{BS}} = -\frac{1}{4S}$, corresponding to an energy $\hbar\omega_{\text{BS}} = 8JS(1-\frac{1}{4S})$ and showing that the bound state disappears in the classical limit $S \to \infty$. For $S=\frac{1}{2}$, the solution of Eq. (5.2.3) or (5.2.5) can be given in analytic form leading to

$$\hbar\omega(K) = 2J\sin^2\frac{K}{2}. \tag{5.2.6}$$

In Fig. 5.1, the two-magnon bound state energy in units of $8JS$ for a LC is shown as function of the reduced wavevector $K/\pi$ for several values of the spin $S$. The light dashed lines represent the bottom and the top of the two-magnon band. The continuous and dotted curves represent the energy of the two-magnon bound state

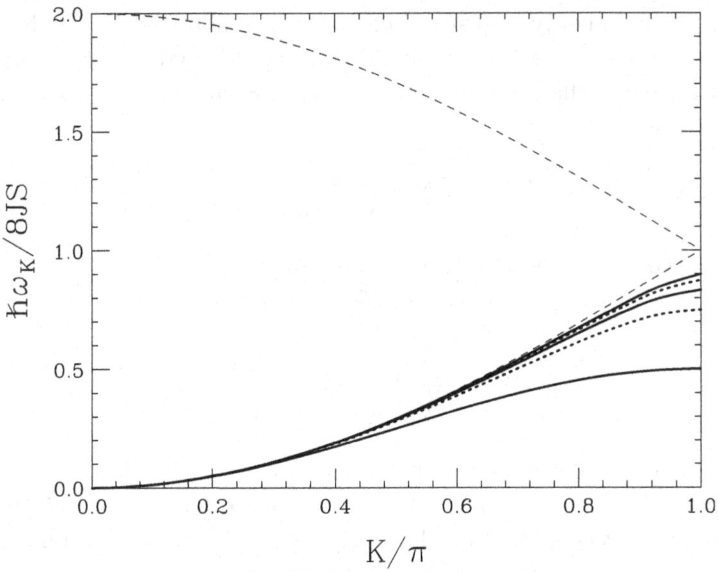


Fig. 5.1. Two-magnon bound state energy versus wavevector for the LC and several values of the spin (alternation of continuous and dotted curves). Going from the lowest to the highest curve the spin assumes the values $S = \frac{1}{2}, 1, \frac{3}{2}, 2, \frac{5}{2}$. The light dashed lines represent the bottom and the top of the two-magnon band.

for $S = \frac{1}{2}, 1, \frac{3}{2}, 2, \frac{5}{2}$ moving from the bottom to the top. In order to prove that the bound state energy is always below the bottom of the two-magnon band, we expand both the energy of the bottom of the two-magnon band and the bound state energy for $K \to 0$. From the expansion of the energy of the bottom of the two-magnon band in powers of $K^2$, we obtain

$$\frac{\hbar\omega_{\text{bottom}}}{8JS} = \frac{K^2}{8} - \frac{K^4}{384} + \frac{K^6}{46080} + \cdots . \tag{5.2.7}$$

Using Eq. (5.2.5) and assuming that $x = -1 + AK^2 + B(S)K^4 + C(S)K^6 + \cdots$, we obtain a polynomial in $K^2$: imposing that the coefficients of such polynomial are zero we obtain

$$A = \frac{1}{8}, \quad B = -\frac{3+4S^2}{1536S^2}, \quad C = -\frac{45 - 180S + 120S^2 - 16S^4}{737280S^4}. \tag{5.2.8}$$

The difference between the energy of the bottom of the two-magnon band and the energy of the bound state in units of $8JS$ is

$$\Delta(S) = \frac{\hbar\left(\omega_{\text{bottom}} - \omega_{\text{BS}}\right)}{8JS} = \frac{K^4}{512S^2} + \frac{\left(3 - 12S + 8S^2\right)K^6}{49152S^4}. \tag{5.2.9}$$

As one can see from Eq. (5.2.9), the difference $\Delta(S)$ is positive for any $S$ for $K \to 0$, so that the bound state energy is below the two-magnon band for any wavevector $K \neq 0$.

The eigenstate belonging to the eigenvalue $\hbar\omega_{\rm BS}$ is given by Eqs. (5.1.1) and (5.2.2)

$$|K\rangle = \frac{G_K}{2S}\sum_q \frac{\cos q}{x_{\rm BS}+\cos\frac{K}{2}\cos q} a^+_{\frac{K}{2}+q} a^+_{\frac{K}{2}-q}|0\rangle. \tag{5.2.10}$$

It is interesting to write the eigenstate $|K\rangle$ in terms of localized spin deviations using the Fourier transform given by Eq. (2.2.14): one has

$$|K\rangle = \frac{G_K}{2S}\sum_{l,m} e^{i\frac{K}{2}(l+m)}\frac{1}{N}\sum_q e^{iq(l-m)}\frac{\cos q}{x_{\rm BS}+\cos\frac{K}{2}\cos q} a^+_l a^+_m|0\rangle. \tag{5.2.11}$$

In the thermodynamic limit, the sum over $q$ in Eq. (5.2.11) can be transformed into an integral like

$$\frac{1}{\pi}\int_0^\pi dq \frac{\cos[q(l-m)]\cos q}{x_{\rm BS}+\cos\frac{K}{2}\cos q} = \frac{1}{\cos\frac{K}{2}}[\delta_{l,m} - I_K(l-m)] \tag{5.2.12}$$

where[3]

$$\begin{aligned} I_K(n) &= \frac{1}{\pi}\int_0^\pi dq \frac{\cos nq}{1+\frac{\cos\frac{K}{2}}{x_{\rm BS}}\cos q} \\ &= \frac{|x_{\rm BS}|}{\sqrt{x_{\rm BS}^2-\cos^2\frac{K}{2}}}\left[\frac{|x_{\rm BS}|-\sqrt{x_{\rm BS}^2-\cos^2\frac{K}{2}}}{\cos\frac{K}{2}}\right]^{|n|} \\ &= \frac{|x_{\rm BS}|}{\sqrt{x_{\rm BS}^2-\cos^2\frac{K}{2}}} e^{-\frac{|n|}{\xi}} \end{aligned} \tag{5.2.13}$$

with

$$\xi = \frac{1}{\ln\left(\frac{\cos\frac{K}{2}}{|x_{\rm BS}|-\sqrt{x_{\rm BS}^2-\cos^2\frac{K}{2}}}\right)}. \tag{5.2.14}$$

Replacing Eq. (5.2.13) into Eq. (5.2.11), one obtains

$$|K\rangle = \frac{G_K}{2S}\sum_{l,m} e^{i\frac{K}{2}(l+m)}\frac{1}{\cos\frac{K}{2}}\left[\delta_{l,m} - \frac{|x_{\rm BS}|}{\sqrt{x_{\rm BS}^2-\cos^2\frac{K}{2}}}e^{-\frac{|l-m|}{\xi}}\right] a^+_l a^+_m|0\rangle \tag{5.2.15}$$

that is an exponential decay of the wave function with the distance between the two localized deviations. This means that only states with two deviations close to each other contribute significantly to the bound state wave function.

For $S = 1/2$, the analytic solution $x_{\rm BS} = -\frac{1}{2}(1+\cos^2\frac{K}{2})$ was obtained in Eq. (5.2.6). Replacing such result in Eq. (5.2.15) and noticing that $(a_l^+)^2|0\rangle = 0$ the eigenstate of the bound state becomes

$$|K\rangle = -G_K \sum_{l\neq m} e^{i\frac{K}{2}(l+m)} \frac{1}{\cos\frac{K}{2}} \frac{3+\cos K}{1-\cos K} e^{-\frac{|l-m|}{\xi}} a_l^+ a_m^+ |0\rangle, \tag{5.2.16}$$

where

$$\xi = \frac{1}{\ln\left(\frac{1}{\cos\frac{K}{2}}\right)}. \tag{5.2.17}$$

Equation (5.2.16) looks very similar to Eq. (2.22) of Oguchi.[38] The light discrepancy comes from the way of computing the integral (5.2.13). Indeed, in the paper of Oguchi,[38] the integral (5.2.13) was evaluated by changing the upper limit of integration from $\pi$ to $\infty$ since the author was interested in the behaviour of the amplitude of the eigenstate for large $|l-m|$. This approximation does not change the qualitative conclusion about the exponential decay of the amplitude but slightly changes the quantitative result. Note that for $K \to 0$, the correlation length goes to infinity, so that the decay of the amplitude with the distance is very slow and the two deviations may be located at an arbitrary distance. On the contrary, for $K = \pi$, the correlation length goes to zero implying that the two deviations are located on NN sites. Going from for $K = 0$ to $K = \pi$, the correlation length decays monotonically from $\infty$ to 0.

## 5.3. Bound States in 2D

In a SQ lattice, the bound states are obtained as solutions of Eq. (5.1.4) for which

$$\begin{aligned} B_{xx}(\boldsymbol{K},\omega) &= \frac{1}{\pi^2}\int_0^\pi dq_x \int_0^\pi dq_y \frac{\cos q_x\left(\cos\frac{K_x}{2} - \cos q_x\right)}{x + \cos\frac{K_x}{2}\cos q_x + \cos\frac{K_y}{2}\cos q_y} \\ &= \mp\frac{1}{\pi}\int_0^\pi dq \frac{\cos q\left(\cos\frac{K_x}{2} - \cos q\right)}{\sqrt{\left(x+\cos\frac{K_x}{2}\cos q\right)^2 - \cos^2\frac{K_y}{2}}}, \end{aligned} \tag{5.3.1}$$

$$\begin{aligned} B_{xy}(\boldsymbol{K},\omega) &= \frac{1}{\pi^2}\int_0^\pi dq_x \int_0^\pi dq_y \frac{\cos q_x\left(\cos\frac{K_y}{2} - \cos q_y\right)}{x + \cos\frac{K_x}{2}\cos q_x + \cos\frac{K_y}{2}\cos q_y} \\ &= \frac{\cos\frac{K_y}{2}}{\cos\frac{K_x}{2}} \pm \frac{1}{\cos\frac{K_x}{2}}\frac{1}{\pi}\int_0^\pi dq \frac{\left(x+\cos\frac{K_y}{2}\cos q\right)\left(\cos\frac{K_y}{2} - \cos q\right)}{\sqrt{\left(x+\cos\frac{K_y}{2}\cos q\right)^2 - \cos^2\frac{K_x}{2}}}, \end{aligned} \tag{5.3.2}$$

$$\begin{aligned} B_{yx}(\boldsymbol{K},\omega) &= \frac{1}{\pi^2}\int_0^\pi dq_x \int_0^\pi dq_y \frac{\cos q_y\left(\cos\frac{K_x}{2} - \cos q_x\right)}{x + \cos\frac{K_x}{2}\cos q_x + \cos\frac{K_y}{2}\cos q_y} \\ &= \frac{\cos\frac{K_x}{2}}{\cos\frac{K_y}{2}} \pm \frac{1}{\cos\frac{K_y}{2}}\frac{1}{\pi}\int_0^\pi dq \frac{\left(x+\cos\frac{K_x}{2}\cos q\right)\left(\cos\frac{K_x}{2} - \cos q\right)}{\sqrt{\left(x+\cos\frac{K_x}{2}\cos q\right)^2 - \cos^2\frac{K_y}{2}}}, \end{aligned} \tag{5.3.3}$$

$$B_{yy}(\boldsymbol{K},\omega) = \frac{1}{\pi^2}\int_0^\pi dq_x \int_0^\pi dq_y \frac{\cos q_y\left(\cos\frac{K_y}{2} - \cos q_y\right)}{x + \cos\frac{K_x}{2}\cos q_x + \cos\frac{K_y}{2}\cos q_y}$$

$$= \mp\frac{1}{\pi}\int_0^\pi dq \frac{\cos q\left(\cos\frac{K_y}{2} - \cos q\right)}{\sqrt{\left(x + \cos\frac{K_y}{2}\cos q\right)^2 - \cos^2\frac{K_x}{2}}}, \tag{5.3.4}$$

where the upper sign has to be assumed for $-2 < x < -\cos\frac{K_x}{2} - \cos\frac{K_y}{2}$, corresponding to energies below the two-magnon band and the lower sign has to be assumed for $x > \cos\frac{K_x}{2} + \cos\frac{K_y}{2}$, corresponding to energies above the two-magnon band. Remember that $x = \frac{\hbar\omega}{8JS} - 2$. One integration in Eqs. (5.3.1)–(5.3.4) has been performed taking advantage from the relationship[3]

$$\frac{1}{\pi}\int_0^\pi d\xi \frac{1}{a + \cos\xi} = \pm\frac{1}{\sqrt{a^2 - 1}} \tag{5.3.5}$$

for $a > 1$ (upper sign) and $a < -1$ (lower sign), respectively. The integrals (5.3.1)–(5.3.4) may be expressed in terms of elliptic integrals,[39] however, because of the complexity of the final formula we choose to perform the last integration numerically. It should be noted that for a general point in the BZ, the integrals (5.3.1)–(5.3.4) are all distinct and the equation for the bound states (5.1.4) writes

$$\left(1 - \frac{B_{xx}}{2S}\right)\left(1 - \frac{B_{yy}}{2S}\right) - \frac{B_{xy}\,B_{yx}}{4S^2} = 0. \tag{5.3.6}$$

To check the reliability of the numerical calculation we focus on some points of the BZ for which integrals (5.3.1)–(5.3.4) can be written in a simple analytic form. For $\boldsymbol{K} = (\pi,\pi)$, for instance, one has $B_{xx} = B_{yy} = -\frac{1}{2x}$ and $B_{xy} = B_{yx} = 0$ so that Eq. (5.3.6) becomes

$$\left(1 + \frac{1}{4Sx}\right)^2 = 0. \tag{5.3.7}$$

Equation (5.3.7) implies that at the ZC two degenerate bound states exist with energy

$$\hbar\omega_{\mathrm{BS}}(\pi,\pi) = 16JS\left(1 - \frac{1}{8S}\right). \tag{5.3.8}$$

Along the line $(K,\pi)$, Eqs. (5.3.1)–(5.3.4) reduce to

$$B_{xx} = \frac{1}{\pi}\int_0^\pi dq \frac{\cos q\left(\cos\frac{K}{2} - \cos q\right)}{x + \cos\frac{K}{2}\cos q} = \left(1 + \frac{x}{\cos^2\frac{K}{2}}\right)\left(1 \pm \frac{x}{\sqrt{x^2 - \cos^2\frac{K}{2}}}\right), \tag{5.3.9}$$

$$B_{yy} = -\frac{1}{2\pi}\int_0^\pi dq \frac{1}{x + \cos\frac{K}{2}\cos q} = \pm\frac{1}{2}\frac{1}{\sqrt{x^2 - \cos^2\frac{K}{2}}} \tag{5.3.10}$$

and $B_{xy} = B_{yx} = 0$. The bound state equation (5.3.6) splits into the couple of equations

$$1 - \frac{1}{2S}\left(1 + \frac{x}{\cos^2\frac{K}{2}}\right)\left(1 \pm \frac{x}{\sqrt{x^2 - \cos^2\frac{K}{2}}}\right) = 0 \qquad (5.3.11)$$

and

$$1 \mp \frac{1}{4S}\frac{1}{\sqrt{x^2 - \cos^2\frac{K}{2}}} = 0. \qquad (5.3.12)$$

Equation (5.3.11) coincides with Eqs. (5.2.3) and (5.2.4), so that we conclude that only one solution exists for $-1 < x < -\cos\frac{K}{2}$ and no solution exists for $x > \cos\frac{K}{2}$. For $S = 1/2$, the solution of Eq. (5.3.11) (upper sign) is $x_{\rm BS} = -\frac{1}{2}(1 + \cos^2\frac{K}{2})$ leading to the bound state energy

$$\hbar\omega_{\rm BS}^{(1)}(K, \pi) = 2J\left(3 - \cos^2\frac{K}{2}\right). \qquad (5.3.13)$$

Another bound state is obtained from Eq. (5.3.12) by choosing the upper sign while no solution is obtained by the same equation by choosing the lower sign. The energy of this second bound state is given by

$$\hbar\omega_{\rm BS}^{(2)}(K, \pi) = 16JS\left(1 - \frac{1}{2}\sqrt{\frac{1}{16S^2} + \cos^2\frac{K}{2}}\right). \qquad (5.3.14)$$

From Eqs. (5.3.11) and (5.3.12), one sees that two bound states exist for any wavevector $\boldsymbol{K} = (K, \pi)$ below the two-magnon band. The same conclusion holds for $\boldsymbol{K} = (\pi, K)$. The two-magnon bound states of wavevector $\boldsymbol{K} = (\pi, K)$ are shown in the central panel of Fig. 5.2. The light dashed lines represent the top and bottom of the two-magnon band.

Now, we look for the bound states for $\boldsymbol{K} = (K, 0)$. In this case, the integrals (5.3.1)–(5.3.4) are evaluated numerically and the zeros of Eq. (5.3.6) are obtained. Only the choice of the upper sign leads to a solution so that one can conclude that the existence of the bound states is limited to the energy region below the two-magnon band. In the right panel of Fig. 5.2, one can see that only one bound state is found for $\boldsymbol{K} = (K, 0)$ and the same result is obtained for $\boldsymbol{K} = (0, K)$. The bound state energy differs from the bottom of the two-magnon band of less than $16JS \times 10^{-7}$ for $K < 0.7\pi$. This is the reason why the bound state appears indistinguishable from the bottom of the two-magnon band in the right panel of Fig. 5.2. To check that the bound state is really below the bottom of the two-magnon band for any $K$, we try to evaluate analytically the difference between the bottom of the two-magnon band and the bound state energy. To do this, assume $x_{\rm BS} = -1 - \cos\frac{K}{2} - \epsilon$ which leads to a bound state energy $\hbar\omega_{\rm BS} = \hbar\omega_{\rm bottom} - 8JS\epsilon$. Putting this value of $x_{\rm BS}$ into Eqs. (5.3.1)–(5.3.4), in the limit $\epsilon \to 0$ we obtain

$$B_{xx} = -\frac{1 - \cos\frac{K}{2}}{2\pi\sqrt{\cos\frac{K}{2}}}\ln\epsilon + B_{xx}^{\rm reg}, \qquad (5.3.15)$$

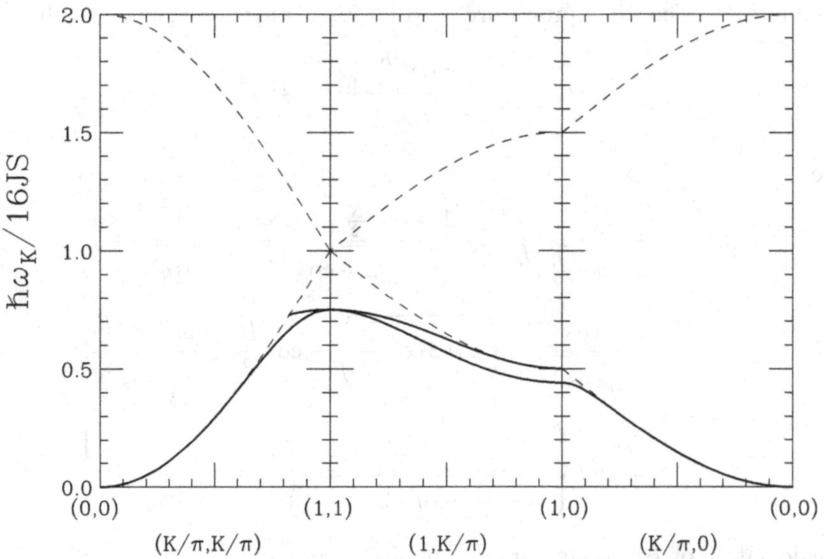


Fig. 5.2. Two-magnon bound state energy versus wavevector along three high symmetry directions $(K, K)$, $(\pi, K)$ and $(K, 0)$ for a SQ lattice with $S = 1/2$. The light dashed lines represent the top and the bottom of the two-magnon band. The continuous heavy curves represent the bound states.

where

$$B_{xx}^{\mathrm{reg}} = -\frac{1}{\pi\sqrt{\cos\frac{K}{2}}}\int_0^\pi dq \frac{\cos q\sqrt{1-\cos q}}{\sqrt{2+\cos\frac{K}{2}(1-\cos q)}} + \frac{1-\cos\frac{K}{2}}{\pi\sqrt{\cos\frac{K}{2}}}\ln\left(\pi\sqrt{2\cos\frac{K}{2}}\right)$$

$$+\frac{1-\cos\frac{K}{2}}{\pi\sqrt{\cos\frac{K}{2}}}\int_0^\pi dq\left\{\frac{\cos q}{\sqrt{(1-\cos q)\left[2+\cos\frac{K}{2}(1-\cos q)\right]}} - \frac{1}{q}\right\}. \quad (5.3.16)$$

As one can see from Eq. (5.3.15), $B_{xx}$ diverges logarithmically when the energy of the bound state meets the energy of the bottom of the two-magnon band. This divergence is due to the fact that the integrand diverges as $\frac{1}{q}$ for $q \to 0$. The "regular" contribution given by Eq. (5.3.16) is obtained by subtracting the divergent contribution (first term of the series expansion in $q$ of the integrand) and evaluating the difference for $\epsilon = 0$. This approach leads to the third term of Eq. (5.3.16). The "singular" part is then obtained by evaluating directly the integral of the term subtracted for $\epsilon \neq 0$. The second term of Eq. (5.3.16) comes from the integral of the singular part evaluated at $q = \pi$ and taking the limit $\epsilon \to 0$. The above procedure is not required to evaluate $B_{xy}$ because it remains finite for $\epsilon = 0$ since the vanishing of the denominator for $q \to 0$ is compensated by a concomitant vanishing of the numerator. The result is

$$B_{xy} = \frac{1}{\cos\frac{K}{2}}\left[1 - \frac{1}{\pi}\int_0^\pi dq \frac{\left(1+\cos\frac{K}{2}-\cos q\right)\sqrt{1-\cos q}}{\sqrt{2\cos\frac{K}{2}+1-\cos q}}\right]. \quad (5.3.17)$$

To evaluate $B_{yx}$ the same procedure used for $B_{xx}$ is necessary and one obtains

$$B_{yx} = -\frac{1-\cos\frac{K}{2}}{2\pi\sqrt{\cos\frac{K}{2}}}\ln\epsilon + B_{yx}^{\rm reg}, \tag{5.3.18}$$

where

$$\begin{aligned} B_{yx}^{\rm reg} &= -\frac{1}{\pi\sqrt{\cos\frac{K}{2}}}\int_0^\pi dq \frac{\left[1+\cos\frac{K}{2}(1-\cos q)\right]\sqrt{1-\cos q}}{\sqrt{2+\cos\frac{K}{2}(1-\cos q)}} \\ &+ \frac{1-\cos\frac{K}{2}}{\pi\sqrt{\cos\frac{K}{2}}}\ln\left(\pi\sqrt{2\cos\frac{K}{2}}\right) + \cos\frac{K}{2} + \frac{1-\cos\frac{K}{2}}{\pi\sqrt{\cos\frac{K}{2}}} \\ &\times \int_0^\pi dq\left\{\frac{1+\cos\frac{K}{2}(1-\cos q)}{\sqrt{(1-\cos q)\left[2+\cos\frac{K}{2}(1-\cos q)\right]}} - \frac{1}{q}\right\}. \end{aligned} \tag{5.3.19}$$

The calculation of $B_{yy}$ is similar to that of $B_{xy}$ and gives

$$B_{yy} = -\frac{1}{\pi}\int_0^\pi dq \frac{\cos q\sqrt{1-\cos q}}{\sqrt{2\cos\frac{K}{2}+1-\cos q}}. \tag{5.3.20}$$

Replacing Eqs. (5.3.15)–(5.3.20) into Eq. (5.3.6), we obtain the equation for the bound state near the bottom of the two-magnon band along the $(1,0)$-direction

$$\frac{1-\cos\frac{K}{2}}{4\pi S\sqrt{\cos\frac{K}{2}}}\ln\epsilon\left(1-\frac{B_{yy}-B_{xy}}{2S}\right) + \left(1-\frac{B_{xx}^{\rm reg}}{2S}\right)\left(1-\frac{B_{yy}}{2S}\right) - \frac{1}{4S^2}B_{xy}B_{yx}^{\rm reg} = 0. \tag{5.3.21}$$

The solution of Eq. (5.3.21) is

$$\epsilon = \exp\left[-\frac{4\pi S\sqrt{\cos\frac{K}{2}}}{1-\cos\frac{K}{2}}f(K,S)\right] \tag{5.3.22}$$

where

$$f(K,S) = \frac{\left(1-\frac{B_{xx}^{\rm reg}}{2S}\right)\left(1-\frac{B_{yy}}{2S}\right) - \frac{B_{xy}B_{yx}^{\rm reg}}{4S^2}}{1-\frac{B_{yy}-B_{xy}}{2S}}. \tag{5.3.23}$$

In the limit of small wavevectors from Eqs. (5.3.16), (5.3.17), (5.3.19) and (5.3.20), one obtains

$$B_{xx}^{\rm reg} = B_{yy} = -\frac{1}{\pi}\int_0^\pi dq\cos q\left(\frac{1-\cos q}{3-\cos q}\right)^{1/2} + O(K^2) \tag{5.3.24}$$

and

$$B_{yx}^{\rm reg} = B_{xy} = 1 - \frac{1}{\pi}\int_0^\pi dq\,(2-\cos q)\left(\frac{1-\cos q}{3-\cos q}\right)^{1/2} + O(K^2). \tag{5.3.25}$$

Then Eq. (5.3.23) becomes

$$f(K,S) = 1 - \frac{1}{2S}\left[1 - \frac{2}{\pi}\int_0^\pi dq \left(\frac{1-\cos q}{3-\cos q}\right)^{1/2}\right] + O(K^2) = 1 + O(K^2). \tag{5.3.26}$$

The integral in Eq. (5.3.26) is easily evaluated by using the change of variable $\cos\frac{q}{2} = \xi$ and its value is $\frac{\pi}{2}$. Then

$$\epsilon \simeq \exp\left[-\frac{32\pi S}{K^2}\right]. \tag{5.3.27}$$

The relationship (5.3.27) proves that for small wavevectors, the energy of the bound state is below the bottom of the two-magnon bound states by an exponentially small amount which explains why the bound state and the bottom of the two-magnon band are indistinguishable in the right panel of Fig. 5.2.

Let us conclude this section looking for the bound states with $\boldsymbol{K} = (K,K)$. From Eqs. (5.3.1)–(5.3.4) one obtains $B_{xx} = B_{yy}$, $B_{xy} = B_{yx}$ so that the bound state equation (5.3.6) splits into the couple of equations

$$1 - \frac{1}{2S}(B_{xx} + B_{xy}) = 0 \tag{5.3.28}$$

and

$$1 - \frac{1}{2S}(B_{xx} - B_{xy}) = 0 \tag{5.3.29}$$

where

$$B_{xx} + B_{xy} = 1 \pm \frac{x}{\pi\cos\frac{K}{2}}\int_0^\pi dq \frac{\cos\frac{K}{2} - \cos q}{\sqrt{\left(x + \cos\frac{K}{2}\cos q\right)^2 - \cos^2\frac{K}{2}}} \tag{5.3.30}$$

and

$$B_{xx} - B_{xy} = -1 \mp \frac{1}{\pi\cos\frac{K}{2}}\int_0^\pi dq \frac{\left(\cos\frac{K}{2} - \cos q\right)\left(x + 2\cos\frac{K}{2}\cos q\right)}{\sqrt{\left(x + \cos\frac{K}{2}\cos q\right)^2 - \cos^2\frac{K}{2}}}. \tag{5.3.31}$$

where the upper (lower) sign holds for $x < -2\cos\frac{K}{2}$ ($x > 2\cos\frac{K}{2}$). The solutions of Eqs. (5.3.28), (5.3.29) exist only for $x < -2\cos\frac{K}{2}$, that is for energies below the two-magnon band. In particular, for $x = -2\cos\frac{K}{2}$ (the value of the bound state energy when it meets the bottom of the two-magnon band), Eq. (5.3.30) becomes

$$B_{xx} + B_{xy} = 1 - \frac{2}{\pi\cos\frac{K}{2}}\int_0^\pi dq \frac{\cos\frac{K}{2} - \cos q}{\sqrt{(3-\cos q)(1-\cos q)}} \to +\infty \tag{5.3.32}$$

for any $K$ so that one solution of Eq. (5.3.28) always exists implying that one bound state exists over the whole (1,1)-direction. Analagously, for $x = -2\cos\frac{K}{2}$, Eq. (5.3.31) gives

$$B_{xx} - B_{xy} = -1 - \frac{1}{\pi\cos\frac{K_c}{2}}\int_0^\pi dq \frac{\left(\cos\frac{K_c}{2} - \cos q\right)(1-\cos q)}{\sqrt{(3-\cos q)(1-\cos q)}} = \frac{1}{\cos\frac{K_c}{2}}\left(\frac{4}{\pi} - 1\right). \tag{5.3.33}$$

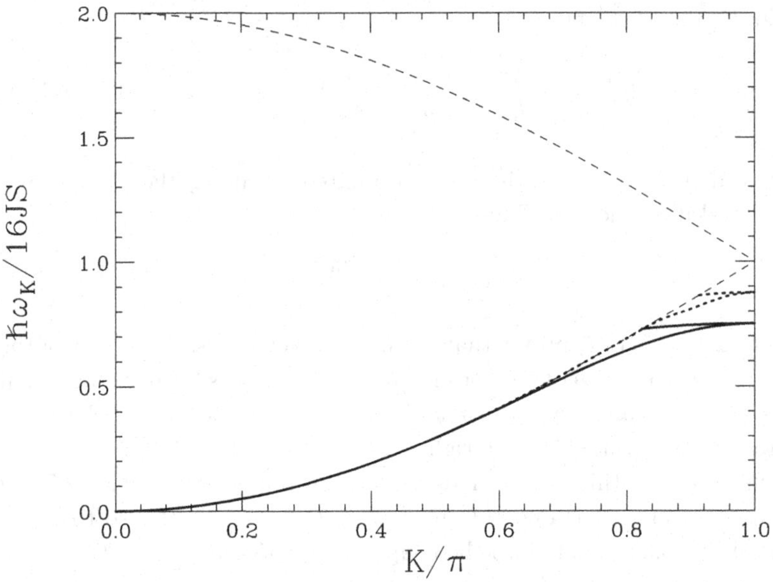


Fig. 5.3. Two-magnon bound state energy versus wavevector along the (1,1)-direction for a SQ lattice. The light dashed lines represent the top and the bottom of the two-magnon band. The continuous and dotted curves represent the two bound states for $S = 1/2$ and 1 respectively.

The integral in Eq. (5.3.33) has been evaluated analytically, performing the change of variable $\xi = \cos\frac{q}{2}$. The replacement of Eq. (5.3.33) into Eq. (5.3.29) shows that one solution exists only for $K > K_c(S)$ where

$$K_c(S) = 2\arccos\left[\frac{1}{2S}\left(\frac{4}{\pi} - 1\right)\right], \tag{5.3.34}$$

so that the existence of a second bound state is restricted to a region close to the ZC as shown in the left panel of Fig. 5.2 for $S = 1/2$. From Eq. (5.3.34), one has $K_c(S) = 0.82381\pi, 0.91275\pi, 0.94194\pi, 0.95648\pi, 0.96519\pi, 0.97100\pi, \ldots$, for $S = \frac{1}{2}, 1, \frac{3}{2}, 2, \frac{5}{2}, 3, \ldots$, respectively, showing that the region of existence of this second bound state decreases, increasing the spin value. In Fig. 5.3, the bound states in a SQ lattice versus the reduced wavevector $\frac{K}{\pi}$ along the $(1,1)$-direction are shown for $S = \frac{1}{2}$ (continuous curve) and 1 (dotted curve) respectively. Two bound states exist for $K > 0.82381\pi$ ($S = \frac{1}{2}$) and for $K > 0.91275\pi$ ($S = 1$). For $K = K_c(S)$, one bound state enters the continuum of the two-magnon band while the other bound state exists for any $K$ and any $S$. Notice that for $K \lesssim 0.7\pi$ ($S = \frac{1}{2}$) and $K \lesssim 0.8\pi$ ($S = 1$), the bound state is so close to the bottom of the two-magnon band that it is indistinguishable from the bottom of the band. An analytic solution of equation (5.3.28) leads to[39]

$$x \simeq -2\cos\frac{K}{2}\left[1 + 8e^{-\pi}\exp\left(-\frac{2\pi S\cos\frac{K}{2}}{1 - \cos\frac{K}{2}}\right)\right]. \tag{5.3.35}$$

Notice that there is a slight difference with Eq. (67) of Wortis[39] where a factor $e^{-\pi}$ is missing. The bound state energy is below the two-magnon band by an amount

$$\Delta(K) = \frac{\hbar\omega_{\rm bottom} - \hbar\omega_{\rm BS}}{16JS} \simeq 8e^{-\pi}\cos\frac{K}{2}\exp\left(-\frac{2\pi S\cos\frac{K}{2}}{1-\cos\frac{K}{2}}\right) \tag{5.3.36}$$

that is less than $2\times10^{-4}$ for $K < \pi/2$ ($S = 1/2$) and less than $2\times10^{-3}$ for $K < 0.8\pi$ ($S = 1$) explaining the failed resolution of the bound states from the bottom of the two-magnon band in Fig. 5.3.

Let us conclude this section by coming back to Fig. 5.2 where the bound states for a SQ lattice with $S = 1/2$ are shown in several directions of the reciprocal space. At least one bound state exists for any $\boldsymbol{K}$ below the two-magnon band. Along the $(1,1)$-direction, one further bound state appears close to the ZC ($0.82381 < K/\pi < 1$). Along the boundary of the Brillouin zone $(\pi, K)$ and $(K, \pi)$, two bound states exist for any wavevector while along the $(1,0)$ and $(0,1)$ directions, only one bound state is found. This scenario is qualitatively the same for other 2D lattice as, for instance, the triangular (TR) lattice.[40]

## 5.4. Bound States in 3D

In a SC lattice the bound states are obtained as solutions of Eq. (5.1.4) with $B_{\alpha\beta}$, $\alpha, \beta = x, y, z$ given by Eq. (5.1.5) and $x = \frac{\hbar\omega}{8JS} - 3$. There are no solution for energy above the top of the two-magnon band. For energies below the bottom of the two-magnon band $\hbar\omega_{\rm bottom} = 24JS[1 - \frac{1}{3}(\cos\frac{K_x}{2} + \cos\frac{K_y}{2} + \cos\frac{K_z}{2})]$, the sums defining $B_{\alpha\beta}$ may be reduced to 2D integrals that can be evaluated numerically. We will investigate in this section the existence of bound states in several directions of the first BZ of the reciprocal lattice.

For $\boldsymbol{K} = (K, K, K)$, the nine $B_{\alpha\beta}$ of Eq. (5.1.5) reduce to $B_{xx}$ and $B_{xy}$: indeed, one has $B_{xx} = B_{yy} = B_{zz}$ and $B_{xy} = B_{yx} = B_{xz} = B_{zx} = B_{yz} = B_{zy}$ where

$$B_{xx} = -\frac{1}{\pi^2}\int_0^\pi dq_x \int_0^\pi dq_y \frac{\cos q_x\left(\cos\frac{K}{2} - \cos q_x\right)}{\sqrt{\left[x + \cos\frac{K}{2}\left(\cos q_x + \cos q_y\right)\right]^2 - \cos^2\frac{K}{2}}} \tag{5.4.1}$$

and

$$B_{xy} = -\frac{1}{\pi^2}\int_0^\pi dq_x \int_0^\pi dq_y \frac{\cos q_x\left(\cos\frac{K}{2} - \cos q_y\right)}{\sqrt{\left[x + \cos\frac{K}{2}\left(\cos q_x + \cos q_y\right)\right]^2 - \cos^2\frac{K}{2}}}. \tag{5.4.2}$$

The bound state equation (5.1.4) reduces to

$$\left[1 - \frac{1}{2S}\left(B_{xx} - B_{xy}\right)\right]^2\left[1 - \frac{1}{2S}\left(B_{xx} + 2B_{xy}\right)\right] = 0 \tag{5.4.3}$$

so that there are two degenerate bound states for

$$\left[1 - \frac{1}{2S}\left(B_{xx} - B_{xy}\right)\right] = 0 \tag{5.4.4}$$

and one bound state for

$$\left[1 - \frac{1}{2S}\left(B_{xx} + 2B_{xy}\right)\right] = 0. \tag{5.4.5}$$

Note that for $K = \pi$, both Eqs. (5.4.4) and (5.4.5) reduce to

$$\left(1 + \frac{1}{4xS}\right) = 0 \tag{5.4.6}$$

so that at the ZC three degenerate bound states exist with energy

$$\hbar\omega_{\rm BS}(\pi,\pi,\pi) = 24JS\left(1 - \frac{1}{12S}\right). \tag{5.4.7}$$

For $K < \pi$, one sees that Eqs. (5.4.4) and (5.4.5) do not always have solution. In particular, the doubly degenerate bound state, solution of Eq. (5.4.4) exists only for $K_c^{(2)} < K < \pi$ and the existence of the single bound state coming from the solution of Eq. (5.4.5) is limited to the region $K_c^{(1)} < K < \pi$, where $K_c^{(2)}$ and $K_c^{(1)}$ can be determined by evaluating the combinations $B_{xx} - B_{yy}$ and $B_{xx} + 2B_{xy}$ for $x = -3\cos\frac{K}{2}$, corresponding to the value of $x$ when the bound state meets the bottom of the two-magnon band. One obtains

$$B_{xx} - B_{xy} = \frac{1}{\pi^2\cos\frac{K}{2}}\int_0^\pi dq_x \int_0^\pi dq_y \frac{\cos q_x\left(\cos q_x - \cos q_y\right)}{\sqrt{\left(3 - \cos q_x - \cos q_y\right)^2 - 1}} = \frac{0.185237}{\cos\frac{K}{2}} \tag{5.4.8}$$

and using Eq. (5.4.4) one finds

$$K_c^{(2)} = 2\arccos\left(\frac{0.185237}{2S}\right) \tag{5.4.9}$$

so that $K_c^{(2)} = 0.88139\pi$ for $S = 1/2$. Then

$$B_{xx} + 2B_{xy} = 3\left(1 - \cos\frac{K}{2}\right)\frac{1}{\pi^2\cos\frac{K}{2}}\int_0^\pi dq_x \int_0^\pi dq_y \frac{\cos q_x}{\sqrt{\left(3 - \cos q_x - \cos q_y\right)^2 - 1}}$$

$$= 0.516386\frac{1 - \cos\frac{K}{2}}{\cos\frac{K}{2}}. \tag{5.4.10}$$

Using Eq. (5.4.5), one obtains

$$K_c^{(1)} = 2\arccos\left(\frac{0.516386}{2S + 0.516386}\right) \tag{5.4.11}$$

so that $K_c^{(1)} = 0.77878\pi$ for $S = 1/2$. The bound states along the $(1,1,1)$-direction are shown in the first panel of Fig. 5.4 along with the bottom and the top of the two-magnon band (dashed curves). The existence of one bound state, solution of Eq. (5.4.5) (lower continuous curve), is restricted to the region $0.77878 < K/\pi < 1$ while the doubly degenerate bound state, solution of Eq. (5.4.4) (higher curve), exists only for $0.88139 < K/\pi < 1$. By increasing the spin $S$, the region of existence of both bound states is further reduced.

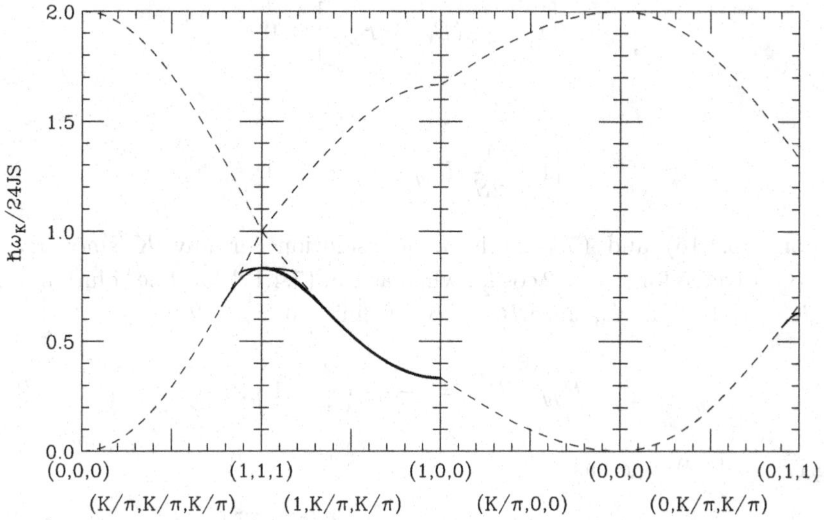


Fig. 5.4. Two-magnon bound state energy versus wavevector along four directions of a SC lattice with $S = 1/2$: $(K, K, K)$ (first panel), $(\pi, K, K)$ (second panel), $(K, 0, 0)$ (third panel) and $(0, K, K)$ (last panel). The dashed lines represent the top and the bottom of the two-magnon band. The continuous curves represent the bound states. The lower heavy curve in the second panel represents a couple of (non-degenerate) bound states with very close energies.

Now, we look for the bound states along the diagonals of the faces of the cube of the BZ. In particular, for $\boldsymbol{K} = (\pi, K, K)$, one has

$$B_{xx} = \frac{1}{2\pi}\int_0^\pi dq \frac{1}{\sqrt{\left(x + \cos\frac{K}{2}\cos q\right)^2 - \cos^2\frac{K}{2}}}, \tag{5.4.12}$$

$$B_{yy} = B_{zz} = -\frac{1}{\pi}\int_0^\pi dq \frac{\cos q\left(\cos\frac{K}{2} - \cos q\right)}{\sqrt{\left(x + \cos\frac{K}{2}\cos q\right)^2 - \cos^2\frac{K}{2}}}, \tag{5.4.13}$$

$B_{xy} = B_{yx} = B_{xz} = B_{zx} = 0$ and

$$B_{yz} = B_{zy} = 1 + \frac{1}{\pi\cos\frac{K}{2}}\int_0^\pi dq \frac{(\cos\frac{K}{2} - \cos q)(x + \cos\frac{K}{2}\cos q)}{\sqrt{(x + \cos\frac{K}{2}\cos q)^2 - \cos^2\frac{K}{2}}}. \tag{5.4.14}$$

The bound state equation (5.1.4) reduces to

$$\left(1 - \frac{1}{2S}B_{xx}\right)\left[1 - \frac{1}{2S}(B_{yy} - B_{yz})\right]\left[1 - \frac{1}{2S}(B_{yy} + B_{yz})\right] = 0 \tag{5.4.15}$$

so that the bound states are the possible solutions of the three equations

$$\left(1 - \frac{1}{2S}B_{xx}\right) = 0, \tag{5.4.16}$$

$$\left[1 - \frac{1}{2S}\left(B_{yy} - B_{yz}\right)\right] = 0 \tag{5.4.17}$$

and

$$\left[1 - \frac{1}{2S}\left(B_{yy} + B_{yz}\right)\right] = 0. \tag{5.4.18}$$

Equations (5.4.16) and (5.4.18) have one solution for any $K$ since $B_{xx}$ and $B_{yy} + B_{yz}$ diverge for $x \to -2\cos\frac{K}{2}$, whereas Eq. (5.4.17) has one solution only for $K_c < K < \pi$. To find $K_c$, since $B_{yy} - B_{yz}$ is finite at $x = -2\cos\frac{K}{2}$

$$B_{yy} - B_{yz} = \frac{1}{\cos\frac{K}{2}}\left(\frac{4}{\pi} - 1\right). \tag{5.4.19}$$

Using Eq. (5.4.17), one has

$$K_c(S) = 2\arccos\left[\frac{1}{2S}\left(\frac{4}{\pi} - 1\right)\right] \tag{5.4.20}$$

and $K_c(\frac{1}{2}) = 0.82381\pi$. In the second panel of Fig. 5.4, the bound states obtained from Eqs. (5.4.16)–(5.4.18) are shown. The solution of Eq. (5.4.17) corresponds to the higher curve. The solutions of Eqs. (5.4.16) and (5.4.18) correspond to the apparently unique lower curve. The bound states, solutions of Eqs. (5.4.16) and (5.4.18), are degenerate only at $K = 0$ and $K = \pi$. However, the energy difference between them for $0 < K < \pi$ reaches its maximum $\hbar\omega = 24JS \times 0.0055$ around $K \simeq 0.8\pi$. The proximity of these two bound states prevents any resolution between them using the scale of Fig. 5.4. Obviously, the same result is obtained for $\boldsymbol{K} = (K, \pi, K)$ and $(K, K, \pi)$.

Now we look for the bound states along the edges of the cube. Assuming $\boldsymbol{K} = (K, 0, 0)$, we have

$$B_{xx} = -\frac{1}{\pi^2}\int_0^\pi dq_x \int_0^\pi dq_y \frac{\cos q_x\left(\cos\frac{K}{2} - \cos q_x\right)}{\sqrt{\left(x + \cos\frac{K}{2}\cos q_x + \cos q_y\right)^2 - 1}}, \tag{5.4.21}$$

$$B_{yy} = B_{zz} = -\frac{1}{\pi^2}\int_0^\pi dq_y \int_0^\pi dq_z \frac{\cos q_y\left(1 - \cos q_y\right)}{\sqrt{\left(x + \cos q_y + \cos q_z\right)^2 - \cos^2\frac{K}{2}}}, \tag{5.4.22}$$

$$B_{xy} = B_{xz} = -\frac{1}{\pi^2}\int_0^\pi dq_x \int_0^\pi dq_y \frac{\cos q_x\left(1 - \cos q_y\right)}{\sqrt{\left(x + \cos\frac{K}{2}\cos q_x + \cos q_y\right)^2 - 1}}, \tag{5.4.23}$$

$$B_{yx} = B_{zx} = -\frac{1}{\pi^2}\int_0^\pi dq_x \int_0^\pi dq_y \frac{\cos q_y\left(\cos\frac{K}{2} - \cos q_x\right)}{\sqrt{\left(x + \cos\frac{K}{2}\cos q_x + \cos q_y\right)^2 - 1}}, \tag{5.4.24}$$

$$B_{yz} = B_{zy} = -\frac{1}{\pi^2}\int_0^\pi dq_y \int_0^\pi dq_z \frac{\cos q_y\,(1-\cos q_z)}{\sqrt{(x+\cos q_y+\cos q_z)^2-\cos^2\frac{K}{2}}}. \tag{5.4.25}$$

The bound state equation (5.1.4) reduces to

$$\left[1-\frac{1}{2S}(B_{yy}-B_{yz})\right]\left\{\left(1-\frac{1}{2S}B_{xx}\right)\left[1-\frac{1}{2S}(B_{yy}+B_{yz})\right]-\frac{1}{2S^2}B_{xy}B_{yx}\right\}=0. \tag{5.4.26}$$

A numerical evaluation of the integrals (5.5.21)–(5.5.25) leads to the conclusion that both equations

$$1-\frac{1}{2S}(B_{yy}-B_{yz})=0 \tag{5.4.27}$$

and

$$\left(1-\frac{1}{2S}B_{xx}\right)\left[1-\frac{1}{2S}(B_{yy}+B_{yz})\right]-\frac{1}{2S^2}B_{xy}B_{yx}=0 \tag{5.4.28}$$

have no solution for any $K$, so that no bound state exists along the $(1,0,0)$-direction as illustrated by the third panel of Fig. 5.4. The same result is obtained for $\boldsymbol{K} = (0,K,0)$ and $(0,0,K)$.

We conclude this section by evaluating the bound states for $\boldsymbol{K} = (0,K,K)$ for which $B_{yy} = B_{zz}$, $B_{xy} = B_{xz}$, $B_{yx} = B_{zx}$ and $B_{yz} = B_{zy}$. The bound state equation (5.1.4) becomes identical to Eq. (5.4.26) but with different $B_{\alpha\beta}$. In particular, we have

$$B_{xx} = -\frac{1}{\pi^2}\int_0^\pi dq_x \int_0^\pi dq_y \frac{\cos q_x\,(1-\cos q_x)}{\sqrt{\left(x+\cos q_x+\cos\frac{K}{2}\cos q_y\right)^2-\cos^2\frac{K}{2}}}, \tag{5.4.29}$$

$$B_{xy} = -\frac{1}{\pi^2}\int_0^\pi dq_x \int_0^\pi dq_y \frac{\cos q_x\left(\cos\frac{K}{2}-\cos q_y\right)}{\sqrt{\left(x+\cos q_x+\cos\frac{K}{2}\cos q_y\right)^2-\cos^2\frac{K}{2}}}, \tag{5.4.30}$$

$$B_{yx} = -\frac{1}{\pi^2}\int_0^\pi dq_x \int_0^\pi dq_y \frac{\cos q_y\,(1-\cos q_x)}{\sqrt{\left(x+\cos q_x+\cos\frac{K}{2}\cos q_y\right)^2-\cos^2\frac{K}{2}}}, \tag{5.4.31}$$

$$B_{yy}-B_{yz} = -\frac{1}{\pi^2}\int_0^\pi dq_y \int_0^\pi dq_z \frac{\cos q_y\,(\cos q_z-\cos q_y)}{\sqrt{\left[x+\cos\frac{K}{2}(\cos q_y+\cos q_z)\right]^2-1}}, \tag{5.4.32}$$

$$B_{yy}+B_{yz} = -\frac{1}{\pi^2}\int_0^\pi dq_y \int_0^\pi dq_z \frac{\cos q_y\left(2\cos\frac{K}{2}-\cos q_y-\cos q_z\right)}{\sqrt{\left[x+\cos\frac{K}{2}(\cos q_y+\cos q_z)\right]^2-1}} \tag{5.4.33}$$

with $-3 < x < -1 - 2\cos\frac{K}{2}$. For $K = \pi$, one obtains $B_{xy} = B_{yx} = 0$,

$$B_{xx} = (1+x)\left(1 + \frac{x}{\sqrt{x^2-1}}\right), \quad B_{yy} - B_{yz} = B_{yy} + B_{yz} = \frac{1}{2\sqrt{x^2-1}} \tag{5.4.34}$$

and Eq. (5.4.26) becomes

$$\left(1 - \frac{1}{4S\sqrt{x^2-1}}\right)^2 \left[1 - \frac{1}{2S}(1+x)\left(1 + \frac{x}{\sqrt{x^2-1}}\right)\right] = 0. \tag{5.4.35}$$

The first factor of Eq. (5.4.35) vanishes for

$$x = -\sqrt{1 + \frac{1}{16S^2}} \tag{5.4.36}$$

while the second factor does not vanish for any $-3 < x < -1$. The energy of the doubly degenerate bound state is

$$\hbar\omega_{\mathrm{BS}}(0,\pi,\pi) = 24JS\left(1 - \frac{1}{3}\sqrt{1 + \frac{1}{16S^2}}\right). \tag{5.4.37}$$

Moving away from $(0,\pi,\pi)$, the degenerate bound state splits into two distinct bound states: the first is the solution of Eq. (5.4.27) and exists for $K_c^{(1)} < K < \pi$ with $K_c^{(1)} = 0.96011\pi$ for $S = 1/2$; the second bound state is the solution of Eq. (5.4.28) and exists for $K_c^{(2)} < K < \pi$ with $K_c^{(2)} = 0.90227\pi$ for $S = 1/2$. The bound states for $\boldsymbol{K} = (0, K, K)$ are shown as continuous curves in the last panel of Fig. 5.4 along with the top and bottom of the two-magnon band (dashed curves). Obviously, the same result is obtained from the equivalent directions $\boldsymbol{K} = (K, 0, K)$ and $(K, K, 0)$.

In summary, the four panels of Fig. 5.4 show that the bound states of a SC isotropic ferromagnet are comprised between 0 and 3 depending on the wavevector $\boldsymbol{K}$. The novelty with respect to the results obtained for the 1D and 2D case is that in 3D, a wide region around $\boldsymbol{K} = (0,0,0)$ exists free of bound states. On the other hand, in a small region around $\boldsymbol{K} = (\pi,\pi,\pi)$, three bound states exist. Regions with one or two bound states are thin shells in the vicinity of the faces of the cube of the BZ. It should be noted that the qualitative shape of the regions with 0, 1, 2, 3 bound states was sketched by Wortis[39] in his fundamental paper *Bound states of two spin waves in the Heisenberg ferromagnet.* Figure 7 of that paper (reproduced by many textbooks on magnetism) is not quantitatively correct. Indeed, the region with zero bound states should be extended to the corners of the cube at the expense of the one bound state region which should be reduced to a thin space between the regions with zero and two bound states, respectively. This misrepresentation was induced by Fig. 6 of the same paper where the two bound states existing for any $K$ along the $(K, K, \pi)$ direction are clearly split off from each other and one of them (the lower in energy) does not merge into the continuum for $K \to 0$ in disagreement with the second panel of Fig. 5.4.

Concluding this section, we summarize the scenario of the two-magnon bound states in an isotropic Heisenberg ferromagnet: in all dimensions, the bound states exist only below the two-magnon band:

i) in 1D (linear chain) only one bound state exists for any wavevector $0 < K < \pi$;
ii) in 2D (square lattice[39] and triangular lattice[40]), one bound state exists for any wavevector, a second bound state exists only in a shell of the BZ near the boundaries;
iii) in 3D (simple cubic,[39] body centred cubic and face centred cubic[41] lattices), no bound state exists in a wide region around the zone centre while three bound states exist in a small region close to the zone corner $\boldsymbol{K} = (\pi, \pi, \pi)$.

## 5.5. Bound States in Anisotropic Ferromagnets

We have obtained in Sec. 1.7 the determinant equation (1.7.17) for a "cubic" anisotropic ferromagnet in the presence of single-ion easy-axis anisotropy ($D > 0$) and exchange easy-axis anisotropy ($J^z \geq J^\perp$)

$$\det \begin{pmatrix} 1 + \frac{D}{4J^\perp S} I_0 & \frac{J^z}{2J^\perp S} I_x & \frac{J^z}{2J^\perp S} I_y & \frac{J^z}{2J^\perp S} I_z \\ \frac{D}{4J^\perp S} A_x & 1 - \frac{1}{2S} B_{xx} & -\frac{1}{2S} B_{xy} & -\frac{1}{2S} B_{xz} \\ \frac{D}{4J^\perp S} A_y & -\frac{1}{2S} B_{yx} & 1 - \frac{1}{2S} B_{yy} & -\frac{1}{2S} B_{yz} \\ \frac{D}{4J^\perp S} A_z & -\frac{1}{2S} B_{zx} & -\frac{1}{2S} B_{zy} & 1 - \frac{1}{2S} B_{zz} \end{pmatrix} = 0, \qquad (5.5.1)$$

where $I_0$, $I_\alpha$, $A_\alpha$ and $B_{\alpha\beta}$ are given by Eqs. (1.7.13)–(1.7.15) and (1.7.18)–(1.7.19). As one can see from Eq. (5.5.1), the exchange anisotropy alone does not change the number of roots of the determinant equation: indeed for $D = 0$, Eq. (5.5.1) reduces to

$$\det \left[ \mathbf{1} - \frac{1}{2S} \mathbf{B}(\boldsymbol{K}, \omega) \right] = 0 \qquad (5.5.2)$$

that looks similar to the determinant equation obtained for the isotropic case given in Eq. (5.1.4). However, the matrix elements differ from the isotropic case because of the factor $J^z/J^\perp$ in $B_{\alpha\beta}$ given by Eq. (1.7.19). In any case, the number of bound states is the same as in the isotropic case even though their energy is changed. On the contrary, the single-ion anisotropy increases the dimension of the determinant entering a new bound state. At the ZC ($K_\alpha = \pi$), one has $I_0 = \frac{1}{x}$, $I_\alpha = 0$, $A_\alpha = 0$ and $B_{\alpha\beta} = -\frac{J^z}{2J^\perp x} \delta_{\alpha,\beta}$ so that Eq. (5.5.1) reduces to

$$\left( 1 + \frac{D}{4J^\perp Sx} \right) \left( 1 + \frac{J^z}{4J^\perp Sx} \right)^3 = 0 \qquad (5.5.3)$$

and from Eq. (1.7.11), the bound state energies become

$$\hbar\omega_{\rm BS}^{(\rm s)}(\pi,\pi,\pi) = 2h + 2D(2S-1) + 8J^{\perp}S - 2D \tag{5.5.4}$$

with $G^0_{\pi,\pi,\pi} = 1$, $G^\alpha_{\pi,\pi,\pi} = 0$ ($\alpha = x, y, z$) and

$$\hbar\omega_{\rm BS}^{(\rm ex)}(\pi,\pi,\pi) = 2h + 2D(2S-1) + 8J^{\perp}S - 2J^z \tag{5.5.5}$$

with $G^0_{\pi,\pi,\pi} = 0$, $G^\alpha_{\pi,\pi,\pi} = 1$ ($\alpha = x, y$ or $z$). The superscripts "s" and "ex" stay for "single-ion" and "exchange" bound states.[42] Using Eqs. (1.7.10) and (1.4.11), one has

$$F_{\boldsymbol{K}}(\boldsymbol{r}) = -\frac{1}{N}\sum_{\boldsymbol{q}}\left(\frac{D}{4J^{\perp}S}G^0_{\boldsymbol{K}} + \frac{J^z}{2J^{\perp}S}\sum_\alpha G^\alpha_{\boldsymbol{K}}\cos q_\alpha\right)\frac{\cos\boldsymbol{q}\cdot\boldsymbol{r}}{x + \sum_\sigma \cos\frac{K_\sigma}{2}\cos q_\sigma} \tag{5.5.6}$$

so that

$$F^{(\rm s)}_{\pi,\pi,\pi}(\boldsymbol{r}) = \frac{1}{N}\sum_{\boldsymbol{q}}\cos\boldsymbol{q}\cdot\boldsymbol{r} = \delta_{\boldsymbol{r},0} \tag{5.5.7}$$

for the single-ion bound state (5.5.4) and

$$F^{(\rm ex)}_{\pi,\pi,\pi}(\boldsymbol{r}) = \frac{2}{N}\sum_{\boldsymbol{q}}\cos\boldsymbol{q}\cdot\boldsymbol{\delta}\cos\boldsymbol{q}\cdot\boldsymbol{r} = \delta_{\boldsymbol{r},\boldsymbol{\delta}} \tag{5.5.8}$$

for the triply degenerate exchange bound state (5.5.5). In Eq. (5.5.8), $\boldsymbol{\delta}$ is a vector connecting two NN spins. The probability of finding the two spin deviations on the same lattice site is obtained from Eq. (1.4.34) that reads

$$P^{(\rm s)}_{\pi,\pi,\pi}(0) = 1 \tag{5.5.9}$$

for the single-ion bound state and

$$P^{(\rm ex)}_{\pi,\pi,\pi}(0) = 0 \tag{5.5.10}$$

for the exchange bound state. On the other hand, the probability of finding the two spin deviations on lattice sites connected by a vector $\boldsymbol{r}$ is obtained from Eq. (1.4.35) that reads

$$P^{(\rm s)}_{\pi,\pi,\pi}(\boldsymbol{r}) = 0 \tag{5.5.11}$$

for the single-ion bound state and

$$P^{(\rm ex)}_{\pi,\pi,\pi}(\boldsymbol{r}) = \delta_{\boldsymbol{r},\boldsymbol{\delta}} \tag{5.5.12}$$

for the exchange bound state. From Eqs. (5.5.9) and (5.5.11), one sees that the single-ion bound state corresponds to a state with two spin deviations on the same site. Analogously from Eqs. (5.5.10) and (5.5.12), one sees that the exchange bound state corresponds to a state with two spin deviations on NN sites. This classification is fully justified at the ZC but it becomes less meaningful as one moves away from the ZC. To see this, we will perform explicit calculations for the 1D case for which Eqs. (5.5.1) and (5.5.6) reduce to

$$f(x,K) \equiv \det \begin{pmatrix} 1+\frac{D}{4J^{\perp}S}I_0 & \frac{J^z}{2J^{\perp}S}\frac{1-xI_0}{\cos\frac{K}{2}} \\ \frac{D}{4J^{\perp}S}\frac{1-xI_0-\frac{J^z}{J^{\perp}}I_0\cos^2\frac{K}{2}}{\cos\frac{K}{2}} & 1-\frac{1}{2S}\left(1+\frac{J^z}{J^{\perp}}\frac{x}{\cos^2\frac{K}{2}}\right)(1-xI_0) \end{pmatrix} = 0 \quad (5.5.13)$$

and

$$F_K(r) = -\frac{1}{\pi}\int_0^{\pi} dq \frac{\cos(r\,q)}{x_{\rm BS}+\cos\frac{K}{2}\cos q}\left(\frac{D}{4J^{\perp}S}G_K^0+\frac{J^z}{2J^{\perp}S}G_K^1\ \cos q\right), \quad (5.5.14)$$

respectively. In Eq. (5.5.14), we have written explicitly $x_{\rm BS}$ instead of $x$ to stress that it is a root of the determinant equation (5.5.13). For $x < -\cos\frac{K}{2}$ (bound states below the two magnon band) one has

$$I_0 = -\frac{1}{\sqrt{x^2-\cos^2\frac{K}{2}}} \quad (5.5.15)$$

and Eq. (5.5.13) has two roots for $K_c < K \le \pi$ and one root for $0 < K \le K_c$ where

$$K_c = 2\arccos\left[\frac{D+2J^z}{4J^{\perp}} - \sqrt{\left(\frac{D+2J^z}{4J^{\perp}}\right)^2 - \frac{D\,J^z}{4(J^{\perp})^2S}}\right]. \quad (5.5.16)$$

Equation (5.5.16) is obtained by evaluating Eq. (5.5.13) for $x = -\cos\frac{K}{2} - \epsilon$ and taking the limit $\epsilon \to 0$. For $\epsilon \to 0$, Eq. (5.5.13) becomes

$$\frac{A}{\sqrt{2\epsilon}} + B + O(\sqrt{\epsilon}) = 0 \quad (5.5.17)$$

where

$$A = \frac{\cos^2\frac{K}{2} - \frac{D+2J^z}{2J^{\perp}}\cos\frac{K}{2} + \frac{D\,J^z}{4(J^{\perp})^2S}}{2S\cos^{3/2}\frac{K}{2}} \quad (5.5.18)$$

and

$$B = \frac{(2S-1)\cos^2\frac{K}{2} + \frac{J^z}{J^{\perp}}\cos\frac{K}{2} - \frac{D\,J^z}{4(J^{\perp})^2S}}{2S\cos^2\frac{K}{2}}. \quad (5.5.19)$$

For $\epsilon \to 0$, the left-hand side of Eq. (5.5.17) diverges to $+\infty$ for $A > 0$ ($K > K_c$) and to $-\infty$ for $A < 0$ ($K < K_c$) where $K_c$ is the root of the equation $A = 0$. For $K = K_c$ ($A = 0$), Eq. (5.5.19) goes to a finite value $B(K_c) < 0$. These results imply that the function $f(x, K)$ of Eq. (5.5.13) crosses twice the $x$-axis for $K > K_c$ starting from 1 when $x \to -\infty$, becoming negative in a region $x_1 < x < x_2$ and diverging to $+\infty$ for $x \to -\cos\frac{K}{2}$. On the contrary, for $K < K_c$ the function $f(x, K)$ crosses the

$x$-axis in one point $x_1$ going from 1 when $x \to -\infty$ to $-\infty$ in the limit $x \to -\cos\frac{K}{2}$. For $K = K_c$, the function $f(x, K_c)$ does not diverge for $x \to -\cos\frac{K_c}{2}$: it goes to the finite value $B(K_c) < 0$ and crosses the $x$-axis only once at the point $x_1$.

For $x > \cos\frac{K}{2}$ (possible bound states above the two-magnon band), one has

$$I_0 = \frac{1}{\sqrt{x^2 - \cos^2\frac{K}{2}}} \tag{5.5.20}$$

and the function $f(x, K)$ never crosses the $x$-axis being a monotonic function of $x$ going from $+\infty$ for $x \to \cos\frac{K}{2}$ to 1 for $x \to \infty$. This means that Eq. (5.5.13) has no solution for $x > \cos\frac{K}{2}$, preventing the presence of bound states above the continuum. From Eqs. (5.2.12) and (5.5.14), one obtains

$$F_K(r) = -\frac{1}{\cos\frac{K}{2}}\left\{\frac{J^z}{2J^\perp S}\delta_{r,0}G_K^1 + \left(\frac{D\cos\frac{K}{2}}{4J^\perp Sx}G_K^0 - \frac{J^z}{2J^\perp S}G_K^1\right)\right.$$
$$\left.\times\frac{1}{\sqrt{1 - \frac{\cos^2\frac{K}{2}}{x_{\mathrm{BS}}^2}}}\left[-\frac{x_{\mathrm{BS}}}{\cos\frac{K}{2}}\left(1 - \sqrt{1 - \frac{\cos^2\frac{K}{2}}{x_{\mathrm{BS}}^2}}\right)\right]^{|r|}\right\}. \tag{5.5.21}$$

The constants $G_K^0$ and $G_K^1$ are the two components of the eigenvector belonging to the eigenvalue $x_{\mathrm{BS}}$ according to the determinant equation (5.5.13). Writing $G_K^0$ and $G_K^1$ in terms of the arbitrary constant $F_K(1)$, Eq. (5.5.21) becomes

$$F_K(0)\left(x_{\mathrm{BS}} + \frac{D}{4J^\perp S}\right) = -F_K(1)\cos\frac{K}{2} \tag{5.5.22}$$

for $r = 0$ and

$$F_K(r) = F_K(1)\left(-\frac{x_{\mathrm{BS}} + \sqrt{x_{\mathrm{BS}}^2 - \cos^2\frac{K}{2}}}{\cos\frac{K}{2}}\right)^{|r|-1} \tag{5.5.23}$$

for $|r| \geq 2$. The probability of finding two spin deviations on the same site (1.4.34) or on different sites (1.4.35) becomes

$$P_K(0) = \frac{\left(1 - \frac{1}{2S}\right)\cos^2\frac{K}{2}\left[1 - \frac{x_{\mathrm{BS}}^2}{\cos^2\frac{K}{2}}\left(1 - \sqrt{1 - \frac{\cos^2\frac{K}{2}}{x_{\mathrm{BS}}^2}}\right)^2\right]}{\left(1 - \frac{1}{2S}\right)\cos^2\frac{K}{2}\left[1 - \frac{x_{\mathrm{BS}}^2}{\cos\frac{K}{2}}\left(1 - \sqrt{1 - \frac{\cos^2\frac{K}{2}}{x_{\mathrm{BS}}^2}}\right)^2\right] + \left(x_{\mathrm{BS}} + \frac{D}{4J^\perp S}\right)^2} \tag{5.5.24}$$

and

$$P_K(r)$$
$$= \frac{\left(x_{\rm BS}+\frac{D}{4J^\perp S}\right)^2\left[1-\frac{x_{\rm BS}^2}{\cos^2\frac{K}{2}}\left(1-\sqrt{1-\frac{\cos^2\frac{K}{2}}{x_{\rm BS}^2}}\right)^2\right]\left[\frac{x_{\rm BS}}{\cos\frac{K}{2}}\left(1-\sqrt{1-\frac{\cos^2\frac{K}{2}}{x_{\rm BS}^2}}\right)\right]^{2(r-1)}}{\left(1-\frac{1}{2S}\right)\cos^2\frac{K}{2}\left[1-\frac{x_{\rm BS}^2}{\cos\frac{K}{2}}\left(1-\sqrt{1-\frac{\cos^2\frac{K}{2}}{x_{\rm BS}^2}}\right)^2\right]+\left(x_{\rm BS}+\frac{D}{4J^\perp S}\right)^2} \tag{5.5.25}$$

for $r \geq 1$. Notice that for $S = 1/2$, the probability (5.5.24) vanishes and the probability (5.5.25) becomes independent of $D$. This result should have been expected since the double occupancy is forbidden for $S = 1/2$ and the single-ion contribution to the anisotropy Hamiltonian reduces to a constant $-D\sum_i (S_i^z)^2 = -\frac{1}{4}DN$ so that $D$ cannot occur in the spin dynamics.

For $K \to \pi$, we may expand Eq. (5.5.13) in powers of $\cos\frac{K}{2}$ and the solutions are given by

$$\begin{aligned} x_{\rm BS}^{(1)} = -\frac{J^z}{4J^\perp S}\Bigg\{1 &+ \frac{4(J^\perp)^2 S\left[(3S-1)J^z - DS\right]}{(J^z)^2(J^z-D)}\cos^2\frac{K}{2} \\ &+ \frac{16(J^\perp)^4 S^2(2S-1)\left[3DJ^zS - D^2S - (4S-1)(J^z)^2\right]}{(J^z)^3(J^z-D)^3}\cos^4\frac{K}{2} \\ &+ O\left(\cos^6\frac{K}{2}\right)\Bigg\} \end{aligned} \tag{5.5.26}$$

and

$$\begin{aligned} x_{\rm BS}^{(2)} = -\frac{D}{4J^\perp S}\Bigg\{1 &- \frac{4(J^\perp)^2 S(2S-1)}{D(J^z-D)}\cos^2\frac{K}{2} \\ &+ \frac{16(J^\perp)^4 S^2(2S-1)\left[D(S-1)+J^zS\right]}{D^2(J^z-D)^3}\cos^4\frac{K}{2} + O\left(\cos^6\frac{K}{2}\right)\Bigg\}. \end{aligned} \tag{5.5.27}$$

The probability of finding two spin deviations on the same site is given by

$$\begin{aligned} P_{K\to\pi}^{(1)}(0) = {} & \frac{8(J^\perp)^2 S(2S-1)}{(J^z-D)^2}\cos^2\frac{K}{2} \\ & - \frac{32(J^\perp)^4 S^2(2S-1)[D^2S - 4DJ^zS + (11S-4)(J^z)^2]}{(J^z)^2(J^z-D)^4}\cos^4\frac{K}{2} \\ & + O\left(\cos^6\frac{K}{2}\right) \end{aligned} \tag{5.5.28}$$

for the bound state given by Eq. (5.5.26) and

$$P^{(2)}_{K\to\pi}(0) = 1 - \frac{2(J^\perp)^2 S(2S-1)}{(J^z-D)^2}\cos^2\frac{K}{2}$$
$$+\frac{4(J^\perp)^4 S^2(2S-1)[D^2(4S-5)+8DJ^zS-2(J^z)^2S]}{D^2(J^z-D)^4}\cos^4\frac{K}{2}$$
$$+O\left(\cos^6\frac{K}{2}\right) \qquad (5.5.29)$$

for the bound state given by Eq. (5.5.27). Obviously, the probability of finding the two spin deviations on distinct lattice sites is $\sum_{r\neq 0} P_K(r) = 1 - P_K(0)$. As one can see from Eqs. (5.5.28) and (5.5.29), the bound state $x^{(1)}_{\mathrm{BS}}$ is an exchange bound state since $P^{(1)}_K(0) \to 0$ for $K \to \pi$ while the bound state $x^{(2)}_{\mathrm{BS}}$, is a single-ion bound state since $P^{(2)}_K(0) \to 1$ for $K \to \pi$. As previously announced, the exchange or single-ion nature of the bound state is rigorously confirmed only for $K = \pi$. Moving away from the zone boundary, a mixing between the exchange and single-ion characters occurs.

In the absence of single-ion anisotropy ($D = 0$), a single bound state exists. In particular, for $S = 1/2$, the energy of the two-magnon bound state is given by

$$\hbar\omega_{\mathrm{BS}} = 2h + 2J^z\left[1 - \left(\frac{J^\perp}{J^z}\right)^2\cos^2\frac{K}{2}\right] \qquad (5.5.30)$$

that in the Heisenberg limit ($J^z = J^\perp$) reduces to the isotropic result given in Eq. (5.2.6) and in the Ising limit ($J^\perp = 0$) reduces to

$$\hbar\omega_{\mathrm{BS}} = 2h + 2J^z. \qquad (5.5.31)$$

The result (5.5.31) represents the energy cost of a couple of NN spin reversals. The bound state (5.5.30) is always below the bottom of the two-magnon band

$$\hbar\omega_{\mathrm{bottom}} = 2h + 4J^z\left(1 - \frac{J^\perp}{J^z}\cos\frac{K}{2}\right) \qquad (5.5.32)$$

and crosses the single-spin wave spectrum

$$\hbar\omega = h + 2J^z\left(1 - \frac{J^\perp}{J^z}\cos k\right) \qquad (5.5.33)$$

at

$$K_{\mathrm{cross}} = \arccos\frac{J^\perp - h\frac{J^z}{J^\perp}}{2J^z - J^\perp}. \qquad (5.5.34)$$

For $K < K_{\mathrm{cross}}$, the single-spin wave excitation is the lowest excited state while for $K > K_{\mathrm{cross}}$, the bound state is the lowest excitation. In the Heisenberg limit, $J^\perp = J^z$, one has $K_{\mathrm{cross}} = \arccos(1 - h/J^z)$ and the bound state becomes the lowest excitation for any $0 < K < \pi$ only when $h = 0$. The result for $S \neq 1/2$ is qualitatively similar to the result obtained for $S = 1/2$ even though only numerical results can be obtained.

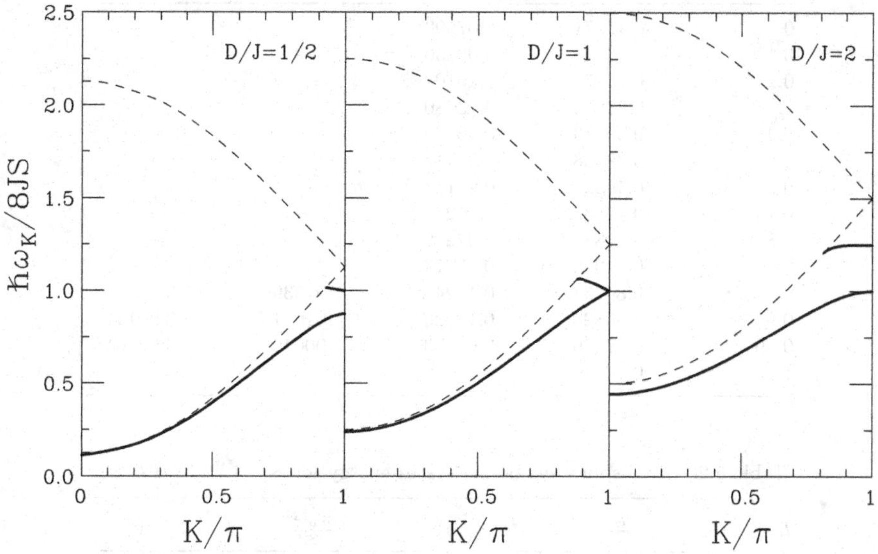


Fig. 5.5. Two-magnon bound state energy versus wavevector for the LC with $S = 1$ and several values of the single-ion anisotropy $D/J = 0.5$, 1 and 2. The light dashed lines represent the bottom and the top of the two-magnon band.

In the absence of exchange anisotropy ($J^\perp = J^z = J$), one bound state exists over the whole BZ and another bound state appears close to the zone boundary. The scenario for $S = 1$ and $D/J = 0.5, 1, 2$ is shown in Fig. 5.5. The wavevector $K_c$ at which the second bound state occurs is given by $K_c/\pi = 0.93008$, 0.87766 and 0.81076 for $D/J = 0.5, 1$ and 2, respectively. The interval $K_c < K < \pi$ of wavevectors in which the second bound state exists increases as the single-ion anisotropy increases. In Tables 5.1–5.3, we give the numerical values of the energy eigenvalues of the bound states together with the associated probability of finding the two spin deviations on the same site for $D/J = 0.5$, 1 and 2, respectively. As one can see from Table 5.1, the bound state (1) existing over the whole BZ has a prevalent exchange character. Indeed, the associated probability of finding the two spin deviations on the same site is less than 1/3 for any $K$ reaching its maximum at $K \simeq 0.8\pi$. On the contrary, the bound state (2) restricted to the region $0.93008\pi < K < \pi$ shows a prevalent single-ion character since $P_K^{(2)} \sim 1$ except when it is very near to $K_c$ where the bound state enter the two-magnon band. An unexpected behaviour is seen in Table 5.2 for $D/J = 1$ where even at the ZB, where the two bound states are degenerate, the pure single-ion or exchange character does not occur. Both bound states are of prevalent single-ion character since they have a probability $P_\pi^{(1)}(0) = P_\pi^{(2)}(0) = 2/3$. Moving from $K = \pi$, both bound states maintain their prevalent single-ion character: The bound state (1) recovers the exchange nature for $K \lesssim 0.8\pi$. Note

Table 5.1. Two-spin wave bound states energy for $S = 1$ and $D/J = 0.5$.

| $K/\pi$ | $\frac{\hbar\omega^{(1)}}{8J}$ | $P_K^{(1)}(0)$ | $\frac{\hbar\omega^{(2)}}{8J}$ | $P_K^{(2)}(0)$ |
|---|---|---|---|---|
| 0 | 0.12271 | 0.07592 | — | — |
| 0.1 | 0.13460 | 0.08230 | — | — |
| 0.2 | 0.16975 | 0.10102 | — | — |
| 0.3 | 0.22662 | 0.13080 | — | — |
| 0.4 | 0.30273 | 0.16960 | — | — |
| 0.5 | 0.39468 | 0.21447 | — | — |
| 0.6 | 0.49825 | 0.26123 | — | — |
| 0.7 | 0.60839 | 0.30273 | — | — |
| 0.8 | 0.71882 | 0.32219 | — | — |
| 0.9 | 0.81955 | 0.25774 | — | — |
| 0.93008 | 0.84453 | 0.19240 | 1.01539 | 0 |
| 0.95 | 0.85824 | 0.13020 | 1.01021 | 0.96141 |
| 0.99 | 0.87426 | 0.00773 | 1.00049 | 0.99805 |
| 1 | 0.875 | 0 | 1 | 1 |

Table 5.2. Two-spin wave bound states energy for $S = 1$ and $D/J = 1$.

| $K/\pi$ | $\frac{\hbar\omega^{(1)}}{8J}$ | $P_K^{(1)}(0)$ | $\frac{\hbar\omega^{(2)}}{8J}$ | $P_K^{(2)}(0)$ |
|---|---|---|---|---|
| 0 | 0.23931 | 0.17961 | — | — |
| 0.1 | 0.25073 | 0.18636 | — | — |
| 0.2 | 0.28447 | 0.20628 | — | — |
| 0.3 | 0.33904 | 0.23843 | — | — |
| 0.4 | 0.41199 | 0.28137 | — | — |
| 0.5 | 1/2 | 1/3 | — | — |
| 0.6 | 0.59901 | 0.39244 | — | — |
| 0.7 | 0.70430 | 0.45686 | — | — |
| 0.8 | 0.81052 | 0.52500 | — | — |
| 0.87766 | 0.88983 | 0.57962 | 1.05902 | 0 |
| 0.9 | 0.91162 | 0.59552 | 1.06029 | 0.70869 |
| 0.95 | 0.95799 | 0.63123 | 1.03569 | 0.69970 |
| 0.99 | 0.99203 | 0.65965 | 1.00773 | 0.67361 |
| 1 | 1 | 2/3 | 1 | 2/3 |

that the expansions (5.5.26)–(5.5.29) for $D = J^{\perp} = J^z$ and $S = 1$ have to be replaced by

$$x_{\mathrm{BS}}^{(1)} = -\frac{1}{4}\left(1 + 2\cos\frac{K}{2} + 2\cos^2\frac{K}{2} - 3\cos^3\frac{K}{2} + \cdots\right), \tag{5.5.35}$$

$$x_{\mathrm{BS}}^{(2)} = -\frac{1}{4}\left(1 - 2\cos\frac{K}{2} + 2\cos^2\frac{K}{2} + 3\cos^3\frac{K}{2} + \cdots\right), \tag{5.5.36}$$

$$P_{K\to\pi}^{(1)}(0) = \frac{2}{3} - \frac{4}{9}\cos\frac{K}{2} - \frac{4}{27}\cos^2\frac{K}{2} + \frac{230}{81}\cos^3\frac{K}{2} + \cdots \tag{5.5.37}$$

and

$$P_{K\to\pi}^{(2)}(0) = \frac{2}{3} + \frac{4}{9}\cos\frac{K}{2} - \frac{4}{27}\cos^2\frac{K}{2} - \frac{230}{81}\cos^3\frac{K}{2} + \cdots. \tag{5.5.38}$$

Table 5.3. Two-spin wave bound states energy for $S = 1$ and $D/J = 2$.

| $K/\pi$ | $\frac{\hbar\omega^{(1)}}{8J}$ | $P_K^{(1)}(0)$ | $\frac{\hbar\omega^{(2)}}{8J}$ | $P_K^{(2)}(0)$ |
|---|---|---|---|---|
| 0 | 0.44331 | 0.44074 | — | — |
| 0.1 | 0.45347 | 0.44795 | — | — |
| 0.2 | 0.48345 | 0.46941 | — | — |
| 0.3 | 0.53166 | 0.50458 | — | — |
| 0.4 | 0.59541 | 0.55279 | — | — |
| 0.5 | 0.67135 | 0.61337 | — | — |
| 0.6 | 0.75458 | 0.68596 | — | — |
| 0.7 | 0.83921 | 0.77038 | — | — |
| 0.8 | 0.91723 | 0.86479 | — | — |
| 0.81076 | 0.92479 | 0.87522 | 1.20711 | 0 |
| 0.85 | 0.95023 | 0.91270 | 1.23675 | 0.23507 |
| 0.9 | 0.97660 | 0.95618 | 1.24756 | 0.14936 |
| 0.95 | 0.99392 | 0.98805 | 1.24985 | 0.04583 |
| 0.99 | 0.99975 | 0.99951 | 1.25000 | 0.00197 |
| 1 | 1 | 1 | 1.25 | 0 |

From Table 5.3, we see that the bound state (1) existing over the whole BZ, has a prevalent single-ion character while the bound state (2) restricted to $0.81076\pi < K < \pi$ has a prevalent exchange character. This may be explained noticing that for large single-ion anisotropy ($D > J$), the lowest bound state becomes the single-ion one at zone boundary and it extends over the whole BZ.

The single-ion anisotropy also enters a new bound state in 2D and 3D. Numerical results were performed for a SC lattice[42,43] with $J^\perp = J^z = J$ and $S = 1$ along the high symmetry direction $\boldsymbol{K} = (K, K, K)$: a new "single-ion" bound state appears in addition to the three "exchange" bound states existing in the isotropic case (see the first panel of Fig. 5.4). For small anisotropy, let us say $D \lesssim J$, the four bound states exist in a restricted region close to the ZC. As the anisotropy increases, the "single-ion" bound state exists over a wider region of wavevectors extending towards $K = 0$. As for the 1D case, for a single-ion anisotropy $D > J$, the single-ion bound state is lower than the exchange bound state since at the ZC, the state with two deviations on the same site (single-ion bound state) has a lower energy: the energy cost of such a state with a spin reversal is only due to the exchange energy; indeed for $S = 1$, the anisotropy energy does not change when the spin is reversed since for $S_i^z = \pm 1$, one has $-D(S_i^z)^2 = -D$. The reduced dispersion of the spectrum observed for large anisotropy confirms that the excitation is of Ising-type. For very large anisotropy $D \sim 10 - 20J$, the single-ion bound state extends over the whole range $0 < K < \pi$. This peculiarity is the most striking discrepancy with respect to the isotropic case where the existence of bound states around the zone centre is prevented. For large uniaxial anisotropy, the "single-ion" bound state may become the lowest excitation crossing the single-spin wave energy spectrum. The occurrence of a bound state below the two-magnon band improves the chance of an experimental check.

In 2D, the bound states are evaluated for the TR lattice in the isotropic case[40] and for the anisotropic model[43] with $S = 1$, $D = 7.41J^z$, $J^\perp = 0.71J^z$, a choice of parameters used to describe the actual compound $FeCl_2$, a triangular anisotropic ferromagnet in which the antiferromagnetic coupling between planes is an order of magnitude smaller than the ferromagnetic coupling in the triangular planes. The 2D case is qualitatively similar to the 1D case since at least one bound state exists for any wavevector. Results qualitatively similar to the SQ lattice shown in Fig. 5.2 are obtained in the TR isotropic ferromagnet.[40] The single-ion anisotropy enters one more bound state: The most important effect, however, is that the anisotropy splits the bound state existing over the whole BZ in the isotropic case, well below the two-magnon band with an increased chance of experimental check.

The study of the bound states has also been extended to the case of Heisenberg Hamiltonian with NNN interaction. In the case of a LC with NNN interaction,[44] the result is similar to that obtained in Section 5.2 until the NNN interaction becomes negative and large enough to change the ferromagnetic ground state into a non-collinear ground state.

Calculation of three-magnon bound states[45] have been performed in a SQ lattice with $S = 1/2$ for $\boldsymbol{K} = (K, K)$ for an isotropic and anisotropic Hamiltonian. The existence of four three-magnon bound states was proved in a region close to the ZC ($K \gtrsim 0.76\pi$).

An interesting question is about the experimental measure of the bound states. Since a direct measure of the bound state seems to be hopeless, one more direct approach seems to test the effect of the presence of the bound state on the one-magnon excitation that can be studied directly by means of the inelastic neutron scattering (INS). As we have seen in Section 4.5, Fig. 4.8, the presence of a bound state induces a sort of resonance in the magnon spectrum in proximity of the ZC. In particular, such a resonance occurs at a wavevector corresponding to the crossing of the simple spin wave energy spectrum with the damped bound state energy inside the two-magnon band.

# Chapter 6

# PERTURBATION THEORY IN PLANAR FERROMAGNETS

## 6.1. Bogoliubov Transformation

Until now we have studied isotropic or uniaxial ferromagnetic systems for which the ground state is known exactly. In this chapter, we will study the *planar* ferromagnet in which the single-ion easy-plane anisotropy changes dramatically the scenario. First of all, the ground state of the planar ferromagnet is not known exactly, then the bilinear boson Hamiltonian is no longer diagonal but a Bogoliubov transformation has to be used to diagonalize it. Even worse, the DM spin-boson transformation[15,16] leads to a bilinear Hamiltonian that is non-hermitian violating one of the postulates of the quantum mechanics. This fact also occurred in the isotropic ferromagnet, but only when the interaction between spin waves was accounted for, and in Section 3.4, we showed that this drawback does not affect the physical quantities at low temperature. Moreover, two other inconsistencies occur: the *kinematical consistency*, that is, the disappearance of any anisotropy dependence of the spin wave spectrum for $S = \frac{1}{2}$ is lost as well as the existence of a Goldstone mode at $\boldsymbol{k} = 0$ as required by the invariance property of the Hamiltonian under rotation about the $z$-axis. This intriguing puzzle was tackled[46] some time ago: only a very careful treatment of the terms coming from the perturbation expansion in $\frac{1}{S}$ allows to overcome any physical inconsistence. The detailed procedure to arrive at a correct spin wave spectrum may be useful to make the reader more familiar with the many-body theory in unorthodox systems. The many-body approach to this problem is similar to that used to treat the $^4$He case[47] even though the absence of the Bose–Einstein condensation makes the problem more difficult.

Let us consider the exchange Hamiltonian

$$\mathcal{H} = -J\sum_{i,\boldsymbol{\delta}} \boldsymbol{S}_i \cdot \boldsymbol{S}_{i+\boldsymbol{\delta}} + D\sum_i (S_i^z)^2, \tag{6.1.1}$$

where $i$ runs over all sites of a Bravais lattice; $J > 0$ is the ferromagnetic exchange integral between a spin located on the lattice site $i$ and its NN. For $D > 0$, the single-ion anisotropy favours the $xy$-plane and the Hamiltonian (6.1.1) is invariant under rotation about the $z$-axis. This symmetry implies the existence of a Goldstone mode[48] or a "soft mode" at $\boldsymbol{k}=0$ in the spin wave excitation spectrum. Moreover, for $S=\frac{1}{2}$ the single-ion anisotropy reduces to a constant so that any dependence of the spin wave spectrum on the anisotropy $D$ has to disappear (kinematical consistency). Without loss of generality, we assume that $x$ is the quantization axis in the $xy$-plane. By using the DM spin-boson transformation

$$\begin{aligned} S_i^x &= S - a_i^+ a_i, \\ S_i^y &= \frac{\sqrt{2S}}{2}\left(a_i + a_i^+ - \frac{1}{2S}a_i^+ a_i a_i\right), \\ S_i^z &= \frac{\sqrt{2S}}{2i}\left(a_i - a_i^+ - \frac{1}{2S}a_i^+ a_i a_i\right) \end{aligned} \tag{6.1.2}$$

and the Fourier transforms (2.2.14) together with a normal ordering of the Bose operators that is putting all creation Bose operators $a_{\boldsymbol{k}}^+$ on the left of the Bose destruction operators $a_{\boldsymbol{k}}$, one obtains

$$\mathcal{H} = E_0 + \mathcal{H}_2 + \mathcal{H}_4 + \mathcal{H}_6 \tag{6.1.3}$$

where

$$E_0 = -zJS^2N + \frac{1}{2}DSN. \tag{6.1.4}$$

The bilinear Hamiltonian is

$$\mathcal{H}_2 = \mathcal{H}_0 + \mathcal{H}_2^{\mathrm{NH}} \tag{6.1.5}$$

where

$$\mathcal{H}_0 = 2zJS\sum_{\boldsymbol{q}}\left[(\omega_{\boldsymbol{q}}+\delta)a_{\boldsymbol{q}}^+ a_{\boldsymbol{q}} - \frac{1}{2}\delta(a_{\boldsymbol{q}}a_{-\boldsymbol{q}} + a_{\boldsymbol{q}}^+ a_{-\boldsymbol{q}}^+)\right] \tag{6.1.6}$$

with

$$\omega_{\boldsymbol{q}} = 1-\gamma_{\boldsymbol{q}}, \quad d = \frac{D}{2zJ}, \quad \delta = d\left(1-\frac{1}{2S}\right) \tag{6.1.7}$$

is the hermitian bilinear Hamiltonian and

$$\mathcal{H}_2^{\mathrm{NH}} = -\frac{zJd}{2}\sum_{\boldsymbol{q}} a_{\boldsymbol{q}}^+ a_{-\boldsymbol{q}}^+ \tag{6.1.8}$$

is the "non-hermitian" bilinear Hamiltonian. The four-operator Hamiltonian is

$$\mathcal{H}_4 = \sum_{\boldsymbol{q}_1,\boldsymbol{q}_2,\boldsymbol{q}_3,\boldsymbol{q}_4} \delta_{\boldsymbol{q}_1+\boldsymbol{q}_2,\boldsymbol{q}_3+\boldsymbol{q}_4}\left[V_{\boldsymbol{q}_1,\boldsymbol{q}_2,\boldsymbol{q}_3,\boldsymbol{q}_4}\, a_{\boldsymbol{q}_1}^+ a_{\boldsymbol{q}_2}^+ a_{\boldsymbol{q}_3} a_{\boldsymbol{q}_4} + D^{(1)} a_{\boldsymbol{q}_1}^+ a_{-\boldsymbol{q}_2} a_{\boldsymbol{q}_3} a_{\boldsymbol{q}_4}\right] \tag{6.1.9}$$

with

$$V_{\boldsymbol{q}_1,\boldsymbol{q}_2,\boldsymbol{q}_3,\boldsymbol{q}_4} = -\frac{zJ}{4N}[\gamma_{\boldsymbol{q}_1-\boldsymbol{q}_3} + \gamma_{\boldsymbol{q}_1-\boldsymbol{q}_4} + \gamma_{\boldsymbol{q}_2-\boldsymbol{q}_3} + \gamma_{\boldsymbol{q}_2-\boldsymbol{q}_4} - 2(\gamma_{\boldsymbol{q}_1}+\gamma_{\boldsymbol{q}_2}) + 4d] \tag{6.1.10}$$

and

$$D^{(1)} = \frac{zJ\delta}{N}. \tag{6.1.11}$$

The six-operator Hamiltonian is

$$\mathcal{H}_6 = \sum_{q_1,q_2,q_3,q_4,q_5,q_6} D^{(2)}\delta_{q_1+q_2+q_3,q_4+q_5+q_6}\, a^+_{q_1}a^+_{q_2}a_{-q_3}a_{q_4}a_{q_5}a_{q_6} \tag{6.1.12}$$

with

$$D^{(2)} = -\frac{zJd}{4SN^2}. \tag{6.1.13}$$

Unlike the isotropic ferromagnet, the bilinear Hamiltonian (6.1.5) of the planar ferromagnet is no longer diagonal. The merit of the DM transformation is in generating only a finite number of interaction potentials even though the boson Hamiltonian so obtained is no longer hermitian. The non-hermiticity of the bilinear Hamiltonian (6.1.5) prevents its diagonalization using the Bogoliubov transformation. To overcome this unpleasant feature we limit ourselves to diagonalize the hermitian bilinear Hamiltonian (6.1.6) and treating the non-hermitian bilinear Hamiltonian (6.1.8) as a perturbation. We introduce the Bogoliubov transformation

$$a_{\boldsymbol{q}} = l_{\boldsymbol{q}}(\alpha_{\boldsymbol{q}} - x_{\boldsymbol{q}}\alpha^+_{-\boldsymbol{q}}), \quad a^+_{\boldsymbol{q}} = l_{\boldsymbol{q}}(\alpha^+_{\boldsymbol{q}} - x_{\boldsymbol{q}}\alpha_{-\boldsymbol{q}}) \tag{6.1.14}$$

where $l_{\boldsymbol{q}}$ and $x_{\boldsymbol{q}}$ are scalar functions of the wavevector $\boldsymbol{q}$. The requirement that $\alpha_{\boldsymbol{q}}$ and $\alpha^+_{\boldsymbol{q}}$ are Bose operators satisfying the same commutation rules of $a_{\boldsymbol{q}}$ and $a^+_{\boldsymbol{q}}$ leads to the relationships

$$l_{\boldsymbol{q}} = l_{-\boldsymbol{q}} = l^*_{\boldsymbol{q}}, \quad x_{\boldsymbol{q}} = x_{-\boldsymbol{q}} = x^*_{\boldsymbol{q}}, \quad l^2_{\boldsymbol{q}}(1 - x^2_{\boldsymbol{q}}) = 1. \tag{6.1.15}$$

Equations in (6.1.15) are simultaneously satisfied assuming

$$l_{\boldsymbol{q}} = \cosh\phi_{\boldsymbol{q}}, \quad x_{\boldsymbol{q}} = \tanh\phi_{\boldsymbol{q}}. \tag{6.1.16}$$

By means of Eq. (6.1.14), the Hamiltonian (6.1.6) becomes

$$\begin{aligned}\mathcal{H}_0 = 2zJS\sum_{\boldsymbol{q}}\Big\{ & [l^2_{\boldsymbol{q}}(1 + x^2_{\boldsymbol{q}})(\omega_{\boldsymbol{q}} + \delta) + 2\delta l^2_{\boldsymbol{q}}x_{\boldsymbol{q}}]\alpha^+_{\boldsymbol{q}}\alpha_{\boldsymbol{q}} \\ & - [2l^2_{\boldsymbol{q}}x_{\boldsymbol{q}}(\omega_{\boldsymbol{q}} + \delta) + \delta l^2_{\boldsymbol{q}}(1 + x^2_{\boldsymbol{q}})]\frac{1}{2}(\alpha_{\boldsymbol{q}}\alpha_{-\boldsymbol{q}} + \alpha^+_{\boldsymbol{q}}\alpha^+_{-\boldsymbol{q}}) \\ & + l^2_{\boldsymbol{q}}x^2_{\boldsymbol{q}}(\omega_{\boldsymbol{q}} + \delta) + \delta l^2_{\boldsymbol{q}}x_{\boldsymbol{q}}\Big\}.\end{aligned} \tag{6.1.17}$$

Obviously, the Hamiltonian (6.1.17) can be diagonalized choosing

$$2l^2_{\boldsymbol{q}}x_{\boldsymbol{q}}(\omega_{\boldsymbol{q}} + \delta) + \delta l^2_{\boldsymbol{q}}(1 + x^2_{\boldsymbol{q}}) = 0 \tag{6.1.18}$$

or, using Eq. (6.1.16),

$$\tanh 2\phi_{\boldsymbol{q}} = -\frac{\delta}{\omega_{\boldsymbol{q}} + \delta}. \tag{6.1.19}$$

Using the relationships between the hyperbolic functions

$$l_{\boldsymbol{q}}^2 = \cosh^2 \phi_{\boldsymbol{q}} = \frac{1}{2}\left(\frac{1}{\sqrt{1-\tanh^2 2\phi_{\boldsymbol{q}}}} + 1\right), \tag{6.1.20}$$

$$l_{\boldsymbol{k}}^2 x_{\boldsymbol{q}}^2 = \sinh^2 \phi_{\boldsymbol{q}} = \frac{1}{2}\left(\frac{1}{\sqrt{1-\tanh^2 2\phi_{\boldsymbol{q}}}} - 1\right), \tag{6.1.21}$$

$$2l_{\boldsymbol{q}}^2 x_{\boldsymbol{q}} = \sinh 2\phi_{\boldsymbol{q}} = \frac{\tanh 2\phi_{\boldsymbol{q}}}{\sqrt{1-\tanh^2 2\phi_{\boldsymbol{q}}}} \tag{6.1.22}$$

and Eq. (6.1.19), the Hamiltonian (6.1.6) becomes

$$\mathcal{H}_0 = 2zJS\sum_{\boldsymbol{q}} \sqrt{\omega_{\boldsymbol{q}}(\omega_{\boldsymbol{q}} + 2\delta)}\alpha_{\boldsymbol{q}}^+\alpha_{\boldsymbol{q}} + zJSN[X - Y + \delta(X+Z)], \tag{6.1.23}$$

where

$$X = \frac{1}{N}\sum_{\boldsymbol{q}} 2l_{\boldsymbol{q}}^2 x_{\boldsymbol{q}}^2 = \frac{1}{N}\sum_{\boldsymbol{q}} \frac{\omega_{\boldsymbol{q}} + \delta}{\sqrt{\omega_{\boldsymbol{q}}(\omega_{\boldsymbol{q}} + 2\delta)}} - 1, \tag{6.1.24}$$

$$Y = \frac{1}{N}\sum_{\boldsymbol{q}} 2l_{\boldsymbol{q}}^2 x_{\boldsymbol{q}}^2 \gamma_{\boldsymbol{q}} = \frac{1}{N}\sum_{\boldsymbol{q}} \frac{\omega_{\boldsymbol{q}} + \delta}{\sqrt{\omega_{\boldsymbol{q}}(\omega_{\boldsymbol{q}} + 2\delta)}}\gamma_{\boldsymbol{q}}, \tag{6.1.25}$$

$$Z = \frac{1}{N}\sum_{\boldsymbol{q}} 2l_{\boldsymbol{q}}^2 x_{\boldsymbol{q}} = -\delta\frac{1}{N}\sum_{\boldsymbol{q}} \frac{1}{\sqrt{\omega_{\boldsymbol{q}}(\omega_{\boldsymbol{q}} + 2\delta)}} \tag{6.1.26}$$

and

$$T = \frac{1}{N}\sum_{\boldsymbol{q}} 2l_{\boldsymbol{q}}^2 x_{\boldsymbol{q}} \gamma_{\boldsymbol{q}} = -\delta\frac{1}{N}\sum_{\boldsymbol{q}} \frac{\gamma_{\boldsymbol{q}}}{\sqrt{\omega_{\boldsymbol{q}}(\omega_{\boldsymbol{q}} + 2\delta)}}. \tag{6.1.27}$$

As one can see from Eq. (6.1.23), the choice of $\mathcal{H}_0$ as "unperturbed" Hamiltonian leads to a spin wave spectrum kinematically consistent with a soft mode at $\boldsymbol{k} = 0$. Note that the Bogoliubov transformation (6.1.14) enters a first correction to the "classical" ground state $E_0$ given by Eq. (6.1.4). Replacing Eq. (6.1.14) into Eqs. (6.1.9) and (6.1.12), after normal ordering the Hamiltonian (6.1.3) becomes

$$\mathcal{H} = E_0 + \Delta E_0 + \mathcal{H}_0 + \mathcal{H}_2' + \mathcal{H}_4' + \mathcal{H}_6', \tag{6.1.28}$$

where $E_0$ is given by Eq. (6.1.4) and

$$\Delta E_0 = zJSN\left\{X - Y + \delta(X+Z) - \frac{1}{4S}[-dZ + (X-Y)^2 + T(T-Z) + d(Z^2+2X^2) + 3\delta XZ] + \frac{3d}{32S^2}Z(Z^2+4X^2)\right\}. \quad (6.1.29)$$

In Eq. (6.1.29), the terms linear, quadratic and cubic in $X, Y, Z$ and $T$ come from the normal ordering of $\mathcal{H}_2$, $\mathcal{H}_4$ and $\mathcal{H}_6$, respectively. The unperturbed Hamiltonian is

$$\mathcal{H}_0 = 2zJS\sum_{q}\sqrt{\omega_q(\omega_q+2\delta)}\alpha_q^+\alpha_q, \quad (6.1.30)$$

and the remaining part of the bilinear Hamiltonian coming from the Hamiltonian (6.1.8) and from the normal ordering of Hamiltonians (6.1.9) and (6.1.12) becomes

$$\mathcal{H}_2' = zJ\sum_{k}\left(-P_q\alpha_q^+\alpha_q + \frac{1}{2}Q_q\alpha_q^+\alpha_{-q}^+ + \frac{1}{2}R_q\alpha_q\alpha_{-q}\right) \quad (6.1.31)$$

where

$$P_q = \frac{1}{\sqrt{\omega_q(\omega_q+2\delta)}}[A_q(\omega_q+\delta) - \delta B_q], \quad (6.1.32)$$

$$Q_q = \frac{1}{\sqrt{\omega_q(\omega_q+2\delta)}}[-\delta A_q + B_q(\omega_q+\delta)] - C_q, \quad (6.1.33)$$

$$R_q = \frac{1}{\sqrt{\omega_q(\omega_q+2\delta)}}[-\delta A_q + B_q(\omega_q+\delta)] + C_q \quad (6.1.34)$$

with

$$A_q = (X-Y)\omega_q + \frac{1}{2}\delta(3Z+4X) + \frac{d}{2S}X(2-3Z), \quad (6.1.35)$$

$$B_q = -\frac{1}{2}(Z-T)\gamma_q - \frac{1}{2}T\omega_q - \frac{\delta}{2}(1-2Z-3X) - \frac{d}{16S}(4-8Z+9Z^2+12X^2) \quad (6.1.36)$$

and

$$C_k = \frac{1}{2}(Z-T)\gamma_q - \frac{1}{2}T\omega_q + \frac{\delta}{2}(1+3X) + \frac{d}{16S}(4-3Z^2-12X^2). \quad (6.1.37)$$

The four-operator Hamiltonian $\mathcal{H}_4'$ is given by

$$\mathcal{H}_4' = \sum_{q_1,q_2,q_3,q_4} l_{q_1}l_{q_2}l_{q_3}l_{q_4}\delta_{q_1+q_2,q_3+q_4}(\Phi^{(1)}_{q_1,q_2,q_3,q_4}\alpha_{q_1}^+\alpha_{q_2}^+\alpha_{q_3}\alpha_{q_4} + \Phi^{(2)}_{q_1,q_2,-q_3,q_4}\alpha_{q_1}^+\alpha_{q_2}^+\alpha_{-q_3}^+\alpha_{q_4} + \Phi^{(3)}_{q_1,-q_2,q_3,q_4}\alpha_{q_1}^+\alpha_{-q_2}\alpha_{q_3}\alpha_{q_4} + \Phi^{(4)}_{-q_1,-q_2,q_3,q_4}\alpha_{-q_1}\alpha_{-q_2}\alpha_{q_3}\alpha_{q_4} + \Phi^{(5)}_{q_1,q_2,-q_3,-q_4}\alpha_{q_1}^+\alpha_{q_2}^+\alpha_{-q_3}^+\alpha_{-q_4}^+), \quad (6.1.38)$$

where

$$\begin{aligned}
\Phi^{(1)}_{q_1,q_2,q_3,q_4} &= V_{q_1,q_2,q_3,q_4} + x_{q_1}x_{q_3}V_{q_2,-q_3,-q_1,q_4} + x_{q_1}x_{q_4}V_{q_2,-q_4,-q_1,q_3} \\
&\quad + x_{q_2}x_{q_3}V_{q_1,-q_3,-q_2,q_4} + x_{q_2}x_{q_4}V_{q_1,-q_4,-q_2,q_3} \\
&\quad + x_{q_1}x_{q_2}x_{q_3}x_{q_4}V_{q_4,q_3,q_2,q_1} \\
&\quad - \frac{3}{2}D^{(1)}[x_{q_1} + x_{q_2} + x_{q_1}x_{q_2}(x_{q_3} + x_{q_4})] \\
&\quad + 3D^{(3)}Z(1 + x_{q_1}x_{q_2} + x_{q_1}x_{q_3} + x_{q_1}x_{q_4} + x_{q_2}x_{q_3} + x_{q_2}x_{q_4} \\
&\quad + x_{q_1}x_{q_2}x_{q_3}x_{q_4}) \\
&\quad + 6D^{(3)}X[x_{q_1} + x_{q_2} + x_{q_1}x_{q_2}(x_{q_3} + x_{q_4})], \qquad (6.1.39)
\end{aligned}$$

$$\begin{aligned}
\Phi^{(2)}_{q_1,q_2,-q_3,q_4} &= -\frac{2}{3}(x_{q_1}V_{q_2,-q_3,-q_1,q_4} + x_{q_2}V_{q_1,-q_3,-q_2,q_4} + x_{q_3}V_{q_1,q_2,q_3,q_4} \\
&\quad + x_{q_1}x_{q_2}x_{q_4}V_{q_4,q_3,q_2,q_1} + x_{q_1}x_{q_3}x_{q_4}V_{q_4,-q_2,-q_3,q_1} \\
&\quad + x_{q_2}x_{q_3}x_{q_4}V_{q_4,-q_1,-q_3,q_2}) \\
&\quad + D^{(1)}(x_{q_1}x_{q_2} + x_{q_1}x_{q_3} + x_{q_2}x_{q_3} + x_{q_1}x_{q_2}x_{q_3}x_{q_4}) \\
&\quad - 2D^{(3)}Z(x_{q_1} + x_{q_2} + x_{q_3} + x_{q_1}x_{q_2}x_{q_3} + x_{q_1}x_{q_2}x_{q_4} \\
&\quad + x_{q_1}x_{q_3}x_{q_4} + x_{q_2}x_{q_3}x_{q_4}) \\
&\quad - 4D^{(3)}X(x_{q_1}x_{q_2} + x_{q_1}x_{q_3} + x_{q_2}x_{q_3} + x_{q_1}x_{q_2}x_{q_3}x_{q_4}), \qquad (6.1.40)
\end{aligned}$$

$$\begin{aligned}
\Phi^{(3)}_{q_1,-q_2,q_3,q_4} &= -\frac{2}{3}(x_{q_2}V_{q_1,q_2,q_3,q_4} + x_{q_3}V_{q_1,-q_3,-q_2,q_4} + x_{q_4}V_{q_1,-q_4,-q_2,q_3} \\
&\quad + x_{q_1}x_{q_2}x_{q_3}V_{q_3,-q_2,-q_4,q_1} + x_{q_1}x_{q_2}x_{q_4}V_{q_4,-q_2,-q_3,q_1} \\
&\quad + x_{q_1}x_{q_3}x_{q_4}V_{q_4,q_3,q_2,q_1}) \\
&\quad + D^{(1)}(1 + x_{q_1}x_{q_2} + x_{q_1}x_{q_3} + x_{q_1}x_{q_4}) \\
&\quad - 2D^{(3)}Z(x_{q_1} + x_{q_2} + x_{q_3} + x_{q_4} + x_{q_1}x_{q_2}x_{q_3} + x_{q_1}x_{q_2}x_{q_4} \\
&\quad + x_{q_1}x_{q_3}x_{q_4}) \\
&\quad - 4D^{(3)}X(1 + x_{q_1}x_{q_2} + x_{q_1}x_{q_3} + x_{q_1}x_{q_4}), \qquad (6.1.41)
\end{aligned}$$

$$\begin{aligned}
\Phi^{(4)}_{-q_1,-q_2,q_3,q_4} &= \frac{1}{6}(x_{q_1}x_{q_2}V_{q_1,q_2,q_3,q_4} + x_{q_1}x_{q_3}V_{q_1,-q_3,-q_2,q_4} + x_{q_1}x_{q_4}V_{q_1,-q_4,-q_2,q_3} \\
&\quad + x_{q_2}x_{q_3}V_{q_2,-q_3,-q_1,q_4} + x_{q_2}x_{q_4}V_{q_2,-q_4,-q_1,q_3} + x_{q_3}x_{q_4}V_{q_4,q_3,q_2,q_1}) \\
&\quad - \frac{1}{4}D^{(1)}(x_{q_1} + x_{q_2} + x_{q_3} + x_{q_4}) \\
&\quad + \frac{1}{2}D^{(3)}Z(1 + x_{q_1}x_{q_2} + x_{q_1}x_{q_3} + x_{q_1}x_{q_4} + x_{q_2}x_{q_3} + x_{q_2}x_{q_4} \\
&\quad + x_{q_3}x_{q_4}) + D^{(3)}X(x_{q_1} + x_{q_2} + x_{q_3} + x_{q_4}) \qquad (6.1.42)
\end{aligned}$$

and

$$\Phi^{(5)}_{q_1,q_2,-q_3,-q_4} = \frac{1}{6}(x_{q_1}x_{q_2}V_{q_4,q_3,q_2,q_1} + x_{q_1}x_{q_3}V_{q_4,-q_2,-q_3,q_1} + x_{q_1}x_{q_4}V_{q_3,-q_2,-q_4,q_1}$$
$$+ x_{q_2}x_{q_3}V_{q_4,-q_1,-q_3,q_2} + x_{q_2}x_{q_4}V_{q_3,-q_1,-q_4,q_2} + x_{q_3}x_{q_4}V_{q_1,q_2,q_3,q_4})$$
$$- \frac{1}{4}D^{(1)}[x_{q_1}x_{q_2}(x_{q_3} + x_{q_4}) + x_{q_3}x_{q_4}(x_{q_1} + x_{q_2})]$$
$$+ \frac{1}{2}D^{(3)}Z(x_{q_1}x_{q_2} + x_{q_1}x_{q_3} + x_{q_1}x_{q_4} + x_{q_2}x_{q_3} + x_{q_2}x_{q_4}$$
$$+ x_{q_3}x_{q_4} + x_{q_1}x_{q_2}x_{q_3}x_{q_4}) + D^{(3)}X[x_{q_1}x_{q_2}(x_{q_3} + x_{q_4})$$
$$+ x_{q_3}x_{q_4}(x_{q_1} + x_{q_2})] \tag{6.1.43}$$

with

$$D^{(3)} = \frac{zJd}{4SN}. \tag{6.1.44}$$

The potentials $\Phi$ are written in a symmetric even though heavy form which is convenient in the perturbation expansion since it allows to group many terms in a single Feynman diagram with an appropriated multiplicity. For instance, all perturbation terms containing the potentials $\Phi^{(1)}$ that differ only by the exchange of the first and second or the third and fourth subscripts give the same contribution so that a multiplicity 4 can be introduced retaining only one term. Analogously, the potentials $\Phi^{(2)}$ and $\Phi^{(3)}$ are invariant under any permutation of the first, second and third or second, third and fourth subscripts, respectively. The potentials $\Phi^{(4)}$ and $\Phi^{(5)}$ are invariant under any of the 24 permutations of their subscripts. Note that $D^{(1)}$ given by Eq. (6.1.11) is kinematically consistent whereas $D^{(3)}$ given by Eq. (6.1.44) is not kinematically consistent.

The six-operator Hamiltonian $\mathcal{H}'_6$ is given by

$$\mathcal{H}'_6 = \sum_{q_1,q_2,q_3,q_4,q_5,q_6} l_{q_1}l_{q_2}l_{q_3}l_{q_4}l_{q_5}l_{q_6}\delta_{q_1+q_2+q_3,q_4+q_5+q_6}$$
$$\times (\Psi^{(1)}_{q_1,q_2,q_3,q_4,q_5,q_6}\alpha^+_{q_1}\alpha^+_{q_2}\alpha^+_{q_3}\alpha_{q_4}\alpha_{q_5}\alpha_{q_6}$$
$$+ \Psi^{(2)}_{q_1,q_2,q_3,-q_4,q_5,q_6}\alpha^+_{q_1}\alpha^+_{q_2}\alpha^+_{q_3}\alpha^+_{-q_4}\alpha_{q_5}\alpha_{q_6}$$
$$+ \Psi^{(3)}_{q_1,q_2,-q_3,q_4,q_5,q_6}\alpha^+_{q_1}\alpha^+_{q_2}\alpha_{-q_3}\alpha_{q_4}\alpha_{q_5}\alpha_{q_6}$$
$$+ \Psi^{(4)}_{q_1,q_2,q_3,-q_4,-q_5,q_6}\alpha^+_{q_1}\alpha^+_{q_2}\alpha^+_{q_3}\alpha^+_{-q_4}\alpha^+_{-q_5}\alpha_{q_6}$$
$$+ \Psi^{(5)}_{q_1,-q_2,-q_3,q_4,q_5,q_6}\alpha^+_{q_1}\alpha_{-q_2}\alpha_{-q_3}\alpha_{q_4}\alpha_{q_5}\alpha_{q_6}$$
$$+ \Psi^{(6)}_{-q_1,-q_2,-q_3,q_4,q_5,q_6}\alpha_{-q_1}\alpha_{-q_2}\alpha_{-q_3}\alpha_{q_4}\alpha_{q_5}\alpha_{q_6}$$
$$+ \Psi^{(7)}_{q_1,q_2,q_3,-q_4,-q_5,-q_6}\alpha^+_{q_1}\alpha^+_{q_2}\alpha^+_{q_3}\alpha^+_{-q_4}\alpha^+_{-q_5}\alpha^+_{-q_6}) \tag{6.1.45}$$

where

$$\Psi^{(1)}_{q_1,q_2,q_3,q_4,q_5,q_6} = -\frac{4}{3}D^{(2)}[x_{q_1} + x_{q_2} + x_{q_3} + (x_{q_1}x_{q_2} + x_{q_1}x_{q_3} + x_{q_2}x_{q_3}) \times (x_{q_4} + x_{q_5} + x_{q_6}) + x_{q_1}x_{q_2}x_{q_3}(x_{q_4}x_{q_5} + x_{q_4}x_{q_6} + x_{q_5}x_{q_6})], \tag{6.1.46}$$

$$\Psi^{(2)}_{q_1,q_2,q_3,-q_4,q_5,q_6} = D^{(2)}[x_{q_1}x_{q_2} + x_{q_1}x_{q_3} + x_{q_1}x_{q_4} + x_{q_2}x_{q_3} + x_{q_2}x_{q_4} + x_{q_3}x_{q_4} + (x_{q_1}x_{q_2}x_{q_3} + x_{q_1}x_{q_2}x_{q_4} + x_{q_1}x_{q_3}x_{q_4} + x_{q_2}x_{q_3}x_{q_4}) \times (x_{q_5} + x_{q_6}) + x_{q_1}x_{q_2}x_{q_3}x_{q_4}x_{q_5}x_{q_6}], \tag{6.1.47}$$

$$\Psi^{(3)}_{q_1,q_2,-q_3,q_4,q_5,q_6} = D^{(2)}[1 + (x_{q_1} + x_{q_2})(x_{q_3} + x_{q_4} + x_{q_5} + x_{q_6}) + x_{q_1}x_{q_2}(x_{q_3}x_{q_4} + x_{q_3}x_{q_5} + x_{q_3}x_{q_6} + x_{q_4}x_{q_5} + x_{q_4}x_{q_6} + x_{q_5}x_{q_6})], \tag{6.1.48}$$

$$\Psi^{(4)}_{q_1,q_2,q_3,-q_4,-q_5,q_6} = -\frac{2}{5}D^{(2)}[x_{q_1}x_{q_2}x_{q_3} + x_{q_1}x_{q_2}x_{q_4} + x_{q_1}x_{q_2}x_{q_5} + x_{q_1}x_{q_3}x_{q_4} + x_{q_1}x_{q_3}x_{q_5} + x_{q_1}x_{q_4}x_{q_5} + x_{q_2}x_{q_3}x_{q_4} + x_{q_2}x_{q_3}x_{q_5} + x_{q_2}x_{q_4}x_{q_5} + x_{q_3}x_{q_4}x_{q_5} + (x_{q_1}x_{q_2}x_{q_3}x_{q_4} + x_{q_1}x_{q_2}x_{q_3}x_{q_5} + x_{q_1}x_{q_2}x_{q_4}x_{q_5} + x_{q_1}x_{q_3}x_{q_4}x_{q_5} + x_{q_2}x_{q_3}x_{q_4}x_{q_5})x_{q_6}], \tag{6.1.49}$$

$$\Psi^{(5)}_{q_1,-q_2,-q_3,q_4,q_5,q_6} = -\frac{2}{5}D^{(2)}[x_{q_2} + x_{q_3} + x_{q_4} + x_{q_5} + x_{q_6} + x_{q_1}(x_{q_2}x_{q_3} + x_{q_2}x_{q_4} + x_{q_2}x_{q_5} + x_{q_2}x_{q_6} + x_{q_3}x_{q_4} + x_{q_3}x_{q_5} + x_{q_3}x_{q_6} + x_{q_4}x_{q_5} + x_{q_4}x_{q_6} + x_{q_5}x_{q_6})], \tag{6.1.50}$$

$$\Psi^{(6)}_{-q_1,-q_2,-q_3,q_4,q_5,q_6} = \frac{1}{15}D^{(2)}(x_{q_1}x_{q_2} + x_{q_1}x_{q_3} + x_{q_1}x_{q_4} + x_{q_1}x_{q_5} + x_{q_1}x_{q_6} + x_{q_2}x_{q_3} + x_{q_2}x_{q_4} + x_{q_2}x_{q_5} + x_{q_2}x_{q_6} + x_{q_3}x_{q_4} + x_{q_3}x_{q_5} + x_{q_3}x_{q_6} + x_{q_4}x_{q_5} + x_{q_4}x_{q_6} + x_{q_5}x_{q_6}) \tag{6.1.51}$$

and

$$\Psi^{(7)}_{q_1,q_2,q_3,-q_4,-q_5,-q_6} = \frac{1}{15}D^{(2)}(x_{q_1}x_{q_2}x_{q_3}x_{q_4} + x_{q_1}x_{q_2}x_{q_3}x_{q_5} + x_{q_1}x_{q_2}x_{q_3}x_{q_6} + x_{q_1}x_{q_2}x_{q_4}x_{q_5} + x_{q_1}x_{q_2}x_{q_4}x_{q_6} + x_{q_1}x_{q_2}x_{q_5}x_{q_6} + x_{q_1}x_{q_3}x_{q_4}x_{q_5} + x_{q_1}x_{q_3}x_{q_4}x_{q_6} + x_{q_1}x_{q_3}x_{q_5}x_{q_6} + x_{q_1}x_{q_4}x_{q_5}x_{q_6} + x_{q_2}x_{q_3}x_{q_4}x_{q_5} + x_{q_2}x_{q_3}x_{q_4}x_{q_6} + x_{q_2}x_{q_3}x_{q_5}x_{q_6} + x_{q_2}x_{q_4}x_{q_5}x_{q_6} + x_{q_3}x_{q_4}x_{q_5}x_{q_6}). \tag{6.1.52}$$

Like the potentials $\Phi$, the potentials $\Psi$ also have been written in a symmetric form convenient to group them in a reduced number of Feynman diagrams with appropriate multiplicity. For instance, $\Psi^{(6)}$ and $\Psi^{(7)}$ are invariant under all the $6! = 720$ permutations of their subscripts.

## 6.2. The Dyson Matrix Equation

Analogously to what we have done for the isotropic ferromagnet in Chapter 4, we introduce the temperature Green function in order to perform a perturbation expansion in the interaction Hamiltonian. For the planar ferromagnet Hamiltonian, however, it is convenient to introduce the temperature *matrix* Green function, the elements of which are given by

$$\mathcal{G}_{\boldsymbol{k}}^{11}(\tau) = -\langle T[\alpha_{\boldsymbol{k}}^{\mathrm{H}}(\tau)\alpha_{\boldsymbol{k}}^{+}]\rangle, \tag{6.2.1}$$

$$\mathcal{G}_{\boldsymbol{k}}^{12}(\tau) = -\langle T[\alpha_{\boldsymbol{k}}^{\mathrm{H}}(\tau)\alpha_{-\boldsymbol{k}}]\rangle, \tag{6.2.2}$$

$$\mathcal{G}_{\boldsymbol{k}}^{21}(\tau) = -\langle T[\alpha_{-\boldsymbol{k}}^{+\mathrm{H}}(\tau)\alpha_{\boldsymbol{k}}^{+}]\rangle \tag{6.2.3}$$

and

$$\mathcal{G}_{\boldsymbol{k}}^{22}(\tau) = -\langle T[\alpha_{-\boldsymbol{k}}^{+\mathrm{H}}(\tau)\alpha_{-\boldsymbol{k}}]\rangle = \mathcal{G}_{-\boldsymbol{k}}^{11}(-\tau), \tag{6.2.4}$$

where $\tau = it$ is an imaginary time and $T$ is the ordering time operator that puts the operators within the square brackets in such a way that the time decreases going from the first to the last. The superscript H stays for "Heisenberg picture". The introduction of the matrix Green function dates back to the paper by Beliaev[47] for the liquid helium and of Nambu[49] for the superconductivity and applied to the antiferromagnet by Harris, Kumar, Halpering and Hohenberg.[50] According to Eq. (4.1.18), we can write a perturbation expansion for each of the four temperature Green functions (6.2.1)–(6.2.4)

$$\mathcal{G}_{\boldsymbol{k}}^{11}(\tau) = -\sum_{n=0}^{\infty}\left(-\frac{1}{\hbar}\right)^n \frac{1}{n!}\int_0^{\beta\hbar} d\tau_1 \cdots$$
$$\times \int_0^{\beta\hbar} d\tau_n \langle T[\mathcal{H}'(\tau_1)\cdots\mathcal{H}'(\tau_n)\alpha_{\boldsymbol{k}}(\tau)\alpha_{\boldsymbol{k}}^{+}]\rangle_0^{\mathrm{c}}, \tag{6.2.5}$$

$$\mathcal{G}_{\boldsymbol{k}}^{12}(\tau) = -\sum_{n=0}^{\infty}\left(-\frac{1}{\hbar}\right)^n \frac{1}{n!}\int_0^{\beta\hbar} d\tau_1 \cdots$$
$$\times \int_0^{\beta\hbar} d\tau_n \langle T[\mathcal{H}'(\tau_1)\cdots\mathcal{H}'(\tau_n)\alpha_{\boldsymbol{k}}(\tau)\alpha_{-\boldsymbol{k}}]\rangle_0^{\mathrm{c}}, \tag{6.2.6}$$

$$\mathcal{G}_{\boldsymbol{k}}^{21}(\tau) = -\sum_{n=0}^{\infty}\left(-\frac{1}{\hbar}\right)^n \frac{1}{n!}\int_0^{\beta\hbar} d\tau_1 \cdots$$
$$\times \int_0^{\beta\hbar} d\tau_n \langle T[\mathcal{H}'(\tau_1)\cdots\mathcal{H}'(\tau_n)\alpha_{-\boldsymbol{k}}^{+}(\tau)\alpha_{\boldsymbol{k}}^{+}]\rangle_0^{\mathrm{c}} \tag{6.2.7}$$

and

$$\mathcal{G}_{\boldsymbol{k}}^{22}(\tau) = \mathcal{G}_{-\boldsymbol{k}}^{11}(-\tau), \tag{6.2.8}$$

where the subscript "0" means that the average is performed over the statistical ensemble with the unperturbed Hamiltonian $\mathcal{H}_0$ given by (6.1.30) and the superscript "c" means that only contributions corresponding to connected diagrams have to be retained in the sum. The interaction Hamiltonian $\mathcal{H}'$ is given by the sum

$$\mathcal{H}' = \mathcal{H}_2' + \mathcal{H}_4' + \mathcal{H}_6', \tag{6.2.9}$$

where $\mathcal{H}_2'$, $\mathcal{H}_4'$ and $\mathcal{H}_6'$ are given by Eqs. (6.1.31), (6.1.38) and (6.1.45), respectively. The unperturbed temperature Green functions are given by

$$\mathcal{G}_{\boldsymbol{k}}^{(0)}(\tau) = -\langle T[\alpha_{\boldsymbol{k}}(\tau)\alpha_{\boldsymbol{k}}^{+}]\rangle = \begin{cases} -(1+n_{\boldsymbol{k}})e^{-\epsilon_{\boldsymbol{k}}\tau} & \text{for } \tau > 0, \\ -n_{\boldsymbol{k}}e^{-\epsilon_{\boldsymbol{k}}\tau} & \text{for } \tau < 0, \end{cases} \tag{6.2.10}$$

$$-\langle T[\alpha_{\boldsymbol{k}}(\tau)\alpha_{-\boldsymbol{k}}]\rangle = -\langle T[\alpha_{-\boldsymbol{k}}^{+}(\tau)\alpha_{\boldsymbol{k}}^{+}]\rangle = 0, \tag{6.2.11}$$

$$-\langle T[\alpha_{-\boldsymbol{k}}^{+}(\tau)\alpha_{-\boldsymbol{k}}]\rangle = \begin{cases} -n_{\boldsymbol{k}}e^{\epsilon_{\boldsymbol{k}}\tau} & \text{for } \tau > 0, \\ -(1+n_{\boldsymbol{k}})e^{\epsilon_{\boldsymbol{k}}\tau} & \text{for } \tau < 0, \end{cases} \tag{6.2.12}$$

where

$$\hbar\epsilon_{\boldsymbol{k}} = 2zJS\sqrt{\omega_{\boldsymbol{k}}(\omega_{\boldsymbol{k}} + 2\delta)} \tag{6.2.13}$$

is the unperturbed spin wave spectrum. Note that from Eqs. (6.2.10) and (6.2.12), one obtains

$$-\langle T[\alpha_{-\boldsymbol{k}}^{+}(\tau)\alpha_{-\boldsymbol{k}}]\rangle = \mathcal{G}_{-\boldsymbol{k}}^{(0)}(-\tau) = \mathcal{G}_{\boldsymbol{k}}^{(0)}(-\tau) \tag{6.2.14}$$

and the generalized Wick's theorem applied to the thermal averages occurring in Eqs. (6.2.5)–(6.2.7) leads to contractions containing only the unperturbed propagator given by Eq. (6.2.10) similar to the unperturbed propagator of the isotropic ferromagnet given by Eq. (4.1.20). The Fourier transforms (4.1.24) and (4.1.22) become

$$\mathcal{G}_{\boldsymbol{k}}^{\mathrm{ij}}(i\omega_n) = \int_0^{\beta\hbar} d\tau e^{i\omega_n\tau}\mathcal{G}_{\boldsymbol{k}}^{\mathrm{ij}}(\tau) \tag{6.2.15}$$

and

$$\mathcal{G}_{\boldsymbol{k}}^{\mathrm{ij}}(\tau) = \frac{1}{\beta\hbar}\sum_n e^{-i\omega_n\tau}\mathcal{G}_{\boldsymbol{k}}^{\mathrm{ij}}(i\omega_n), \tag{6.2.16}$$

where i,j = 1,2 and $\omega_n = \frac{2\pi n}{\beta\hbar}$. The Fourier transform of the unperturbed temperature Green function coincides with Eq. (4.1.25) replacing the unperturbed spin wave spectrum of the isotropic ferromagnet by the spin wave spectrum of the planar ferromagnet (6.2.13) that is

$$\mathcal{G}_{\boldsymbol{k}}^{(0)}(i\omega_n) = \frac{1}{i\omega_n - \epsilon_{\boldsymbol{k}}}. \tag{6.2.17}$$

The Dyson equation (3.5.2) becomes the *matrix* Dyson equation

$$\begin{pmatrix} \mathcal{G}_{\boldsymbol{k}}^{11}(i\omega_n) & \mathcal{G}_{\boldsymbol{k}}^{12}(i\omega_n) \\ \mathcal{G}_{\boldsymbol{k}}^{21}(i\omega_n) & \mathcal{G}_{\boldsymbol{k}}^{22}(i\omega_n) \end{pmatrix} = \begin{pmatrix} \mathcal{G}_{\boldsymbol{k}}^{(0)}(i\omega_n) & 0 \\ 0 & \mathcal{G}_{\boldsymbol{k}}^{(0)}(i\omega_n) \end{pmatrix}$$
$$+ \begin{pmatrix} \mathcal{G}_{\boldsymbol{k}}^{(0)}(i\omega_n) & 0 \\ 0 & \mathcal{G}_{\boldsymbol{k}}^{(0)}(i\omega_n) \end{pmatrix} \begin{pmatrix} \Sigma_{11}^*(\boldsymbol{k}, i\omega_n) & \Sigma_{12}^*(\boldsymbol{k}, i\omega_n) \\ \Sigma_{21}^*(\boldsymbol{k}, i\omega_n) & \Sigma_{22}^*(\boldsymbol{k}, i\omega_n) \end{pmatrix}$$
$$\times \begin{pmatrix} \mathcal{G}_{\boldsymbol{k}}^{11}(i\omega_n) & \mathcal{G}_{\boldsymbol{k}}^{12}(i\omega_n) \\ \mathcal{G}_{\boldsymbol{k}}^{21}(i\omega_n) & \mathcal{G}_{\boldsymbol{k}}^{22}(i\omega_n) \end{pmatrix} \tag{6.2.18}$$

where $\Sigma_{\rm ij}^*(\boldsymbol{k}, i\omega_n)$ are the proper self-energies with i,j = 1,2 and

$$\Sigma_{22}^*(\boldsymbol{k}, i\omega_n) = \Sigma_{11}^*(-\boldsymbol{k}, -i\omega_n) = \Sigma_{11}^*(\boldsymbol{k}, -i\omega_n), \tag{6.2.19}$$

where the last equality holds for Bravais lattices with a centre of symmetry. The solution of the matrix equation (6.2.18) is then given by

$$\mathcal{G}_{\boldsymbol{k}}^{11}(i\omega_n) = \frac{i\omega_n + \epsilon_{\boldsymbol{k}} + \Sigma_{22}^*(\boldsymbol{k}, i\omega_n)}{F(\boldsymbol{k}, i\omega_n)}, \tag{6.2.20}$$

$$\mathcal{G}_{\boldsymbol{k}}^{12}(i\omega_n) = -\frac{\Sigma_{12}^*(\boldsymbol{k}, i\omega_n)}{F(\boldsymbol{k}, i\omega_n)}, \tag{6.2.21}$$

$$\mathcal{G}_{\boldsymbol{k}}^{21}(i\omega_n) = -\frac{\Sigma_{21}^*(\boldsymbol{k}, i\omega_n)}{F(\boldsymbol{k}, i\omega_n)}, \tag{6.2.22}$$

$$\mathcal{G}_{\boldsymbol{k}}^{22}(i\omega_n) = \frac{-i\omega_n + \epsilon_{\boldsymbol{k}} + \Sigma_{11}^*(\boldsymbol{k}, i\omega_n)}{F(\boldsymbol{k}, i\omega_n)}, \tag{6.2.23}$$

where

$$\begin{aligned} F(\boldsymbol{k}, i\omega_n) &= [i\omega_n - \epsilon_{\boldsymbol{k}} - \Sigma_{11}^*(\boldsymbol{k}, i\omega_n)][i\omega_n + \epsilon_{\boldsymbol{k}} + \Sigma_{11}^*(\boldsymbol{k}, -i\omega_n)] \\ &\quad + \Sigma_{12}^*(\boldsymbol{k}, i\omega_n)\Sigma_{21}^*(\boldsymbol{k}, i\omega_n) \\ &= [i\omega_n - \Sigma_{\rm A}^*(\boldsymbol{k}, i\omega_n)]^2 - [\epsilon_{\boldsymbol{k}} + \Sigma_{\rm S}^*(\boldsymbol{k}, i\omega_n)]^2 \\ &\quad + \Sigma_{12}^*(\boldsymbol{k}, i\omega_n)\Sigma_{21}^*(\boldsymbol{k}, i\omega_n) \end{aligned} \tag{6.2.24}$$

with

$$\Sigma_{\rm A}^*(\boldsymbol{k}, i\omega_n) = \frac{1}{2}[\Sigma_{11}^*(\boldsymbol{k}, i\omega_n) - \Sigma_{22}^*(\boldsymbol{k}, i\omega_n)] \tag{6.2.25}$$

and

$$\Sigma_{\rm S}^*(\boldsymbol{k}, i\omega_n) = \frac{1}{2}[\Sigma_{11}^*(\boldsymbol{k}, i\omega_n) + \Sigma_{22}^*(\boldsymbol{k}, i\omega_n)]. \tag{6.2.26}$$

The poles of the temperature Green functions (6.2.20)–(6.2.23) are the roots of the equation $F(\boldsymbol{k}, i\omega) = 0$. The perturbation theory is meaningful only if the correction to the unperturbed spin wave spectrum is small compared to the unperturbed spectrum $\epsilon_{\boldsymbol{k}}$ so that the argument $i\omega_n$ inside the self-energy contribution may be replaced by $\epsilon_{\boldsymbol{k}}$ and the poles of the temperature Green functions (6.2.20)–(6.2.23) are given by the roots of the equation

$$[\omega - \Sigma_{\rm A}^*(\boldsymbol{k}, \epsilon_{\boldsymbol{k}})]^2 - [\epsilon_{\boldsymbol{k}} + \Sigma_{\rm S}^*(\boldsymbol{k}, \epsilon_{\boldsymbol{k}})]^2 + \Sigma_{12}^*(\boldsymbol{k}, \epsilon_{\boldsymbol{k}})\Sigma_{21}^*(\boldsymbol{k}, \epsilon_{\boldsymbol{k}}) = 0. \tag{6.2.27}$$

## 6.3. First-order Perturbation Theory

In this section, we perform explicitly the calculation of the first-order self-energies $\Sigma_{\rm ij}(\boldsymbol{k})$. As for the isotropic ferromagnet, the first-order self-energies of the planar ferromagnet are independent of the frequency $i\omega_n$. In order to simplify the notations, we neglect everywhere the superscript $^{(1)}$ on both the Green functions and the self-energies indicating the first-order approximation and suppress the asterisk on the *proper* self-energy since at the first-order there is no difference between the self-energy and the proper self-energy. The way followed in Chapter 4 for the isotropic ferromagnet is the same we follow here. The main difference between the isotropic case and the planar one is that in the planar ferromagnet the interaction Hamiltonian is made up of more terms and four temperature Green functions are to be accounted for. The first-order perturbation theory corresponds to the first term of the series expansions (6.2.5)–(6.2.8) that is

$$\mathcal{G}_{\boldsymbol{k}}^{11}(\tau) = \frac{1}{\hbar}\int_0^{\beta\hbar} d\tau_1 \langle T[\mathcal{H}'(\tau_1)\alpha_{\boldsymbol{k}}(\tau)\alpha_{\boldsymbol{k}}^+]\rangle_0^{\rm c}, \tag{6.3.1}$$

$$\mathcal{G}_{\boldsymbol{k}}^{12}(\tau) = \frac{1}{\hbar}\int_0^{\beta\hbar} d\tau_1 \langle T[\mathcal{H}'(\tau_1)\alpha_{\boldsymbol{k}}(\tau)\alpha_{-\boldsymbol{k}}]\rangle_0^{\rm c}, \tag{6.3.2}$$

$$\mathcal{G}_{\boldsymbol{k}}^{21}(\tau) = \frac{1}{\hbar}\int_0^{\beta\hbar} d\tau_1 \langle T[\mathcal{H}'(\tau_1)\alpha_{-\boldsymbol{k}}^+(\tau)\alpha_{\boldsymbol{k}}^+]\rangle_0^{\rm c} \tag{6.3.3}$$

and

$$\mathcal{G}_{\boldsymbol{k}}^{22}(\tau) = \mathcal{G}_{\boldsymbol{k}}^{11}(-\tau), \tag{6.3.4}$$

where $\mathcal{H}'$ is given by Eq. (6.2.9). Each of the Eqs. (6.3.1)–(6.3.3) consists of three contributions corresponding to the three terms occurring in the interaction Hamiltonian $\mathcal{H}'$ given by Eq. (6.2.9). The generalized Wick's theorem illustrated in Chapter 4 and the Fourier transform (6.2.16) enable us to write the contribution of

Eq. (6.3.1) coming from $\mathcal{H}_2'$ in the form

$$\frac{1}{\hbar}\int_0^{\beta\hbar} d\tau_1 \langle T[\mathcal{H}_2'(\tau_1)\alpha_{\boldsymbol{k}}(\tau)\alpha_{\boldsymbol{k}}^+]\rangle_0^c = -\frac{1}{\hbar}zJP_{\boldsymbol{k}}\int_0^{\beta\hbar} d\tau_1 \mathcal{G}_{\boldsymbol{k}}^{(0)}(\tau-\tau_1)\mathcal{G}_{\boldsymbol{k}}^{(0)}(\tau_1)$$
$$= \frac{1}{\beta\hbar}\sum_n e^{-i\omega_n\tau}[\mathcal{G}_{\boldsymbol{k}}^{(0)}(i\omega_n)]^2[\Sigma_{11}(\boldsymbol{k})]_{\mathcal{H}_2'}, \quad (6.3.5)$$

where

$$[\Sigma_{11}(\boldsymbol{k})]_{\mathcal{H}_2'} = -\frac{1}{\hbar}zJP_{\boldsymbol{k}}. \quad (6.3.6)$$

The Feynman diagram corresponding to the first-order self-energy given by Eq. (6.3.6) is represented by the first term of the first line of Fig. 6.1.

The contribution to Eq. (6.3.1) coming from of $\mathcal{H}_4'$ becomes

$$\frac{1}{\hbar}\int_0^{\beta\hbar} d\tau_1 \langle T[\mathcal{H}_4'(\tau_1)\alpha_{\boldsymbol{k}}(\tau)\alpha_{\boldsymbol{k}}^+]\rangle_0^c$$
$$= -\frac{1}{\hbar}\sum_{\boldsymbol{p}} 4l_{\boldsymbol{k}}^2 l_{\boldsymbol{p}}^2 \Phi^{(1)}_{\boldsymbol{k},\boldsymbol{p},\boldsymbol{k},\boldsymbol{p}} \int_0^{\beta\hbar} d\tau_1 \mathcal{G}_{\boldsymbol{k}}^{(0)}(\tau-\tau_1)\mathcal{G}_{\boldsymbol{p}}^{(0)}(0^-)\mathcal{G}_{\boldsymbol{k}}^{(0)}(\tau_1)$$
$$= \frac{1}{\beta\hbar}\sum_n e^{-i\omega_n\tau}[\mathcal{G}_{\boldsymbol{k}}^{(0)}(i\omega_n)]^2[\Sigma_{11}(\boldsymbol{k})]_{\mathcal{H}_4'}, \quad (6.3.7)$$

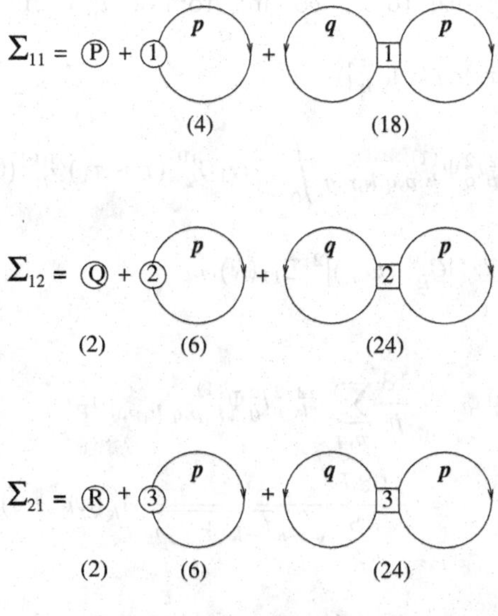


Fig. 6.1. First-order self-energy. The circles labelled by P, Q and R represent the potential interactions of $\mathcal{H}_2'$. The circles containing 1, 2 and 3 represent the potential interactions $\Phi^{(1)}$, $\Phi^{(2)}$ and $\Phi^{(3)}$ of $\mathcal{H}_4'$, respectively. The squares containing 1, 2 and 3 represent the potential interactions $\Psi^{(1)}$, $\Psi^{(2)}$ and $\Psi^{(3)}$ of $\mathcal{H}_6'$, respectively. The numbers in brackets under the diagrams indicate their multiplicity.

where

$$[\Sigma_{11}(\boldsymbol{k})]_{\mathcal{H}'_4} = \frac{4}{\hbar}\sum_{\boldsymbol{p}} l_{\boldsymbol{k}}^2 l_{\boldsymbol{p}}^2 \Phi^{(1)}_{\boldsymbol{k},\boldsymbol{p},\boldsymbol{k},\boldsymbol{p}} n_{\boldsymbol{p}}$$
$$= -\frac{2zJ}{\hbar}\frac{1}{\sqrt{\omega_{\boldsymbol{k}}(\omega_{\boldsymbol{k}}+2\delta)}}[A'_{\boldsymbol{k}}(\omega_{\boldsymbol{k}}+\delta) - B'_{\boldsymbol{k}}\delta] \quad (6.3.8)$$

with

$$A'_{\boldsymbol{k}} = (\lambda_2 + \delta\lambda_1)\omega_{\boldsymbol{k}} + 2d(\lambda_1 + \delta\lambda_0)\left(1 - \frac{3Z}{4S}\right) - \frac{3}{2}\delta^2\lambda_0 + \frac{3d\delta}{2S}\lambda_0 X, \quad (6.3.9)$$

$$B'_{\boldsymbol{k}} = \frac{1}{2}\delta\lambda_1\gamma_{\boldsymbol{k}} + \frac{1}{2}\delta(\lambda_0 - \lambda_1)\omega_{\boldsymbol{k}} + \frac{3}{2}\delta(\lambda_1 + \delta\lambda_0)$$
$$- d\delta\lambda_0\left(1 - \frac{9Z}{8S}\right) - \frac{3d}{2S}X(\lambda_1 + \delta\lambda_0), \quad (6.3.10)$$

$$\lambda_r = \frac{1}{N}\sum_{\boldsymbol{p}} \frac{(\omega_{\boldsymbol{p}})^r n_{\boldsymbol{p}}}{\sqrt{\omega_{\boldsymbol{p}}(\omega_{\boldsymbol{p}} + 2\delta)}} \quad (6.3.11)$$

and

$$n_{\boldsymbol{p}} = \frac{1}{e^{\beta\hbar\epsilon_{\boldsymbol{p}}} - 1}. \quad (6.3.12)$$

To obtain the self-energy (6.3.8) whose Feynman diagram is represented by the second term of the first line of Fig. 6.1, the relationship (4.2.16) is used in which the isotropic unperturbed spectrum $\omega_{\boldsymbol{k}}$ has been replaced by the planar unperturbed spectrum given by Eq. (6.2.13).

The contribution to Eq. (6.3.1) coming from of $\mathcal{H}'_6$ is given by

$$\frac{1}{\hbar}\int_0^{\beta\hbar} d\tau_1 \langle T[\mathcal{H}'_6(\tau_1)\alpha_{\boldsymbol{k}}(\tau)\alpha^+_{\boldsymbol{k}}]\rangle_0^c$$
$$= \frac{18}{\hbar}\sum_{\boldsymbol{p},\boldsymbol{q}} l_{\boldsymbol{k}}^2 l_{\boldsymbol{p}}^2 l_{\boldsymbol{q}}^2 \Psi^{(1)}_{\boldsymbol{k},\boldsymbol{p},\boldsymbol{q},\boldsymbol{k},\boldsymbol{p},\boldsymbol{q}} \int_0^{\beta\hbar} d\tau_1 \mathcal{G}^{(0)}_{\boldsymbol{k}}(\tau - \tau_1)\mathcal{G}^{(0)}_{\boldsymbol{p}}(0^-)\mathcal{G}^{(0)}_{\boldsymbol{q}}(0^-)\mathcal{G}^{(0)}_{\boldsymbol{k}}(\tau_1)$$
$$= \frac{1}{\beta\hbar}\sum_n e^{-i\omega_n\tau}[\mathcal{G}^{(0)}_{\boldsymbol{k}}(i\omega_n)]^2[\Sigma_{11}(\boldsymbol{k})]_{\mathcal{H}'_6} \quad (6.3.13)$$

where

$$[\Sigma_{11}(\boldsymbol{k})]_{\mathcal{H}'_6} = \frac{18}{\hbar}\sum_{\boldsymbol{p},\boldsymbol{q}} l_{\boldsymbol{k}}^2 l_{\boldsymbol{p}}^2 l_{\boldsymbol{q}}^2 \Psi^{(1)}_{\boldsymbol{k},\boldsymbol{p},\boldsymbol{q},\boldsymbol{k},\boldsymbol{p},\boldsymbol{q}} n_{\boldsymbol{p}} n_{\boldsymbol{q}}$$
$$= -\frac{6zJ}{\hbar S}\frac{d}{\sqrt{\omega_{\boldsymbol{k}}(\omega_{\boldsymbol{k}}+2\delta)}}[A''_{\boldsymbol{k}}(\omega_{\boldsymbol{k}}+\delta) - B''_{\boldsymbol{k}}\delta] \quad (6.3.14)$$

with

$$A''_{\boldsymbol{k}} = \delta\lambda_0(\lambda_1 + \delta\lambda_0) \quad (6.3.15)$$

and

$$B''_{\boldsymbol{k}} = -\frac{1}{2}(\lambda_1 + \delta\lambda_0)^2 - \frac{3}{8}\delta^2\lambda_0^2. \quad (6.3.16)$$

The self-energy (6.3.14) is represented by the third diagram of the first line of Fig. 6.1.

The same procedure is followed to evaluate the three contributions generated from Eq. (6.3.2). The contribution coming from $\mathcal{H}_2'$ is

$$\begin{aligned}
&\frac{1}{\hbar}\int_0^{\beta\hbar} d\tau_1 \langle T[\mathcal{H}_2'(\tau_1)\alpha_{\boldsymbol{k}}(\tau)\alpha_{-\boldsymbol{k}}]\rangle_0^c \\
&\quad = \frac{1}{2\hbar} zJ(Q_{\boldsymbol{k}} + Q_{-\boldsymbol{k}}) \int_0^{\beta\hbar} d\tau_1 \mathcal{G}_{\boldsymbol{k}}^{(0)}(\tau - \tau_1)\mathcal{G}_{\boldsymbol{k}}^{(0)}(-\tau_1) \\
&\quad = \frac{1}{\beta\hbar}\sum_n e^{-i\omega_n\tau}\mathcal{G}_{\boldsymbol{k}}^{(0)}(i\omega_n)\mathcal{G}_{\boldsymbol{k}}^{(0)}(-i\omega_n)[\Sigma_{12}(\boldsymbol{k})]_{\mathcal{H}_2'},
\end{aligned} \tag{6.3.17}$$

where

$$[\Sigma_{12}(\boldsymbol{k})]_{\mathcal{H}_2'} = \frac{1}{\hbar} zJQ_{\boldsymbol{k}}. \tag{6.3.18}$$

The corresponding Feynman diagram is given by the first term of the second line of Fig. 6.1.

The contribution to Eq. (6.3.2) coming from of $\mathcal{H}_4'$ is

$$\begin{aligned}
&\frac{1}{\hbar}\int_0^{\beta\hbar} d\tau_1 \langle T[\mathcal{H}_4'(\tau_1)\alpha_{\boldsymbol{k}}(\tau)\alpha_{-\boldsymbol{k}}]\rangle_0^c \\
&\quad = -\frac{6}{\hbar}\sum_{\boldsymbol{p}} l_{\boldsymbol{k}}^2 l_{\boldsymbol{p}}^2 \Phi^{(2)}_{\boldsymbol{k},\boldsymbol{p},-\boldsymbol{k},\boldsymbol{p}} \int_0^{\beta\hbar} d\tau_1 \mathcal{G}_{\boldsymbol{k}}^{(0)}(\tau - \tau_1)\mathcal{G}_{\boldsymbol{p}}^{(0)}(0^-)\mathcal{G}_{\boldsymbol{k}}^{(0)}(-\tau_1) \\
&\quad = \frac{1}{\beta\hbar}\sum_n e^{-i\omega_n\tau}\mathcal{G}_{\boldsymbol{k}}^{(0)}(i\omega_n)\mathcal{G}_{\boldsymbol{k}}^{(0)}(-i\omega_n)[\Sigma_{12}(\boldsymbol{k})]_{\mathcal{H}_4'},
\end{aligned} \tag{6.3.19}$$

where

$$\begin{aligned}
[\Sigma_{12}(\boldsymbol{k})]_{\mathcal{H}_4'} &= \frac{6}{\hbar}\sum_{\boldsymbol{p}} l_{\boldsymbol{k}}^2 l_{\boldsymbol{p}}^2 \Phi^{(2)}_{\boldsymbol{k},\boldsymbol{p},-\boldsymbol{k},\boldsymbol{p}} n_{\boldsymbol{p}} \\
&= \frac{2zJ}{\hbar}\left\{\frac{1}{\sqrt{\omega_{\boldsymbol{k}}(\omega_{\boldsymbol{k}} + 2\delta)}}[-A_{\boldsymbol{k}}'\delta + B_{\boldsymbol{k}}'(\omega_{\boldsymbol{k}} + \delta)] - C_{\boldsymbol{k}}'\right\}
\end{aligned} \tag{6.3.20}$$

with

$$\begin{aligned}
C_{\boldsymbol{k}}' = &-\frac{1}{2}\delta\lambda_1\gamma_{\boldsymbol{k}} + \frac{1}{2}\delta(\lambda_0 - \lambda_1)\omega_{\boldsymbol{k}} + \frac{3}{2}\delta(\lambda_1 + \delta\lambda_0) \\
&- \frac{3d}{2S}X(\lambda_1 + \delta\lambda_0) + \frac{3d\delta}{8S}Z\lambda_0.
\end{aligned} \tag{6.3.21}$$

The Feynman diagram corresponding to Eq. (6.3.20) is given by the second term of the second line of Fig. 6.1.

The contribution to Eq. (6.3.2) coming from of $\mathcal{H}_6'$ is

$$\frac{1}{\hbar}\int_0^{\beta\hbar} d\tau_1 \langle T[\mathcal{H}_6'(\tau_1)\alpha_{\boldsymbol{k}}(\tau)\alpha_{-\boldsymbol{k}}]\rangle_0^{\mathrm{c}}$$

$$= \frac{24}{\hbar}\sum_{\boldsymbol{p},\boldsymbol{q}} l_{\boldsymbol{k}}^2 l_{\boldsymbol{p}}^2 l_{\boldsymbol{q}}^2 \Psi^{(2)}_{\boldsymbol{p},\boldsymbol{q},\boldsymbol{k},-\boldsymbol{k},\boldsymbol{p},\boldsymbol{q}} \int_0^{\beta\hbar} d\tau_1 \mathcal{G}_{\boldsymbol{k}}^{(0)}(\tau-\tau_1)\mathcal{G}_{\boldsymbol{p}}^{(0)}(0^-)\mathcal{G}_{\boldsymbol{q}}^{(0)}(0^-)\mathcal{G}_{\boldsymbol{k}}^{(0)}(-\tau_1)$$

$$= \frac{1}{\beta\hbar}\sum_n e^{-i\omega_n\tau}\mathcal{G}_{\boldsymbol{k}}^{(0)}(i\omega_n)\mathcal{G}_{\boldsymbol{k}}^{(0)}(-i\omega_n)[\Sigma_{12}(\boldsymbol{k})]_{\mathcal{H}_6'}, \tag{6.3.22}$$

where

$$[\Sigma_{12}(\boldsymbol{k})]_{\mathcal{H}_6'} = \frac{24}{\hbar}\sum_{\boldsymbol{p},\boldsymbol{q}} l_{\boldsymbol{k}}^2 l_{\boldsymbol{p}}^2 l_{\boldsymbol{q}}^2 \Psi^{(2)}_{\boldsymbol{p},\boldsymbol{q},\boldsymbol{k},-\boldsymbol{k},\boldsymbol{p},\boldsymbol{q}}\, n_{\boldsymbol{p}}\, n_{\boldsymbol{q}}$$

$$= \frac{6zJ}{\hbar S}d\left\{\frac{1}{\sqrt{\omega_{\boldsymbol{k}}(\omega_{\boldsymbol{k}}+2\delta)}}[-A_{\boldsymbol{k}}''\delta + B_{\boldsymbol{k}}''(\omega_{\boldsymbol{k}}+\delta)] - C_{\boldsymbol{k}}''\right\} \tag{6.3.23}$$

with

$$C_{\boldsymbol{k}}'' = -\frac{1}{2}(\lambda_1+\delta\lambda_0)^2 - \frac{1}{8}\delta^2\lambda_0^2. \tag{6.3.24}$$

The Feynman diagram corresponding to Eq. (6.3.23) is given by the third term of the second line of Fig. 6.1.

As for Eq. (6.3.3), the contribution coming from $\mathcal{H}_2'$ is

$$\frac{1}{\hbar}\int_0^{\beta\hbar} d\tau_1 \langle T[\mathcal{H}_2'(\tau_1)\alpha_{-\boldsymbol{k}}^+(\tau)\alpha_{\boldsymbol{k}}^+]\rangle_0^{\mathrm{c}} = \frac{1}{2\hbar}zJ(R_{\boldsymbol{k}}+R_{-\boldsymbol{k}})$$

$$\times\int_0^{\beta\hbar} d\tau_1 \mathcal{G}_{\boldsymbol{k}}^{(0)}(\tau-\tau_1)\mathcal{G}_{\boldsymbol{k}}^{(0)}(-\tau_1)$$

$$= \frac{1}{\beta\hbar}\sum_n e^{-i\omega_n\tau}\mathcal{G}_{\boldsymbol{k}}^{(0)}(i\omega_n)\mathcal{G}_{\boldsymbol{k}}^{(0)}(-i\omega_n)[\Sigma_{21}(\boldsymbol{k})]_{\mathcal{H}_2'}, \tag{6.3.25}$$

where

$$[\Sigma_{21}(\boldsymbol{k})]_{\mathcal{H}_2'} = \frac{1}{\hbar}zJR_{\boldsymbol{k}}. \tag{6.3.26}$$

The Feynman diagram corresponding to the self-energy (6.3.26) is shown by the first term of the third line of Fig. 6.1.

The contribution coming from of $\mathcal{H}_4'$ is given by

$$\begin{aligned}
&\frac{1}{\hbar}\int_0^{\beta\hbar} d\tau_1 \langle T[\mathcal{H}_4'(\tau_1)\alpha_{-\boldsymbol{k}}^{+}(\tau)\alpha_{\boldsymbol{k}}^{+}]\rangle_0^{\mathrm{c}} \\
&\quad = -\frac{6}{\hbar}\sum_{\boldsymbol{p}} l_{\boldsymbol{k}}^2 l_{\boldsymbol{p}}^2 \Phi^{(3)}_{\boldsymbol{p},-\boldsymbol{k},\boldsymbol{p},\boldsymbol{k}} \int_0^{\beta\hbar} d\tau_1 \mathcal{G}_{\boldsymbol{k}}^{(0)}(\tau-\tau_1)\mathcal{G}_{\boldsymbol{p}}^{(0)}(0^-)\mathcal{G}_{\boldsymbol{k}}^{(0)}(-\tau_1) \\
&\quad = \frac{1}{\beta\hbar}\sum_n e^{-i\omega_n\tau}\mathcal{G}_{\boldsymbol{k}}^{(0)}(i\omega_n)\mathcal{G}_{\boldsymbol{k}}^{(0)}(-i\omega_n)[\Sigma_{21}(\boldsymbol{k})]_{\mathcal{H}_4'}, \qquad (6.3.27)
\end{aligned}$$

where

$$\begin{aligned}
[\Sigma_{21}(\boldsymbol{k})]_{\mathcal{H}_4'} &= \frac{6}{\hbar}\sum_{\boldsymbol{p}} l_{\boldsymbol{k}}^2 l_{\boldsymbol{p}}^2 \Phi^{(3)}_{\boldsymbol{p},-\boldsymbol{k},\boldsymbol{p},\boldsymbol{k}} n_{\boldsymbol{p}} \\
&= \frac{2zJ}{\hbar}\left\{\frac{1}{\sqrt{\omega_{\boldsymbol{k}}(\omega_{\boldsymbol{k}}+2\delta)}}[-A_{\boldsymbol{k}}'\delta + B_{\boldsymbol{k}}'(\omega_{\boldsymbol{k}}+\delta)] + C_{\boldsymbol{k}}'\right\}. \qquad (6.3.28)
\end{aligned}$$

The corresponding Feynman diagram is given by the second term of the third line of Fig. 6.1.

The contribution to Eq. (6.3.3) coming from of $\mathcal{H}_6'$ is

$$\begin{aligned}
&\frac{1}{\hbar}\int_0^{\beta\hbar} d\tau_1 \langle T[\mathcal{H}_6'(\tau_1)\alpha_{-\boldsymbol{k}}^{+}(\tau)\alpha_{\boldsymbol{k}}^{+}]\rangle_0^{\mathrm{c}} \\
&\quad = \frac{24}{\hbar}\sum_{\boldsymbol{p},\boldsymbol{q}} l_{\boldsymbol{k}}^2 l_{\boldsymbol{p}}^2 l_{\boldsymbol{q}}^2 \Psi^{(3)}_{\boldsymbol{p},\boldsymbol{q},-\boldsymbol{k},\boldsymbol{k},\boldsymbol{p},\boldsymbol{q}} \int_0^{\beta\hbar} d\tau_1 \mathcal{G}_{\boldsymbol{k}}^{(0)}(\tau-\tau_1)\mathcal{G}_{\boldsymbol{p}}^{(0)}(0^-)\mathcal{G}_{\boldsymbol{q}}^{(0)}(0^-)\mathcal{G}_{\boldsymbol{k}}^{(0)}(-\tau_1) \\
&\quad = \frac{1}{\beta\hbar}\sum_n e^{-i\omega_n\tau}\mathcal{G}_{\boldsymbol{k}}^{(0)}(i\omega_n)\mathcal{G}_{\boldsymbol{k}}^{(0)}(-i\omega_n)[\Sigma_{21}(\boldsymbol{k})]_{\mathcal{H}_6'}, \qquad (6.3.29)
\end{aligned}$$

where

$$\begin{aligned}
[\Sigma_{21}(\boldsymbol{k})]_{\mathcal{H}_6'} &= \frac{24}{\hbar}\sum_{\boldsymbol{p},\boldsymbol{q}} l_{\boldsymbol{k}}^2 l_{\boldsymbol{p}}^2 l_{\boldsymbol{q}}^2 \Psi^{(3)}_{\boldsymbol{p},\boldsymbol{q},-\boldsymbol{k},\boldsymbol{k},\boldsymbol{p},\boldsymbol{q}}\, n_{\boldsymbol{p}}\, n_{\boldsymbol{q}} \\
&= \frac{6zJ}{\hbar S}d\left\{\frac{1}{\sqrt{\omega_{\boldsymbol{k}}(\omega_{\boldsymbol{k}}+2\delta)}}[-A_{\boldsymbol{k}}''\delta + B_{\boldsymbol{k}}''(\omega_{\boldsymbol{k}}+\delta)] + C_{\boldsymbol{k}}''\right\}. \qquad (6.3.30)
\end{aligned}$$

The corresponding Feynman diagram is given by the third term of the third line of Fig. 6.1. In conclusion the first-order self-energies are

$$\begin{aligned}
\Sigma_{11}(\boldsymbol{k}) = &-\frac{zJ}{\hbar}\frac{1}{\sqrt{\omega_{\boldsymbol{k}}(\omega_{\boldsymbol{k}}+2\delta)}}\left[\left(A_{\boldsymbol{k}} + 2A_{\boldsymbol{k}}' + \frac{6d}{S}A_{\boldsymbol{k}}''\right)\right. \\
&\left.\times(\omega_{\boldsymbol{k}}+\delta) - \left(B_{\boldsymbol{k}} + 2B_{\boldsymbol{k}}' + \frac{6d}{S}B_{\boldsymbol{k}}''\right)\delta\right], \qquad (6.3.31)
\end{aligned}$$

$$\Sigma_{12}(\boldsymbol{k}) = \frac{zJ}{\hbar}\left\{\frac{1}{\sqrt{\omega_{\boldsymbol{k}}(\omega_{\boldsymbol{k}}+2\delta)}}\left[-\left(A_{\boldsymbol{k}}+2A'_{\boldsymbol{k}}+\frac{6d}{S}A''_{\boldsymbol{k}}\right)\delta\right.\right.$$
$$\left.\left.+\left(B_{\boldsymbol{k}}+2B'_{\boldsymbol{k}}+\frac{6d}{S}B''_{\boldsymbol{k}}\right)(\omega_{\boldsymbol{k}}+\delta)\right]-\left(C_{\boldsymbol{k}}+2C'_{\boldsymbol{k}}+\frac{6d}{S}C''_{\boldsymbol{k}}\right)\right\}, \quad (6.3.32)$$

$$\Sigma_{21}(\boldsymbol{k}) = \frac{zJ}{\hbar}\left\{\frac{1}{\sqrt{\omega_{\boldsymbol{k}}(\omega_{\boldsymbol{k}}+2\delta)}}\left[-\left(A_{\boldsymbol{k}}+2A'_{\boldsymbol{k}}+\frac{6d}{S}A''_{\boldsymbol{k}}\right)\delta\right.\right.$$
$$\left.\left.+\left(B_{\boldsymbol{k}}+2B'_{\boldsymbol{k}}+\frac{6d}{S}B''_{\boldsymbol{k}}\right)(\omega_{\boldsymbol{k}}+\delta)\right]+\left(C_{\boldsymbol{k}}+2C'_{\boldsymbol{k}}+\frac{6d}{S}C''_{\boldsymbol{k}}\right)\right\} \quad (6.3.33)$$

and

$$\Sigma_{22}(\boldsymbol{k}) = \Sigma_{11}(\boldsymbol{k}), \quad (6.3.34)$$

where $A_{\boldsymbol{k}}$, $B_{\boldsymbol{k}}$, $C_{\boldsymbol{k}}$, $A'_{\boldsymbol{k}}$, $B'_{\boldsymbol{k}}$, $C'_{\boldsymbol{k}}$, $A''_{\boldsymbol{k}}$, $B''_{\boldsymbol{k}}$ and $C''_{\boldsymbol{k}}$ are given by Eqs. (6.1.35), (6.1.36), (6.1.37), (6.3.9), (6.3.11), (6.3.21), (6.3.15), (6.3.16) and (6.3.24), respectively. From Eqs. (6.3.34), (6.2.25) and (6.2.26), one obtains $\Sigma_{\mathrm{A}} = 0$ and $\Sigma_{\mathrm{S}} = \Sigma_{11}$ so that the first-order spin wave spectrum that can be deduced from Eq. (6.2.27) becomes

$$\omega^2 = [\epsilon_{\boldsymbol{k}} + \Sigma_{11}(\boldsymbol{k})]^2 - \Sigma_{12}(\boldsymbol{k})\Sigma_{21}(\boldsymbol{k})$$
$$= \left(\frac{2zJS}{\hbar}\right)^2\left\{\omega_{\boldsymbol{k}}(\omega_{\boldsymbol{k}}+2\delta) - \frac{1}{S}\left[\left(A_{\boldsymbol{k}}+2A'_{\boldsymbol{k}}+\frac{6d}{S}A''_{\boldsymbol{k}}\right)(\omega_{\boldsymbol{k}}+\delta)\right.\right.$$
$$\left. - \left(B_{\boldsymbol{k}}+2B'_{\boldsymbol{k}}+\frac{6d}{S}B''_{\boldsymbol{k}}\right)\delta\right]$$
$$+ \frac{1}{4S^2}\left[\left(A_{\boldsymbol{k}}+2A'_{\boldsymbol{k}}+\frac{6d}{S}A''_{\boldsymbol{k}}\right)^2 - \left(B_{\boldsymbol{k}}+2B'_{\boldsymbol{k}}+\frac{6d}{S}B''_{\boldsymbol{k}}\right)^2\right.$$
$$\left.\left. + \left(C_{\boldsymbol{k}}+2C'_{\boldsymbol{k}}+\frac{6d}{S}C''_{\boldsymbol{k}}\right)^2\right]\right\}. \quad (6.3.35)$$

As one can see from Eq. (6.3.35), contributions of order 1 (unperturbed spectrum), $\frac{1}{S}$, $\frac{1}{S^2}$ and $\frac{1}{S^3}$ appear inside the curl brackets. Notice that *all* contributions within $\frac{1}{S}$ are contained in Eq. (6.3.35) while not all contributions of the order $\frac{1}{S^2}$ and $\frac{1}{S^3}$ appear because terms of the same order are generated by higher orders of the perturbation expansion. For this reason, Eq. (6.3.35) is consistent only within terms of order $\frac{1}{S}$ so that the first-order spectrum becomes

$$\omega^2 = \left(\frac{2zJS}{\hbar}\right)^2\left\{\omega_{\boldsymbol{k}}(\omega_{\boldsymbol{k}}+2\delta) - \frac{\omega_{\boldsymbol{k}}}{S}[(X-Y)\omega_{\boldsymbol{k}} + \delta(3X+Z-Y+T)\right.$$
$$\left. + 2(\lambda_2+\delta\lambda_1)\omega_{\boldsymbol{k}} + 2\delta\lambda_2 + 4\delta\lambda_1(1+\delta)]\right\}. \quad (6.3.36)$$

In writing Eq. (6.3.36) the "sum rule" $Z - T + \delta(1+X+Z) = 0$ is used. As one can see from Eq. (6.3.36), the first-order spin wave spectrum is kinematically consistent and shows a soft mode at $\boldsymbol{k} = 0$. This result was obtained by Rastelli and Lindgård[18] at $T = 0$ and Balucani *et al.*[51] at $T \neq 0$. These authors did not use the DM spin-boson transformation because in presence of easy-plane anisotropy it leads to a non-hermitian bilinear Hamiltonian as shown by Eq. (6.1.5) that cannot be diagonalized using the Bogoliubov transformation. In particular, these authors used a modified version of the HP spin-boson transformation introduced by Lindgård and Danielsen[52] in 1974: the matching of matrix element (MME) method. The MME method consists on writing the spin operators and their powers in a normal ordered boson expansion where any expansion in $\frac{1}{S}$ is avoided. Rastelli and Lindgård[18] showed that the MME method leads to the same results as those obtained from the normal ordered HP transformation if a "generalized" Bogoliubov transformation[53] is performed. Indeed, the generalized Bogoliubov transformation consists on diagonalizing the bilinear boson Hamiltonian only *after* having normal ordered the higher-order terms of the boson Hamiltonian. As shown in Sec. 6.1, the terms $\mathcal{H}_4$ and $\mathcal{H}_6$ give contributions to both the ground-state energy (6.1.19) and the bilinear Hamiltonian (6.1.31). The generation of higher-order terms in $\frac{1}{S}$ is a general feature of the easy-plane amsotropy independently of which spin-boson transformation is assumed. However, the great advantage of the DM transformation used in this section with respect to the normal ordered HP transformation and the MME method is that the number of interaction potentials is *finite*. Balucani *et al.*,[51] using the MME method, obtained a spectrum similar to that of Eq. (6.3.36) by means of the Green function equation of motion method pushed to second order. Subsequently, Rastelli and Tassi[46] showed that the DM transformation led to the same result performing a first-order perturbation expansion by means of the many-body technique as illustrated in the present section.

Now we perform an explicit calculation of the temperature-independent terms $X, Y, Z, T$ and of the temperature-dependent terms $\lambda_1, \lambda_2$ occurring in the first-order spectrum (6.3.36). We will transform the sums over the wavevector into an integral containing the isotropic ferromagnet density of states by $D(\omega)$ that satisfies the properties

$$\int_0^{\omega_m} d\omega D(\omega) = \frac{1}{N}\sum_{\boldsymbol{q}} 1 = 1, \tag{6.3.37}$$

$$\int_0^{\omega_m} d\omega\, \omega D(\omega) = \frac{1}{N}\sum_{\boldsymbol{q}} (1-\gamma_{\boldsymbol{q}}) = 1 \tag{6.3.38}$$

and

$$\int_0^{\omega_m} d\omega \frac{D(\omega)}{\omega} = \frac{1}{N}\sum_{\boldsymbol{q}} \frac{1}{1-\gamma_{\boldsymbol{q}}} = I_1 \tag{6.3.39}$$

where $\omega_m$ is the maximum spin wave frequency of the isotropic ferromagnet and $I_1$ is the Watson's integral: for a SC lattice, $\omega_m = 2$ and $I_1 = 1.51637$.

Let us begin evaluating $Z$ given by Eq. (6.1.26). Taking advantage from the fact that in the actual compounds the easy-plane anisotropy is generally much smaller than the exchange interaction, we assume that $\delta \ll 1$, so that we may expand $Z$ in ascending powers of $\delta$ limiting ourselves to the first few terms. By the use of the density of states $D(\omega)$ defined by Eqs. (6.3.37)–(6.3.39), Eq. (6.1.26) becomes

$$Z = -\delta \int_0^{\omega_m} \frac{D(\omega)d\omega}{\sqrt{\omega(\omega + 2\delta)}}. \tag{6.3.40}$$

Obviously, we do not know the analytic expression of $D(\omega)$ but we know that for small frequencies the density of states behaves like $D(\omega) = c\sqrt{\omega}$ as a consequence of the quadratic dependence of the dispersion relation of the isotropic ferromagnet on the wavevector. Then we break the integration over the frequency in two parts: in the first integral the frequency range is assumed to be $0 < \omega < \omega_0$, while in the second integral the frequency range is $\omega_0 < \omega < \omega_m$. Assuming

$$\delta \ll \omega_0 \ll \omega_m, \tag{6.3.41}$$

in the first integral we may replace $D(\omega)$ with $c\sqrt{\omega}$ so that the integration can be easily performed and in the second integral we may expand the function in powers of $\delta$ obtaining

$$Z = -\delta \left\{ 2c[\sqrt{\omega_0 + 2\delta} - \sqrt{2\delta}] + \int_{\omega_0}^{\omega_m} d\omega \, \frac{D(\omega)}{\omega} \left( 1 - \frac{\delta}{\omega} + \frac{3}{2}\frac{\delta^2}{\omega^2} + \cdots \right) \right\}. \tag{6.3.42}$$

Under the assumption (6.3.41), we may expand the function inside the square brackets in powers of $\delta$ obtaining

$$Z = -I_1\delta + c(2\delta)^{\frac{3}{2}} - I_2\delta^2 + O(\delta^3), \tag{6.3.43}$$

where

$$I_1 = 2c\sqrt{\omega_0} + \int_{\omega_0}^{\omega_m} d\omega \frac{D(\omega)}{\omega} = \int_0^{\omega_0} d\omega \frac{c\sqrt{\omega}}{\omega} + \int_{\omega_0}^{\omega_m} d\omega \frac{D(\omega)}{\omega} = \int_0^{\omega_m} d\omega \frac{D(\omega)}{\omega} \tag{6.3.44}$$

in agreement with Eq. (6.3.39) and

$$I_2 = \frac{2c}{\sqrt{\omega_0}} - \int_{\omega_0}^{\omega_m} d\omega \frac{D(\omega)}{\omega^2}. \tag{6.3.45}$$

Even though we do not know the numerical value of $I_2$, we have been able to write an expansion of $Z$ in powers of $\delta$ picking the non-analytic term coming from the low frequency square-root dependence of the density of state of the isotropic ferromagnet. In a similar way, we obtain

$$X = \frac{1}{3}c(2\delta)^{\frac{3}{2}} - \frac{1}{2}I_2\delta^2 + O(\delta^3), \tag{6.3.46}$$

$$T = (1 - I_1)\delta + c(2\delta)^{\frac{3}{2}} - (I_1 + I_2)\delta^2 + \frac{2}{3}c(2\delta)^{\frac{5}{2}} + O(\delta^3) \tag{6.3.47}$$

and

$$Y = \frac{1}{3}c(2\delta)^{\frac{3}{2}} - \frac{1}{2}(I_1 + I_2)\delta^2 + \frac{2}{5}c(2\delta)^{\frac{5}{2}} + O(\delta^3). \qquad (6.3.48)$$

As for the temperature-dependent terms $\lambda_1$ and $\lambda_2$ occurring in Eq. (6.3.36), we proceed in a way similar to that used for $Z$. First of all, we transform the sums over the wavevectors into integrals over the frequency making use of the density of states $D(\omega)$. Then we break the integrals in two parts: in the first integral ($0 < \omega < \omega_0$), we replace the density of states by $D(\omega) = c\sqrt{\omega}$ and the occupation number by its expansion

$$n = \frac{1}{e^{\beta\hbar\epsilon} - 1} \simeq \frac{1}{\beta\hbar\epsilon} = \frac{\theta}{\omega(\omega + 2\delta)}, \qquad (6.3.49)$$

where

$$\theta = \frac{k_B T}{2zJS}. \qquad (6.3.50)$$

The approximation (6.3.49) is justified for $\omega_0 \ll \theta$, that is for temperatures such that $D(2S-1) \ll k_BT \ll 2zJS$. In the second integral ($\omega_0 < \omega < \omega_m$), we may expand the integrand function in powers of $\frac{\delta}{\omega}$ since $\frac{\delta}{\omega_0} \ll 1$. By doing so, we obtain

$$\lambda_1 = c\theta \int_0^{\omega_0} d\omega \frac{\sqrt{\omega}}{\omega + 2\delta} + \int_{\omega_0}^{\omega_m} d\omega \frac{D(\omega)}{e^{\frac{\omega}{\theta}} - 1}\left(1 - \frac{\delta}{\omega} - \frac{\delta}{\theta}\frac{e^{\frac{\omega}{\theta}}}{e^{\frac{\omega}{\theta}} - 1} + O(\delta^2)\right). \qquad (6.3.51)$$

Using the relationships

$$\int_0^{\omega_0} d\omega \frac{\sqrt{\omega}}{\omega + 2\delta} = 2\sqrt{\omega_0} - 2\sqrt{2\delta}\arctan\sqrt{\frac{\omega_0}{2\delta}} = 2\sqrt{\omega_0} - \pi\sqrt{2\delta} + O(\delta) \qquad (6.3.52)$$

and

$$2\theta c\sqrt{\omega_0} = \int_0^{\omega_0} d\omega c\sqrt{\omega}\frac{\theta}{\omega} \simeq \int_0^{\omega_0} d\omega \frac{D(\omega)}{e^{\frac{\omega}{\theta}} - 1}, \qquad (6.3.53)$$

Equation (6.3.51) becomes

$$\lambda_1 = \int_0^{\omega_m} d\omega D(\omega) n^{(0)}(\omega) - \pi\theta c\sqrt{2\delta} = \sigma_0 - \pi\theta c\sqrt{2\delta} + O(\delta), \qquad (6.3.54)$$

where

$$\sigma_0 = \frac{1}{N}\sum_{\boldsymbol{q}} n_{\boldsymbol{q}}^{(0)} = \frac{1}{N}\sum_{\boldsymbol{q}} \frac{1}{e^{\frac{\omega_{\boldsymbol{q}}}{\theta}} - 1}. \qquad (6.3.55)$$

In Eqs. (6.3.54) and (6.3.55), $n^{(0)}$ is the Bose factor of the isotropic ferromagnet. Following the same procedure, we obtain for $\lambda_2$

$$\lambda_2 = \sigma_1 - 2\delta\sigma_0 \tag{6.3.56}$$

where

$$\sigma_1 = \frac{1}{N}\sum_{\boldsymbol{q}} \omega_{\boldsymbol{q}} n_{\boldsymbol{q}}^{(0)}. \tag{6.3.57}$$

Replacing Eqs. (6.3.43), (6.3.46)–(6.3.48), (6.3.54) and (6.3.56) into Eq. (6.3.35), one obtains for the renormalized spin wave spectrum

$$\omega^2 = \left(\frac{2zJS}{\hbar}\right)^2 \left\{\omega_{\boldsymbol{k}}(\omega_{\boldsymbol{k}} + 2\delta) + \frac{\delta^2}{2S}\omega_{\boldsymbol{k}}(4I_1 - 2 - I_1\omega_{\boldsymbol{k}}) - 2\frac{\omega_{\boldsymbol{k}}}{S}[(\omega_{\boldsymbol{k}} + \delta)\sigma_1 \right.$$
$$\left. + \delta(2 - \omega_{\boldsymbol{k}})\sigma_0] + \frac{\delta^2}{4S^2}[-1 + (I_1 - 1)\omega_{\boldsymbol{k}}^2] - \frac{2}{S^2}\delta\omega_{\boldsymbol{k}}\sigma_0 + O\left(\delta^{\frac{5}{2}}, \theta\delta^{\frac{3}{2}}\right)\right\}. \tag{6.3.58}$$

Equation (6.3.58) reproduces the first-order spectrum given by Eq. (55) of Rastelli and Tassi[46] except for a minor change in the term proportional to $\sigma_0$. A factor 3 should be replaced by a factor 2 in the last line of Eq. (55). Moreover, the contribution of the spectrum (6.3.58) proportional to $\frac{1}{S}$ recovers Eq. (25) of Balucani *et al.*[51] for the temperature-dependent contribution once the thermal averages $\langle a_{\boldsymbol{p}}^+ a_{\boldsymbol{p}}\rangle$ are expressed in terms of the Bose factors $\langle \alpha_{\boldsymbol{p}}^+ \alpha_{\boldsymbol{p}}\rangle$. Notice that the $\frac{1}{S}$ contribution is correct while the contribution of order $\frac{1}{S^2}$ is manifestly incorrect: indeed, the Goldstone theorem is violated because of the presence of the term $-\frac{\delta^2}{4S^2}$ that remains finite and negative at $\boldsymbol{k} = 0$. As announced, terms of the same order in $\frac{1}{S}$ come from the second-order perturbation theory as we will see in the next section.

## 6.4. Second-Order Perturbation Theory

The only contribution of order $\frac{1}{S^2}$ to the squared renormalized spin wave spectrum of Eq. (6.3.35) coming from the second-order perturbation theory is given by $2\epsilon_{\boldsymbol{k}}\Sigma_{11}^*(\boldsymbol{k}, \epsilon_{\boldsymbol{k}})$ where $\Sigma_{11}^*$ is the second-order proper self-energy evaluated for $i\omega_n = \epsilon_{\boldsymbol{k}}$ coming from the perturbation expansion (6.2.5), that is,

$$\mathcal{G}_{\boldsymbol{k}}^{11}(\tau) = -\frac{1}{2!\hbar^2}\int_0^{\beta\hbar} d\tau_1 \int_0^{\beta\hbar} d\tau_2 \langle T[\mathcal{H}'(\tau_1)\mathcal{H}'(\tau_2)\alpha_{\boldsymbol{k}}(\tau)\alpha_{\boldsymbol{k}}^+]\rangle_0^{\mathrm{c}}, \tag{6.4.1}$$

where $\mathcal{H}'$ is restricted to the sum $\mathcal{H}_2' + \mathcal{H}_4'$, since any product involving $\mathcal{H}_6'$ which is of order $\frac{1}{S^2}$ leads to terms of order at least $\frac{1}{S^3}$ neglected in the present approximation. The term $\mathcal{H}_2'(\tau_1)\mathcal{H}_2'(\tau_2)$ leads to a reducible self-energy so that the terms contributing to the proper self-energy entering the matrix Dyson equation are restricted to the products $\mathcal{H}'_2(\tau_1)\mathcal{H}'_4(\tau_2)$, $\mathcal{H}'_4(\tau_1)\mathcal{H}'_2(\tau_2)$ and $\mathcal{H}'_4(\tau_1)\mathcal{H}'_4(\tau_2)$. The terms $\mathcal{H}'_2(\tau_1)\mathcal{H}'_4(\tau_2)$ and $\mathcal{H}'_4(\tau_1)\mathcal{H}'_2(\tau_2)$ give the same contribution since they are transformed into one another changing the time variables $\tau_1$ and $\tau_2$. This invariance under time permutation is accounted for keeping only one term, for instance $\mathcal{H}'_4(\tau_1)\mathcal{H}'_2(\tau_2)$ and neglecting the factor $\frac{1}{2!}$ in the expansion (6.4.1). Moreover, within the present approximation, the potentials $P, Q, R$ given by Eqs. (6.1.32)–(6.1.37) and $\Phi^{(i)}$ given by Eqs. (6.1.38)–(6.1.43) have to be restricted to their parts independent of $S$ and $D^{(3)} = \frac{zJd}{4SN}$, respectively.

Let us begin evaluating the contribution to the proper self-energy coming from $\mathcal{H}'_4\mathcal{H}'_2$. From Eq. (6.4.1), one obtains

$$\mathcal{G}^{11}_{\boldsymbol{k}}(\tau) = -\frac{1}{\hbar^2}\int_0^{\beta\hbar} d\tau_1 \int_0^{\beta\hbar} d\tau_2 \langle T[\mathcal{H}'_4(\tau_1)\mathcal{H}'_2(\tau_2)\alpha_{\boldsymbol{k}}(\tau)\alpha^+_{\boldsymbol{k}}]\rangle_0^c$$
$$= \frac{1}{\beta\hbar}\sum_n e^{-i\omega_n\tau}[\mathcal{G}^{(0)}_{\boldsymbol{k}}(i\omega_n)]^2[\Sigma^*_{11}(\boldsymbol{k})]_{\mathcal{H}'_4\mathcal{H}'_2}, \tag{6.4.2}$$

where

$$[\Sigma^*_{11}(\boldsymbol{k})]_{\mathcal{H}'_2\mathcal{H}'_4} = 4\beta\frac{zJ}{\hbar}\sum_{\boldsymbol{q}} l^2_{\boldsymbol{k}} l^2_{\boldsymbol{q}}\ \Phi^{(1)}_{\boldsymbol{k},\boldsymbol{q},\boldsymbol{k},\boldsymbol{q}}\ P_{\boldsymbol{q}} n_{\boldsymbol{q}}(1+n_{\boldsymbol{q}})$$
$$-3\frac{zJ}{\hbar^2}\sum_{\boldsymbol{q}} l^2_{\boldsymbol{k}} l^2_{\boldsymbol{q}}\ \Phi^{(2)}_{\boldsymbol{k},\boldsymbol{q},-\boldsymbol{q},\boldsymbol{k}}\ R_{\boldsymbol{q}}\ \frac{1+2n_{\boldsymbol{q}}}{2\epsilon_{\boldsymbol{q}}}$$
$$-3\frac{zJ}{\hbar^2}\sum_{\boldsymbol{q}} l^2_{\boldsymbol{k}} l^2_{\boldsymbol{q}}\ \Phi^{(3)}_{\boldsymbol{k},-\boldsymbol{q},\boldsymbol{q},\boldsymbol{k}}\ Q_{\boldsymbol{q}}\ \frac{1+2n_{\boldsymbol{q}}}{2\epsilon_{\boldsymbol{q}}}. \tag{6.4.3}$$

The three contributions to the second-order proper self-energy (6.4.3) are represented by the Feynman diagrams of the first line of Fig. 6.2. With the help of such diagrams it is easier to write their analytic translation using the same rules given in Chapter 4 for the isotropic ferromagnet. For instance, the third diagram of the first line of Fig. 6.2 corresponds to a product between the potentials $-\frac{zJ}{2}Q_{\boldsymbol{q}}$ and $l^2_{\boldsymbol{k}} l^2_{\boldsymbol{q}}\Phi^{(3)}_{\boldsymbol{k},-\boldsymbol{q},\boldsymbol{q},\boldsymbol{k}}$ times a factor 6 coming from the number of ways of contracting the boson operators that lead to six equivalent contributions since the potential $\Phi^{(3)}$ is invariant under the permutation of the last three subscripts ($3! = 6$ is the multiplicity of the diagram). The sums over internal frequencies are reduced to a single sum over the only frequency remaining after the use of the "frequency conservation" rule applied to the internal lines of the diagram. Finally, using Eqs. (6.2.17) and (4.2.16) one obtains

$$\frac{1}{\beta\hbar}\sum_r \mathcal{G}^{(0)}_{\boldsymbol{q}}(i\omega_r)\mathcal{G}^{(0)}_{\boldsymbol{q}}(-i\omega_r) = \frac{1}{\beta\hbar}\sum_r \frac{1}{i\omega_r-\epsilon_{\boldsymbol{q}}}\,\frac{1}{-i\omega_r-\epsilon_{\boldsymbol{q}}}$$
$$= -\frac{1}{\beta\hbar}\sum_r \frac{1}{2\epsilon_{\boldsymbol{q}}}\left(\frac{1}{i\omega_r-\epsilon_{\boldsymbol{q}}} - \frac{1}{i\omega_r+\epsilon_{\boldsymbol{q}}}\right)$$
$$= \frac{1}{2\epsilon_{\boldsymbol{q}}}[n_{\boldsymbol{q}} - n(-\epsilon_{\boldsymbol{q}})] = \frac{1+2n_{\boldsymbol{q}}}{2\epsilon_{\boldsymbol{q}}}. \tag{6.4.4}$$

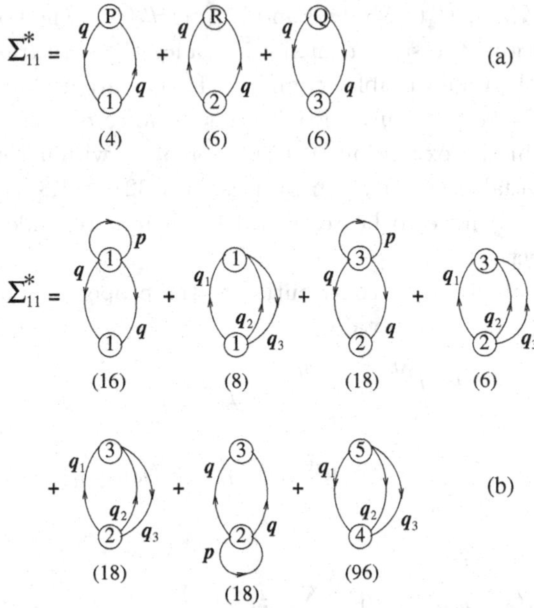

Fig. 6.2. Second-order proper self-energy $[\Sigma_{11}^*]_{\mathcal{H}_4'\mathcal{H}_2'}$ (a) and $[\Sigma_{11}^*]_{\mathcal{H}_4'\mathcal{H}_4'}$ (b). The circles labelled by P, Q and R of (a) represent the potentials occurring in $\mathcal{H}_2'$. The circles containing 1, 2, 3, 4 and 5 in (a) and (b) represent the potentials $\Phi^{(1)}$, $\Phi^{(2)}$, $\Phi^{(3)}$, $\Phi^{(4)}$ and $\Phi^{(5)}$ occurring in $\mathcal{H}_4'$, respectively. The numbers in brackets under the diagrams indicate their multiplicity.

Neglecting terms of order higher than $\frac{\delta^{5/2}}{S^2}$, the proper self-energy (6.4.3) reduces to

$$[\Sigma_{11}^*(\boldsymbol{k})]_{\mathcal{H}_2'\mathcal{H}_4'} = \frac{2zJS}{\hbar}\frac{\delta^2}{4S^2}\frac{1}{\sqrt{\omega_{\boldsymbol{k}}(\omega_{\boldsymbol{k}}+2\delta)}}\left\{\frac{1}{2}[1+2\omega_{\boldsymbol{k}}-(2-I_1)\omega_{\boldsymbol{k}}^2]\right.$$
$$+\frac{\omega_{\boldsymbol{k}}^2}{2\theta}[2(1-2I_1)\rho_1+I_1\,\rho_2]+\sigma_0\omega_{\boldsymbol{k}}[4-I_1-(2-I_1)\omega_{\boldsymbol{k}}]$$
$$\left.+\,\sigma_1[1-(2-I_1)\omega_{\boldsymbol{k}}]\right\}, \tag{6.4.5}$$

with

$$\rho_1 = \frac{1}{N}\sum_{\boldsymbol{q}}\omega_{\boldsymbol{q}} n_{\boldsymbol{q}}^{(0)}(1+n_{\boldsymbol{q}}^{(0)}) \simeq \frac{3}{2}\theta\,\sigma_0, \tag{6.4.6}$$

$$\rho_2 = \frac{1}{N}\sum_{\boldsymbol{q}}\omega_{\boldsymbol{q}}^2 n_{\boldsymbol{q}}^{(0)}(1+n_{\boldsymbol{q}}^{(0)}) \simeq \frac{5}{2}\theta\,\sigma_1 \tag{6.4.7}$$

where $n_{\boldsymbol{q}}^{(0)}$ is the Bose factor of the isotropic model. The relationship between $\rho_r$ and $\sigma_r$ is obtained as follows: in terms of the density of states, the sum over wavevectors is transformed into a sum over frequencies like

$$\rho_r = \int_0^{\omega_m} d\omega D(\omega)\omega^r\frac{e^{\frac{\omega}{\theta}}}{(e^{\frac{\omega}{\theta}}-1)^2} = \theta\int_0^{\omega_m} d\omega[r\omega^{r-1}D(\omega)+\omega^r D'(\omega)]\frac{1}{e^{\frac{\omega}{\theta}}-1}, \tag{6.4.8}$$

where an integration by parts has been performed and the integrated contribution vanishes because the density of states is zero for $\omega = 0$ and $\omega = \omega_m$. At low temperature, the main contribution to the integral comes from the low frequency region where the density of states may be replaced by the function $D(\omega) = c\sqrt{\omega}$. Using the change of variable $\omega \to x = \frac{\omega}{\theta}$ and assuming that the upper limit of the integral $\frac{\omega_m}{\theta}$ goes to infinity, Eq. (6.4.8) becomes

$$\rho_r = c\left(r + \frac{1}{2}\right)\theta^{r+\frac{3}{2}} \int_0^\infty dx \frac{x^{r-\frac{1}{2}}}{e^x - 1} = a_{r+\frac{1}{2}}\, c\, \left(r + \frac{1}{2}\right)\theta^{r+\frac{3}{2}} \tag{6.4.9}$$

with[3]

$$a_r = \Gamma(r)\zeta(r) \tag{6.4.10}$$

where $\Gamma$ and $\zeta$ are the Gamma-function and the Riemann function, respectively. By a similar procedure, we obtain for $\sigma_r$

$$\sigma_r = \int_0^{\omega_m} d\omega D(\omega)\omega^r \frac{1}{e^{\frac{\omega}{\theta}} - 1} = c\,\theta^{r+\frac{3}{2}} \int_0^\infty dx \frac{x^{r+\frac{1}{2}}}{e^x - 1} = a_{r+\frac{3}{2}}\, c\,\theta^{r+\frac{3}{2}}. \tag{6.4.11}$$

From Eqs. (6.4.9) and (6.4.11), we obtain

$$\frac{\rho_{r+1}}{\sigma_r} = \left(r + \frac{3}{2}\right)\theta \tag{6.4.12}$$

so that the temperature-dependent contribution to the self-energy of Eq. (6.4.5) is of the order $\delta^2\theta^{\frac{3}{2}}/S^2$, and consistently with Eq. (6.3.58) has to be neglected. Then the contribution to the renormalized spin wave spectrum coming from the self-energy (6.4.5) becomes

$$2\epsilon_{\boldsymbol{k}}[\Sigma_{11}(\boldsymbol{k})]_{\mathcal{H}_2'\mathcal{H}_4'} = \left(\frac{2zJS}{\hbar}\right)^2 \left\{\frac{\delta^2}{4S^2}[1 + 2\omega_{\boldsymbol{k}} - (2 - I_1)\omega_{\boldsymbol{k}}^2] + O\left(\delta^{\frac{5}{2}}, \delta^2\theta^{\frac{3}{2}}\right)\right\}. \tag{6.4.13}$$

Note that the term $\frac{\delta^2}{4S^2}$ of Eq. (6.4.13) cancels out the analogous term of the first-order spectrum (6.3.58) restoring the soft mode at $\boldsymbol{k} = 0$.

Let us define $[\Sigma_{11}^*(\boldsymbol{k}, i\omega_n)]_{\mathcal{H}_4'\mathcal{H}_4'}$ as the proper self-energy obtained from the perturbation expansion (6.4.1) with $\mathcal{H}' = \mathcal{H}_4'$. It consists of seven contributions represented by the Feynman diagrams classified as (b) in Fig. 6.2. Their anlaytic translation is given by

$$[\Sigma_{11}^*(\boldsymbol{k}, i\omega_n)]_{\mathcal{H}_4'\mathcal{H}_4'} = -\frac{16}{\hbar^2}\sum_{\boldsymbol{p},\boldsymbol{q}} l_{\boldsymbol{k}}^2 l_{\boldsymbol{p}}^2 l_{\boldsymbol{q}}^4\, \Phi^{(1)}_{\boldsymbol{k},\boldsymbol{q},\boldsymbol{k},\boldsymbol{q}} \Phi^{(1)}_{\boldsymbol{q},\boldsymbol{p},\boldsymbol{q},\boldsymbol{p}}\, \beta\hbar\, n_{\boldsymbol{p}} n_{\boldsymbol{q}}(1 + n_{\boldsymbol{q}})$$

$$+\frac{8}{\hbar^2}\sum_{\boldsymbol{q}_1,\boldsymbol{q}_2,\boldsymbol{q}_3} l_{\boldsymbol{k}}^2 l_{\boldsymbol{q}_1}^2 l_{\boldsymbol{q}_2}^2 l_{\boldsymbol{q}_3}^2 \delta_{\boldsymbol{q}_1+\boldsymbol{q}_2,\boldsymbol{q}_3+\boldsymbol{k}}\, \Phi^{(1)}_{\boldsymbol{q}_1,\boldsymbol{q}_2,\boldsymbol{q}_3,\boldsymbol{k}} \Phi^{(1)}_{\boldsymbol{k},\boldsymbol{q}_3,\boldsymbol{q}_2,\boldsymbol{q}_1}$$

$$\times \frac{n_{\boldsymbol{q}_3}(1 + n_{\boldsymbol{q}_1})(1 + n_{\boldsymbol{q}_2}) - n_{\boldsymbol{q}_1} n_{\boldsymbol{q}_2}(1 + n_{\boldsymbol{q}_3})}{i\omega_n - \epsilon_{\boldsymbol{q}_1} - \epsilon_{\boldsymbol{q}_2} + \epsilon_{\boldsymbol{q}_3}}$$

$$- \frac{18}{\hbar^2} \sum_{p,q} l_k^2 l_p^2 l_q^4 \, \Phi^{(2)}_{k,q,-q,k} \Phi^{(3)}_{p,-q,q,p} \, \frac{n_p(1+2n_q)}{2\epsilon_q}$$

$$+ \frac{6}{\hbar^2} \sum_{q_1,q_2,q_3} l_k^2 l_{q_1}^2 l_{q_2}^2 l_{q_3}^2 \delta_{q_1+q_2,q_3+k} \, \Phi^{(2)}_{q_1,q_2,-q_3,k} \Phi^{(3)}_{k,-q_3,q_2,q_1}$$

$$\times \frac{(1+n_{q_1})(1+n_{q_2})(1+n_{q_3}) - n_{q_1}n_{q_2}n_{q_3}}{i\omega_n - \epsilon_{q_1} - \epsilon_{q_2} - \epsilon_{q_3}}$$

$$- \frac{18}{\hbar^2} \sum_{q_1,q_2,q_3} l_k^2 l_{q_1}^2 l_{q_2}^2 l_{q_3}^2 \delta_{q_1+k,q_2+q_3} \, \Phi^{(2)}_{k,q_1,-q_2,q_3} \Phi^{(3)}_{q_3,-q_2,q_1,k}$$

$$\times \frac{n_{q_3}(1+n_{q_1})(1+n_{q_2}) - n_{q_1}n_{q_2}(1+n_{q_3})}{i\omega_n + \epsilon_{q_1} + \epsilon_{q_2} - \epsilon_{q_3}}$$

$$- \frac{18}{\hbar^2} \sum_{p,q} l_k^2 l_p^2 l_q^4 \, \Phi^{(2)}_{p,q,-q,p} \Phi^{(3)}_{k,-q,q,k} \, \frac{n_p(1+2n_q)}{2\epsilon_q}$$

$$- \frac{96}{\hbar^2} \sum_{q_1,q_2,q_3} l_k^2 l_{q_1}^2 l_{q_2}^2 l_{q_3}^2 \delta_{q_1+q_2,q_3+k}$$

$$\times \Phi^{(4)}_{-q_1,-q_2,q_3,k} \Phi^{(5)}_{k,q_3,-q_2,-q_1}$$

$$\times \frac{(1+n_{q_1})(1+n_{q_2})(1+n_{q_3}) - n_{q_1}n_{q_2}n_{q_3}}{i\omega_n + \epsilon_{q_1} + \epsilon_{q_2} + \epsilon_{q_3}}. \tag{6.4.14}$$

The lowest order in temperature is obtained by neglecting in Eq. (6.4.14) all terms involving products of two or more Bose factors labelled by different wavevectors. By doing so, we obtain

$$[\Sigma_{11}^*(k, i\omega_n)]_{\mathcal{H}_4'\mathcal{H}_4'} = \frac{6}{\hbar^2} \sum_{q_1,q_2,q_3} l_k^2 l_{q_1}^2 l_{q_2}^2 l_{q_3}^2 \delta_{q_1+q_2,q_3+k}$$

$$\times \left[ \frac{\Phi^{(2)}_{q_1,q_2,-q_3,k} \Phi^{(3)}_{k,-q_3,q_2,q_1}}{i\omega_n - \epsilon_{q_1} - \epsilon_{q_2} - \epsilon_{q_3}} \right.$$

$$\left. - 16 \frac{\Phi^{(4)}_{-q_1,-q_2,q_3,k} \Phi^{(5)}_{k,q_3,-q_2,-q_1}}{i\omega_n + \epsilon_{q_1} + \epsilon_{q_2} + \epsilon_{q_3}} \right]$$

$$+ \frac{8}{\hbar^2} \sum_{q_1,q_2,q_3} l_k^2 l_{q_1}^2 l_{q_2}^2 l_{q_3}^2 \delta_{q_1+q_2,q_3+k}$$

$$\times \frac{\Phi^{(1)}_{q_1,q_2,q_3,k} \Phi^{(1)}_{k,q_3,q_2,q_1}}{i\omega_n - \epsilon_{q_1} - \epsilon_{q_2} + \epsilon_{q_3}} n_{q_3}$$

$$+ \frac{18}{\hbar^2} \sum_{q_1,q_2,q_3} l_k^2 l_{q_1}^2 l_{q_2}^2 l_{q_3}^2 \delta_{q_1+q_2,q_3+k}$$

$$\times\left[\frac{\Phi^{(2)}_{q_1,q_2,-q_3,k}\Phi^{(3)}_{k,-q_3,q_2,q_1}}{i\omega_n-\epsilon_{q_1}-\epsilon_{q_2}-\epsilon_{q_3}}-\frac{\Phi^{(2)}_{k,q_3,-q_2,q_1}\Phi^{(3)}_{q_1,-q_2,q_3,k}}{i\omega_n-\epsilon_{q_1}+\epsilon_{q_2}+\epsilon_{q_3}}\right.$$

$$\left.-16\frac{\Phi^{(4)}_{-q_1,-q_2,q_3,k}\Phi^{(5)}_{k,q_3,-q_2,-q_1}}{i\omega_n+\epsilon_{q_1}+\epsilon_{q_2}+\epsilon_{q_3}}\right]n_{q_1}$$

$$-\frac{9}{\hbar^2}\sum_{p,q}l_k^2l_p^2l_q^4\big[\Phi^{(2)}_{k,q,-q,k}\Phi^{(3)}_{p,-q,q,p}$$

$$+\Phi^{(2)}_{p,q,-q,p}\Phi^{(3)}_{k,-q,q,k}\big]\frac{n_p}{\epsilon_q}. \qquad (6.4.15)$$

Now, we select the terms of Eq. (6.4.15) consistent with the spectrum (6.3.58) and (6.4.13), that is, we neglect all contributions higher than $\delta^2/S^2$ in the temperature-independent terms and all contributions higher than $\theta^{\frac{5}{2}}$ or $\delta\theta^{\frac{3}{2}}$ in the temperature-dependent terms. Accordingly, in Eqs. (6.3.43) and (6.3.46)–(6.3.48) we may claim that

$$\frac{1}{N}\sum_q l_q^2\sim\frac{1}{N}\sum_q l_q^2\omega_q=O(1), \qquad (6.4.16)$$

$$\frac{1}{N}\sum_q l_q^2x_q\sim\frac{1}{N}\sum_q l_q^2x_q\omega_q=O(\delta) \qquad (6.4.17)$$

and

$$\sum_q l_k^2x_q^2=O\left(\delta^{\frac{3}{2}}\right),\quad \sum_q l_k^2x_q^2\omega_q=O(\delta^2). \qquad (6.4.18)$$

Equations (6.4.16)–(6.4.18) allow us to establish the general rule that the powers of $\delta$ increase according to the number of factors $x_q$ associated to $l_q^2$. This remark leads to the conclusion that the main anisotropy contributions come from the terms generated by the products $\Phi^{(i)}\Phi^{(j)}$ having the minimum number of $x_q$. Consistently with the order $\frac{1}{S^2}$, we can write Eq. (6.1.10) in the form

$$V_{q_1,q_2,q_3,q_4}=-\frac{zJ}{2N}(v_{q_1,q_2,q_3,q_4}+2\delta), \qquad (6.4.19)$$

where

$$v_{q_1,q_2,q_3,q_4}=\frac{1}{2}(\gamma_{q_1-q_3}+\gamma_{q_1-q_4}+\gamma_{q_2-q_3}+\gamma_{q_2-q_4})-(\gamma_{q_1}+\gamma_{q_2}). \qquad (6.4.20)$$

Using Eqs. (6.1.38)–(6.1.43), the product of potentials occurring under the sum of the temperature-independent term of Eq. (6.4.15) may be replaced by

$$\Phi^{(2)}_{q_1,q_2,-q_3,k}\Phi^{(3)}_{k,-q_3,q_2,q_1}=\left(\frac{zJ}{2N}\right)^2\left\{\frac{4}{3}[x_{q_1}^2\; v_{q_2,-q_3,-q_1,k}\; v_{k,-q_1,-q_3,q_2}\right.$$

$$+\;2x_{q_1}x_{q_2}\; v_{q_2,-q_3,-q_1,k}\; v_{k,-q_2,-q_3,q_1}]$$

$$\left.+\;4\delta x_{q_1}v_{q_2,-q_3,-q_1,k}+O\left(\delta^{\frac{5}{2}}\right)\right\} \qquad (6.4.21)$$

and

$$\Phi^{(4)}_{-q_1,-q_2,q_3,k}\Phi^{(5)}_{k,q_3,-q_2,-q_1} = \left(\frac{zJ}{N}\right)^2 O(\delta^3). \tag{6.4.22}$$

Replacing Eq. (6.4.21) in the temperature-independent part of the proper self-energy (6.4.15), we obtain

$$\begin{aligned}[\Sigma^*_{11}(\boldsymbol{k},\epsilon_{\boldsymbol{k}},T=0)]_{\mathcal{H}'_4\mathcal{H}'_4} &= \frac{2zJS}{\hbar}\frac{\delta^2}{8S^2}\frac{1}{N^2}\sum_{q_1,q_2,q_3}\delta_{q_1+q_2,q_3+k} \\ &\times\left[\frac{1}{\omega^2_{q_1}}v_{q_2,-q_3,-q_1,k}\,v_{k,-q_1,-q_3,q_2}\right. \\ &+\frac{2}{\omega_{q_1}\omega_{q_2}}v_{q_2,-q_3,-q_1,k}\,v_{k,-q_2,-q_3,q_1} \\ &\left.-\frac{6}{\omega_{q_1}}v_{q_2,-q_3,-q_1,k}\right]\frac{1}{\omega_k-\omega_{q_1}-\omega_{q_2}-\omega_{q_3}}.\end{aligned} \tag{6.4.23}$$

As for the temperature-dependent part of the proper self-energy (6.4.15), we can see that the terms of order $\theta^{\frac{5}{2}}$ and $\delta\theta^{\frac{3}{2}}$ come from the contribution containing $\Phi^{(1)}$ that gives

$$\begin{aligned}[\Sigma^*_{11}(\boldsymbol{k},\epsilon_{\boldsymbol{k}},T\neq 0)]_{\mathcal{H}'_4\mathcal{H}'_4} &= \frac{2zJS}{\hbar}\frac{1}{2S^2}\left[2\delta\sigma_0+\frac{1}{N^2}\sum_{q_1,q_2,q_3}\delta_{q_1+q_2,q_3+k}\right. \\ &\left.\times\frac{v_{q_1,q_2,q_3,k}\,v_{k,q_3,q_2,q_1}}{\omega_k-\omega_{q_1}-\omega_{q_2}+\omega_{q_3}}n^{(0)}_{q_3}+O(\delta^2\theta)\right].\end{aligned} \tag{6.4.24}$$

Equation (6.4.24) reduces to the isotropic result (3.4.25) for $\delta=0$ and it is of order $\theta^{\frac{5}{2}}$. All other terms in Eq. (6.4.15) give higher-order contributions. The renormalized second-order spin wave spectrum is the sum of (6.3.58), (6.4.13), (6.4.23) and (6.4.24), where the last two equations have to be multiplied by $2\epsilon_{\boldsymbol{k}}$. The result is

$$\begin{aligned}\omega^2 = \left(\frac{2zJS}{\hbar}\right)^2\omega_k\Bigg\{&\omega_k+2\delta+\frac{\delta^2}{2S}(4I_1-2-I_1\omega_k)+\frac{\delta^2}{4S^2}[2+(2I_1-3)\omega_k] \\ &+\frac{\delta^2}{4S^2}\frac{1}{N^2}\sum_{q_1,q_2,q_3}\delta_{q_1+q_2,q_3+k}\frac{v_{q_2,-q_3,-q_1,k}}{\omega_{q_1}}\left(\frac{v_{k,-q_1,-q_3,q_2}}{\omega_{q_1}}\right. \\ &\left.+2\frac{v_{k,-q_2,-q_3,q_1}}{\omega_{q_2}}-6\right)\frac{1}{\omega_k-\omega_{q_1}-\omega_{q_2}-\omega_{q_3}} \\ &-\frac{2}{S}[(\omega_k+\delta)\sigma_1+\delta(2-\omega_k)\sigma_0]+\frac{2\delta}{S^2}\sigma_0 \\ &+\frac{1}{S^2N^2}\sum_{q_1,q_2,q_3}\delta_{q_1+q_2,q_3+k}\frac{v_{q_1,q_2,q_3,k}\,v_{k,q_3,q_2,q_1}}{\omega_k-\omega_{q_1}-\omega_{q_2}+\omega_{q_3}}n^{(0)}_{q_3}\Bigg\}.\end{aligned} \tag{6.4.25}$$

Equation (6.4.25) coincides with Eq. (60) of Rastelli and Tassi[46] except for minor changes that amend Eq. (60). As one can see, the renormalized spectrum (6.4.25) shows kinematical consistency and satisfies the Goldstone theorem.

In conclusion, we have shown that the DM spin-boson transformation leads to the same results obtained from the HP transformation[18] and MME method[51] if all contributions of the same order in $\frac{1}{S}$ are carefully taken into account: the apparent discrepencies are simply due to the treatment of terms not homogeneous in the perturbation expansion. The spectrum (6.4.25) was obtained after a systematic use of the many-body theory. This approach, even though very heavy, can be worked out. On the contrary, the method of the Green function equation of motion illustrated in Chapter 3 much more hardly could be pushed to the same order of approximation.

# Chapter 7

# SPIN WAVES IN NON-COLLINEAR SYSTEMS

## 7.1. Local Axis Transformation and Boson Hamiltonian

Until now, we have studied ferromagnetic systems where the spins are parallel in the ground state. As we have remarked in Chapter 1, other kinds of magnetic order have been observed in actual compounds as antiferromagnets, spiral and helix magnets. In this chapter, we will study Hamiltonian models that can give non-collinear configurations in the ground state. A non-collinear ground state can be obtained simply from the Heisenberg Hamiltonian (1.2.5) if the exchange interaction is not limited to NN spins.[54] Let us consider the exchange Hamiltonian

$$\mathcal{H} = -\sum_{i,\boldsymbol{\delta}_\alpha} J_\alpha \boldsymbol{S}_i \cdot \boldsymbol{S}_{i+\boldsymbol{\delta}_\alpha}, \tag{7.1.1}$$

where $i$ runs over all sites of a Bravais lattice; $J_\alpha$ is the exchange integral between a spin located on the lattice site $i$ and the shell of NN ($\alpha = 1$), NNN ($\alpha = 2$), third nearest neighbours (TNN) ($\alpha = 3$) and so on. In Eq. (7.1.1), $\boldsymbol{\delta}_\alpha$ is the set of vectors connecting the site $i$ with its NN, NNN, TNN,... for $\alpha = 1, 2, 3, \ldots$, respectively. Let's assume that $x, y, z$ are the crystal axes and introduce a reference system of local axes $\xi, \eta, \zeta$, one for each lattice site. Suppose that each local axis $\zeta$ associated with the lattice site $i$ lies on the surface of a cone characterized by an apex angle $\theta$ (independent of $i$), where $\theta$ is the angle between the crystal axis $z$ and the local axis $\zeta$. Then, suppose that the angle between the projections into the crystal $xy$-plane of the local axis $\zeta$ at the lattice sites $i$ and $j$ is $\boldsymbol{Q} \cdot (\boldsymbol{r}_i - \boldsymbol{r}_j)$ where $\boldsymbol{Q}$ is the helix wavevector. The more general relation between the components of the spin $\boldsymbol{S}_i$ along the crystal and local axes is given by

$$\begin{pmatrix} S_i^x \\ S_i^y \\ S_i^z \end{pmatrix} = \begin{pmatrix} \cos\theta\cos\phi_i & -\sin\phi_i & \sin\theta\cos\phi_i \\ \cos\theta\sin\phi_i & \cos\phi_i & \sin\theta\sin\phi_i \\ -\sin\theta & 0 & \cos\theta \end{pmatrix} \begin{pmatrix} S_i^\xi \\ S_i^\eta \\ S_i^\zeta \end{pmatrix}, \tag{7.1.2}$$

where $\phi_i = \boldsymbol{Q} \cdot \boldsymbol{r}_i + \phi$. Note that if the spin was a classic vector of magnitude $S$ directed along the $\zeta$-axis, Eq. (7.1.2) should become

$$\begin{aligned} S_i^x &= S \sin\theta \cos(\boldsymbol{Q} \cdot \boldsymbol{r}_i + \phi), \\ S_i^y &= S \sin\theta \sin(\boldsymbol{Q} \cdot \boldsymbol{r}_i + \phi), \\ S_i^z &= S \cos\theta \end{aligned} \tag{7.1.3}$$

corresponding to the usual representation of a vector $\boldsymbol{S}$ in spherical coordinates. Using the transformation (7.1.2) and writing the operators $S_i^\xi$ and $S_i^\eta$ in terms of the raising and lowering operators $S_i^+$ and $S_i^-$ given by Eq. (1.3.3), the Hamiltonian (7.1.1) becomes

$$\begin{aligned} \mathcal{H} = -\sum_{i,\boldsymbol{\delta}_\alpha} J_\alpha \Big\{ & (\sin^2\theta \cos \boldsymbol{Q} \cdot \boldsymbol{\delta}_\alpha + \cos^2\theta)\, S_i^\zeta S_{i+\boldsymbol{\delta}_\alpha}^\zeta \\ & + \frac{1}{4} \sin^2\theta (1 - \cos \boldsymbol{Q} \cdot \boldsymbol{\delta}_\alpha)(S_i^+ S_{i+\boldsymbol{\delta}_\alpha}^+ + S_i^- S_{i+\boldsymbol{\delta}_\alpha}^-) \\ & + \frac{1}{4}[(1 + \cos^2\theta) \cos \boldsymbol{Q} \cdot \boldsymbol{\delta}_\alpha + \sin^2\theta - 2\, i \cos\theta \sin \boldsymbol{Q} \cdot \boldsymbol{\delta}_\alpha] S_i^+ S_{i+\boldsymbol{\delta}_\alpha}^- \\ & + \frac{1}{4}[(1 + \cos^2\theta) \cos \boldsymbol{Q} \cdot \boldsymbol{\delta}_\alpha + \sin^2\theta + 2\, i \cos\theta \sin \boldsymbol{Q} \cdot \boldsymbol{\delta}_\alpha] S_i^- S_{i+\boldsymbol{\delta}_\alpha}^+ \\ & - \frac{1}{2} \sin\theta [\cos\theta (1 - \cos \boldsymbol{Q} \cdot \boldsymbol{\delta}_\alpha) - i \sin \boldsymbol{Q} \cdot \boldsymbol{\delta}_\alpha](S_i^\zeta S_{i+\boldsymbol{\delta}_\alpha}^+ + S_i^- S_{i+\boldsymbol{\delta}_\alpha}^\zeta) \\ & - \frac{1}{2} \sin\theta [\cos\theta (1 - \cos \boldsymbol{Q} \cdot \boldsymbol{\delta}_\alpha) + i \sin \boldsymbol{Q} \cdot \boldsymbol{\delta}_\alpha](S_i^\zeta S_{i+\boldsymbol{\delta}_\alpha}^- + S_i^+ S_{i+\boldsymbol{\delta}_\alpha}^\zeta) \Big\}, \end{aligned} \tag{7.1.4}$$

where the parameters $\theta$ and $\boldsymbol{Q}$ have to be determined by a minimization of the ground-state energy. Note that the angle $\phi$ does not appear in the Hamiltonian (7.1.4). The presence of the angle $\theta$ in Eq. (7.1.4) is a consequence of the broken symmetry entered by the choice of the $z$-axis as the cone axis, while the absence of the angle $\phi$ reflects the invariance of the Hamiltonian under rotation about the $z$-axis. Obviously, the choice of the $z$-axis as the cone axis is arbitrary for the isotropic Heisenberg Hamiltonian (7.1.1). Only the anisotropic terms entered by the crystal field may fix the cone axis and the phase angle $\phi$. By means of the transformation to the local axes given by Eq. (7.1.2), we can investigate the occurrence of non-collinear configurations. Replacing the local spin operators $S_i^+, S_i^-$ and $S_i^\zeta$ by the boson creation and destruction operators using the spin-boson DM transformation given by (2.1.7), and using the Fourier transforms (2.2.14), one obtains

$$\mathcal{H} = E_0 + \sum_{n=1}^{6} \mathcal{H}_n, \tag{7.1.5}$$

where

$$E_0 = -\sum_{\boldsymbol{\delta}_\alpha} J_\alpha S^2 N (\sin^2\theta \cos \boldsymbol{Q} \cdot \boldsymbol{\delta}_\alpha + \cos^2\theta) \tag{7.1.6}$$

is the zeroth-order ground-state energy which is the "exact" ground-state energy only in the "classical limit" $S \to \infty$. Except for the isotropic ferromagnet the ground state of the Hamiltonian (7.1.1) is generally not known. An example was given in Chapter 6 where we have studied the planar ferromagnet. In the present case, a series of contributions ordered in powers of $\frac{1}{S}$ come from the interaction Hamiltonians $\mathcal{H}_n$ when a systematic perturbative theory is performed:

$$\mathcal{H}_1 = S\sqrt{2SN}\cos\theta\sin\theta\sum_{\boldsymbol{\delta}_\alpha} J_\alpha(1-\cos\boldsymbol{Q}\cdot\boldsymbol{\delta}_\alpha)(a_0+a_0^+) \tag{7.1.7}$$

is the linear term that will disappear when the minimum conditions of $E_0$ with respect to $\theta$ and $\boldsymbol{Q}$ will be taken into account:

$$\mathcal{H}_2 = \sum_{\boldsymbol{k}} A_{\boldsymbol{k}} a_{\boldsymbol{k}}^+ a_{\boldsymbol{k}} + \frac{1}{2}\sum_{\boldsymbol{k}} B_{\boldsymbol{k}}(a_{\boldsymbol{k}}a_{-\boldsymbol{k}} + a_{\boldsymbol{k}}^+ a_{-\boldsymbol{k}}^+) \tag{7.1.8}$$

with

$$A_{\boldsymbol{k}} = \sum_{\boldsymbol{\delta}_\alpha} 2J_\alpha S\Big\{ \sin^2\theta\cos\boldsymbol{Q}\cdot\boldsymbol{\delta}_\alpha + \cos^2\theta - \frac{1}{2}\cos\boldsymbol{k}\cdot\boldsymbol{\delta}_\alpha[\sin^2\theta + (1+\cos^2\theta)\cos\boldsymbol{Q}\cdot\boldsymbol{\delta}_\alpha] + \cos\theta\sin\boldsymbol{k}\cdot\boldsymbol{\delta}_\alpha\sin\boldsymbol{Q}\cdot\boldsymbol{\delta}_\alpha\Big\} \tag{7.1.9}$$

and

$$B_{\boldsymbol{k}} = -\sum_{\boldsymbol{\delta}_\alpha} J_\alpha S\sin^2\theta\cos\boldsymbol{k}\cdot\boldsymbol{\delta}_\alpha(1-\cos\boldsymbol{Q}\cdot\boldsymbol{\delta}_\alpha) \tag{7.1.10}$$

is the bilinear Hamiltonian that describes the system in the absence of interaction between the spin waves. From Eq. (7.1.9), one sees that $A_{\boldsymbol{k}} \neq A_{-\boldsymbol{k}}$. The remaining terms come from the interaction between the spin waves:

$$\mathcal{H}_3 = -\frac{1}{N}\sum_{\boldsymbol{k}_1,\boldsymbol{k}_2,\boldsymbol{k}_3}\left[\frac{1}{2}\sqrt{2SN}\cos\theta\sin\theta\sum_{\boldsymbol{\delta}_\alpha} J_\alpha(1-\cos\boldsymbol{Q}\cdot\boldsymbol{\delta}_\alpha) + C_{\boldsymbol{k}_3}\right] \times \delta_{\boldsymbol{k}_1,\boldsymbol{k}_2+\boldsymbol{k}_3}\, a_{\boldsymbol{k}_1}^+ a_{\boldsymbol{k}_2} a_{\boldsymbol{k}_3} - \frac{1}{N}\sum_{\boldsymbol{k}_1,\boldsymbol{k}_2,\boldsymbol{k}_3} C_{\boldsymbol{k}_1}\,\delta_{\boldsymbol{k}_1+\boldsymbol{k}_2,\boldsymbol{k}_3}\, a_{\boldsymbol{k}_1}^+ a_{\boldsymbol{k}_2}^+ a_{\boldsymbol{k}_3} \tag{7.1.11}$$

with

$$C_{\boldsymbol{k}} = \sqrt{2SN}\sin\theta\sum_{\boldsymbol{\delta}_\alpha} J_\alpha[\cos\theta\cos\boldsymbol{k}\cdot\boldsymbol{\delta}_\alpha(1-\cos\boldsymbol{Q}\cdot\boldsymbol{\delta}_\alpha) + \sin\boldsymbol{k}\cdot\boldsymbol{\delta}_\alpha\sin\boldsymbol{Q}\cdot\boldsymbol{\delta}_\alpha], \tag{7.1.12}$$

$$\mathcal{H}_4 = -\frac{1}{N}\sum_{\boldsymbol{k}_1,\boldsymbol{k}_2,\boldsymbol{k}_3,\boldsymbol{k}_4} V_{\boldsymbol{k}_1,\boldsymbol{k}_2,\boldsymbol{k}_3,\boldsymbol{k}_4}\,\delta_{\boldsymbol{k}_1+\boldsymbol{k}_2,\boldsymbol{k}_3+\boldsymbol{k}_4}\, a_{\boldsymbol{k}_1}^+ a_{\boldsymbol{k}_2}^+ a_{\boldsymbol{k}_3} a_{\boldsymbol{k}_4} - \frac{1}{2SN}\sum_{\boldsymbol{k}_1,\boldsymbol{k}_2,\boldsymbol{k}_3,\boldsymbol{k}_4} B_{\boldsymbol{k}_4}\,\delta_{\boldsymbol{k}_1,\boldsymbol{k}_2+\boldsymbol{k}_3+\boldsymbol{k}_4}\, a_{\boldsymbol{k}_1}^+ a_{\boldsymbol{k}_2} a_{\boldsymbol{k}_3} a_{\boldsymbol{k}_4} \tag{7.1.13}$$

with

$$V_{k_1,k_2,k_3,k_4} = \sum_{\delta_\alpha} J_\alpha \Big\{ \cos[(k_2 - k_4) \cdot \delta_\alpha](\sin^2\theta \cos Q \cdot \delta_\alpha + \cos^2\theta)$$

$$-\frac{1}{2}\cos k_1 \cdot \delta_\alpha[(1+\cos^2\theta)\cos Q \cdot \delta_\alpha + \sin^2\theta]$$

$$+ \cos\theta \sin k_1 \cdot \delta_\alpha \sin Q \cdot \delta_\alpha \Big\}, \tag{7.1.14}$$

$$\mathcal{H}_5 = \frac{1}{2S^2N} \sum_{k_1,k_2,k_3,k_4,k_5} C_{k_1-k_3}\, \delta_{k_1+k_2,k_3+k_4+k_5}\, a^+_{k_1} a^+_{k_2} a_{k_3} a_{k_4} a_{k_5} \tag{7.1.15}$$

and

$$\mathcal{H}_6 = \frac{1}{8S^2N^2} \sum_{k_1,k_2,k_3,k_4,k_5,k_6} B_{k_1-k_5-k_6}\, \delta_{k_1+k_2,k_3+k_4+k_5+k_6}\, a^+_{k_1} a^+_{k_2} a_{k_3} a_{k_4} a_{k_5} a_{k_6}. \tag{7.1.16}$$

As noted in Chapters 2 and 6, the DM transformation generates only a finite number of interaction potentials even though the boson Hamiltonian so obtained is no longer hermitian. Let us perform the same steps on the anisotropic Hamiltonian

$$\mathcal{H}_{\text{an}} = -h \sum_i S_i^z + D \sum_i (S_i^z)^2 \tag{7.1.17}$$

that accounts for an external uniform magnetic field directed along the $z$-axis $h = g\mu_B H$ and a single-ion easy-plane $D > 0$ anisotropy that forces the spins in the $xy$-plane. One obtains

$$\mathcal{H}_{\text{an}} = E_0^{\text{an}} + \sum_{n=1}^{6} \mathcal{H}_n^{\text{an}}, \tag{7.1.18}$$

where

$$E_0^{\text{an}} = DS^2N\left(1 - \frac{1}{2S}\right)\cos^2\theta + \frac{1}{2}DSN - hSN\cos\theta, \tag{7.1.19}$$

$$\mathcal{H}_1^{\text{an}} = -\frac{1}{2}\sqrt{2SN}\sin\theta[D(2S-1)\cos\theta - h](a_0 + a_0^+), \tag{7.1.20}$$

$$\mathcal{H}_2^{\text{an}} = \left[DS\left(1 - \frac{1}{2S}\right)(1 - 3\cos^2\theta) + h\cos\theta\right]\sum_k a_k^+ a_k$$

$$+\frac{1}{2}DS\left(1 - \frac{1}{2S}\right)\sin^2\theta \sum_k (a_k a_{-k} + a_k^+ a_{-k}^+), \tag{7.1.21}$$

$$\mathcal{H}_2^{\rm NH} = \frac{D}{4}\sin^2\theta\sum_{\boldsymbol{k}} a^+_{\boldsymbol{k}} a^+_{-\boldsymbol{k}}, \tag{7.1.22}$$

$$\begin{aligned}\mathcal{H}_3^{\rm an} &= D\sqrt{2SN}\cos\theta\sin\theta\frac{1}{N}\sum_{\boldsymbol{k}_1,\boldsymbol{k}_2,\boldsymbol{k}_3}\delta_{\boldsymbol{k}_1+\boldsymbol{k}_2,\boldsymbol{k}_3} a^+_{\boldsymbol{k}_1} a^+_{\boldsymbol{k}_2} a_{\boldsymbol{k}_3}\\ &+\sqrt{2SN}\sin\theta\frac{1}{N}\sum_{\boldsymbol{k}_1,\boldsymbol{k}_2,\boldsymbol{k}_3}\left[D\left(1-\frac{3}{4S}\right)\cos\theta-\frac{h}{4S}\right]\delta_{\boldsymbol{k}_1,\boldsymbol{k}_2+\boldsymbol{k}_3}\, a^+_{\boldsymbol{k}_1} a_{\boldsymbol{k}_2} a_{\boldsymbol{k}_3},\end{aligned} \tag{7.1.23}$$

$$\begin{aligned}\mathcal{H}_4^{\rm an} &= -\frac{D}{2}\left(1-3\cos^2\theta\right)\frac{1}{N}\sum_{\boldsymbol{k}_1,\boldsymbol{k}_2,\boldsymbol{k}_3,\boldsymbol{k}_4}\delta_{\boldsymbol{k}_1+\boldsymbol{k}_2,\boldsymbol{k}_3+\boldsymbol{k}_4}\, a^+_{\boldsymbol{k}_1} a^+_{\boldsymbol{k}_2} a_{\boldsymbol{k}_3} a_{\boldsymbol{k}_4}\\ &-\sin^2\theta\frac{D}{2N}\left(1-\frac{1}{2S}\right)\sum_{\boldsymbol{k}_1,\boldsymbol{k}_2,\boldsymbol{k}_3,\boldsymbol{k}_4}\delta_{\boldsymbol{k}_1,\boldsymbol{k}_2+\boldsymbol{k}_3+\boldsymbol{k}_4}\, a^+_{\boldsymbol{k}_1} a_{\boldsymbol{k}_2} a_{\boldsymbol{k}_3} a_{\boldsymbol{k}_4},\end{aligned} \tag{7.1.24}$$

$$\mathcal{H}_5^{\rm an} = -\sqrt{2SN}\cos\theta\sin\theta\frac{D}{4S^2N}\sum_{\boldsymbol{k}_1,\boldsymbol{k}_2,\boldsymbol{k}_3,\boldsymbol{k}_4,\boldsymbol{k}_5}\delta_{\boldsymbol{k}_1+\boldsymbol{k}_2,\boldsymbol{k}_3+\boldsymbol{k}_4+\boldsymbol{k}_5}\, a^+_{\boldsymbol{k}_1} a^+_{\boldsymbol{k}_2} a_{\boldsymbol{k}_3} a_{\boldsymbol{k}_4} a_{\boldsymbol{k}_5} \tag{7.1.25}$$

and

$$\mathcal{H}_6^{\rm an} = \sin^2\theta\frac{D}{8SN^2}\sum_{\boldsymbol{k}_1,\boldsymbol{k}_2,\boldsymbol{k}_3,\boldsymbol{k}_4,\boldsymbol{k}_5,\boldsymbol{k}_6}\delta_{\boldsymbol{k}_1+\boldsymbol{k}_2,\boldsymbol{k}_3+\boldsymbol{k}_4+\boldsymbol{k}_5+\boldsymbol{k}_6}\, a^+_{\boldsymbol{k}_1} a^+_{\boldsymbol{k}_2} a_{\boldsymbol{k}_3} a_{\boldsymbol{k}_4} a_{\boldsymbol{k}_5} a_{\boldsymbol{k}_6}. \tag{7.1.26}$$

Note that the anisotropic Hamiltonian (7.1.18) is independent of $\boldsymbol{Q}$ and $\phi$. Consequently, the helix wavevector $\boldsymbol{Q}$ is determined only by the exchange Hamiltonian (7.1.5). Moreover, the $\phi$ angle cannot be selected by the anisotropic Hamiltonian (7.1.18) which is invariant under rotation about the $z$-axis. As we have seen in Chapter 6, the DM spin-boson transformation leads to a non-hermitian bilinear Hamiltonian (7.1.22). As for the isotropic contribution, however, the non-hermiticity appears only in the interaction potentials between spin waves. In Chapter 6, we have seen that a careful treatment of the terms of the same order in $\frac{1}{S}$ prevents any unphysical effect due to the non-hermiticity of the bilinear Hamiltonian. According to Chapter 6, we choose the unperturbed Hamiltonian as the hermitian contribution of the Hamiltonian bilinear $\mathcal{H}_2+\mathcal{H}_2^{\rm an}$ with $\mathcal{H}_2$ and $\mathcal{H}_2^{\rm an}$ given by Eqs. (7.1.8) and (7.1.21), respectively. A novelty with respect to the Hamiltonian of the planar ferromagnet is the presence of linear terms $\mathcal{H}_1+\mathcal{H}_1^{\rm an}$ containing a single Bose operator and of cubic terms $\mathcal{H}_3+\mathcal{H}_3^{\rm an}$ containing three Bose operators. While the linear term disappears when the minimum conditions are taken into account, the cubic terms do not vanish. Moreover, the cubic terms treated at the second-order in the $\frac{1}{S}$ perturbation expansion give a contribution of the same order of the quartic term $\mathcal{H}_4+\mathcal{H}_4^{\rm an}$ treated at the first-order. Like for the planar ferromagnet, a careful treatment of the perturbative terms has to be done.

## 7.2. Harmonic Approximation and Bogoliubov Transformation

Let us begin diagonalizing the isotropic bilinear Hamiltonian (7.1.8). From Eq. (7.1.9), we see that $A_{\boldsymbol{k}} \neq A_{-\boldsymbol{k}}$ so that it is convenient to write the Hamiltonian (7.1.8) in the more symmetric form

$$\mathcal{H}_2 = \frac{1}{2}\sum_{\boldsymbol{k}}[A_{\boldsymbol{k}}a^+_{\boldsymbol{k}}a_{\boldsymbol{k}} + A_{-\boldsymbol{k}}a^+_{-\boldsymbol{k}}a_{-\boldsymbol{k}} + B_{\boldsymbol{k}}(a_{\boldsymbol{k}}a_{-\boldsymbol{k}} + a^+_{\boldsymbol{k}}a^+_{-\boldsymbol{k}})] \tag{7.2.1}$$

before introducing the Bogoliubov transformation (6.1.14)

$$a_{\boldsymbol{k}} = l_{\boldsymbol{k}}(\alpha_{\boldsymbol{k}} - x_{\boldsymbol{k}}\alpha^+_{-\boldsymbol{k}}), \quad a^+_{\boldsymbol{k}} = l_{\boldsymbol{k}}(\alpha^+_{\boldsymbol{k}} - x_{\boldsymbol{k}}\alpha_{-\boldsymbol{k}}) \tag{7.2.2}$$

where

$$l^2_{\boldsymbol{k}}(1 - x^2_{\boldsymbol{k}}) = 1. \tag{7.2.3}$$

By means of the Bogoliubov transformation (7.2.2), the Hamiltonian (7.2.1) becomes

$$\begin{aligned}\mathcal{H}_2 = \sum_{\boldsymbol{k}} l^2_{\boldsymbol{k}}\{&[A^{(\mathrm{s})}_{\boldsymbol{k}} + A^{(\mathrm{a})}_{\boldsymbol{k}} + (A^{(\mathrm{s})}_{\boldsymbol{k}} - A^{(\mathrm{a})}_{\boldsymbol{k}})x^2_{\boldsymbol{k}} - 2B_{\boldsymbol{k}}x_{\boldsymbol{k}}]\alpha^+_{\boldsymbol{k}}\alpha_{\boldsymbol{k}} \\ &+ [-2A^{(\mathrm{s})}_{\boldsymbol{k}}x_{\boldsymbol{k}} + B_{\boldsymbol{k}}(1 + x^2_{\boldsymbol{k}})]\frac{1}{2}(\alpha_{\boldsymbol{k}}\alpha_{-\boldsymbol{k}} + \alpha^+_{\boldsymbol{k}}\alpha^+_{-\boldsymbol{k}}) + A^{(\mathrm{s})}_{\boldsymbol{k}}x^2_{\boldsymbol{k}} - B_{\boldsymbol{k}}x_{\boldsymbol{k}}\}\end{aligned} \tag{7.2.4}$$

where

$$\begin{aligned}A^{(\mathrm{s})}_{\boldsymbol{k}} = \frac{1}{2}(A_{\boldsymbol{k}} + A_{-\boldsymbol{k}}) = \sum_{\boldsymbol{\delta}_\alpha} 2J_\alpha S\Big\{&\sin^2\theta\cos\boldsymbol{Q}\cdot\boldsymbol{\delta}_\alpha + \cos^2\theta \\ &-\frac{1}{2}\cos\boldsymbol{k}\cdot\boldsymbol{\delta}_\alpha[\sin^2\theta + (1+\cos^2\theta)\cos\boldsymbol{Q}\cdot\boldsymbol{\delta}_\alpha]\Big\}\end{aligned} \tag{7.2.5}$$

and

$$A^{(\mathrm{a})}_{\boldsymbol{k}} = \frac{1}{2}(A_{\boldsymbol{k}} - A_{-\boldsymbol{k}}) = \cos\theta\sum_{\boldsymbol{\delta}_\alpha} 2J_\alpha S\sin\boldsymbol{k}\cdot\boldsymbol{\delta}_\alpha\sin\boldsymbol{Q}\cdot\boldsymbol{\delta}_\alpha. \tag{7.2.6}$$

The Hamiltonian (7.2.4) is diagonalized if $x_{\boldsymbol{k}}$ satisfies the equation

$$-2A^{(\mathrm{s})}_{\boldsymbol{k}}x_{\boldsymbol{k}} + B_{\boldsymbol{k}}(1 + x^2_{\boldsymbol{k}}) = 0. \tag{7.2.7}$$

The solutions of Eq. (7.2.7) are

$$x_{\boldsymbol{k}} = \frac{A^{(\mathrm{s})}_{\boldsymbol{k}} \pm \sqrt{A^{(\mathrm{s})2}_{\boldsymbol{k}} - B^2_{\boldsymbol{k}}}}{B_{\boldsymbol{k}}}. \tag{7.2.8}$$

Replacing the solutions given by Eqs. (7.2.8) in Eq. (7.2.3), one can see that the solution with the + sign leads to a $l^2_{\boldsymbol{k}} < 0$ so that it has to be rejected. On the contrary, the solution with the − sign leads to

$$l_k^2 = \frac{1}{2}\left(\frac{A_k^{(s)}}{\sqrt{A_k^{(s)2} - B_k^2}} + 1\right), \tag{7.2.9}$$

$$l_k^2 x_k = \frac{1}{2}\frac{B_k}{\sqrt{A_k^{(s)2} - B_k^2}}, \tag{7.2.10}$$

$$l_k^2 x_k^2 = \frac{1}{2}\left(\frac{A_k^{(s)}}{\sqrt{A_k^{(s)2} - B_k^2}} - 1\right) \tag{7.2.11}$$

and the Hamiltonian (7.2.4) becomes

$$\mathcal{H}_2 = \sum_k \left(\sqrt{A_k^{(s)2} - B_k^2} + A_k^{(a)}\right)\alpha_k^+\alpha_k + \frac{1}{2}\sum_k \left(\sqrt{A_k^{(s)2} - B_k^2} - A_k^{(s)}\right). \tag{7.2.12}$$

By means of the definitions

$$S_k = A_k^{(s)} + B_k = \sum_{\delta_\alpha} 2J_\alpha S[\sin^2\theta(\cos \boldsymbol{Q}\cdot\boldsymbol{\delta}_\alpha - \cos \boldsymbol{k}\cdot\boldsymbol{\delta}_\alpha) + \cos^2\theta(1 - \cos \boldsymbol{k}\cdot\boldsymbol{\delta}_\alpha \cos \boldsymbol{Q}\cdot\boldsymbol{\delta}_\alpha)] \tag{7.2.13}$$

and

$$D_k = A_k^{(s)} - B_k = \sum_{\delta_\alpha} 2J_\alpha S[\cos \boldsymbol{Q}\cdot\boldsymbol{\delta}_\alpha(1 - \cos \boldsymbol{k}\cdot\boldsymbol{\delta}_\alpha) + \cos^2\theta(1 - \cos \mathbf{Q}\cdot\boldsymbol{\delta}_\alpha)] \tag{7.2.14}$$

the Hamiltonian (7.2.12) becomes

$$\mathcal{H}_2 = \Delta E_0 + \sum_k \left(\sqrt{S_k D_k} + A_k^{(a)}\right)\alpha_k^+\alpha_k \tag{7.2.15}$$

where

$$\Delta E_0 = \frac{1}{2}\sum_k \left(\sqrt{S_k D_k} - A_k^{(s)}\right) \tag{7.2.16}$$

is the first quantum correction of order $\frac{1}{S}$ with respect to the classical ground-state energy $E_0$ given by Eq. (7.1.6).

Taking the anisotropic bilinear Hamiltonian (7.1.21) into account, Eqs. (7.2.13) and (7.2.14) do not change if $S_k$ and $D_k$ are replaced by

$$S_k^{\rm an} = S_k + 2DS\left(1 - \frac{1}{2S}\right)(1 - 2\cos^2\theta) + h\cos\theta \tag{7.2.17}$$

and

$$D_k^{\rm an} = D_k - 2DS\left(1 - \frac{1}{2S}\right)\cos^2\theta + h\cos\theta, \tag{7.2.18}$$

respectively.

## 7.3. Ground-State Configurations

The zero-order approximation assumes that the ground-state energy is the absolute minimum of $E_G = E_0 + E_0^{an}$ with $E_0$ and $E_0^{an}$ given by Eqs. (7.1.6) and (7.1.19), respectively. The absolute minimum of $E_G$ has to be looked for among the simultaneous solutions of the equations $\frac{\partial E_G}{\partial \theta} = 0$ and $\frac{\partial E_G}{\partial Q_i} = 0$ with $i = x, y, z$ that is

$$\sin\theta \left\{ \cos\theta \left[ \sum_{\boldsymbol{\delta}_\alpha} 2J_\alpha S(1 - \cos \boldsymbol{Q}\cdot\boldsymbol{\delta}_\alpha) - D(2S-1) \right] + h \right\} = 0 \qquad (7.3.1)$$

and

$$\sin^2\theta \sum_{\boldsymbol{\delta}_\alpha} J_\alpha(\boldsymbol{\delta}_\alpha \cdot \mathbf{u}_i) \sin \boldsymbol{Q}\cdot\boldsymbol{\delta}_\alpha = 0, \quad i = x, y, z, \qquad (7.3.2)$$

where $\mathbf{u}_i$ are the unit vectors along the $i = x, y, z$ axes. Note that Eq. (7.3.1) coincides with the vanishing of the linear term $\mathcal{H}_1 + \mathcal{H}_1^{an}$ where $\mathcal{H}_1$ and $\mathcal{H}_1^{an}$ are given by Eqs. (7.1.7) and (7.1.20), respectively. This general property confirms that linear terms in the boson Hamiltonian disappear when the ground state corresponds to an energy minimum. The minimum obtained from Eq. (7.3.1) is

$$\sin\theta = 0 \qquad (7.3.3)$$

for $h > h_c$ and

$$\cos\theta = \frac{h}{h_c} \qquad (7.3.4)$$

for $h \le h_c$ where

$$h_c = D(2S-1) - \sum_{\boldsymbol{\delta}_\alpha} 2J_\alpha S(1 - \cos \boldsymbol{Q}\cdot\boldsymbol{\delta}_\alpha). \qquad (7.3.5)$$

The solution (7.3.3) corresponds to the "saturated ferromagnetic" phase with all spins directed along the $z$-axis that is stable only in presence of magnetic fields strong enough. The solution (7.3.4) corresponds to a "cone structure" with an apex angle depending on the magnetic field that reduces to the saturated ferromagnetic phase for $h = h_c$ and to a regular "helix structure" with all spins lying in the $xy$-plane for $h = 0$. For $0 \le h \le h_c$, the spin wave spectrum is given by

$$\hbar\omega_{\boldsymbol{k}} = \sqrt{S_{\boldsymbol{k}}^{an} D_{\boldsymbol{k}}^{an}} + A_{\boldsymbol{k}}^{(a)} \qquad (7.3.6)$$

with

$$S_{\boldsymbol{k}}^{an} = \sum_{\boldsymbol{\delta}_\alpha} 2J_\alpha S \left[ \cos \boldsymbol{Q}\cdot\boldsymbol{\delta}_\alpha - \cos \boldsymbol{k}\cdot\boldsymbol{\delta}_\alpha + \left(\frac{h}{h_c}\right)^2 \cos \boldsymbol{k}\cdot\boldsymbol{\delta}_\alpha \,(1 - \cos \boldsymbol{Q}\cdot\boldsymbol{\delta}_\alpha) \right]$$

$$+ D(2S-1)\left[1 - \left(\frac{h}{h_c}\right)^2\right], \qquad (7.3.7)$$

$$D_{\boldsymbol{k}}^{an} = \sum_{\boldsymbol{\delta}_\alpha} 2J_\alpha S \cos \boldsymbol{Q}\cdot\boldsymbol{\delta}_\alpha (1 - \cos \boldsymbol{k}\cdot\boldsymbol{\delta}_\alpha) \qquad (7.3.8)$$

and

$$A_{\boldsymbol{k}}^{(\mathrm{a})} = \frac{h}{h_c} \sum_{\boldsymbol{\delta}_\alpha} 2J_\alpha S \sin \boldsymbol{k} \cdot \boldsymbol{\delta}_\alpha \sin \boldsymbol{Q} \cdot \boldsymbol{\delta}_\alpha. \tag{7.3.9}$$

Note that the spectrum (7.3.6) shows a soft mode at $\boldsymbol{k} = 0$ even in presence of an external magnetic field since Eq. (7.3.8) vanishes in that limit. The existence of the soft mode at $\boldsymbol{k} = 0$ is not restricted to the "unperturbed" spectrum (7.3.6) obtained from the bilinear Hamiltonian describing a collection of non-interacting spin waves. The soft mode remains also when the interactions between the spin waves are taken into account since it reflects the invariance of the total Hamiltonian under rotation about the magnetic field direction. Indeed, the existence of this soft mode was proved, in general, making use of a generalization of the Goldstone theorem.[55] For $h = 0$, the apex angle of the cone is $\frac{\pi}{2}$ and the ground-state configuration is a regular helix in which all spins rotate in the $xy$-plane with a pitch determined by the solutions of Eq. (7.3.2). The presence of a single-ion easy-plane anisotropy prevents the existence of a soft mode at $\boldsymbol{k} = \boldsymbol{Q}$ even for $h = 0$ since Eq. (7.3.7) vanishes only in the isotropic limit ($D = h = 0$). In this case, the existence of a soft mode at $\boldsymbol{k} = \boldsymbol{Q}$ reflects a more general property of the system[55] corresponding to the invariance of the isotropic Hamiltonian under a rotation of the plane where the spins lie. The spiral configurations ($0 < \theta < \frac{\pi}{2}$) observed in some rare earths[9,56] in the absence of an external magnetic field can be explained only by adding crystal field terms of higher order to the Heisenberg Hamiltonian. For Hamiltonians with single-ion easy-plane anisotropy, the regular helix phases are the only possible configurations. This theoretical expectation is confirmed by the experimental data on some transition metal halides like[57] $NiBr_2$ or[58] $NiI_2$ and $CoI_2$.

In order to find the helix wavevector, equations in (7.3.2) have to be investigated: They correspond to D equations to be solved simultaneously where D is the lattice dimensionality and they depend on the exchange interactions only.

From now on, we restrict to the isotropic case with the minimum of $E_0$ corresponding to $\theta = \frac{\pi}{2}$. The spin wave spectrum (7.3.6) reduces to

$$\hbar\omega_{\boldsymbol{k}} = \sqrt{S_{\boldsymbol{k}} D_{\boldsymbol{k}}} \tag{7.3.10}$$

since $A_{\boldsymbol{k}}^{(\mathrm{a})}$ given by Eq. (7.2.6) vanishes for $\theta = \frac{\pi}{2}$. Equations (7.2.12) and (7.2.13) become

$$S_{\boldsymbol{k}} = \sum_{\boldsymbol{\delta}_\alpha} 2J_\alpha S(\cos \boldsymbol{Q} \cdot \boldsymbol{\delta}_\alpha - \cos \boldsymbol{k} \cdot \boldsymbol{\delta}_\alpha) \tag{7.3.11}$$

and

$$D_{\boldsymbol{k}} = \sum_{\boldsymbol{\delta}_\alpha} 2J_\alpha S \cos \boldsymbol{Q} \cdot \boldsymbol{\delta}_\alpha (1 - \cos \boldsymbol{k} \cdot \boldsymbol{\delta}_\alpha). \tag{7.3.12}$$

From Eqs. (7.3.10)–(7.3.12), it is direct to see that there are two *soft modes* at $\boldsymbol{k} = 0$ and $\boldsymbol{k} = \boldsymbol{Q}$. Classically, the two soft modes correspond to a "uniform" rotation of the spins in the $xy$-plane ($\boldsymbol{k} = 0$) and to a rigid rotation of the $xy$-plane around any straight line lying in the plane ($\boldsymbol{k} = \boldsymbol{Q}$).

Let us conclude this section evaluating the average value of the spin

$$\begin{aligned}\langle S_i^x\rangle &= \langle S_i^\zeta\rangle \cos(\boldsymbol{Q}\cdot\boldsymbol{r}_i + \phi),\\ \langle S_i^y\rangle &= \langle S_i^\zeta\rangle \sin(\boldsymbol{Q}\cdot\boldsymbol{r}_i + \phi),\\ \langle S_i^z\rangle &= 0,\end{aligned} \tag{7.3.13}$$

where

$$\begin{aligned}\langle S_i^\zeta\rangle &= S - \langle a_i^+ a_i\rangle = S\left(1 - \frac{1}{NS}\sum_{\boldsymbol{k}}\langle a_{\boldsymbol{k}}^+ a_{\boldsymbol{k}}\rangle\right)\\ &= S\left(1 - \frac{1}{2NS}\sum_{\boldsymbol{k}}\frac{A_{\boldsymbol{k}} - \hbar\omega_{\boldsymbol{k}}}{\hbar\omega_{\boldsymbol{k}}} - \frac{1}{NS}\sum_{\boldsymbol{k}}\frac{A_{\boldsymbol{k}}}{\hbar\omega_{\boldsymbol{k}}}\langle\alpha_{\boldsymbol{k}}^+\alpha_{\boldsymbol{k}}\rangle\right),\end{aligned} \tag{7.3.14}$$

where $\hbar\omega_{\boldsymbol{k}}$ is given by Eq. (7.3.10) and $A_{\boldsymbol{k}}$ is given by Eq. (7.1.9) with $\theta = \frac{\pi}{2}$. As one can see, the average value of the spin is reduced with respect to its saturation value $S$ even at zero temperature due to the existence of quantum fluctuations. The non saturation at zero temperature is called *spin reduction*..

## 7.4. Néel Antiferromagnet

The exact ground-state wave function and the exact ground-state energy of an antiferromagnet are not known except for the LC with $S = \frac{1}{2}$ for which the famous solutions of Bethe[10] in 1931 for the wave function and of Hulthén[11] in 1938 for the energy are available. In this section, we will obtain the approximate antiferromagnetic ground state and the antiferromagnetic spin waves as a special case of a helix configuration. This procedure is not conventional but it is very simple. Let us consider the Hamiltonian (7.1.1) with a NN exchange interaction that is $J_\alpha = J < 0$ for $\alpha = 1$ and $J_\alpha = 0$ otherwise. Moreover, $\boldsymbol{\delta}_\alpha = \boldsymbol{\delta}$ for $\alpha = 1$ and zero otherwise. The classical ground-state energy $E_0$ obtained from Eqs. (7.1.6) with $\theta = \frac{\pi}{2}$ is

$$E_0 = |J|S^2N\sum_{\boldsymbol{\delta}}\cos\boldsymbol{Q}\cdot\boldsymbol{\delta} = z|J|S^2N\,\gamma_{\boldsymbol{Q}} \tag{7.4.1}$$

with $\gamma_{\boldsymbol{Q}}$ given by Eq. (2.2.17). We write explicitly the energy (7.4.1) for some loose packed lattices like the SQ lattice

$$E_0^{\rm SQ} = 2|J|S^2N(\cos aQ_x + \cos aQ_y), \tag{7.4.2}$$

the SC lattice

$$E_0^{\rm SC} = 2|J|S^2N(\cos aQ_x + \cos aQ_y + \cos aQ_z), \tag{7.4.3}$$

the BCC lattice

$$E_0^{\rm BCC} = 8|J|S^2N\cos\frac{aQ_x}{2}\cos\frac{aQ_y}{2}\cos\frac{aQ_z}{2} \tag{7.4.4}$$

and the tetragonal (T) lattice

$$E_0^{\rm T} = 2|J|S^2N(\cos aQ_x + \cos aQ_y) + 2|J'|S^2N\cos cQ_z, \qquad (7.4.5)$$

where $J$ and $J'$ are the NN antiferromagnetic exchange interactions in the $xy$-plane and out-of-plane, respectively. For these lattices, the helix wavevector corresponding to the absolute minimum of Eq. (7.4.1) is

$$\boldsymbol{Q}^{\rm SQ} = \left(\frac{\pi}{a}, \frac{\pi}{a}\right), \quad \boldsymbol{Q}^{\rm SC} = \left(\frac{\pi}{a}, \frac{\pi}{a}, \frac{\pi}{a}\right),$$

$$\boldsymbol{Q}^{\rm BCC} = \left(\frac{2\pi}{a}, \frac{2\pi}{a}, \frac{2\pi}{a}\right), \quad \boldsymbol{Q}^{\rm T} = \left(\frac{\pi}{a}, \frac{\pi}{a}, \frac{\pi}{c}\right) \qquad (7.4.6)$$

so that, in general, the antiferromagnetic ground state for loose packed lattices corresponds to

$$\boldsymbol{Q} = \frac{1}{2}(\boldsymbol{a}^* + \boldsymbol{b}^* + \boldsymbol{c}^*), \quad \gamma_{\boldsymbol{Q}} = -1. \qquad (7.4.7)$$

where $\boldsymbol{a}^*, \boldsymbol{b}^*, \boldsymbol{c}^*$ are the reciprocal lattice basic vectors. For instance, the reciprocal triad of the BCC lattice is

$$\boldsymbol{a}^* = \frac{2\pi}{a}(\mathbf{u}_x + \mathbf{u}_y), \quad \boldsymbol{b}^* = \frac{2\pi}{a}(\mathbf{u}_y + \mathbf{u}_z), \quad \boldsymbol{c}^* = \frac{2\pi}{a}(\mathbf{u}_z + \mathbf{u}_x). \qquad (7.4.8)$$

The classical ground-state spin configuration can be easily obtained from Eq. (7.3.13) replacing $\langle S_i^\zeta \rangle$ with $S$. Since the generic lattice vector is given by

$$\boldsymbol{r}_{l,m,n} = l\boldsymbol{a} + m\boldsymbol{b} + n\boldsymbol{c}, \qquad (7.4.9)$$

the generic spin vector located at the lattice site $(l, m, n)$ becomes

$$\boldsymbol{S}_{l,m,n} = S\cos\left[(l+m+n)\pi + \phi\right]\mathbf{u}_x + S\sin\left[(l+m+n)\pi + \phi\right]\mathbf{u}_y \qquad (7.4.10)$$

where Eq. (7.4.7) has been used. In Fig. 7.1, we show the classical ground-state configuration of the SQ lattice obtained from Eq. (7.4.10) with $n = 0$ and $\phi = \frac{\pi}{2}$. The kind of order shown in Fig. 7.1 is common to all *Néel antiferromagnets* in which the NN spins of a given spin are antiparallel to that spin. It should be noted that all the lattices considered in this section satisfy the condition $\mathbf{Q} \cdot \boldsymbol{\delta} = \pi$ that is peculiar of the Néel antiferromagnetic order. Making use of Eq. (7.4.7), the spin wave spectrum (7.3.10) becomes

$$\hbar\omega_{\boldsymbol{k}} = 2z|J|S\sqrt{1-\gamma_{\boldsymbol{k}}^2}, \qquad (7.4.11)$$

where $\gamma_{\boldsymbol{k}}$ is given by Eq. (2.2.17). The spectrum (7.4.11) vanishes linearly with $k$ in the long wavelength limit. In particular, for the cubic lattices one has

$$\hbar\omega_{\boldsymbol{k}} \sim 2|J|S\sqrt{2z}(ak) \qquad (7.4.12)$$

and the free energy becomes

$$F = E_{\rm GS} + k_BTN\frac{v_c}{8\pi^3}\int_{\rm BZ} d^3\boldsymbol{k}\ln(1 - e^{-\beta\hbar\omega_{\boldsymbol{k}}})$$

$$= E_{\rm GS} + k_BTN\frac{m}{2\pi^2}\int_0^\infty k^2dk\ln(1 - e^{-\beta 2|J|S\sqrt{2z}k}), \qquad (7.4.13)$$

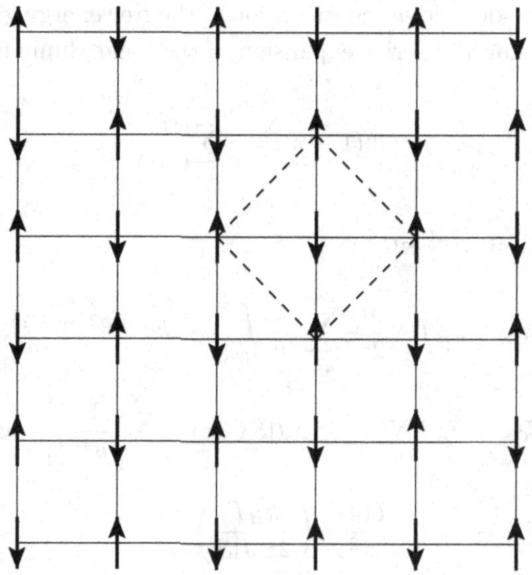

Fig. 7.1. Ground-state spin configuration for a SQ antiferromagnet. The dashed line represents the magnetic unit cell.

where $m = 2$ for the SC and $m = 1$ for the BCC lattice and

$$E_{\text{GS}} = E_0 + \Delta E_0 = -z|J|S^2 N\left(1 + \frac{c}{S}\right) \tag{7.4.14}$$

where

$$c = \frac{1}{N}\sum_{\boldsymbol{k}}\left(1 - \sqrt{1-\gamma_{\boldsymbol{k}}^2}\right) \tag{7.4.15}$$

is the first-order contribution to the zero-point energy. The numerical value[59] of $c$ is 0.097 and 0.073 for the SC and BCC lattices, respectively. Equation (7.4.15) evaluated for a SQ lattice gives[59] $c = 0.158$ showing that the zero-point energy increases as the dimensionality decreases. This general trend is confirmed by evaluating Eq. (7.4.15) for the antiferromagnetic LC

$$c = \frac{1}{\pi}\int_0^{\pi} dk(1 - \sin k) = 1 - \frac{2}{\pi} = 0.363. \tag{7.4.16}$$

The exact ground-state energy[11] of the antiferromagnetic LC with $S = \frac{1}{2}$ is

$$E_{\text{GS}} = \frac{1}{2}|J|N(1 - 4\ln 2) = -\frac{1}{2}|J|N\ 1.773 \tag{7.4.17}$$

while the first-order result (7.4.14) for $S = \frac{1}{2}$ and $c$ given by Eq. (7.4.16) is

$$E_{\text{GS}} = -\frac{1}{2}|J|N(1 + 2c) = -\frac{1}{2}|J|N\ 1.727 \tag{7.4.18}$$

pointing out that the first-order result is not so bad.

The temperature-dependent contribution of the free energy given by Eq. (7.4.13) can be evaluated using the series expansion of the logarithmic function

$$\ln(1-x) = -\sum_{n=1}^{\infty} \frac{x^n}{n} \tag{7.4.19}$$

by means of which Eq. (7.4.13) becomes

$$\begin{aligned} F &= E_{\text{GS}} - k_B T N \frac{m}{2\pi^2} \sum_{n=1}^{\infty} \frac{1}{n} \int_0^{\infty} k^2 dk e^{-n\beta 2|J|S\sqrt{2z}\, k} \\ &= E_{\text{GS}} - k_B T N \frac{m}{2\pi^2} (\beta 2|J|S\sqrt{2z})^{-3} \sum_{n=1}^{\infty} \frac{1}{n^4} \int_0^{\infty} dx x^2 e^{-x} \\ &= E_{\text{GS}} - k_B T N \frac{\zeta(4)}{\pi^2} \eta \left( \frac{k_B T}{2z|J|S} \right)^3 \end{aligned} \tag{7.4.20}$$

where $\zeta(4) = \frac{\pi^4}{90}$; $\eta = 6\sqrt{3}$ and 8 for the SC and BCC lattices, respectively. The internal energy becomes

$$U = -T^2 \frac{\partial}{\partial T} \left( \frac{F}{T} \right) = E_{\text{GS}} + 2z|J|SN \frac{3\zeta(4)}{\pi^2} \eta \left( \frac{k_B T}{2z|J|S} \right)^4. \tag{7.4.21}$$

The heat capacity becomes

$$C = \frac{\partial U}{\partial T} = k_B N \frac{12\zeta(4)}{\pi^2} \eta \left( \frac{k_B T}{2z|J|S} \right)^3. \tag{7.4.22}$$

As one can see from Eq. (7.4.22), the heat capacity of an antiferromagnet is proportional to $T^3$ at low temperature. This behaviour differs from that obtained for a ferromagnet $C \sim T^{3/2}$ given by Eq. (2.4.8). The reason is the different dispersion relation of the spin wave spectra in the long wavelength limit: linear for an antiferromagnet, quadratic for a ferromagnet. Experimentally, it is difficult to distinguish between the vibrational and magnetic contributions to the heat capacity of an antiferromagnet since both contributions are of the same order in temperature ($T^3$) in the low temperature limit. However, one can distinguish between the vibrational and magnetic contribution measuring the heat capacity of the sample *above* the Néel temperature $T_N$ where the magnetic order disappears. Since, generally the Néel temperature is much lower than the Debye temperature, the cubic contribution to the heat capacity measured above $T_N$ is entirely due to the lattice vibrations. Then, measuring the heat capacity of the sample below $T_N$ and subtracting the vibrational contribution obtained from the "high" temperature experiment, one can distinguish between the magnetic and vibrational contribution.

The average value of the local spin at low temperature is

$$\langle S_i^\zeta \rangle = S\left(1 - \frac{c'}{2S} - \frac{1}{NS}\sum_k \frac{n_k}{\sqrt{1-\gamma_k^2}}\right), \tag{7.4.23}$$

where

$$c' = \frac{1}{N}\sum_k \left(\frac{1}{\sqrt{1-\gamma_k^2}} - 1\right) \tag{7.4.24}$$

and

$$n_k = \frac{1}{e^{\frac{2z|J|S}{k_BT}\sqrt{1-\gamma_k^2}} - 1} \simeq \frac{1}{e^{\frac{2|J|S}{k_BT}\sqrt{2z}ak} - 1}. \tag{7.4.25}$$

The spin reduction $c'$ that is the deviation from the saturation value $S$ at zero temperature is given[59] by 0.156 and 0.119 for SC and BCC lattices, respectively. Like the zero-point energy also the spin reduction increases as the dimensionality decreases: for instance one finds[59] $c' = 0.394$ for a SQ lattice. For the LC, the sum in Eq. (7.4.24) diverges logarithmically pointing out that LRO is lost even at $T = 0$. Using the long wavelength approximation (7.4.25) in Eq. (7.4.23), the low temperature thermal average of the local spin becomes

$$\langle S_i^\zeta \rangle = S\left[1 - \frac{c'}{2S} - \frac{\zeta(2)}{2\pi^2 S}\eta\left(\frac{k_BT}{2z|J|S}\right)^2\right]. \tag{7.4.26}$$

The low temperature deviation from the zero temperature value of the local spin is proportional to $T^2$ to be compared with the deviation proportional to $T^{3/2}$ given by Eq. (2.4.11) for a ferromagnet. Also in this case the different low temperature behaviour of the ferromagnetic magnetization from the antiferromagnetic "sublattice" magnetization (the correct order parameter of the antiferromagnet since the total magnetization is always zero) is due to the different dispersion relation of the ferromagnetic and antiferromagnetic spin wave spectrum in the long wavelength limit..

For quasi-2D lattices characterized by a stacking of planes in which the exchange interaction between the spins in the plane $|J|$ is much larger than the exchange interaction between the NN spins belonging to different planes $|J'|$ the heat capacity is proportional to $T^2$ for $\sqrt{JJ'} < k_BT < |J|$ while the usual 3D behaviour $C \sim T^3$ is recovered for $k_BT < \sqrt{JJ'}$. Note that in many actual compounds this behaviour has been experimentally tested. In particular, for the $BX_2$ compounds,[60] where B is a 3$d$-transition metal like vanadium (V), manganese (Mn), iron (Fe), cobalt (Co) or nickel (Ni) and X is a halogen like bromine (Br), chlorine (Cl) or iodine (I) the in-plane exchange interaction is two order of magnitude greater than the out-of-plane exchange coupling. Moreover, in some of the "high-$T_c$ superconductors" (HTCS)[61] like $La_{2-x}B_xCuO_{4-y}$ with B=Ca, Sr or Ba, the ratio $j' = |J'/J|$ is of the order $10^{-3} - 10^{-4}$.

The study of the temperature dependence of the sublattice magnetization for quasi-2D antiferromagnets leads to an interesting non-analytic behaviour in a region

$\sqrt{JJ'} < k_BT < |J|$ in analogy with that found in the ferromagnetic case given by Eq. (2.5.15). For T lattices with $j' \ll 1$ well describing some of the HTCS, Eq. (7.4.23) becomes

$$\langle S_i^\varsigma \rangle = S\left[1 - \frac{c'}{2S} - \frac{1}{2NS}\sum_{\boldsymbol{k}}\left(\sqrt{\frac{s_{\boldsymbol{k}}}{d_{\boldsymbol{k}}}} + \sqrt{\frac{d_{\boldsymbol{k}}}{s_{\boldsymbol{k}}}}\right)n_{\boldsymbol{k}}\right] \tag{7.4.27}$$

where $c' = 0.394$ is the value of the SQ lattice and

$$s_{\boldsymbol{k}} = \left[1 + \frac{1}{2}(\cos k_x + \cos k_y)\right] + \frac{1}{2}j'(1 + \cos k_z) \tag{7.4.28}$$

and

$$d_{\boldsymbol{k}} = \left[1 - \frac{1}{2}(\cos k_x + \cos k_y)\right] + \frac{1}{2}j'(1 - \cos k_z) \tag{7.4.29}$$

with $k_x$ and $k_y$ measured in units of in-plane lattice space $a$ while $k_z$ is measured in units of the out-of-plane distance $c$. For $k_BT < |J|$, the thermal contribution of Eq. (7.4.27) becomes

$$\begin{aligned}
\Delta M(T) &= \frac{1}{2NS}\sum_{\boldsymbol{k}}\left(\sqrt{\frac{s_{\boldsymbol{k}}}{d_{\boldsymbol{k}}}} + \sqrt{\frac{d_{\boldsymbol{k}}}{s_{\boldsymbol{k}}}}\right)n_{\boldsymbol{k}} \\
&\simeq \frac{1}{S(2\pi)^2}\int_{-\pi}^{\pi} dk_z \int_0^{\infty} k_\perp dk_\perp \left(\frac{1}{\epsilon_{\boldsymbol{k}}^-}\frac{1}{e^{\frac{8|J|S\epsilon_{\boldsymbol{k}}^-}{k_BT}} - 1} + \frac{1}{\epsilon_{\boldsymbol{k}}^+}\frac{1}{e^{\frac{8|J|S\epsilon_{\boldsymbol{k}}^+}{k_BT}} - 1}\right) \\
&= -\frac{1}{S(2\pi)^2}\frac{k_BT}{8|J|S}\int_{-\pi}^{\pi} dk_z\left[\ln\left(1 - e^{-\frac{8|J|S}{k_BT}\sqrt{j'(1-\cos k_z)}}\right)\right. \\
&\quad \left. + \ln\left(1 - e^{-\frac{8|J|S}{k_BT}\sqrt{j'(1+\cos k_z)}}\right)\right],
\end{aligned} \tag{7.4.30}$$

where $k_\perp^2 = k_x^2 + k_y^2$, $\epsilon_{\boldsymbol{k}}^\pm = \sqrt{k_\perp^2/2 + j'(1 \pm \cos k_z)}$. For $k_BT < \sqrt{JJ'}$, the exponential functions occurring in the arguments of the logarithmic functions in Eq. (7.4.30) can be replaced by their expansions around $k_z = 0$ (the first), and $k_z = \pi$ (the second one). Then the logarithmic functions so obtained can be expanded in their turn to give

$$\begin{aligned}
\Delta M(T) &= \frac{1}{S(2\pi)^2}\frac{k_BT}{8|J|S}4\sum_{n=1}^{\infty}\frac{1}{n}\int_0^{\infty} dk_z e^{-\frac{8|J|S}{k_BT}n\sqrt{\frac{j'}{2}}k_z} \\
&= \frac{\sqrt{2}\zeta(2)}{S\pi^2}\left(\frac{k_BT}{8|J|S}\right)^2\frac{1}{\sqrt{j'}}.
\end{aligned} \tag{7.4.31}$$

Note the singular dependence on the coupling between planes $j'$ which causes the thermal contribution to be orders of magnitude greater than for conventional 3D systems.

For $\sqrt{JJ'} < k_BT < |J|$, the exponential functions in Eq. (7.4.30) can be expanded in powers of their arguments obtaining

$$\Delta M(T) = -\frac{1}{S(2\pi)^2}\frac{k_BT}{8|J|S}2\int_0^\pi dk_z \ln\left[\left(\frac{8|J|S}{k_BT}\right)^2 j' \sin k_z\right]$$

$$= \frac{1}{2\pi S}\frac{k_BT}{8|J|S}\ln\left[\left(\frac{k_BT}{8|J|S}\right)^2 \frac{2}{j'}\right]. \tag{7.4.32}$$

A new singular dependence on $j'$ occurs at intermediate temperature. Equations (7.4.31) and (7.4.32) point out that at finite temperature the 2D antiferromagnet does not have LRO. However, any perturbation like the coupling between planes or easy-axis anisotropy in the plane restores the LRO.

## 7.5. Antiferromagnetism in Close-Packed Lattices

In the close-packed lattices like the triangular (TR), hexagonal (H), face centred cubic (FCC) and rhombohedral (R) lattice, the minimum of the classical energy (7.4.1) is no longer given by $\boldsymbol{Q}\cdot\boldsymbol{\delta} = \pi$. For instance, in the TR lattice, assuming a reference system in which the NNs are located at $\boldsymbol{\delta} = (\pm a, 0)$, $(\pm\frac{a}{2}, \pm\frac{\sqrt{3}}{2}a)$, the classical energy (7.4.1) becomes

$$E_0^{\mathrm{TR}} = 2|J|S^2N\left(\cos aQ_x + 2\cos\frac{aQ_x}{2}\cos\frac{\sqrt{3}}{2}aQ_y\right). \tag{7.5.1}$$

The absolute minimum has to be looked for among the simultaneous solutions of the two equations $\frac{\partial E_0}{\partial Q_x} = 0$ and $\frac{\partial E_0}{\partial Q_y} = 0$, that is

$$\sin\frac{aQ_x}{2}\left(2\cos\frac{aQ_x}{2} + \cos\frac{\sqrt{3}}{2}aQ_y\right) = 0 \tag{7.5.2}$$

and

$$\cos\frac{aQ_x}{2}\sin\frac{\sqrt{3}}{2}aQ_y = 0, \tag{7.5.3}$$

respectively. The solution of Eqs. (7.5.2) and (7.5.3) correspon̄ding to the absolute minimum is

$$\boldsymbol{Q}^{\mathrm{TR}} = \left(\frac{4\pi}{3a}, 0\right) = \frac{1}{3}(\boldsymbol{a}^* + \boldsymbol{b}^*), \tag{7.5.4}$$

where

$$\boldsymbol{a}^* = \frac{2\pi}{a}\left(\mathbf{u}_x + \frac{1}{\sqrt{3}}\mathbf{u}_y\right), \quad \boldsymbol{b}^* = \frac{2\pi}{a}\left(\mathbf{u}_x - \frac{1}{\sqrt{3}}\mathbf{u}_y\right) \tag{7.5.5}$$

are the basic vectors of the reciprocal lattice. The energy (7.5.1) becomes

$$E_0^{\mathrm{TR}} = -3|J|S^2N. \tag{7.5.6}$$

The corresponding spin configuration is obtained from Eq. (7.3.13) where the generic lattice vector is given by

$$\boldsymbol{r}_{l,m} = \frac{a}{2}(l+m)\mathbf{u}_x + \frac{\sqrt{3}}{2}a(l-m)\mathbf{u}_y \tag{7.5.7}$$

with $l, m$ integers. The spin vector localized at the site $(l, m)$ is given by

$$\boldsymbol{S}_{l,m} = S\cos\left[\frac{2\pi}{3}(l+m)+\phi\right]\mathbf{u}_x + S\sin\left[\frac{2\pi}{3}(l+m)+\phi\right]\mathbf{u}_y. \tag{7.5.8}$$

From Eq. (7.5.8) one can see that the spins lie in the $xy$-plane and form a helix with a pitch of 120°. In Fig. 7.2, the ground-state configuration is shown for $\phi = 0$: the Néel antiferromagnetic configuration is not supported by the TR lattice. The physical reason is that if one supposes that two sublattices are antiparallel, *each* spin of the third sublattice could assume any direction since it is surrounded by three spins up and three spins down. For such a configuration the LRO is absent even at $T = 0$. This scenario is found in the TR Ising antiferromagnet where the scalar product of the Heisenberg Hamiltonian (7.1.1) is replaced by the product of the $z$-components of the spins. Indeed, Wannier[62] in 1950 showed that the TR Ising antiferromagnet has a *finite* entropy per spin at zero temperature $\frac{S(0)}{k_B N} = 0.5583$ and the LRO is absent even at $T = 0$. On the contrary, due to the rotation degree of freedom of the spins in the Heisenberg Hamiltonian (7.1.1), the minimum energy is realized by a spin configuration in which the spins of the three compenetrated sublattices rotate by an angle of 120° passing from one sublattice to the other and the LRO at $T = 0$ is recovered. The TR antiferromagnet is an example where the *frustration* by lattice structure prevents the Néel antiferromagnetic configuration. The same configuration characterizes the order in the triangular planes of the H lattice while the order along the $c$-axis is determined by the sign of the NN out-of-plane exchange interaction $J'$: antiferromagnetic or ferromagnetic chains are obtained for $J' < 0$ or $J' > 0$, respectively. In the H lattice, the LRO is recovered at finite temperature by a mechanism similar to that seen in Eqs. (7.4.31) and (7.4.32).

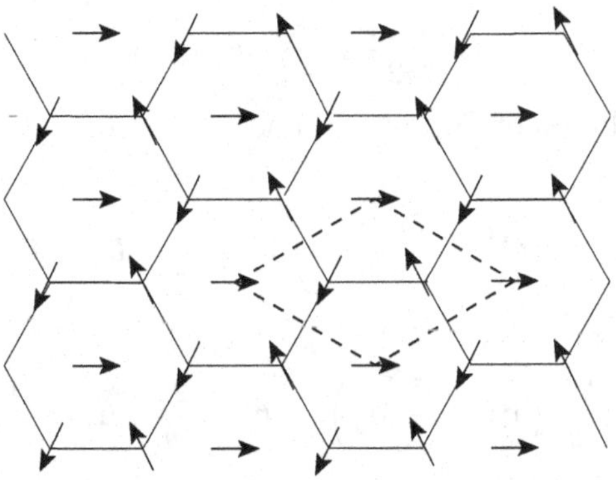

Fig. 7.2. Ground-state spin configuration for a TR antiferromagnet. The dashed line represents the magnetic unit cell.

The FCC lattice for which $\boldsymbol{\delta} = (\pm\frac{a}{2}, \pm\frac{a}{2}, 0)$, $(\pm\frac{a}{2}, 0, \pm\frac{a}{2})$ and $(0, \pm\frac{a}{2}, \pm\frac{a}{2})$ enters another type of frustration. The ground-state energy (7.4.1) becomes

$$E_0^{\mathrm{FCC}} = 4|J|S^2N\left(\cos\frac{aQ_x}{2}\cos\frac{aQ_y}{2} + \cos\frac{aQ_x}{2}\cos\frac{aQ_z}{2} + \cos\frac{aQ_y}{2}\cos\frac{aQ_z}{2}\right) \tag{7.5.9}$$

and the equations for the minimum $\frac{\partial E_0^{\mathrm{FCC}}}{\partial Q_i} = 0$ with $i = x, y, z$ become

$$\sin\frac{aQ_x}{2}\left(\cos\frac{aQ_y}{2} + \cos\frac{aQ_z}{2}\right) = 0, \tag{7.5.10}$$

$$\sin\frac{aQ_y}{2}\left(\cos\frac{aQ_x}{2} + \cos\frac{aQ_z}{2}\right) = 0 \tag{7.5.11}$$

and

$$\sin\frac{aQ_z}{2}\left(\cos\frac{aQ_x}{2} + \cos\frac{aQ_y}{2}\right) = 0. \tag{7.5.12}$$

In this case, the absolute minimum is no longer given by a unique wavevector or, more precisely, by the star of symmetric wavevectors that lead to the same spin configuration. In this case, the absolute minimum corresponds to an *infinite* continuous variety of wavevectors given by

$$\boldsymbol{Q}^{\mathrm{FCC}} = \left(\frac{2\pi}{a}\xi, 0, \frac{2\pi}{a}\right) = \frac{1}{2}\xi(\boldsymbol{a}^* + \boldsymbol{c}^*) + \frac{1}{2}(\boldsymbol{b}^* + \boldsymbol{c}^*), \tag{7.5.13}$$

where $\xi$ is arbitrary and

$$\boldsymbol{a}^* = \frac{2\pi}{a}(\mathbf{u}_x + \mathbf{u}_y - \mathbf{u}_z), \quad \boldsymbol{b}^* = \frac{2\pi}{a}(\mathbf{u}_x - \mathbf{u}_y + \mathbf{u}_z), \quad \boldsymbol{c}^* = \frac{2\pi}{a}(-\mathbf{u}_x + \mathbf{u}_y + \mathbf{u}_z) \tag{7.5.14}$$

are the basic vectors of the reciprocal lattice. Obviously, the helix wavevectors obtained by cyclic permutations of $Q_x, Q_y$ and $Q_z$ lead to the same ground-state energy given by

$$E_0^{\mathrm{FCC}} = -4|J|S^2N. \tag{7.5.15}$$

Since Eq. (7.5.15) is independent of $\xi$, the classical ground-state energy of the FCC antiferromagnet is *infinitely* degenerate. Also the ground-state of the FCC Ising antiferromagnet[63] is highly degenerate ($2^{N^{1/3}}$) and LRO disappears even at $T = 0$. However, unlike the TR Ising antiferromagnet in the FCC Ising antiferromagnet the entropy per spin vanishes at $T = 0$.

The spin wave spectrum of the FCC Heisenberg antiferromagnet is obtained from Eqs. (7.3.10)–(7.3.12). Assuming to measure the lengths in units of the lattice space $a$, one has

$$\hbar\omega_{\boldsymbol{k}} = 8|J|S\sqrt{s_{\boldsymbol{k}} d_{\boldsymbol{k}}} \tag{7.5.16}$$

where

$$s_{\boldsymbol{k}} = 1 + \cos x \cos y + \cos x \cos z + \cos y \cos z \tag{7.5.17}$$

and

$$d_{\boldsymbol{k}} = 1 - \cos y \cos z + \cos(\pi\xi) \cos x\,(\cos y - \cos z) \tag{7.5.18}$$

with $x = \frac{ak_x}{2}$, $y = \frac{ak_y}{2}$ and $z = \frac{ak_z}{2}$. As one can see from Eq. (7.5.18), the spin wave spectrum has a line of soft modes corresponding to the $x$-axis ($k_y = k_z = 0$). The presence of a *soft line* in the spin wave spectrum causes the divergence of the third term of Eq. (7.3.14) pointing out that the LRO is absent at any finite temperature. It is interesting to understand how the classical scenario is modified when quantum fluctuations are taken into account. To this end, we evaluate the first quantum correction to the classical ground-state energy given by Eq. (7.2.16). For the FCC lattice one obtains

$$\Delta E_0^{\mathrm{FCC}} = 4|J|SN[I^{\mathrm{FCC}}(\xi) - 1] \tag{7.5.19}$$

where

$$I^{\mathrm{FCC}}(\xi) = \frac{1}{\pi^3}\int_0^\pi dx \int_0^\pi dy \int_0^\pi dz \sqrt{s_{\boldsymbol{k}} d_{\boldsymbol{k}}}\ . \tag{7.5.20}$$

As one can see from Eqs. (7.5.20), (7.5.17) and (7.5.18), the first quantum correction to the classical ground state is a function of $\xi$ through the term $d_{\boldsymbol{k}}$ appearing in the spin wave spectrum. The minimum of the first-order ground-state energy corresponds to the minimum of $I^{\mathrm{FCC}}(\xi)$ and the ground-state configuration is obtained replacing the value of $\xi$ that minimizes $I^{\mathrm{FCC}}(\xi)$ into Eq. (7.5.13). A numerical evaluation of the integral (7.5.20) shows that $I^{\mathrm{FCC}}(\xi)$ is minimum for $\xi = 0$ and $\xi = 1$ [$I^{\mathrm{FCC}}(0) = I^{\mathrm{FCC}}(1) = 0.75584$] and maximum for $\xi = \frac{1}{2}$ [$I^{\mathrm{FCC}}(\frac{1}{2}) = 0.79661$]. Consequently, the zero-point energy has the absolute minimum at $Q_x = 0$ and $Q_x = \frac{2\pi}{a}$. The corresponding helix wavevectors become

$$\boldsymbol{Q}_{\mathrm{GS}}^{\mathrm{FCC}} = \left(0, 0, \frac{2\pi}{a}\right) \quad \text{and} \quad \boldsymbol{Q}_{\mathrm{GS}}^{\mathrm{FCC}} = \left(\frac{2\pi}{a}, 0, \frac{2\pi}{a}\right). \tag{7.5.21}$$

From the generic vector of the FCC lattice

$$\boldsymbol{r}_{l,m,n} = \frac{a}{2}(l+m)\mathbf{u}_x + \frac{a}{2}(l+n)\mathbf{u}_y + \frac{a}{2}(m+n)\mathbf{u}_z \tag{7.5.22}$$

with $l, m, n$ integers and from the first helix wavevector of Eq. (7.5.21), one obtains the ground-state configuration of the FCC lattice

$$\boldsymbol{S}_{l,m,n} = S\cos[\pi(m+n) + \phi]\mathbf{u}_x + S\sin[\pi(m+n) + \phi]\mathbf{u}_y. \tag{7.5.23}$$

This configuration corresponds to a stacking of alternating ferromagnetic planes perpendicular to the $z$-axis as shown in Fig. 7.3. This kind of antiferromagnetic order is classified as "antiferromagnetic order of type AF1". From the second helix

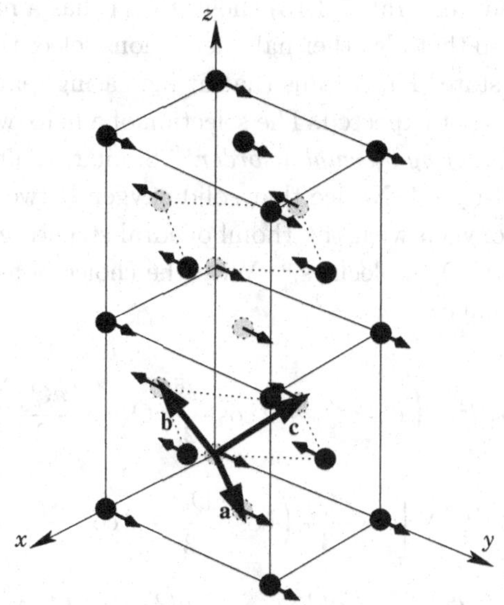


Fig. 7.3. FCC antiferromagnet. Ground-state configuration obtained accounting for the first-order quantum correction. The thick arrows represent the basic lattice vectors of the FCC lattice.

wavevector of Eq. (7.5.21), one obtains a ground-state configuration similar to the previous one. The only difference is that the ferromagnetic planes are perpendicular to the $y$-axis. From Eq. (7.5.23), one sees that the spins make an angle $\phi$ (arbitrary) with the $x$-axis even though the configuration shown in Fig. 7.3 corresponds to the choice $\phi = \frac{\pi}{2}$. The fact that the spins lie in the $xy$-plane is due to the broken symmetry entered by the initial choice of the cone axis parallel to the $z$-axis. In other words, the calculations performed in this chapter are reminiscent of a single-ion easy-plane anisotropy $D \to 0^+$. Physically, only the anisotropy entered by the crystal field can select the angles $\theta$ and $\phi$ in actual compounds. For a configuration as shown in Fig. 7.3, the lattice anisotropy is expected to choose between the high symmetry values $\phi = 0$ (1,0,0), [or $\phi = \frac{\pi}{2}$ (0,1,0)] and $\phi = \frac{\pi}{4}$ (1,1,0) according to $x$ ($y$) are easy-axis or a hard-axis. The antiferromagnetic order of type AF1 was observed in $UO_2$ by neutron-diffraction[64] below $T_N \simeq 31\,K$. The magnetic uranium ions $U^{4+}$ are located on the nodes of a FCC lattice and their magnetic moments ($\sim 1.8\mu_B$) lie within the alternating ferromagnetic sheets: the orientation of the antiferromagnetic axis in the $xy$-plane could not be singled out, however, because of the multidomain structure of the sample.

We have seen that in the FCC antiferromagnet the quantum fluctuations destroy the infinite degeneracy present in classic approximation, leading to an ordered ground state. This fact is referred as *order by quantum disorder.*[65] Another interesting question is about the thermal contribution to the free energy as given by the temperature-dependent term of Eq. (7.4.13). Since the spin wave spectrum is a function of $\xi$ through $d_{\boldsymbol{k}}$ given by Eq. (7.5.18), the question is: which helix wavevectors are selected by thermal fluctuations? A numerical evaluation of the temperature-dependent integral (7.4.13) shows that it has a minimum at $\xi = 0$ and a maximum at $\xi = \frac{1}{2}$ so that also thermal fluctuations select the helix wavevector of the quantum ground state. This means that at increasing temperature, a change of the helix wavevector is not expected. The selection of a helix wavevector by thermal fluctuations is called *order by thermal disorder.*[66] Similar results are obtained for the R lattice[66] shown in Fig. 7.4. Notice that solid oxygen between 24 and 44 K shows a phase known as $\beta$-oxygen with the rhombohedral structure and direct exchange interaction between the $O_2$ molecules.[67] With the choice of axes shown in Fig. 7.4, the energy (7.4.1) becomes

$$E_0^{\rm R} = 2|J|S^2N\left(\cos aQ_y + 2\cos\frac{\sqrt{3}}{2}aQ_x\cos\frac{aQ_y}{2}\right)$$
$$-2J'S^2N\left[\cos\frac{cQ_z}{3}\left(\cos\frac{aQ_x}{\sqrt{3}} + 2\cos\frac{aQ_x}{2\sqrt{3}}\cos\frac{aQ_y}{2}\right)\right.$$
$$\left.+\sin\frac{cQ_z}{3}\left(\sin\frac{aQ_x}{\sqrt{3}} - 2\sin\frac{aQ_x}{2\sqrt{3}}\cos\frac{aQ_y}{2}\right)\right], \qquad (7.5.24)$$

where $J < 0$ and $J'$ are the exchange integrals between the NN in the plane and out-of-plane, respectively (Fig. 7.4). Defining the following quantities

$$e_0 = \frac{E_0^{\rm R}}{6|J|S^2N}, \quad x_0 = \frac{\sqrt{3}}{2}aQ_x, \quad y_0 = \frac{aQ_y}{2}, \quad z_0 = \frac{cQ_z}{3}, \quad j' = \frac{J'}{|J|}, \qquad (7.5.25)$$

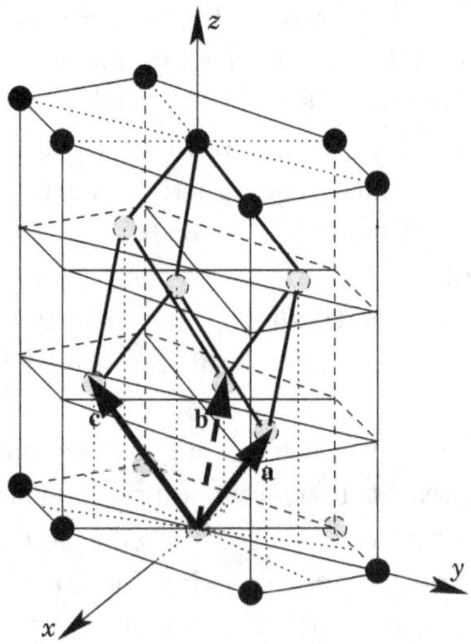


Fig. 7.4. Rhombohedral lattice with the lattice vectors of the primitive cell.

Equation (7.5.24) becomes

$$e_0 = \frac{1}{3}\left\{\cos 2y_0 + 2\cos x_0 \cos y_0 - j'\left[\cos z_0 \left(\cos\frac{2x_0}{3}\right.\right.\right.$$

$$\left.\left.\left. + 2\cos\frac{x_0}{3}\cos y_0\right) + \sin z_0\left(\sin\frac{2x_0}{3} - 2\sin\frac{x_0}{3}\cos y_0\right)\right]\right\}. \quad (7.5.26)$$

The minimum conditions $\frac{\partial e_0}{\partial x_0} = 0$, $\frac{\partial e_0}{\partial y_0} = 0$ and $\frac{\partial e_0}{\partial z_0} = 0$ lead to the equations

$$\sin x_0 \cos y_0 \left(1 - \frac{j'\cos z_0}{\cos\frac{2x_0}{3} + 2\cos\frac{x_0}{3}\cos y_0}\right) = 0, \quad (7.5.27)$$

$$\sin y_0\,(2\cos y_0 + \cos x_0)\left(1 - \frac{j'\cos z_0}{\cos\frac{2x_0}{3} + 2\cos\frac{x_0}{3}\cos y_0}\right) = 0 \quad (7.5.28)$$

and

$$\tan z_0 = \frac{2\sin\frac{x_0}{3}\left(\cos\frac{x_0}{3} - \cos y_0\right)}{\cos\frac{2x_0}{3} + 2\cos\frac{x_0}{3}\cos y_0}, \quad (7.5.29)$$

respectively. Using the trigonometric relationship

$$\cos z_0 = \pm\frac{1}{\sqrt{1+\tan^2 z_0}},$$

Eqs. (7.5.26)–(7.5.28) become

$$\sin x_0 \cos y_0 \left(1 - \frac{|j'|}{\sqrt{1 + 4\cos x_0 \cos y_0 + 4\cos^2 y_0}}\right) = 0, \quad (7.5.30)$$

$$\sin y_0\,(2\cos y_0 + \cos x_0)\left(1 - \frac{|j'|}{\sqrt{1 + 4\cos x_0 \cos y_0 + 4\cos^2 y_0}}\right) = 0 \quad (7.5.31)$$

and

$$\cos z_0 = \pm\frac{\cos\frac{2x_0}{3} + 2\cos\frac{x_0}{3}\cos y_0}{\sqrt{1 + 4\cos x_0\cos y_0 + 4\cos^2 y_0}} \quad (7.5.32)$$

with

$$e_0 = \frac{1}{3}\left(\cos 2y_0 + 2\cos x_0\cos y_0 - |j'|\sqrt{1 + 4\cos x_0\cos y_0 + 4\cos^2 y_0}\right). \quad (7.5.33)$$

The upper and lower signs in Eq. (7.5.32) correspond to $j' > 0$ (ferromagnetic out-of-plane exchange interaction) and $j' < 0$ (antiferromagnetic out-of-plane exchange interaction), respectively. Notice that Eqs. (7.5.30), (7.5.31) and (7.5.33) are independent of the sign of $j'$ while the sign of $j'$ is important in Eq. (7.5.32) that is crucial to obtain the configuration corresponding to the minimum energy. The absolute minimum of the classical energy corresponds to

$$x_0 = 0, \quad y_0 = 0, \quad e_0 = 1 - |j'| \quad (7.5.34)$$

with $z_0 = \pi$ or 0 for $j' < -3$ or $j' > 3$, respectively. For $-3 \leq j' \leq 3$, the minimum is no longer an *isolated point* but it consists of a *continuous line* given by the vanishing of the common term in Eqs. (7.5.30) and (7.5.31), that is,

$$4\cos^2\frac{Q_y}{2} + 4\cos\frac{\sqrt{3}}{2}Q_x\cos\frac{Q_y}{2} + 1 - j'^2 = 0. \tag{7.5.35}$$

The corresponding $Q_z$ is given by the equation

$$\cos\frac{Q_z}{3} = \frac{1}{j'}\left(\cos\frac{Q_x}{\sqrt{3}} + 2\cos\frac{Q_x}{2\sqrt{3}}\cos\frac{Q_y}{2}\right), \tag{7.5.36}$$

where $Q_x$ and $Q_y$ are solutions of Eq. (7.5.35). The corresponding classical energy is given by

$$e_0 = -\frac{1}{2} - \frac{1}{6}j'^2. \tag{7.5.37}$$

This infinitely degenerate minimum in the classical energy leads to a *degenerate helix* (DH).[66] The DH given by Eq. (7.5.35) is shown in Fig. 7.5 for $0 < |j'| < 1$ and in Fig. 7.6 for $1 < |j'| < 3$. Notice that the projection of the DH line onto the $Q_xQ_y$-plane is invariant under the change of sign of $j'$ as shown by the left panels of Figs. 7.5 and 7.6 while the projection onto the $Q_xQ_z$-plane is reversed by the change of sign of $j'$: compare the central ($j' > 0$) with the panel on the right ($j' < 0$) of Figs. 7.5 and 7.6. The curve (7.5.35) reduces to a circle of equation

$$Q_x^2 + \left(Q_y - \frac{4\pi}{3}\right)^2 = \frac{4}{3}j'^2 \tag{7.5.38}$$

for $|j'| \to 0$ and to a circle of equation

$$Q_x^2 + Q_y^2 = \frac{2}{3}(9 - j'^2) \tag{7.5.39}$$

for $|j'| \to 3$. For $|j'| = 0$ and $|j'| = 3$, Eqs. (7.5.38) and (7.5.39) reduce to the points $(0, \frac{4\pi}{3})$ and $(0, 0)$, respectively. For $|j'| \to 1^-$, the curve given by Eq. (7.5.35) reduces to an equilateral triangle with barycentre in $(0, \frac{4\pi}{3})$ while for $|j'| \to 1^+$, the curve given by Eq. (7.5.35) reduces to a hexagon centred in the origin.

## 7.6. Order by Quantum and Thermal Disorder

In the previous section, we have seen that a R lattice shows an infinite degeneracy in the (classical) ground state for $|J'/J| < 3$. In this section, we will see how the quantum and thermal fluctuations destroy this infinite degeneracy. This can be done evaluating explicitly the free energy (7.4.13) for the R lattice assuming that the helix wavevector belongs to the DH given by Eqs. (7.5.35) and (7.5.36). By use

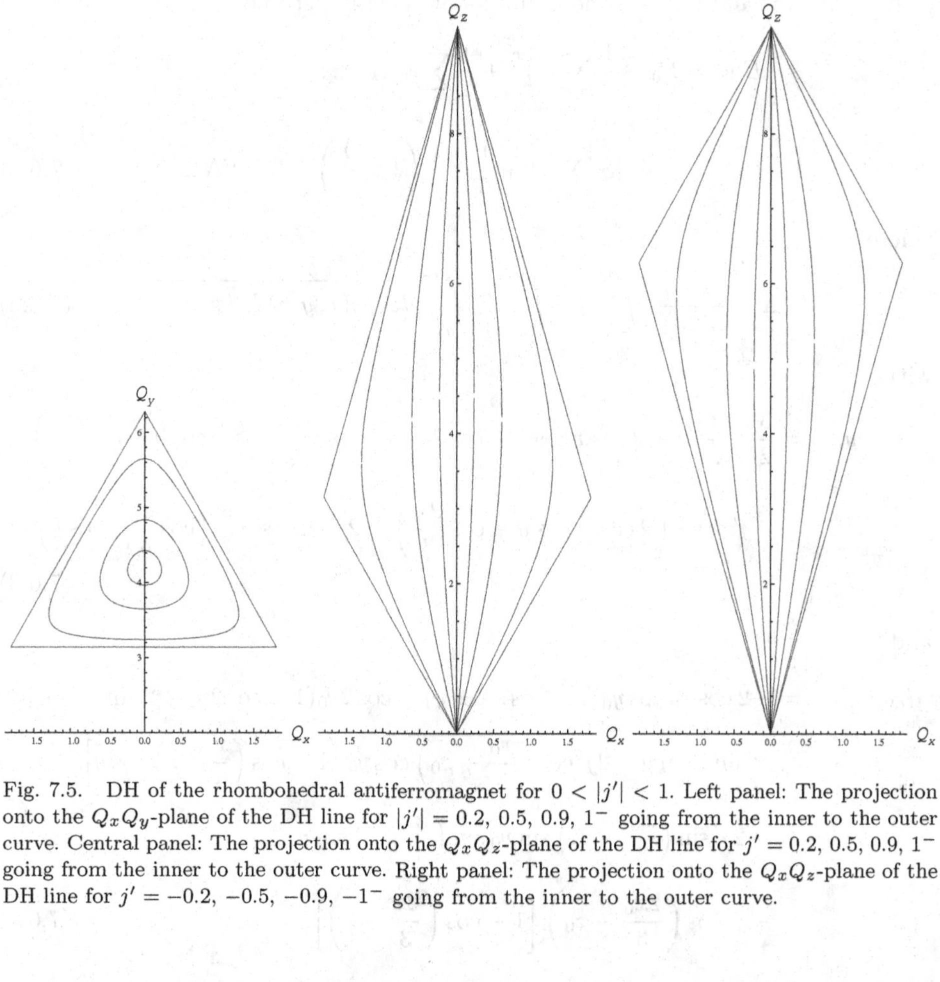


Fig. 7.5. DH of the rhombohedral antiferromagnet for $0 < |j'| < 1$. Left panel: The projection onto the $Q_xQ_y$-plane of the DH line for $|j'| = 0.2, 0.5, 0.9, 1^-$ going from the inner to the outer curve. Central panel: The projection onto the $Q_xQ_z$-plane of the DH line for $j' = 0.2, 0.5, 0.9, 1^-$ going from the inner to the outer curve. Right panel: The projection onto the $Q_xQ_z$-plane of the DH line for $j' = -0.2, -0.5, -0.9, -1^-$ going from the inner to the outer curve.

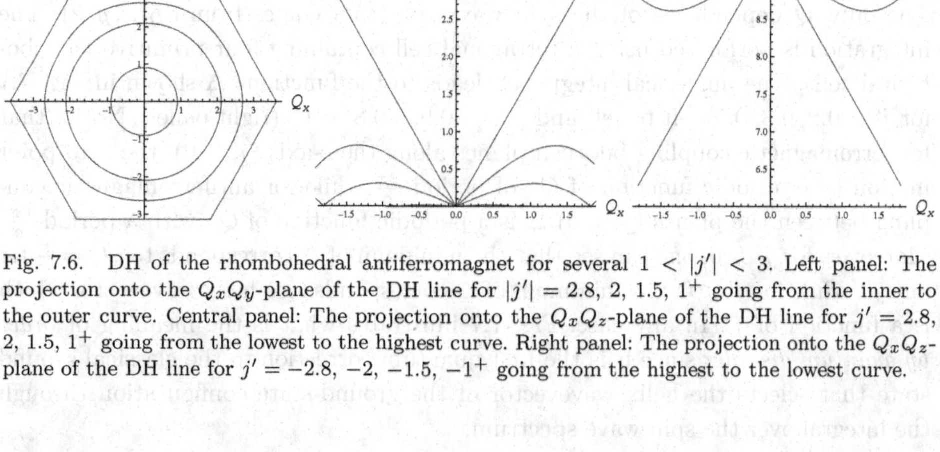


Fig. 7.6. DH of the rhombohedral antiferromagnet for several $1 < |j'| < 3$. Left panel: The projection onto the $Q_xQ_y$-plane of the DH line for $|j'| = 2.8, 2, 1.5, 1^+$ going from the inner to the outer curve. Central panel: The projection onto the $Q_xQ_z$-plane of the DH line for $j' = 2.8, 2, 1.5, 1^+$ going from the lowest to the highest curve. Right panel: The projection onto the $Q_xQ_z$-plane of the DH line for $j' = -2.8, -2, -1.5, -1^+$ going from the highest to the lowest curve.

of Eqs. (7.2.16) and (7.5.37) the ground-state energy becomes

$$E_{\rm GS} = E_0 - \frac{1}{2}\sum_{\boldsymbol{k}} A_{\boldsymbol{k}}^{(s)} + \frac{1}{2}\sum_{\boldsymbol{k}} \hbar\omega_{\boldsymbol{k}}$$

$$= -3|J|S^2N\left(1+\frac{1}{3}j'^2\right)\left(1+\frac{1}{S}\right) + 2|J|SN\Delta, \tag{7.6.1}$$

where

$$\Delta = \frac{1}{2(2\pi)^3}\int_{-2\pi}^{2\pi} dx \int_{-\pi}^{\pi} dy \int_0^{2\pi} dz \sqrt{s(x,y,z)d_{\boldsymbol{Q}}(x,y,z)} \tag{7.6.2}$$

with

$$s(x,y,z) = \frac{3}{2} + \frac{1}{2}j'^2 + 2\cos x\cos y + \cos 2y - j'\cos z\left(2\cos\frac{x}{3}\cos y + \cos\frac{2x}{3}\right)$$
$$- j'\cos z\left(2\cos\frac{x}{3}\cos y + \cos\frac{2x}{3}\right) - 2j'\sin z\sin\frac{x}{3}\left(\cos\frac{x}{3} - \cos y\right) \tag{7.6.3}$$

and

$$d_{\boldsymbol{Q}}(x,y,z) = -2\cos x_0\cos y_0(1-\cos x\cos y) - \cos 2y_0(1-\cos 2y) + 2\sin x_0\sin y_0$$
$$\times \sin x\sin y + 2j'\cos\left(\frac{x_0}{3}+z_0\right)\cos y_0\left[1-\cos\left(\frac{x}{3}+z\right)\cos y\right]$$
$$- 2j'\sin\left(\frac{x_0}{3}+z_0\right)\sin y_0\sin\left(\frac{x}{3}+z\right)\sin y$$
$$+ j'\cos\left(\frac{2x_0}{3}-z_0\right)\left[1-\cos\left(\frac{2x}{3}-z\right)\right]. \tag{7.6.4}$$

In Eqs. (7.6.3) and (7.6.4), $x, y, z$ mean $x = \frac{\sqrt{3}}{2}ak_x$, $y = \frac{1}{2}ak_y$ and $z = \frac{1}{3}ck_z$. Notice that $s(x,y,z)$ is independent of $\boldsymbol{Q}$ ($x_0 = \frac{\sqrt{3}}{2}aQ_x$, $y_0 = \frac{1}{2}aQ_y$ and $z_0 = \frac{1}{3}cQ_z$). The only $\boldsymbol{Q}$ dependence of the spin wave spectrum comes from $d_{\boldsymbol{Q}}(x,y,z)$. The integration is performed using a tetragonal cell containing four primitive rhombohedral cells. The numerical integration leads to the functions $\Delta$ shown in Fig. 7.7 for $j' = 0.9, 0.8, 0.5$ (left panel) and $j' = -0.9, -0.8, -0.5$ (right panel). Notice that for ferromagnetic coupling between planes along the $c$-axis ($j' > 0$), the zero-point motion is a periodic function of $Q_z$ of period $\frac{2\pi}{c}$ while for antiferromagnetic coupling between the planes ($j' < 0$) $\Delta$ is a periodic function of $Q_z$ with a period $\frac{6\pi}{c}$. Moreover, for $j' > 0$, one can see that the minimum of $\Delta$ corresponds to $Q_z = \frac{\pi}{c}$ for any $j'$, while for $j' < 0$, the minimum of $\Delta$ corresponds to a helix wavevector which is a function of $j'$. In any case, Fig. 7.7 illustrates what is the meaning of *order by quantum disorder* since it is the first quantum correction to the classical ground state that selects the helix wavevector of the ground-state configuration through the integral over the spin wave spectrum.

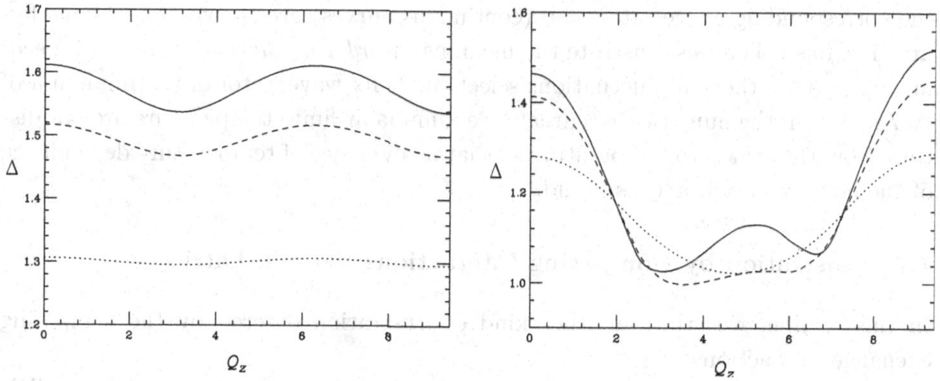


Fig. 7.7. First quantum correction to the ground-state energy of a rhombohedral lattice as function of $Q_z$ moving along the DH helix. Left panel: $j' = 0.9$ (continuous curve), $j' = 0.8$ (dashed curve) and $j' = 0.5$ (dotted curve). Right panel: $j' = -0.9$ (continuous curve), $j' = -0.8$ (dashed curve) and $j' = -0.5$ (dotted curve).

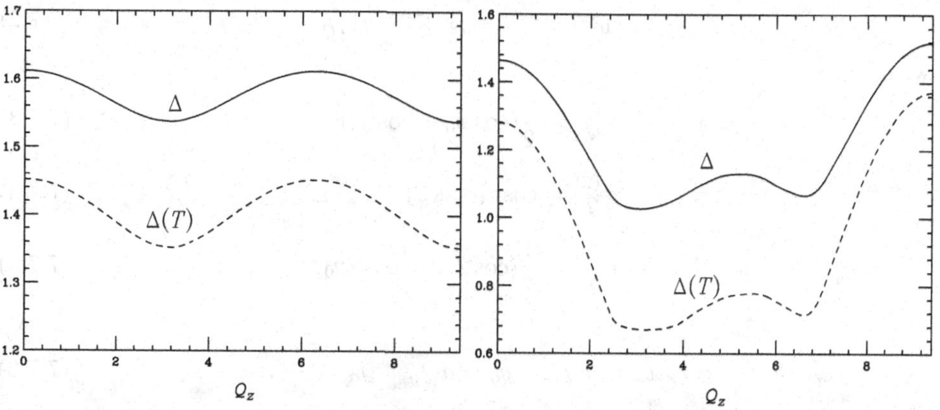


Fig. 7.8. Free energy as function of $Q_z$ moving along the DH line for $T = 0$ (continuous curves) and $\frac{k_BT}{4|J|S} = 0.5$ (dashed curves). Left panel: $j' = 0.9$. Right panel: $j' = -0.9$.

The free energy of the R lattice in the harmonic approximation reads

$$F = E_{\rm GS} + k_BT\sum_k \ln(1 - e^{-\beta\hbar\omega_k})$$

$$= -3|J|S^2N\left(1 + \frac{1}{3}j'^2\right)\left(1 + \frac{1}{S}\right) + 2|J|SN\Delta(T), \qquad (7.6.5)$$

where

$$\Delta(T) = \Delta + \frac{\tau}{(2\pi)^3}\int_{-2\pi}^{2\pi} dx \int_{-\pi}^{\pi} dy \int_0^{2\pi} dz\ \ln\left(1 - e^{-\frac{1}{\tau}\sqrt{s(x,y,z)d_Q(x,y,z)}}\right), \qquad (7.6.6)$$

with $\tau = \frac{k_BT}{4|J|S}$ and $\Delta$ given by Eq. (7.6.2). The numerical integration of Eq. (7.6.6) is shown in Fig. 7.8 for $j' = \pm 0.9$ and $\tau = 0.5$ (dashed curves). For comparison, the corresponding curves at $T = 0$ (continuous curves) are shown in the same figure. The dashed curves illustrate the meaning of *order by thermal disorder.* Indeed, in this case the thermal fluctuations select the helix wavevector of the minimum of $\Delta(T)$. Within the numerical accuracy, the minima at finite temperature are not distinct from those at zero temperature so that no evidence of temperature dependence of the helix wavevector $\boldsymbol{Q}$ is found.

## 7.7. Frustration by Competing Interactions: Square Lattice

In this section, we study another kind of frustration entered by the competing exchange interactions.

The ground-state energy of a SQ lattice with exchange interactions up to TNN is given by

$$E_0^{\mathrm{SQ}} = 4J_1 S^2 N e_0 \tag{7.7.1}$$

where $J_1 > 0$ and

$$e_0 = -\gamma_{\boldsymbol{Q}}^{(1)} - j_2\gamma_{\boldsymbol{Q}}^{(2)} - j_3\gamma_{\boldsymbol{Q}}^{(3)} \tag{7.7.2}$$

with

$$\gamma_{\boldsymbol{Q}}^{(1)} = \frac{1}{2}(\cos x_0 + \cos y_0), \tag{7.7.3}$$

$$\gamma_{\boldsymbol{Q}}^{(2)} = \cos x_0 \cos y_0, \tag{7.7.4}$$

$$\gamma_{\boldsymbol{Q}}^{(3)} = \frac{1}{2}(\cos 2x_0 + \cos 2y_0) \tag{7.7.5}$$

and

$$x_0 = aQ_x, \quad y_0 = aQ_y, \quad j_\alpha = \frac{J_\alpha}{J_1}. \tag{7.7.6}$$

The minimum conditions are given by

$$\sin x_0(1 + 2j_2\cos y_0 + 4j_3\cos x_0) = 0, \tag{7.7.7}$$

$$\sin y_0(1 + 2j_2\cos x_0 + 4j_3\cos y_0) = 0. \tag{7.7.8}$$

The minimum configurations in the $j_2j_3$-plane are illustrated in Fig. 7.9. The $j_2j_3$-plane is divided in four regions and the corresponding energies and helix wavevectors are given by

**F**: ferromagnetic phase

$$x_0 = y_0 = 0, \quad e_0^{\mathrm{F}} = -1 - j_2 - j_3. \tag{7.7.9}$$

**AF$_1$**: antiferromagnetic collinear order in which the spins along the $(1, 0)$ lines are parallel

$$x_0 = 0, \quad y_0 = \pi, \quad e_0^{\mathrm{AF}_1} = j_2 - j_3. \tag{7.7.10}$$

Obviously, the symmetric phase with $x_0 = \pi$ and $y_0 = 0$ in which the parallel spins are along the $(0, 1)$ lines are degenerate with Eq. (7.7.10).

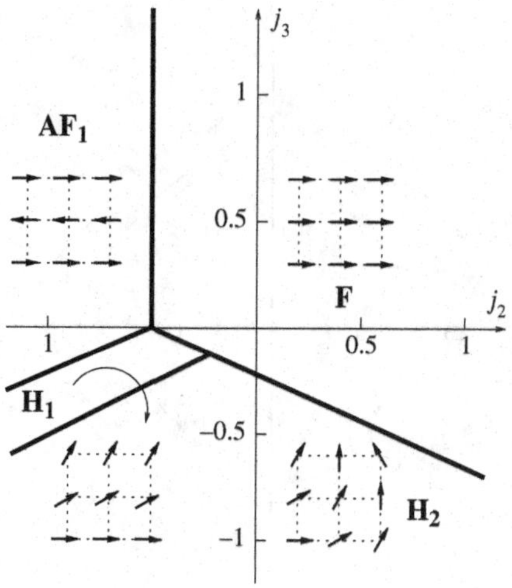


Fig. 7.9. Ground-state configurations of a SQ lattice with competing interactions $j_\alpha = \frac{J_\alpha}{J_1}$ and $J_1 > 0$.

$\mathbf{H_1}$: non-collinear order in which the spins are parallel along the $(1, 0)$ direction and make an angle $y_0$ moving along the $(0, 1)$ direction

$$x_0 = 0, \quad y_0 = \arccos\left(-\frac{1+2j_2}{4j_3}\right), \quad e_0^{\mathrm{H_1}} = -\frac{1}{2} + \frac{(1+2j_2)^2}{16j_3}. \tag{7.7.11}$$

The same energy is obtained by exchanging $x_0$ and $y_0$. In this case, the spins are parallel along the $(0, 1)$ direction and make an angle $x_0$ along the $(1, 0)$ direction. In general, this magnetic order is *incommensurate* with the lattice periodicity.

$\mathbf{H_2}$: non-collinear order in which the spins are parallel along the $(1, -1)$ direction and make an angle $x_0$ moving along the $(1, 1)$ direction

$$x_0 = y_0 = \arccos\left[-\frac{1}{2(j_2+2j_3)}\right], \quad e_0^{\mathrm{H_2}} = j_3 + \frac{1}{4(j_2+2j_3)}. \tag{7.7.12}$$

Obviously, another energy minimum exists for $x_0 = -y_0$. In this case, the parallel spins are along the $(1, 1)$ direction and make an angle $x_0$ along the $(1, -1)$ direction.

The boundaries between the different regions shown in Fig. 7.9 are lines of continuous (second order) phase transitions except the lines between regions $\mathbf{AF_1} - \mathbf{F}$ where the transition is first order and $\mathbf{H_1} - \mathbf{H_2}$ which is a *disorder* line. The equations of the separation lines between the different regions are

$\mathbf{AF_1} - \mathbf{F} : j_2 = -\frac{1}{2}$,
$\mathbf{F} - \mathbf{H_1}$ and $\mathbf{F} - \mathbf{H_2} : j_3 = -\frac{1}{4}(1+2j_2)$,
$\mathbf{AF_1} - \mathbf{H_1} : j_3 = \frac{1}{4}(1+2j_2)$,
$\mathbf{H_1} - \mathbf{H_2} : j_3 = \frac{1}{2}j_2$.

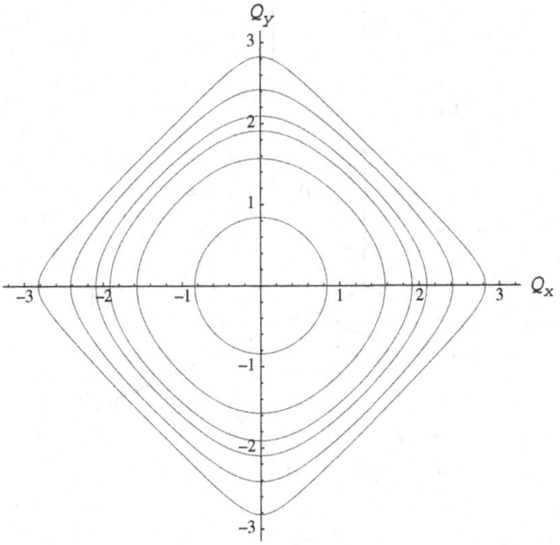


Fig. 7.10. Infinite degeneration line of the SQ lattice in the reciprocal space for several $j_2$ ($j_3 = \frac{1}{2}j_2$). From the inner to the outer curve: $j_2 = -0.3, -0.5, -0.75, -1, -2, -10$, respectively.

Notice that for $j_3 = \frac{1}{2}j_2$ Eqs. (7.7.7) and (7.7.8) are simultaneously satisfied assuming

$$\cos x_0 + \cos y_0 = -\frac{1}{2j_2} \tag{7.7.13}$$

so that the minima form a line in the reciprocal space given by

$$Q_y = \arccos\left(-\frac{1}{2j_2} - \cos Q_x\right) \tag{7.7.14}$$

with an infinitely degenerate ground-state energy given by

$$e_0^{\mathrm{DH}} = \frac{1}{2}j_2 + \frac{1}{8j_2}. \tag{7.7.15}$$

The DH given by Eq. (7.7.14) is shown in Fig. 7.10 for $j_2 = -0.3, -0.5, -0.75, -1, -2, -10$ going from the inner to the outer curve. Notice that for $j_2 = -0.25$, the DH shrinks to the point (0,0), and for $j_2 \to \infty$, it becomes a square given by $Q_y = \pm(\pi - |Q_x|)$ with $-\pi < Q_x < \pi$. The infinite degeneracy of the ground state implies absence of LRO even at $T = 0$. This infinite degeneracy, however, is destroyed by quantum fluctuations or by the addition of interactions to spins further. Indeed, the first quantum correction to the classical ground state is given by

$$E_{\mathrm{GS}}^{\mathrm{DH}} = 4J_1S^2N\left[\left(\frac{1}{2}j_2 + \frac{1}{8j_2}\right)\left(1 + \frac{1}{S}\right) + \frac{\Delta}{S}\right], \tag{7.7.16}$$

where

$$\Delta = \frac{1}{(2\pi)^2}\int_{-\pi}^{\pi} dx \int_{-\pi}^{\pi} dy \sqrt{s(x,y)d_Q(x,y)} \tag{7.7.17}$$

with

$$s(x,y) = -\frac{1}{2}j_2 - \frac{1}{8j_2} - \frac{1}{2}(\cos x + \cos y) - j_2 \cos x \cos y$$
$$-\frac{j_2}{4}(\cos 2x + \cos 2y), \tag{7.7.18}$$
$$d_Q(x,y) = \frac{1}{2}[\cos x_0(1-\cos x) + \cos y_0(1-\cos y)] + j_2 \cos x_0 \cos y_0$$
$$\times (1 - \cos x \cos y) - j_2 \sin x_0 \sin y_0 \sin x \sin y$$
$$+\frac{j_2}{4}[(\cos 2x_0)(1-\cos 2x) + \cos 2y_0(1-\cos 2y)] \tag{7.7.19}$$

and

$$y_0 = \arccos\left(-\frac{1}{2j_2} - \cos x_0\right). \tag{7.7.20}$$

The numerical evaluation of the 2D integral in Eq. (7.7.17) shows that the minima of $\Delta$ occur at $x_0 = 0$, $y_0 = \arccos(-\frac{1}{2j_2} - 1)$ and $x_0 = \arccos(-\frac{1}{2j_2} - 1)$, $y_0 = 0$ corresponding to the helix $\mathbf{H_1}$ while the maxima of $\Delta$ are located at $x_0 = \pm y_0$ where $x_0 = \arccos(-\frac{1}{4j_2})$ corresponding to the helix $\mathbf{H_2}$. This result points out that the disorder line $j_3 = \frac{1}{2}j_2$ becomes a first-order transition line between the $\mathbf{H_1}$ and $\mathbf{H_2}$ phases. More precisely, the existence region of the helix $\mathbf{H_1}$ shown in Fig. 7.9 is expected to widen at the expense of the existence region of the helix $\mathbf{H_2}$ since the gap between the energies of the helices $\mathbf{H_2}$ and $\mathbf{H_1}$ increases with $|j_2|$. The LRO at $T = 0$ is recovered on the "new" boundary line $\mathbf{H_1} - \mathbf{H_2}$ which becomes a first-order transition line. One can say that the first-order transition between the phases $\mathbf{H_1}$ and $\mathbf{H_2}$ is driven by quantum fluctuations.

A similar effect is obtained introducing a fourth nearest neighbour (FNN) exchange interaction $j_4 = \frac{2J_4}{J_1}$ in the classical ground-state energy. Indeed, $j_4$ does not change the qualitative feature of the phase diagram of Fig. 7.9. The main change is the disappearance of the disorder line between the helices $\mathbf{H_1}$ and $\mathbf{H_2}$ replaced by a first-order transition line. The quantitative effect of the FNN interaction is to shift the boundary lines as shown in Fig. 7.11 for $j_4 = 0.1$ (a) and $j_4 = -0.1$ (b), respectively. The thin lines in both figures correspond to the case $j_4 = 0$ shown in Fig. 7.9. The points $\mathrm{P}_1(-0.475, -0.1375)$ and $\mathrm{P}_2(-0.025, -0.1125)$ are triple points where two second-order transition lines and one first-order transition line meet.

The phase diagram at $T = 0$ of the T lattice with competing interactions in the $xy$-plane is the same as that shown in Figs. 7.9 and 7.11. Indeed, the ground-state energy of the T lattice may be obtained from the SQ energy (7.7.2) by the simple addition of the out-of-plane interaction term $j'\gamma'_Q$ with $j' = J'/J_1$ and $\gamma'_Q = \cos z_0$ ($z_0 = cQ_z$). The stacking of planes along the (0,0,1) is ordered ferromagnetically ($z_0 = 0$) or antiferromagnetically ($z_0 = \pi$) according to $j' > 0$ or $j' < 0$, respectively. For the T lattice, the presence of the disorder line $j_3 = \frac{1}{2}j_2$ implies the absence of LRO at any finite temperature.

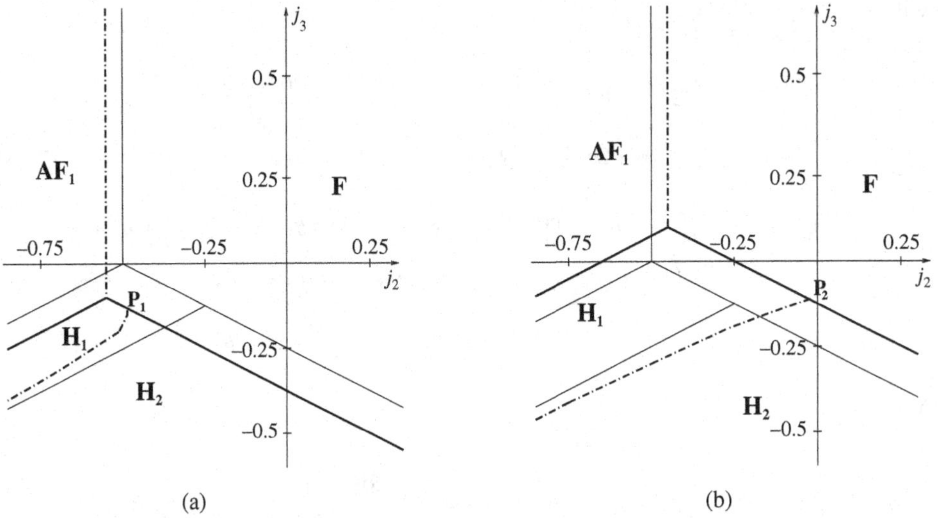


Fig. 7.11. Ground-state configurations of a SQ lattice with competing interactions $j_\alpha = \frac{z_\alpha J_\alpha}{z_1 J_1}$ and $J_1 > 0$: thick lines (a) $j_4 = 0.1$; (b) $j_4 = -0.1$; thin lines $j_4 = 0$. The dashed lines mean first-order transition lines.

For $J_1 < 0$, the classical ground-state energy of the SQ lattice becomes

$$E_0^{\mathrm{SQ}} = 4|J_1|S^2 N e_0 \tag{7.7.21}$$

where

$$e_0 = \gamma_{\boldsymbol{Q}}^{(1)} + j_2\gamma_{\boldsymbol{Q}}^{(2)} + j_3\gamma_{\boldsymbol{Q}}^{(3)}. \tag{7.7.22}$$

Notice that from the definition $j_\alpha = J_\alpha/J_1$ with $\alpha = 2, 3$, the positive $j_\alpha$ correspond to both $J_\alpha$ and $J_1$ antiferromagnetic interactions. The phase diagram at $T = 0$ is shown in Fig. 7.12. The $j_2 j_3$-plane is divided in four regions whose helix wavevectors and energies are given by

**AF**: Néel antiferromagnetic phase

$$x_0 = y_0 = \pi, \quad e_0^{\mathrm{AF}} = -1 + j_2 + j_3. \tag{7.7.23}$$

$\mathbf{AF_1}$: antiferromagnetic collinear order coincident with (7.7.10)

$$x_0 = 0, y_0 = \pi, \quad e_0^{\mathrm{AF}_1} = -j_2 + j_3. \tag{7.7.24}$$

$\mathbf{H_1}$: non-collinear order in which the spins are antiparallel along the $(1, 0)$ direction and make an angle $\pm y_0$ along the $(0, 1)$ direction

$$x_0 = \pi, \quad y_0 = \arccos\left(-\frac{1-2j_2}{4j_3}\right), \quad e_0^{\mathrm{H}_1} = -\frac{1}{2} - \frac{(1-2j_2)^2}{16j_3}. \tag{7.7.25}$$

$\mathbf{H_2}$: non-collinear order similar to Eq. (7.7.12)

$$x_0 = \pm y_0 = \arccos\left[-\frac{1}{2(j_2+2j_3)}\right], \quad e_0^{\mathrm{H}_2} = -j_3 - \frac{1}{4(j_2+2j_3)}. \tag{7.7.26}$$

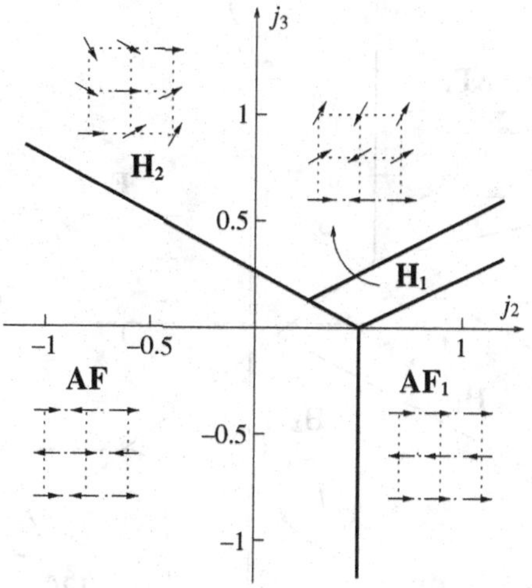


Fig. 7.12. Ground-state configurations of a SQ lattice with competing interactions $j_\alpha = \frac{J_\alpha}{J_1}$ and $J_1 < 0$.

The boundaries between the different regions shown in Fig. 7.12 are

$\mathbf{AF_1} - \mathbf{AF}$: $j_2 = \frac{1}{2}$,
$\mathbf{AF} - \mathbf{H_1}$ and $\mathbf{AF} - \mathbf{H_2}$: $j_3 = \frac{1}{4}(1 - 2j_2)$,
$\mathbf{AF_1} - \mathbf{H_1}$: $j_3 = -\frac{1}{4}(1 - 2j_2)$,
$\mathbf{H_1} - \mathbf{H_2}$: $j_3 = \frac{1}{2}j_2$.

The invariance of the energy under the exchange of $x_0$ with $y_0$ leads to the symmetric configurations obtained by the change of $x_0$ and $y_0$. Moreover, the disorder line $j_3 = \frac{1}{2}j_2$ remains also in the case $J_1 < 0$.

## 7.8. Frustration by Competing Interactions: Triangular Lattice

The ground-state energy of a TR lattice with $J_1 > 0$, NNN and TNN exchange interactions is given by

$$E_0^{\mathrm{TR}} = 6J_1 S^2 N\, e_0, \tag{7.8.1}$$

where

$$e_0 = -\gamma_Q^{(1)} - j_2\gamma_Q^{(2)} - j_3\gamma_Q^{(3)} \tag{7.8.2}$$

with

$$\gamma_Q^{(1)} = \frac{1}{3}(\cos 2y_0 + 2\cos x_0 \cos y_0), \tag{7.8.3}$$

$$\gamma_Q^{(2)} = \frac{1}{3}(\cos 2x_0 + 2\cos x_0 \cos 3y_0), \tag{7.8.4}$$

$$\gamma_Q^{(3)} = \frac{1}{3}(\cos 4y_0 + 2\cos 2x_0 \cos 2y_0) \tag{7.8.5}$$

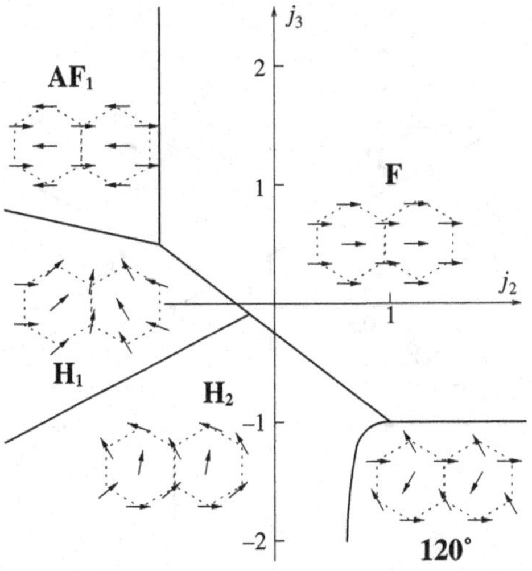


Fig. 7.13. Ground-state configurations of a TR lattice with competing interactions $j_\alpha = \frac{J_\alpha}{J_1}$ and $J_1 > 0$.

and

$$x_0 = \frac{\sqrt{3}}{2} a Q_x, \quad y_0 = \frac{1}{2} a Q_y, \quad j_\alpha = \frac{J_\alpha}{J_1}. \tag{7.8.6}$$

The minimum conditions are given by

$$\sin x_0 [\cos y_0 + j_2(2\cos x_0 + \cos 3y_0) + 4 j_3 \cos x_0 \cos 2y_0] = 0, \tag{7.8.7}$$

$$\begin{aligned} &\sin y_0 [2\cos y_0 + \cos x_0 + 3 j_2 \cos x_0 (4\cos^2 y_0 - 1) \\ &+ 4 j_3 \cos y_0 (2\cos 2y_0 - \cos 2x_0)] = 0. \end{aligned} \tag{7.8.8}$$

The minimum energy configurations in the $j_2 j_3$-plane are illustrated in Fig. 7.13. The $j_2 j_3$-plane is divided in five regions and the corresponding energies and helix wavevectors are given by

**F**: ferromagnetic phase

$$x_0 = y_0 = 0, \quad e_0^{\rm F} = -1 - j_2 - j_3. \tag{7.8.9}$$

**AF$_1$**: antiferromagnetic collinear order with alternating lines of NNN parallel spins

$$x_0 = \pi, \quad y_0 = 0, \quad e_0^{\rm AF_1} = \frac{1}{3}(1 + j_2 - 3 j_3). \tag{7.8.10}$$

**H$_1$**: non-collinear order in which lines of NN spins are parallel while the spins of the adjacent NN line form an angle $x_0$:

$$y_0 = 0, \quad x_0 = \arccos\left[-\frac{1 + j_2}{2(j_2 + 2j_3)}\right], \quad e_0^{\rm H_1} = \frac{1}{2} j_2 + \frac{(1 - 2j_3)^2}{6(j_2 + 2j_3)}. \tag{7.8.11}$$

Two equivalent configurations are characterized by

$$x_0' = \arccos\left[\sqrt{\frac{-1+j_2+4j_3}{4(j_2+2j_3)}}\right], \quad y_0' = \pm x_0'. \tag{7.8.12}$$

One can verify that $2\cos^2 x_0' - 1 = \cos x_0$ so that $x_0' = x_0/2$ and the configurations obtained from Eq. (7.8.12) coincide with that shown in Fig. 7.13 given by Eq. (7.8.11) rotating the TR lattice by $\pm\frac{\pi}{3}$. In general, this magnetic order is *incommensurate* with the lattice periodicity.

$\mathbf{H_2}$: non-collinear order in which the spins are parallel along a NNN line and the spins in the adjacent NNN line form an angle $y_0$:

$$x_0 = 0, \quad y_0 = \arccos\left[\frac{2j_3 - 3j_2 - \sqrt{(2j_3+3j_2)^2 - 8j_3}}{8j_3}\right],$$

$$\begin{aligned} e_0^{\mathrm{H_2}} &= -\frac{1}{3}(2\cos^2 y_0 + 2\cos y_0 - 1) - \frac{1}{3}j_2(8\cos^3 y_0 - 6\cos y_0 + 1) \\ &\quad - \frac{1}{3}j_3(8\cos^4 y_0 - 4\cos^2 y_0 - 1). \end{aligned} \tag{7.8.13}$$

Also in this case, two equivalent solutions exist

$$x_0' = \pm 3y_0', \quad y_0' = \arccos\left[\frac{10j_3 - 3j_2 - \sqrt{(3j_2+2j_3)^2 - 8j_3}}{16j_3}\right]^{\frac{1}{2}}. \tag{7.8.14}$$

One can verify that $y_0' = y_0/2$ and the two configurations given by Eq. (7.8.14) are obtained from that shown in Fig. 7.13 rotating the T lattice of an angle $\pm\frac{\pi}{3}$.

**120°**: a commensurate phase in which the spins of the magnetic cell form an angle $\frac{2\pi}{3}$ each other

$$x_0 = 0, \quad y_0 = \frac{2\pi}{3}, \quad e_0^{120^0} = \frac{1}{2}(1 - 2j_2 + j_3). \tag{7.8.15}$$

The boundaries between the different regions shown in Fig. 7.13 are lines of continuous (second order) phase transitions except the lines between regions $\mathbf{AF_1} - \mathbf{F}$, $\mathbf{120°} - \mathbf{F}$ and $\mathbf{120°} - \mathbf{H_2}$ where the transition is first order and $\mathbf{H_1} - \mathbf{H_2}$ which is a disorder line. The equations of the separation lines between the different regions are

$\mathbf{AF_1} - \mathbf{F}$: $j_2 = -1$,
$\mathbf{F} - \mathbf{H_1}$ and $\mathbf{F} - \mathbf{H_2}$: $j_3 = -\frac{1}{4}(1 + 3j_2)$,
$\mathbf{AF_1} - \mathbf{H_1}$: $j_3 = \frac{1}{4}(1 - j_2)$,
$\mathbf{H_1} - \mathbf{H_2}$: $j_3 = \frac{1}{2}j_2$.
$\mathbf{120°} - \mathbf{F}$: $j_3 = -1$.
$\mathbf{120°} - \mathbf{H_2}$: $j_3 = \frac{j_2^2}{1-2j_2}$.

For $j_3 = \frac{1}{2}j_2$ Eqs. (7.8.7) and (7.8.8) are simultaneously satisfied assuming

$$1 + j_2(4\cos^2 y_0 + 4\cos x_0 \cos y_0 - 3) = 0 \tag{7.8.16}$$

so that a DH of equation

$$Q_y = 2\arccos\left[\frac{1}{2}\left(-\cos\frac{\sqrt{3}}{2}Q_x + \sqrt{3 + \cos^2\frac{\sqrt{3}}{2}Q_x - \frac{1}{j_2}}\right)\right] \tag{7.8.17}$$

occurs for any $j_2$ belonging to the line $j_3 = \frac{1}{2}j_2$. In Fig. 7.14, the DH lines are shown for $j_2 = -0.3, -0.5, -0.75, -1, -2, -10$ going from the inner to the outer curve. Notice that for $j_2 = -0.2$, the DH shrinks to the point $(0, 0)$. All the DH curves shown in Fig. 7.14 appear similar to circles. For instance, for $j_2 \to \infty$ the distance from the origin of the DH curve is in the range $2.09440 < |\boldsymbol{Q}| < 2.10557$.

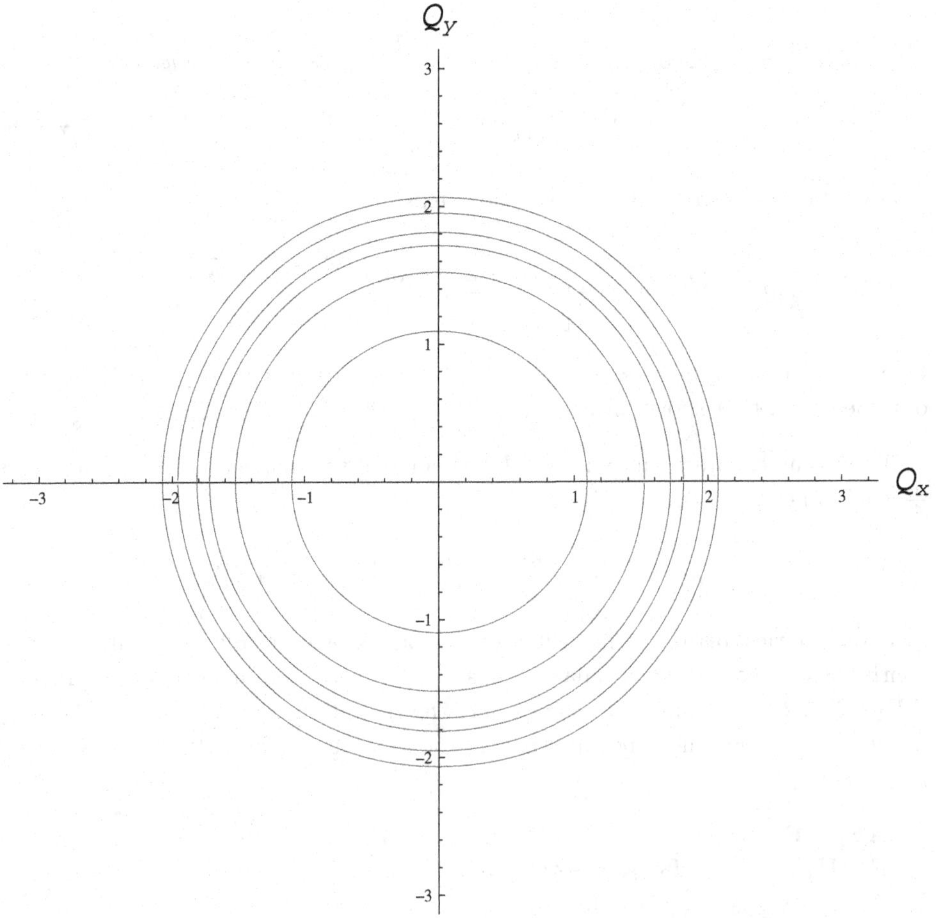


Fig. 7.14. Infinite degeneration line of the TR lattice in the reciprocal space for several $j_2$ ($j_3 = \frac{1}{2}j_2$). From the inner to the outer curve: $j_2 = -0.3, -0.5, -0.75, -1, -2, -10$, respectively.

Similarly, in the TR lattice as well as the SQ lattice, the quantum fluctuations destroy the DH. Indeed, the ground-state energy becomes

$$E_{\rm GS}^{\rm DH} = 6J_1S^2N\left\{\left[\frac{j_2}{2}+\frac{(1-j_2)^2}{12j_2}\right]\left(1+\frac{1}{S}\right)+\frac{\Delta}{S}\right\}, \tag{7.8.18}$$

where

$$\Delta = \frac{1}{(2\pi)^2}\int_{-\pi}^{\pi}dx\int_{-\pi}^{\pi}dy\sqrt{s(x,y)d_{\mathbf{Q}}(x,y)} \tag{7.8.19}$$

with

$$\begin{aligned} s(x,y) = &-\frac{1}{2}j_2-\frac{(1-j_2)^2}{12j_2}-\frac{1}{3}(\cos 2y+2\cos x\cos y)\\ &-\frac{1}{3}j_2(\cos 2x+2\cos x\cos 3y)-\frac{1}{6}j_2(\cos 4y+2\cos 2x\cos 2y), \end{aligned} \tag{7.8.20}$$

$$\begin{aligned} d_{\mathbf{Q}}(x,y) = &\frac{1}{3}[\cos 2y_0(1-\cos 2y)+2\cos x_0\cos y_0(1-\cos x\cos y)]\\ &+\frac{1}{3}j_2[\cos 2x_0(1-\cos 2x)+2\cos x_0\cos 3y_0(1-\cos x\cos 3y)]\\ &+\frac{1}{6}j_2[\cos 4y_0(1-\cos 4y)+2\cos 2x_0\cos 2y_0(1-\cos 2x\cos 2y)]\\ &-\frac{2}{3}\sin x_0\sin y_0\sin x\sin y-\frac{2}{3}j_2\sin x_0\sin 3y_0\sin x\sin 3y\\ &-\frac{1}{3}j_2\sin 2x_0\sin 2y_0\sin 2x\sin 2y \end{aligned} \tag{7.8.21}$$

and

$$y_0 = \arccos\left(-\frac{1}{2}\cos x_0+\frac{1}{2}\sqrt{3+\cos^2 x_0-\frac{1}{j_2}}\right). \tag{7.8.22}$$

For $|j_2| < 1.29$, the minima of the function $\Delta$ given by Eq. (7.8.19) are located along the directions $Q_x = 0$ and $Q_y = \pm\sqrt{3}Q_x$ corresponding to the six equivalent helices $\mathbf{H_1}$ and the maxima are located along the directions $Q_x = 0$ and $Q_y = \pm\frac{1}{\sqrt{3}}Q_x$ corresponding to the six equivalent helices $\mathbf{H_2}$. For $|j_2| > 2.4$, the maxima and minima are exchanged and the configuration corresponding to the helix $\mathbf{H_2}$ becomes stable. However, an unexpected event occurs for $1.29 < |j_2| < 2.4$ where the quantum fluctuations select neither the helix $\mathbf{H_1}$ nor the helix $\mathbf{H_2}$. New minima appear for $|j_2| = 1.29$ around the wavevectors of the $\mathbf{H_1}$ configuration that become maxima. By increasing $|j_2|$, the new minima move toward the neighbourhood of the wavevectors corresponding to the helix $\mathbf{H_2}$ and collapse into these wavevectors for $|j_2| = 2.4$. The above results imply that the existence region of the helix $\mathbf{H_1}$ widens for $|j_2| < 1.29$ and shrinks for $|j_2| > 2.4$ while a *new* phase appears for $1.29 < |j_2| < 2.4$ not expected from the classical result. This is a novelty that has never appeared in literature. The DH is also destroyed by adding the FNN exchange interaction $j_4 = \frac{2J_4}{J_1}$ to the classical ground-state energy of the TR lattice. The new phase

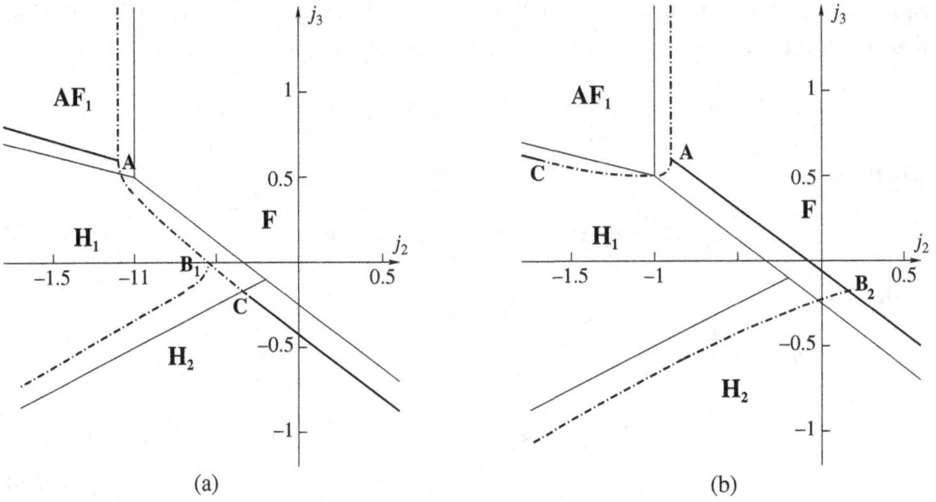


Fig. 7.15. Ground-state configurations of a TR lattice with competing interactions $j_\alpha = \frac{z_\alpha J_\alpha}{z_1 J_1}$ and $J_1 > 0$: thick lines (a) $j_4 = 0.1$; (b) $j_4 = -0.1$; thin lines $j_4 = 0$. The dashed-dotted lines mean first-order transition lines.

diagrams are shown in Fig. 7.15 (a) for $j_4 = 0.1$ and (b) for $j_4 = -0.1$, respectively. As one can see, the new phase diagrams are very rich. The dashed-dotted lines indicate first-order phase transitions while continuous lines mean second-order phase transitions. The thin lines refer to the case $j_4 = 0$ as shown in Fig. 7.13. As for the SQ lattice, a first-order transition line between the phases $\mathbf{H_1}$ and $\mathbf{H_2}$ replaces the DH (disorder) line. A variety of multicritical points appears in Fig. 7.15. The points A are the meeting of two first-order and one second-order transition lines. The point $\mathrm{B_1}$ is the meeting of three first-order transition lines while the point $\mathrm{B_2}$ is the meeting of two second-order and one first-order transition line. The points C are points in which a first-order transition changes to a second-order transition. Even though from Fig. 7.15(a) is not evident, the tangent of the curves meeting at the triple point $\mathrm{B_1}(-0.5132, -0.0355)$ is the same.

The phase diagram at $T = 0$ of the H lattice with competing interactions in the basal ($xy$) plane is the same as that shown in Figs. 7.13 and 7.15. Indeed, the ground-state energy of the H lattice may be obtained from the TR energy (7.8.2) adding the out-of-plane interaction $j'\gamma'_Q$ with $j' = J'/J_1$ and $\gamma'_Q = \cos z_0$ ($z_0 = cQ_z$). The stacking of planes along the $(0, 0, 1)$ direction is ordered ferromagnetically ($z_0 = 0$) or antiferromagnetically ($z_0 = \pi$) according to $j' > 0$ or $j' < 0$, respectively. For the H lattice, the presence of the disorder line $j_3 = \frac{1}{2}j_2$ implies the absence of LRO at any finite temperature.

For $J_1 < 0$, the classical ground-state energy of the TR lattice becomes

$$E_0^{\mathrm{TR}} = 6|J_1|S^2 N e_0, \tag{7.8.23}$$

where

$$e_0 = \gamma_Q^{(1)} + j_2\gamma_Q^{(2)} + j_3\gamma_Q^{(3)} \tag{7.8.24}$$

with $\gamma_Q^{(\alpha)}$ given by Eqs. (7.8.3)–(7.8.5). The phase diagram at $T = 0$ is shown in Figs. 7.16–7.17. Fig. 7.17 is a magnification of the region delimited by the rectangle shown in Fig. 7.16. The $j_2j_3$-plane is divided in six regions and the corresponding energies and helix wavevectors are given by:

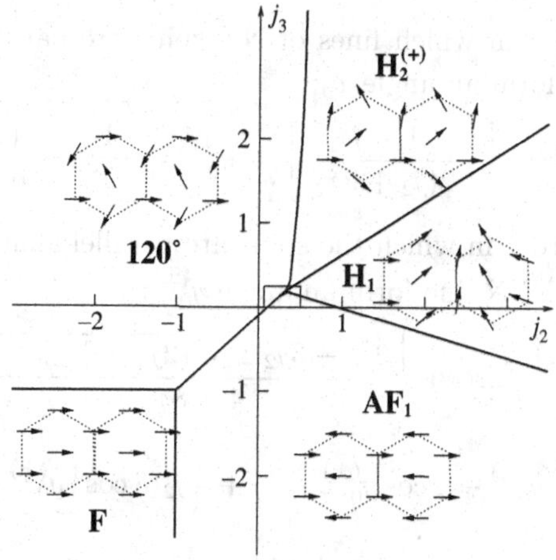


Fig. 7.16. Ground-state configurations of a TR lattice with competing interactions $j_\alpha = \frac{J_\alpha}{J_1}$ and $J_1 < 0$. The configurations inside the small rectangle with $\frac{1}{10} < j_2 < \frac{1}{2}$ and $0 < j_3 < \frac{1}{4}$ are shown in detail in Fig. 7.17.

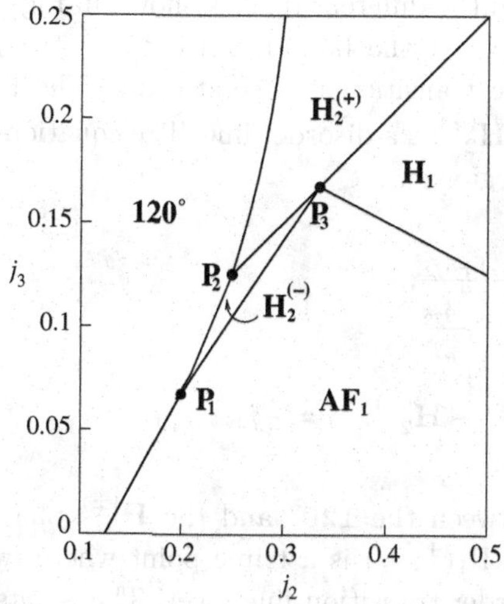


Fig. 7.17. Ground-state configurations of a TR lattice with competing interactions $j_\alpha = \frac{J_\alpha}{J_1}$ and $J_1 < 0$ for $\frac{1}{10} < j_2 < \frac{1}{2}$ and $0 < j_3 < \frac{1}{4}$.

F: ferromagnetic phase

$$x_0 = y_0 = 0, \quad e_0^F = 1 + j_2 + j_3. \tag{7.8.25}$$

**$\mathbf{AF_1}$**: antiferromagnetic collinear order with alternating lines of NNN parallel spins

$$x_0 = \pi, \quad y_0 = 0, \quad e_0^{\mathrm{AF_1}} = -\frac{1}{3}(1 + j_2 - 3j_3). \tag{7.8.26}$$

In this configuration, four of the six NN spins of a given spin are antiparallel and two are parallel.

**$\mathbf{H_1}$**: non-collinear order in which lines of NN spins are parallel while the spins of the adjacent NN line form an angle $x_0$:

$$y_0 = 0, \quad x_0 = \arccos\left[-\frac{1 + j_2}{2(j_2 + 2j_3)}\right], \quad e_0^{\mathrm{H_1}} = -\frac{1}{2}j_2 - \frac{(1 - 2j_3)^2}{6(j_2 + 2j_3)}. \tag{7.8.27}$$

**$\mathbf{H_2^{(\pm)}}$**: non-collinear order in which the spins are parallel along a NNN line and the spins in the adjacent NNN line form an angle $y_0^{(\pm)}$:

$$x_0 = 0, \quad y_0^{(\pm)} = \arccos\left[\frac{2j_3 - 3j_2 \pm \sqrt{(2j_3 + 3j_2)^2 - 8j_3}}{8j_3}\right],$$

$$e_0^{\mathrm{H_2^{(\pm)}}} = \frac{1}{3}(2\cos^2 y_0^{(\pm)} + 2\cos y_0^{(\pm)} - 1) + \frac{1}{3}j_2(8\cos^3 y_0^{(\pm)} - 6\cos y_0^{(\pm)} + 1)$$
$$+ \frac{1}{3}j_3(8\cos^4 y_0^{(\pm)} - 4\cos^2 y_0^{(\pm)} - 1). \tag{7.8.28}$$

**120°**: commensurate phase with

$$x_0 = 0, \quad y_0 = \frac{2\pi}{3}, \quad l_0^{120°} = -\frac{1}{2}(1 - 2j_2 + j_3). \tag{7.8.29}$$

The boundaries between the different regions shown in Figs. 7.16 and 7.17 are first-order transition lines except the boundaries between the phases $\mathbf{AF_1} - \mathbf{H_1}$ and $\mathbf{AF_1} - \mathbf{H_2^{(-)}}$ where the transition is second order. The line between the phases $\mathbf{H_1} - \mathbf{H_2^{(+)}}$ and $\mathbf{H_2^{(-)}} - \mathbf{H_2^{(+)}}$ is a disorder line. The equations of the separation lines between the different regions are

$\mathbf{AF_1} - \mathbf{F}$: $j_2 = -1$,
$\mathbf{AF_1} - \mathbf{120°}$: $j_3 = \frac{-1+8j_2}{9}$,
$\mathbf{AF_1} - \mathbf{H_2^{(-)}}$: $j_3 = \frac{-1+9j_2}{12}$,
$\mathbf{AF_1} - \mathbf{H_1}$: $j_3 = \frac{1-j_2}{4}$,
$\mathbf{H_2^{(+)}} - \mathbf{H_1}$ and $\mathbf{H_2^{(+)}} - \mathbf{H_2^{(-)}}$: $j_3 = \frac{1}{2}j_2$,
$\mathbf{120°} - \mathbf{F}$: $j_3 = -1$.

The boundary line between the **120°** and the $\mathbf{H_2^{(\pm)}}$ phases has to be evaluated numerically. The point $\mathrm{P}_1(\frac{1}{5}, \frac{1}{15})$ is a triple point where two first-order transition lines and one second-order transition line meet. The points $\mathrm{P}_2(\frac{1}{4}, \frac{1}{8})$ and $\mathrm{P}_3(\frac{1}{3}, \frac{1}{6})$ lie on the disorder line.

Concluding this section, it is worthwhile noticing that many magnetic compounds are made up of a staking of triangular planes. For instance, the members of the $ABX_3$ and $BX_2$ families where A is an alkali metal, B a transition metal and X a halogen[68] show a magnetic behaviour satisfactorily described by the triangular model discussed in this section.

## 7.9. Frustration by Competing Interaction: Honeycomb Lattice

In this section, we will study the classical ground state of a HON lattice. The novelty with respect to the SQ and TR lattices is that the HON lattice is not a Bravais lattice, that is, the primitive cell contains two non-equivalent lattice points so that the Hamiltonian (7.1.1) becomes

$$\mathcal{H} = -J_1 \sum_{i,\boldsymbol{\delta}_1} \left( \boldsymbol{S}_i^{(a)} \cdot \boldsymbol{S}_{i+\boldsymbol{\delta}_1}^{(b)} + \boldsymbol{S}_i^{(b)} \cdot \boldsymbol{S}_{i-\boldsymbol{\delta}_1}^{(a)} \right) - J_2 \sum_{i,\boldsymbol{\delta}_2} \left( \boldsymbol{S}_i^{(a)} \cdot \boldsymbol{S}_{i+\boldsymbol{\delta}_2}^{(a)} \right.$$
$$\left. + \boldsymbol{S}_i^{(b)} \cdot \boldsymbol{S}_{i+\boldsymbol{\delta}_2}^{(b)} \right) - J_3 \sum_{i,\boldsymbol{\delta}_3} \left( \boldsymbol{S}_i^{(a)} \cdot \boldsymbol{S}_{i+\boldsymbol{\delta}_3}^{(b)} + \boldsymbol{S}_i^{(b)} \cdot \boldsymbol{S}_{i-\boldsymbol{\delta}_3}^{(a)} \right), \tag{7.9.1}$$

where $a$ and $b$ label the two non-equivalent atoms in the unit cell and $i$ labels the lattice sites of *one* sublattice; $\boldsymbol{\delta}_1 = (\pm\frac{\sqrt{3}}{2}a, \frac{1}{2}a)$, $(0, -a)$; $\boldsymbol{\delta}_2 = (\pm\frac{\sqrt{3}}{2}a, \pm\frac{3}{2}a)$, $(\pm\sqrt{3}a, 0)$; $\boldsymbol{\delta}_3 = (\pm\sqrt{3}a, -a)$, $(0, 2a)$ are the lattice vectors joining the spin $i$ with its three NN, six NNN and three TNN, respectively. With respect to the treatment of Section 7.1, we introduce an additional degree of freedom:[69] we consider the possibility that the non-equivalent spins in the unit cell make an angle $\alpha$. The rotation matrix between cartesian and local axis is assumed to be of the form (7.1.2) with $\phi_i^{(a)} = \boldsymbol{Q} \cdot \boldsymbol{r}_i + \phi$ for the $a$-atoms and $\phi_i^{(b)} = \boldsymbol{Q} \cdot \boldsymbol{r}_i + \phi + \alpha$ for the $b$-atoms. Performing the following transformation to Bose operators

$$\begin{cases} S_i^{(a)\xi} = \dfrac{\sqrt{2S}}{2} \left( a_i + a_i^+ \right), \\ S_i^{(a)\eta} = \dfrac{\sqrt{2S}}{2i} \left( a_i - a_i^+ \right), \\ S_i^{(a)\zeta} = S - a_i^+ a_i, \end{cases} \qquad \begin{cases} S_i^{(b)\xi} = \dfrac{\sqrt{2S}}{2} \left( b_i + b_i^+ \right), \\ S_i^{(b)\eta} = \dfrac{\sqrt{2S}}{2i} \left( b_i - b_i^+ \right), \\ S_i^{(b)\zeta} = S - b_i^+ b_i, \end{cases} \tag{7.9.2}$$

we obtain the bilinear Hamiltonian

$$\mathcal{H}_2 = E_0 + \sum_{\boldsymbol{k}} A_{\boldsymbol{k}} (a_{\boldsymbol{k}}^+ a_{\boldsymbol{k}} + b_{\boldsymbol{k}}^+ b_{\boldsymbol{k}}) + \frac{1}{2} \sum_{\boldsymbol{k}} B_{\boldsymbol{k}} (a_{\boldsymbol{k}} a_{-\boldsymbol{k}} + a_{\boldsymbol{k}}^+ a_{-\boldsymbol{k}}^+ + b_{\boldsymbol{k}} b_{-\boldsymbol{k}} + b_{\boldsymbol{k}}^+ b_{-\boldsymbol{k}}^+)$$
$$+ \sum_{\boldsymbol{k}} (C_{\boldsymbol{k}}^* a_{\boldsymbol{k}} b_{-\boldsymbol{k}} + C_{\boldsymbol{k}} a_{\boldsymbol{k}}^+ b_{-\boldsymbol{k}}^+) + \sum_{\boldsymbol{k}} (D_{\boldsymbol{k}}^* a_{\boldsymbol{k}} b_{\boldsymbol{k}}^+ + D_{\boldsymbol{k}} a_{\boldsymbol{k}}^+ b_{\boldsymbol{k}}), \tag{7.9.3}$$

where

$$E_0 = -6 J_1 S^2 N \left\{ \sin^2\theta \, \frac{1}{3} \sum_{\boldsymbol{\delta}_1} \cos(\boldsymbol{Q} \cdot \boldsymbol{\delta}_1 + \alpha) + \cos^2\theta \right.$$
$$+ j_2 \left( \sin^2\theta \, \frac{1}{6} \sum_{\boldsymbol{\delta}_2} \cos \boldsymbol{Q} \cdot \boldsymbol{\delta}_2 + \cos^2\theta \right)$$

$$+j_3\left[\sin^2\theta\,\frac{1}{3}\sum_{\boldsymbol{\delta}_3}\cos(\boldsymbol{Q}\cdot\boldsymbol{\delta}_3+\alpha)+\cos^2\theta\right]\Bigg\}, \tag{7.9.4}$$

$$A_{\boldsymbol{k}}=6J_1S\Bigg\{\sin^2\theta\,\frac{1}{3}\sum_{\boldsymbol{\delta}_1}\cos(\boldsymbol{Q}\cdot\boldsymbol{\delta}_1+\alpha)+\cos^2\theta$$

$$+j_2\left(\sin^2\theta\,\frac{1}{6}\sum_{\boldsymbol{\delta}_2}\cos\boldsymbol{Q}\cdot\boldsymbol{\delta}_2+\cos^2\theta\right)$$

$$+j_3\left[\sin^2\theta\frac{1}{3}\sum_{\boldsymbol{\delta}_3}\cos(\boldsymbol{Q}\cdot\boldsymbol{\delta}_3+\alpha)+\cos^2\theta\right]-\frac{1}{12}j_2\Bigg[\sin^2\theta\sum_{\boldsymbol{\delta}_2}\cos\boldsymbol{k}\cdot\boldsymbol{\delta}_2$$

$$+\frac{1}{2}(1+\cos\theta)^2\sum_{\boldsymbol{\delta}_2}\cos(\boldsymbol{k}+\boldsymbol{Q})\cdot\boldsymbol{\delta}_2+\frac{1}{2}(1-\cos\theta)^2\sum_{\boldsymbol{\delta}_2}\cos(\boldsymbol{k}-\boldsymbol{Q})\cdot\boldsymbol{\delta}_2\Bigg]\Bigg\}, \tag{7.9.5}$$

$$B_{\boldsymbol{k}}=-\frac{1}{2}J_1S\;j_2\sin^2\theta\sum_{\boldsymbol{\delta}_2}\cos\boldsymbol{k}\cdot\boldsymbol{\delta}_2(1-\cos\boldsymbol{Q}\cdot\boldsymbol{\delta}_2), \tag{7.9.6}$$

$$C_{\boldsymbol{k}}=-J_1S\sin^2\theta\Bigg\{\sum_{\boldsymbol{\delta}_1}e^{i\boldsymbol{k}\cdot\boldsymbol{\delta}_1}[1-\cos(\boldsymbol{Q}\cdot\boldsymbol{\delta}_1+\alpha)]$$

$$+\,j_3\sum_{\boldsymbol{\delta}_3}e^{i\boldsymbol{k}\cdot\boldsymbol{\delta}_3}[1-\cos(\boldsymbol{Q}\cdot\boldsymbol{\delta}_3+\alpha)]\Bigg\} \tag{7.9.7}$$

and

$$D_{\boldsymbol{k}}=-J_1S\Bigg\{\sum_{\boldsymbol{\delta}_1}e^{i\boldsymbol{k}\cdot\boldsymbol{\delta}_1}[\sin^2\theta+(1+\cos^2\theta)\cos(\boldsymbol{Q}\cdot\boldsymbol{\delta}_1+\alpha)$$

$$+\,2i\cos\theta\;\sin(\boldsymbol{Q}\cdot\boldsymbol{\delta}_1+\alpha)]$$

$$+j_3\sum_{\boldsymbol{\delta}_3}e^{i\boldsymbol{k}\cdot\boldsymbol{\delta}_3}[\sin^2\theta+(1+\cos^2\theta)\cos(\boldsymbol{Q}\cdot\boldsymbol{\delta}_3+\alpha)$$

$$+\,2i\cos\theta\;\sin(\boldsymbol{Q}\cdot\boldsymbol{\delta}_3+\alpha)]\Bigg\}, \tag{7.9.8}$$

where $j_2=2\frac{J_2}{J_1}$ and $j_3=\frac{J_3}{J_1}$. The minimization of the classical ground-state energy $E_0$ given by Eq. (7.9.4) with respect to $\theta$, $\alpha$ and $\boldsymbol{Q}$ leads to the phase diagrams shown in Figs. 7.18 and 7.19 for $J_1>0$ and $J_1<0$, respectively.

Assuming $J_1>0$ and introducing the reduced variables

$$e_0=\frac{E_0}{6J_1S^2N},\quad x_0=\frac{\sqrt{3}}{2}aQ_x,\quad y_0=\frac{1}{2}aQ_y \tag{7.9.9}$$

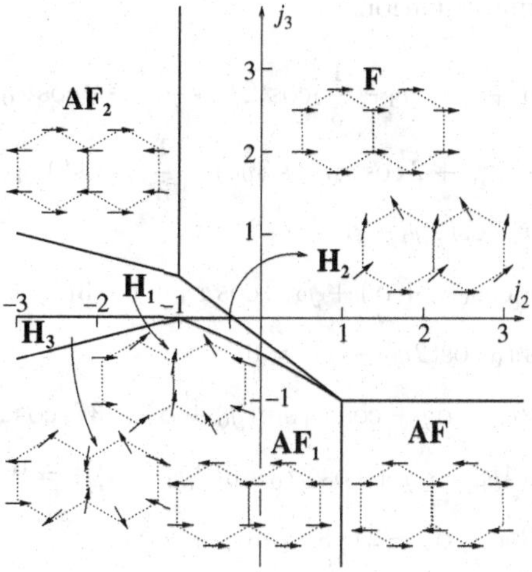


Fig. 7.18. Ground-state configurations of a HON lattice with competing interactions $j_\alpha = \frac{z_\alpha J_\alpha}{z_1 J_1}$ and $J_1 > 0$.

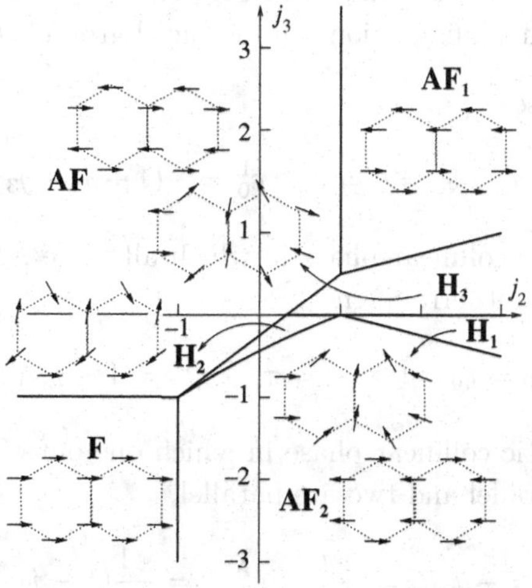


Fig. 7.19. Ground-state configurations of a HON lattice with competing interactions $j_\alpha = \frac{z_\alpha J_\alpha}{z_1 J_1}$ and $J_1 < 0$.

Equation (7.9.4) becomes

$$\begin{aligned} e_0 = &-\frac{1}{3}\sin^2\theta\{\cos(2y_0-\alpha)+2\cos x_0\cos(y_0+\alpha) \\ &+j_2(\cos 2x_0+2\cos x_0\cos 3y_0) \\ &+j_3[\cos(4y_0+\alpha)+2\cos 2x_0\cos(2y_0-\alpha)]\} \\ &-(1+j_2+j_3)\cos^2\theta. \end{aligned} \tag{7.9.10}$$

Imposing that the first partial derivatives of $e_0$ with respect to $\theta, x_0, y_0$ and $\alpha$ vanish, one obtains the following equations

$$\sin\theta\cos\theta\{1+j_2+j_3-\frac{1}{3}\left[\cos(2y_0-\alpha)+2\cos x_0\cos(y_0+\alpha)\right]$$
$$-\frac{1}{3}j_2(\cos 2x_0+2\cos x_0\cos 3y_0)-\frac{1}{3}j_3[\cos(4y_0+\alpha)$$
$$+2\cos 2x_0\cos(2y_0-\alpha)]\}=0, \tag{7.9.11}$$

$$\sin^2\theta\sin x_0[\cos(y_0+\alpha)+j_2(2\cos x_0+\cos 3y_0)$$
$$+4j_3\cos x_0\cos(2y_0-\alpha)]=0, \tag{7.9.12}$$

$$\sin^2\theta\{\sin(2y_0-\alpha)+\cos x_0\sin(y_0+\alpha)+3j_2\cos x_0\sin 3y_0$$
$$+2j_3[\sin(4y_0+\alpha)+\cos 2x_0\sin(2y_0-\alpha)]\}=0, \tag{7.9.13}$$

$$\sin^2\theta\{\sin(2y_0-\alpha)-2\cos x_0\sin(y_0+\alpha)$$
$$+j_3[2\cos 2x_0\sin(2y_0-\alpha)-\sin(4y_0+\alpha)]\}=0. \tag{7.9.14}$$

Equation (7.9.11) gives $\theta=\frac{\pi}{2}$ and the minimum energy configurations obtained from Eqs. (7.9.12)–(7.9.14) are shown in Fig. 7.18. The $j_2j_3$-plane is divided into seven regions: the spin configurations and reduced ground-state energy are

**F**: ferromagnetic phase

$$x_0=y_0=\alpha=0, \quad e_0^{\mathrm{F}}=-(1+j_2+j_3). \tag{7.9.15}$$

**AF**: antiferromagnetic collinear phase in which all three NN spins of a given spin are antiparallel (Néel antiferromagnet)

$$x_0=y_0=0, \quad \alpha=\pi, \quad e_0^{\mathrm{AF}}=1-j_2+j_3. \tag{7.9.16}$$

**AF$_1$**: antiferromagnetic collinear phase in which one out of the three NN spins of a given spin is antiparallel and two are parallel

$$x_0=0, \quad y_0=\frac{\pi}{3}, \quad \alpha=-\frac{\pi}{3}, \quad e_0^{\mathrm{AF}_1}=-\frac{1}{3}(1-j_2-3j_3). \tag{7.9.17}$$

**AF$_2$**: antiferromagnetic collinear phase in which two of the three NN spins of a given spin are antiparallel and one is parallel

$$x_0=\pi, \quad y_0=\alpha=0, \quad e_0^{\mathrm{AF}_2}=\frac{1}{3}(1+j_2-3j_3). \tag{7.9.18}$$

**H$_1$**: helical phase with zero angle between the two non-equivalent spins

$$x_0=\arccos\left[-\frac{1+j_2}{2(j_2+2j_3)}\right], \quad y_0=\alpha=0,$$
$$e_0^{\mathrm{H}_1}=\frac{1}{2}j_2+\frac{(1-2j_3)^2}{6(j_2+2j_3)}. \tag{7.9.19}$$

**H$_2$**: helical phase with an angle between the two non-equivalent spins different from 0 and $\pi$

$$x_0 = 0,$$

$$y_0 = \frac{1}{3}\arccos\left\{-\frac{1}{2j_3(1+2j_3)}\left[1+3j_3+j_2(1+j_3)\sqrt{\frac{(1+j_3)(1-2j_3)}{j_2^2-j_3(1+2j_3)}}\right]\right\},$$

$$\alpha = -y_0 + \arctan\frac{(1+j_3)\sin 3y_0}{2+(1+3j_3)\cos 3y_0},$$

$$e_0^{\mathrm{H}_2} = -\frac{1}{3}\{\cos(2y_0-\alpha)+2\cos(y_0+\alpha)+j_2(1+2\cos 3y_0)$$
$$+j_3[\cos(4y_0+\alpha)+2\cos(2y_0-\alpha)]\}. \tag{7.9.20}$$

**H$_3$**: helical phase with an angle $\pi$ between the two non-equivalent spins

$$x_0 = \arccos\left[\frac{1-j_2}{2(j_2-2j_3)}\right],\ y_0 = 0, \quad \alpha = \pi,$$

$$e_0^{\mathrm{H}_3} = \frac{1}{2}j_2 + \frac{(1-2j_3)^2}{6(j_2-2j_3)}. \tag{7.9.21}$$

The boundaries between the different regions shown in Fig. 7.18 are first-order transition lines except the boundaries between the phases **F**$-$**H$_1$**, **F**$-$**H$_2$**, **AF$_1$**$-$**H$_2$**, **AF$_1$** $-$ **H$_3$** and **AF$_2$** $-$ **H$_1$** where the transition is second order. The transition line between the **H$_1$** and **H$_2$**, **H$_3$** phases is a disorder line. Indeed, for $j_3 = 0$ and $j_2 < -\frac{1}{3}$ the ground state is infinitely degenerate: the equation of the DH line is

$$Q_y = \frac{2}{3a}\arccos\left[\frac{1-j_2^2(1+4\cos^2\frac{\sqrt{3}}{2}aQ_x)}{4j_2^2\cos\frac{\sqrt{3}}{2}aQ_x}\right]. \tag{7.9.22}$$

The DH line reduces to the point $(0,0)$ at $j_2 = -\frac{1}{3}$, it forms closed curves around the origin for $-1 < j_2 < -\frac{1}{3}$. These closed curves become circles for $j_2 \to -\frac{1}{3}$ and a rhombus for $j_2 \to -1$. For $j_2 < -1$, the closed curves are pseudo-triangle centred around the point $(0, \frac{4\pi}{3\sqrt{3}a})$ and the symmetric points obtained rotating by 60, 120, 180, 240, 300 degrees. The equations of the separation lines between the different regions are

**F** $-$ **AF**: $j_3 = -1$,
**F** $-$ **AF$_2$**: $j_2 = -1$,
**F** $-$ **H$_1$** and **F** $-$ **H$_2$**: $j_3 = -\frac{1+3j_2}{4}$,
**AF** $-$ **AF$_1$**: $j_2 = 1$,
**AF$_1$** $-$ **H$_2$**: $j_3 = \frac{1}{8}(1-3j_2-\sqrt{9j_2^2+10j_2+17})$,
**AF$_1$** $-$ **H$_3$**: $j_3 = \frac{1+j_2}{4}$,
**AF$_2$** $-$ **H$_1$**: $j_3 = \frac{1-j_2}{4}$,
**H$_1$** $-$ **H$_2$** and **H$_1$** $-$ **H$_3$**: $j_3 = 0$.

For $J_1 < 0$, the phase diagram at $T = 0$ is shown in Fig. 7.19. In the present case, the reduced quantities are the same as those given in Eq. (7.9.9) except that $J_1$ has to be replaced by $|J_1|$. The $j_2j_3$-plane is divided into seven regions: the spin configurations and reduced ground-state energies are

**F**: ferromagnetic phase

$$x_0 = y_0 = \alpha = 0, \quad e_0^{\mathrm{F}} = 1 + j_2 + j_3. \tag{7.9.23}$$

**AF**: Néel antiferromagnetic collinear phase

$$x_0 = y_0 = 0, \quad \alpha = \pi, \quad e_0^{\mathrm{AF}} = -1 + j_2 - j_3. \tag{7.9.24}$$

$\mathbf{AF_1}$: antiferromagnetic collinear phase in which one out of the three NN spins of a given spin is antiparallel and two are parallel

$$x_0 = 0, \quad y_0 = \frac{\pi}{3}, \quad \alpha = -\frac{\pi}{3}, \quad e_0^{\mathrm{AF}_1} = \frac{1}{3}(1 - j_2 - 3j_3). \tag{7.9.25}$$

$\mathbf{AF_2}$: antiferromagnetic collinear phase in which two of the three NN spins of a given spin are antiparallel and one is parallel

$$x_0 = \pi, \quad y_0 = \alpha = 0, \quad e_0^{\mathrm{AF}_2} = -\frac{1}{3}(1 + j_2 - 3j_3). \tag{7.9.26}$$

$\mathbf{H_1}$: helical phase with zero angle between the two non-equivalent spins

$$x_0 = \arccos\left[-\frac{1 + j_2}{2(j_2 + 2j_3)}\right], \quad y_0 = \alpha = 0,$$
$$e_0^{\mathrm{H}_1} = -\frac{1}{2}j_2 - \frac{(1 - 2j_3)^2}{6(j_2 + 2j_3)}. \tag{7.9.27}$$

$\mathbf{H_2}$: helical phase with an angle $\alpha \neq 0, \pi$ between the two non-equivalent spins

$$x_0 = 0,$$
$$y_0 = \frac{1}{3}\arccos\left\{-\frac{1}{2j_3(1 + 2j_3)}\left[1 + 3j_3 + j_2(1 + j_3)\sqrt{\frac{(1 + j_3)(1 - 2j_3)}{j_2^2 - j_3(1 + 2j_3)}}\right]\right\},$$
$$\alpha = -y_0 + \arctan\frac{(1 + j_3)\sin 3y_0}{2 + (1 + 3j_3)\cos 3y_0},$$
$$e_0^{\mathrm{H}_2} = \frac{1}{3}\{\cos(2y_0 - \alpha) + 2\cos(y_0 + \alpha) + j_2(1 + 2\cos 3y_0)$$
$$+ j_3\left[\cos(4y_0 + \alpha) + 2\cos(2y_0 - \alpha)\right]\}. \tag{7.9.28}$$

$\mathbf{H_3}$: helical phase with an angle $\pi$ between the two non-equivalent spins

$$x_0 = \arccos\left[\frac{1 - j_2}{2(j_2 - 2j_3)}\right], \quad y_0 = 0, \quad \alpha = \pi,$$
$$e_0^{\mathrm{H}_3} = -\frac{1}{2}j_2 - \frac{(1 - 2j_3)^2}{6(j_2 - 2j_3)}. \tag{7.9.29}$$

The boundaries between the different regions shown in Fig. 7.19 are first-order transition lines except the boundaries between the phases $\mathbf{AF-H_2}$, $\mathbf{AF-H_3}$, $\mathbf{AF_1-H_3}$, $\mathbf{AF_2-H_1}$ and $\mathbf{AF_2-H_2}$ where the transition is second order. The $\mathbf{H_3-H_2}$ and $\mathbf{H_3-H_1}$ boundary line ($j_3=0$) is a disorder line. The equations of the separation lines between the different regions are

$\mathbf{AF-F}$: $j_3=-1$,
$\mathbf{AF-H_2}$ and $\mathbf{AF-H_3}$: $j_3=-\frac{1-3j_2}{4}$,
$\mathbf{AF-AF_1}$: $j_2=1$,
$\mathbf{AF_1-H_3}$: $j_3=\frac{1+j_2}{4}$,
$\mathbf{AF_2-H_2}$: $j_3=\frac{1}{8}(1+3j_2-\sqrt{9j_2^2-10j_2+17})$,
$\mathbf{AF_2-H_1}$: $j_3=\frac{1-j_2}{4}$,
$\mathbf{AF_2-F}$: $j_2=-1$,
$\mathbf{H_3-H_2}$ and $\mathbf{H_3-H_1}$: $j_3=0$.

To obtain the spin wave spectrum of the HON lattice with competing interaction, we have to diagonalize the harmonic Hamiltonian (7.9.3). To do this, we use a procedure different from that used in Sec. 7.2. Instead of making use of the Bogoliubov transformation (7.2.2), we introduce the more general transformation

$$\alpha_{\boldsymbol{k}}^{(i)}=u_{\boldsymbol{k}}^{(i)}a_{\boldsymbol{k}}-l_{\boldsymbol{k}}^{(i)}a_{-\boldsymbol{k}}^{+}+v_{\boldsymbol{k}}^{(i)}b_{\boldsymbol{k}}-m_{\boldsymbol{k}}^{(i)}b_{-\boldsymbol{k}}^{+} \tag{7.9.30}$$

in order to transform the Hamiltonian (7.9.3) into the diagonal Hamiltonian

$$\mathcal{H}_2=E_0'+\sum_i\sum_{\boldsymbol{k}}\hbar\omega_{\boldsymbol{k}}^{(i)}\alpha_{\boldsymbol{k}}^{(i)+}\alpha_{\boldsymbol{k}}^{(i)} \tag{7.9.31}$$

The number of the new operators $\alpha_{\boldsymbol{k}}^{(i)}$ will be established by the determinant equation generated by the transformation (7.9.30) and by the requirement of stability, that is the frequences $\omega_{\boldsymbol{k}}^{(i)}$ appearing in Eq. (7.9.31) have to be positive. From Eqs. (7.9.30), one obtains directly the transformations for $\alpha_{-\boldsymbol{k}}^{+}$ assuming that $u_{-\boldsymbol{k}}^{(i)}=u_{\boldsymbol{k}}^{(i)*}$, $l_{-\boldsymbol{k}}^{(i)}=l_{\boldsymbol{k}}^{(i)*}$, $v_{-\boldsymbol{k}}^{(i)}=v_{\boldsymbol{k}}^{(i)*}$ and $m_{-\boldsymbol{k}}^{(i)}=m_{\boldsymbol{k}}^{(i)*}$. Moreover, the condition that the new operators $\alpha_{\boldsymbol{k}}^{(i)}$, as well as $a_{\boldsymbol{k}}$ and $b_{\boldsymbol{k}}$ are Bose operators, implies that

$$|u_{\boldsymbol{k}}^{(i)}|^2-|l_{\boldsymbol{k}}^{(i)}|^2+|v_{\boldsymbol{k}}^{(i)}|^2-|m_{\boldsymbol{k}}^{(i)}|^2=1. \tag{7.9.32}$$

The commutators of the operators $a_{\boldsymbol{k}}$, $a_{-\boldsymbol{k}}^{+}$, $b_{\boldsymbol{k}}$ and $b_{-\boldsymbol{k}}^{+}$ occurring in Eqs. (7.9.30) with the Hamiltonian (7.9.3) give

$$[a_{\boldsymbol{k}},\mathcal{H}_2]=A_{\boldsymbol{k}}a_{\boldsymbol{k}}+B_{\boldsymbol{k}}a_{-\boldsymbol{k}}^{+}+C_{\boldsymbol{k}}b_{-\boldsymbol{k}}^{+}+D_{\boldsymbol{k}}b_{\boldsymbol{k}}, \tag{7.9.33}$$

$$[a_{-\boldsymbol{k}}^{+},\mathcal{H}_2]=-A_{\boldsymbol{k}}a_{-\boldsymbol{k}}^{+}-B_{\boldsymbol{k}}a_{\boldsymbol{k}}-C_{\boldsymbol{k}}b_{\boldsymbol{k}}-D_{\boldsymbol{k}}b_{-\boldsymbol{k}}^{+}, \tag{7.9.34}$$

$$[b_{\boldsymbol{k}},\mathcal{H}_2]=A_{\boldsymbol{k}}b_{\boldsymbol{k}}+B_{\boldsymbol{k}}b_{-\boldsymbol{k}}^{+}+C_{\boldsymbol{k}}^*a_{-\boldsymbol{k}}^{+}+D_{\boldsymbol{k}}^*a_{\boldsymbol{k}} \tag{7.9.35}$$

and

$$[b_{-\boldsymbol{k}}^{+},\mathcal{H}_2]=-A_{\boldsymbol{k}}b_{-\boldsymbol{k}}^{+}-B_{\boldsymbol{k}}b_{\boldsymbol{k}}-C_{\boldsymbol{k}}^*a_{\boldsymbol{k}}-D_{\boldsymbol{k}}^*a_{-\boldsymbol{k}}^{+}, \tag{7.9.36}$$

so that the commutator of the operator $\alpha_k^{(i)}$ with the Hamiltonian (7.9.3) becomes

$$\begin{aligned}[\alpha_k^{(i)}, \mathcal{H}_2] &= (A_k u_k^{(i)} + B_k l_k^{(i)} + D_k^* v_k^{(i)} + C_k^* m_k^{(i)}) a_k \\ &\quad + (B_k u_k^{(i)} + A_k l_k^{(i)} + C_k^* v_k^{(i)} + D_k^* m_k^{(i)}) a_{-k}^+ \\ &\quad + (D_k u_k^{(i)} + C_k l_k^{(i)} + A_k v_k^{(i)} + B_k m_k^{(i)}) b_k \\ &\quad + (C_k u_k^{(i)} + D_k l_k^{(i)} + B_k v_k^{(i)} + A_k m_k^{(i)}) b_{-k}^+. \end{aligned} \tag{7.9.37}$$

The coefficients of the transformation (7.9.30) that bring the Hamiltonian (7.9.3) into the diagonal Hamiltonian (7.9.31) are determined comparing the commutator of $\alpha_k^{(i)}$ with the Hamiltonian (7.9.31), that is

$$[\alpha_k^{(i)}, \mathcal{H}_2] = \hbar\omega_k^{(i)} \alpha_k^{(i)} = \hbar\omega_k^{(i)} (u_k^{(i)} a_k - l_k^{(i)} a_{-k}^+ + v_k^{(i)} b_k - m_k^{(i)} b_{-k}^+) \tag{7.9.38}$$

with Eq. (7.9.37). By equating Eqs. (7.9.37) and (7.9.38) one obtains a linear homogeneous system in the coefficients that can be written in a matrix form as

$$\begin{pmatrix} A_k - \hbar\omega_k^{(i)} & B_k & D_k^* & C_k^* \\ B_k & A_k + \hbar\omega_k^{(i)} & C_k^* & D_k^* \\ D_k & C_k & A_k - \hbar\omega_k^{(i)} & B_k \\ C_k & D_k & B_k & A_k + \hbar\omega_k^{(i)} \end{pmatrix} \begin{pmatrix} u_k^{(i)} \\ l_k^{(i)} \\ v_k^{(i)} \\ m_k^{(i)} \end{pmatrix} = 0. \tag{7.9.39}$$

The non-trivial solutions of this matrix equation are obtained by imposing that the determinant of the square matrix appearing in Eq. (7.9.39) is zero. So doing one obtains a biquadratic equation in $\omega_k^{(i)}$ the solutions of which are

$$\begin{aligned}[\hbar\omega_k^{(i)}]^2 &= A_k^2 - B_k^2 - |C_k|^2 + |D_k|^2 \\ &\quad \mp \sqrt{4|A_k D_k - B_k C_k|^2 + (C_k D_k^* - C_k^* D_k)^2}. \end{aligned} \tag{7.9.40}$$

The square root of Eq. (7.9.38) leads to four frequencies for $\omega_k^{(i)}$ two of which are positive. The two negative roots have to be rejected for stability reasons and the two positive roots give the spin wave spectra of the bilinear Hamiltonian (7.9.31), that is,

$$\begin{aligned}\hbar\omega_k^{(\mp)} &= \Big[A_k^2 - B_k^2 - |C_k|^2 + |D_k|^2 \\ &\quad \mp \sqrt{4|A_k D_k - B_k C_k|^2 + (C_k D_k^* - C_k^* D_k)^2}\Big]^{\frac{1}{2}}. \end{aligned} \tag{7.9.41}$$

The coefficients of the transformation (7.9.30) are given by

$$\begin{aligned} l_k^{(\mp)} &= \frac{u_k^{(\mp)}}{\Delta_k^{(\mp)}} \Big[A_k (C_k^* D_k + C_k D_k^*) - 2B_k |C_k|^2 + (C_k^* D_k - C_k D_k^*)\, \hbar\omega_k^{(\mp)} \\ &\quad \mp B_k \sqrt{4|A_k D_k - B_k C_k|^2 + (C_k D_k^* - C_k^* D_k)^2}\Big], \end{aligned} \tag{7.9.42}$$

$$v_k^{(\mp)} = \frac{u_k^{(\mp)}}{\Delta_k^{(\mp)}}\Big[2A_kB_kC_k - D_k(2A_k^2 - |C_k|^2) - C_k^2D_k^* - 2(A_kD_k - B_kC_k)\hbar\omega_k^{(\mp)}$$

$$\pm D_k\sqrt{4|A_kD_k - B_kC_k|^2 + (C_kD_k^* - C_k^*D_k)^2}\Big], \tag{7.9.43}$$

$$m_k^{(\mp)} = \frac{u_k^{(\mp)}}{\Delta_k^{(\mp)}}\Big[2A_kB_kD_k - C_k(2B_k^2 - |D_k|^2) - C_k^*D_k^2$$

$$\mp C_k\sqrt{4|A_kD_k - B_kC_k|^2 + (C_kD_k^* - C_k^*D_k)^2}\Big] \tag{7.9.44}$$

with

$$\Delta_k^{(\mp)} = -2A_k|D_k|^2 + B_k(C_k^*D_k + C_kD_k^*)$$

$$\pm(A_k + \hbar\omega_k^{(\mp)})\sqrt{4|A_kD_k - B_kC_k|^2 + (C_kD_k^* - C_k^*D_k)^2}. \tag{7.9.45}$$

The value of $|u_k^{(\mp)}|$ is determined by the condition (7.9.32). The "inverse" transformations of Eqs. (7.9.30) that is the transformation from $a$ and $b$ to $\alpha^{(i)}$ $(i = 1, 2)$ is given by

$$a_k = u_k^{(-)*}\alpha_k^{(1)} + l_k^{(-)*}\alpha_{-k}^{(1)+} + u_k^{(+)*}\alpha_k^{(2)} + l_k^{(+)*}\alpha_{-k}^{(2)+} \tag{7.9.46}$$

and

$$b_k = v_k^{(-)*}\alpha_k^{(1)} + m_k^{(-)*}\alpha_{-k}^{(1)+} + v_k^{(+)*}\alpha_k^{(2)} + m_k^{(+)*}\alpha_{-k}^{(2)+}. \tag{7.9.47}$$

Using Eqs. (7.9.46) and (7.9.47) and (7.9.3), one can obtain the zero-point energy. From Eq. (7.9.31), one obtains

$$E_0' = E_0 + \Delta E_0, \tag{7.9.48}$$

where

$$\Delta E_0 = -\sum_k \hbar\omega_k^{(-)}(|l_k^{(-)}|^2 + |m_k^{(-)}|^2) - \sum_k \hbar\omega_k^{(+)}(|l_k^{(+)}|^2 + |m_k^{(+)}|^2). \tag{7.9.49}$$

## 7.10. Neutron Scattering Cross-Section for a Helimagnet

In this section, we will evaluate the neutron cross-section for a system with non-collinear order. The starting point is the magnetic neutron cross-section given by Eq. (3.1.1) and (3.1.2). To simplify the calculations, we will restrict ourselves to the harmonic approximation so that we disregard damping and renormalization of the spin waves. The first step consists of writing the structure factor (3.1.2) in terms of the spin components along the local axis. To do this, we make use of the rotation matrix (7.1.2) with $\theta = \frac{\pi}{2}$. In this way, the spins lie in the $xy$-plane. We begin evaluating the correlation function $\langle S_i^x S_j^x(t)\rangle$, that is,

$$\langle S_i^x S_j^x(t)\rangle = \sin \boldsymbol{Q}\cdot\boldsymbol{r}_i \sin \boldsymbol{Q}\cdot\boldsymbol{r}_j \langle S_i^\eta S_j^\eta(t)\rangle + \cos \boldsymbol{Q}\cdot\boldsymbol{r}_i \cos \boldsymbol{Q}\cdot\boldsymbol{r}_j \langle S_i^\zeta S_j^\zeta(t)\rangle. \tag{7.10.1}$$

For a Bravais lattice within the harmonic approximation, we write the local spin correlation functions in terms of Bose operators using the first system appearing in Eq. (7.9.2)

$$\begin{aligned}\langle S_i^\eta S_j^\eta(t)\rangle &= -\frac{S}{2}\langle a_i a_j(t) - a_i a_j^+(t) - a_i^+ a_j(t) + a_i^+ a_j^+(t)\rangle \\ &= -\frac{S}{2N}\sum_{\boldsymbol{q}} e^{i\boldsymbol{q}\cdot(\boldsymbol{r}_i-\boldsymbol{r}_j)}\langle a_{\boldsymbol{q}} a_{-\boldsymbol{q}}(t) - a_{\boldsymbol{q}} a_{\boldsymbol{q}}^+(t) - a_{\boldsymbol{q}}^+ a_{\boldsymbol{q}}(t) + a_{\boldsymbol{q}}^+ a_{-\boldsymbol{q}}^+(t)\rangle.\end{aligned} \tag{7.10.2}$$

The last line of Eq. (7.10.2) is obtained by the property of the thermal average of the boson operators

$$\langle a_{\boldsymbol{q}} a_{\boldsymbol{q}'}(t)\rangle = \delta_{\boldsymbol{q},-\boldsymbol{q}'}\langle a_{\boldsymbol{q}} a_{-\boldsymbol{q}}(t)\rangle, \quad \langle a_{\boldsymbol{q}}^+ a_{\boldsymbol{q}'}(t)\rangle = \delta_{\boldsymbol{q},\boldsymbol{q}'}\langle a_{\boldsymbol{q}}^+ a_{\boldsymbol{q}}(t)\rangle. \tag{7.10.3}$$

The use of the Bogoliubov transformation (7.2.2) allows us to diagonalize the bilinear Hamiltonian (7.2.1) obtaining the harmonic Hamiltonian (7.2.12) with $A_{\boldsymbol{k}}^{(\mathrm{a})} = 0$. The time evolution of the operators $\alpha$ is then given by Eqs. (3.1.14). From Eq. (7.2.2) and (3.1.14), one obtains

$$\langle a_{\boldsymbol{q}} a_{-\boldsymbol{q}}(t)\rangle = \langle a_{\boldsymbol{q}}^+ a_{-\boldsymbol{q}}^+(t)\rangle = -l_{\boldsymbol{q}}^2 x_{\boldsymbol{q}}[(1+n_{\boldsymbol{q}})e^{i\omega_{\boldsymbol{q}}t} + n_{\boldsymbol{q}}e^{-i\omega_{\boldsymbol{q}}t}], \tag{7.10.4}$$

$$\langle a_{\boldsymbol{q}} a_{\boldsymbol{q}}^+(t)\rangle = l_{\boldsymbol{q}}^2[(1+n_{\boldsymbol{q}})e^{i\omega_{\boldsymbol{q}}t} + x_{\boldsymbol{q}}^2\, n_{\boldsymbol{q}}e^{-i\omega_{\boldsymbol{q}}t}], \tag{7.10.5}$$

$$\langle a_{\boldsymbol{q}}^+ a_{\boldsymbol{q}}(t)\rangle = l_{\boldsymbol{q}}^2[n_{\boldsymbol{q}}e^{-i\omega_{\boldsymbol{q}}t} + x_{\boldsymbol{q}}^2(1+n_{\boldsymbol{q}})e^{i\omega_{\boldsymbol{q}}t}], \tag{7.10.6}$$

where

$$n_{\boldsymbol{q}} = \frac{1}{e^{\beta\hbar\omega_{\boldsymbol{q}}} - 1} \tag{7.10.7}$$

with $\hbar\omega_{\boldsymbol{q}}$ given by Eqs. (7.3.10)–(7.3.12). By the use of Eqs. (7.10.4)–(7.10.6), the correlation function (7.10.2) becomes

$$\langle S_i^\eta S_j^\eta(t)\rangle = \frac{S}{2N}\sum_{\boldsymbol{q}} e^{i\boldsymbol{q}\cdot(\boldsymbol{r}_i-\boldsymbol{r}_j)}\sqrt{\frac{S_{\boldsymbol{q}}}{D_{\boldsymbol{q}}}}[(1+n_{\boldsymbol{q}})e^{i\omega_{\boldsymbol{q}}t} + n_{\boldsymbol{q}}e^{-i\omega_{\boldsymbol{q}}t}], \tag{7.10.8}$$

where $S_{\boldsymbol{q}}$ and $D_{\boldsymbol{q}}$ are given by Eqs. (7.3.11) and (7.3.12), respectively. In a similar way, one obtains

$$\langle S_i^\xi S_j^\xi(t)\rangle = \frac{S}{2N}\sum_{\boldsymbol{q}} e^{i\boldsymbol{q}\cdot(\boldsymbol{r}_i-\boldsymbol{r}_j)}\sqrt{\frac{D_{\boldsymbol{q}}}{S_{\boldsymbol{q}}}}\left[(1+n_{\boldsymbol{q}})e^{iw_{\boldsymbol{q}}t} + n_{\boldsymbol{q}}e^{-iw_{\boldsymbol{q}}t}\right] \tag{7.10.9}$$

$$\langle S_i^\zeta S_j^\zeta(t)\rangle = S^2 - \frac{2S}{N}\sum_{\boldsymbol{q}} l_{\boldsymbol{q}}^2[n_{\boldsymbol{q}} + x_{\boldsymbol{q}}^2(1+n_{\boldsymbol{q}})] \simeq \langle S_i^\zeta\rangle^2. \tag{7.10.10}$$

The last equality of Eq. (7.10.10) is obtained from Eq. (7.3.14). Notice that within the harmonic approximation the correlation $\langle S_i^\zeta S_j^\zeta(t)\rangle$ is time independent. Replacing Eqs. (7.10.8) and (7.10.10) into Eq. (7.10.1), one obtains

$$\langle S_i^x S_j^x(t)\rangle = \sin \boldsymbol{Q}\cdot\boldsymbol{r}_i \sin \boldsymbol{Q}\cdot\boldsymbol{r}_j \frac{S}{2N}\sum_{\boldsymbol{q}} e^{i\boldsymbol{q}\cdot(\boldsymbol{r}_i-\boldsymbol{r}_j)} \times\sqrt{\frac{S_{\boldsymbol{q}}}{D_{\boldsymbol{q}}}}[(1+n_{\boldsymbol{q}})e^{i\omega_{\boldsymbol{q}}t}+n_{\boldsymbol{q}}e^{-i\omega_{\boldsymbol{q}}t}] + \cos \boldsymbol{Q}\cdot\boldsymbol{r}_i \ \cos \boldsymbol{Q}\cdot\boldsymbol{r}_j \ \langle S_i^\zeta\rangle^2. \tag{7.10.11}$$

Similarly, one has

$$\langle S_i^y S_j^y(t)\rangle = \cos \boldsymbol{Q}\cdot\boldsymbol{r}_i \ \cos \boldsymbol{Q}\cdot\boldsymbol{r}_j \frac{S}{2N}\sum_{\boldsymbol{q}} e^{i\boldsymbol{q}\cdot(\boldsymbol{r}_i-\boldsymbol{r}_j)} \times\sqrt{\frac{S_{\boldsymbol{q}}}{D_{\boldsymbol{q}}}}[(1+n_{\boldsymbol{q}})e^{i\omega_{\boldsymbol{q}}t}+n_{\boldsymbol{q}}e^{-i\omega_{\boldsymbol{q}}t}] + \sin \boldsymbol{Q}\cdot\boldsymbol{r}_i \sin \boldsymbol{Q}\cdot\boldsymbol{r}_j\langle S_i^\zeta\rangle^2 \tag{7.10.12}$$

and

$$\langle S_i^z S_j^z(t)\rangle = \frac{S}{2N}\sum_{\boldsymbol{q}} e^{i\boldsymbol{q}\cdot(\boldsymbol{r}_i-\boldsymbol{r}_j)}\sqrt{\frac{D_{\boldsymbol{q}}}{S_{\boldsymbol{q}}}}[(1+n_{\boldsymbol{q}})e^{i\omega_{\boldsymbol{q}}t}+n_{\boldsymbol{q}}e^{-i\omega_{\boldsymbol{q}}t}]. \tag{7.10.13}$$

All other correlation functions do not contribute to the neutron scattering cross-section since the sum over $i$ and $j$ and the exponential factor in Eq. (3.1.2) imply that $S^{\alpha\beta}(\boldsymbol{K},\omega)=0$ for $\alpha\neq\beta$. For $\boldsymbol{Q}\neq 0$ and $\boldsymbol{Q}=\frac{\boldsymbol{G}}{2}$, the only non-zero terms occurring in the structure factor are $S^{xx}(\boldsymbol{K},\omega) = S^{yy}(\boldsymbol{K},\omega) = S_\perp(\vec{\boldsymbol{K}},\omega)$ and $S^{zz}(\boldsymbol{K},\omega) = S_\parallel(\vec{\boldsymbol{K}},\omega)$ where

$$S_\perp(\boldsymbol{K},\omega) = \frac{SN}{8\hbar}\left\{\sqrt{\frac{S_{\boldsymbol{K}-\boldsymbol{Q}}}{D_{\boldsymbol{K}-\boldsymbol{Q}}}}[(1+n_{\boldsymbol{K}-\boldsymbol{Q}})\delta(\omega-\omega_{\boldsymbol{K}-\boldsymbol{Q}})+n_{\boldsymbol{K}-\boldsymbol{Q}}\delta(\omega+\omega_{\boldsymbol{K}-\boldsymbol{Q}})] + \sqrt{\frac{S_{\boldsymbol{K}+\boldsymbol{Q}}}{D_{\boldsymbol{K}+\boldsymbol{Q}}}}[(1+n_{\boldsymbol{K}+\boldsymbol{Q}})\delta(\omega-\omega_{\boldsymbol{K}-\boldsymbol{Q}})+n_{\boldsymbol{K}+\boldsymbol{Q}}\delta(\omega+\omega_{\boldsymbol{K}+\boldsymbol{Q}})]\right\} + \frac{(N\langle S_i^\zeta\rangle)^2}{4\hbar}(\delta_{\boldsymbol{K},\boldsymbol{Q}}+\delta_{\boldsymbol{K},-\boldsymbol{Q}})\delta(\omega) \tag{7.10.14}$$

and

$$S_\parallel(\boldsymbol{K},\omega) = \frac{SN}{2\hbar}\sqrt{\frac{D_{\boldsymbol{K}}}{S_{\boldsymbol{K}}}}[(1+n_{\boldsymbol{K}})\delta(\omega-\omega_{\boldsymbol{K}})+n_{\boldsymbol{K}}\delta(\omega+\omega_{\boldsymbol{K}})]. \tag{7.10.15}$$

The $\delta$-functions with an argument containing the frequency come from the integration over the time present in the structure factor of Eq. (3.1.2). Then the neutron cross-section (3.1.1) becomes

$$\frac{d^2\sigma}{d\Omega dE'} = r_0^2 \frac{k'}{k} \left[\frac{1}{2} gF(\boldsymbol{K})\right]^2 e^{-2W(\boldsymbol{K})} \left[\left(1+\frac{K_z^2}{K^2}\right) S_\perp(\boldsymbol{K},\omega) + \left(1-\frac{K_z^2}{K^2}\right) S_\parallel(\boldsymbol{K},\omega)\right], \tag{7.10.16}$$

where $S_\perp(\boldsymbol{K},\omega)$ and $S_\parallel(\boldsymbol{K},\omega)$ are given by Eqs. (7.10.14) and (7.10.15), respectively. Obviously, the term within the square brackets in the neutron cross-section (7.10.16) is invariant if one adds any reciprocal lattice vector $\boldsymbol{G}$ to the scattering wavevector $\boldsymbol{K}$.

The elastic neutron scattering comes from the last term of $S_\perp(\boldsymbol{K},\omega)$ in Eq. (7.10.14). Such a term containing $\delta(\omega)$ is located at $\boldsymbol{K} = \pm\boldsymbol{Q}$ and its intensity is proportional to the square of the order parameter times a geometric factor going from 1 to 2 according to the scattering wavevector $\boldsymbol{K}$ is parallel or perpendicular to the plane in which the spins lie. This contrasts with what happens in collinear systems like ferromagnets or antiferromagnets where the geometric factor vanishes when the scattering wavevector is parallel to the direction of the order parameter. Notice that the location of the elastic peaks gives directly the helix wavevector of the non-collinear phase.

As for the inelastic (one-magnon) neutron scattering cross-section, there are contributions from the creation of a magnon at the expense of the neutron energy corresponding to the terms proportional to $\delta(\omega - \omega_{\boldsymbol{K}})$ and also present at zero temperature, and contributions from the destruction of a magnon which gives its energy to the neutron corresponding to the terms proportional to $\delta(\omega + \omega_{\boldsymbol{K}})$ and present only at non-zero temperature. In any case, for any scattering wavevector $\boldsymbol{K}$, one sees *three* inelastic peaks in correspondence to the frequencies

$$\omega_{\boldsymbol{K}-\boldsymbol{Q}}, \quad \omega_{\boldsymbol{K}+\boldsymbol{Q}}, \quad \omega_{\boldsymbol{K}} \tag{7.10.17}$$

the intensity of which are proportional to

$$\sqrt{\frac{S_{\boldsymbol{K}-\boldsymbol{Q}}}{D_{\boldsymbol{K}-\boldsymbol{Q}}}}, \quad \sqrt{\frac{S_{\boldsymbol{K}+\boldsymbol{Q}}}{D_{\boldsymbol{K}+\boldsymbol{Q}}}}, \quad \sqrt{\frac{D_{\boldsymbol{K}}}{S_{\boldsymbol{K}}}}, \tag{7.10.18}$$

respectively. Notice that from the position of these peaks it is possible to obtain the whole energy spectrum of the spin waves in a non-collinear magnet.

The existence of a DH in the "classical" ground-state energy of the SQ and TR lattices for $j_3 = \frac{1}{2}j_2$, of the HON lattice for $j_3 = 0$, of the FCC lattice for any $j'$ and of the R lattice for $|j'| < 3$, should imply that the isolated elastic peaks become a *ridge* located on the DH line.[70] However, we have seen that the quantum fluctuations destroy the DH line of the "classical" ground state selecting well-defined order wavevectors $\boldsymbol{Q}$ out of the DH manifold (order by quantum disorder). At the same time, the soft line in the magnon spectrum corresponding to the DH is expected to be replaced by a "rippled soft line", accounting for the contributions to the spectrum from a careful treatment of the terms of the same order in $\frac{1}{S}$ obtained from the normal ordering of the anharmonic terms of the boson Hamiltonian. This challenging calculation is expected to reduce the continuous line of soft modes in the classical spin wave spectrum to a set of discrete soft modes corresponding to the wavevectors $\boldsymbol{k} = \boldsymbol{Q}$ even though the existence of a line of low lying energy excitations may survive. At finite temperature, the modes belonging to the rippled line are thermally populated and a ridge in correspondence of such a line should appear in a quasi-elastic neutron scattering (QENS) experiment.

An analogous scenario was observed[71] in the quasi-2D Heisenberg antiferromagnet $K_2NiF_4$ ($S=1$) where the in-plane antiferromagnetic exchange interaction $J_{\rm NN} = -9.68\,{\rm meV}$ is more than 270 times the inter-plane exchange interaction $J'$ and a small uniaxial (Ising) anisotropy $g\mu_B H_A = 0.073\,{\rm meV}$ determines the 2D Ising character of the phase transition at $T_N = 97.2\,{\rm K}$. In this case, the role of the rippled soft line is played by the low energy branch in the spin wave spectrum along the $(0,0,1)$ direction that causes the appearance of a ridge in the QENS experiment at temperatures $T = 0.8 - 0.9T_N$ in addition to the conventional elastic Bragg peaks. The ridge intensity increases as the temperature increases up to $T_N$. In the paramagnetic phase where the Bragg peaks have been disappeared, the ridge intensity survives up to $2T_N$ even though its intensity decreases with temperature. In $K_2NiF_4$, the presence of a "Bragg ridge" in addition to "Bragg peaks" is ascribed to the quasi-2D magnetic structure of the system confirmed by the absence of dispersion in the spin wave spectrum along the $(0,0,1)$ direction. Indeed, no dispersion of the low lying energy branch at $\hbar\omega(0,0,k_z) \simeq \sqrt{g\mu_B H_A\, 8|J_{\rm NN}|S} = 2.37\,{\rm meV}$ (at $T = 5\,{\rm K}$) was observed in the INS experiment[71] with an energy resolution of $\sim 0.1\,{\rm meV}$.

The existence of a Bragg ridge profile in the QENS cross-section was observed[72,73] in a polycrystal of solid oxygen in both the monoclinic $\alpha$ ($0 < T < 24\,{\rm K}$) and rhombohedral $\beta$ ($24 < T < 44\,{\rm K}$) phases. Simultaneous Bragg peaks and ridge-like profile appeared in the QENS cross-section[73] on the $\alpha$-oxygen with an energy resolution of $\Delta E = 0.06\,{\rm meV}$ while only a similar ridge profile without Bragg peaks was observed in the experiment on the $\beta$-oxygen. As for the ordered $\alpha$-phase, the cross-section profile is ascribed to the low energy magnon branch $10 < \hbar\omega/k_B < 45$ K due to the weak inter-plane exchange interaction[74] in agreement with the $K_2NiF_4$ scenario. As for the $\beta$-phase, one expects the existence of low lying energy branches along both the $(0,0,1)$ direction ($0 < \hbar\omega/k_B < 8\,{\rm K}$) and the rippled soft line $\simeq (0, \frac{4\pi}{3a}, k_z)$ ($\hbar\omega/k_B \simeq 0.5\,{\rm K}$) due to the rhombohedral structure of the $\beta$-oxygen.[74] Unfortunately, the two contributions cannot be separated owing to the limited energy resolution of the QENS experiment.[73] A support to the simultaneous presence of these two low lying energy branches in the spin wave spectrum is the absence of LRO in the $\beta$-phase of the solid oxygen.

A ridge-like structure with Bragg peaks was observed[75] in the rhombohedral $S = \frac{3}{2}$ Heisenberg antiferromagnet $CuCrO_2$ for $T < 25\,\mathrm{K}$. This experiment might be pointed out as a strong evidence for the existence of a quasi-DH in the rhombohedral lattice in addition to a low-lying branch related to the weak inter-plane coupling.

# Chapter 8

# SPIN WAVES IN MULTILAYERS

## 8.1. Spin Green Functions and Random Phase Approximation

In the previous chapters, we have studied the spin waves in *infinite* systems that is in perfect periodic lattices. On the contrary, in this chapter we will study the spin waves in *finite* systems describing magnetic multilayers or spin waves in a semi-infinite system describing the effect of a surface on a magnetic system. The main point is the loss of periodicity in the direction perpendicular to the magnetic layers. According to Chapter 3, we define the retarded Green function[24]

$$G_{l,m}(t) \equiv \langle\langle S_l^+; S_m^- \rangle\rangle = -i\theta(t)\langle [S_l^+(t), S_m^-] \rangle \tag{8.1.1}$$

and its time Fourier transform

$$G_{l,m}(\omega + i\epsilon) \equiv \langle\langle S_l^+; S_m^- \rangle\rangle_{\omega+i\epsilon} = \int_{-\infty}^{+\infty} dt e^{i\omega t} G_{l,m}(t). \tag{8.1.2}$$

The equation of motion of the time Green function (8.1.1) is

$$\frac{d}{dt}\langle\langle S_l^+; S_m^- \rangle\rangle = -2\, i\delta(t)\delta_{l,m}\langle S_l^z \rangle + \left\langle\left\langle -\frac{i}{\hbar}[S_l^+, \mathcal{H}]; S_m^- \right\rangle\right\rangle, \tag{8.1.3}$$

where the commutation rule $[S_l^+, S_m^-] = 2S_l^z \delta_{l,m}$ has been used and $\mathcal{H}$ is the Heisenberg Hamiltonian

$$\mathcal{H} = -\sum_{j,\delta} J_\delta \left[ S_j^z S_{j+\delta}^z + \frac{1}{2}(S_j^+ S_{j+\delta}^- + S_j^- S_{j+\delta}^+) \right] \tag{8.1.4}$$

with $J_\delta > 0$. The equation of motion of the Green function (8.1.2) is then

$$\hbar\omega\langle\langle S_l^+; S_m^- \rangle\rangle_\omega = 2\hbar\delta_{l,m}\langle S_l^z \rangle + \sum_\delta 2J_\delta \langle\langle S_l^+ S_{l+\delta}^z - S_{l+\delta}^+ S_l^z; S_m^- \rangle\rangle_\omega. \tag{8.1.5}$$

The infinite chain of equations for the various Green functions is made finite by a convenient decoupling. In particular, the random phase approximation[28] (RPA) consists of decoupling the Green function occurring in the last term of Eq. (8.1.5) as

$$\langle\langle S_l^+ S_{l+\delta}^z - S_{l+\delta}^+ S_l^z; S_m^-\rangle\rangle_\omega \simeq \langle S_l^z\rangle [G_{l,m}(\omega) - G_{l+\delta,m}(\omega)] \qquad (8.1.6)$$

where one neglects the correlations between $S^+$ on one lattice site and $S^z$ on another lattice site. This approximation becomes exact at zero temperature because $S_l^z|0\rangle = S|0\rangle$ where $S$ is a $c$-number. The RPA allows to write Eq. (8.1.5) in the form

$$\left(\hbar\omega - \langle S_l^z\rangle \sum_\delta 2J_\delta\right) G_{l,m}(\omega) + \langle S_l^z\rangle \sum_\delta 2J_\delta\, G_{l+\delta,m}(\omega) = 2\hbar\delta_{l,m}\langle S_l^z\rangle. \qquad (8.1.7)$$

## 8.2. Multilayers

Starting from Eq. (8.1.7), we will obtain the spin wave spectrum for a stacking of ferromagnetic planes (multilayer). Suppose, we pile up the layers along the $z$-axis. Assuming PBC in the $xy$-planes, we may define a spatial Fourier transform as

$$G_{l,m}(\omega) = \frac{1}{\sqrt{N_s}} \sum_{\boldsymbol{k}_\parallel} e^{i\boldsymbol{k}_\parallel\cdot(\boldsymbol{r}_i - \boldsymbol{r}_j)} G_{n,n'}(\boldsymbol{k}_\parallel, \omega) \qquad (8.2.1)$$

where $l = \boldsymbol{r}_i + n\mathbf{u}_z$ and $m = \boldsymbol{r}_j + n'\mathbf{u}_z$ in which $n, n'$ label the planes and $\boldsymbol{r}_i$, $\boldsymbol{r}_j$ are 2D lattice vectors within the $xy$-planes parallel to the lower ($n = 1$) and upper ($n = N$) surface of the multilayer; $\boldsymbol{k}_\parallel$ is a wavevector in the $k_x k_y$-plane and $N_s$ is the number of the spins belonging to a plane of the multilayer. Suppose that the exchange interaction between NN spins in the $xy$-planes is $2J_\parallel$ and the exchange interaction between NN spins belonging to adjacent planes is $2J_\perp$. Such a model is called[76] "free surface model" since there is no difference between the exchange interactions between the NN spins on the surfaces and the exchange interactions between NN spins in the layers of the bulk. This model could be appropriated to describe a film of $N$ layers. The Heisenberg Hamiltonian for such a film is

$$\mathcal{H} = -\sum_{s=1}^{N} \sum_{j\delta_\parallel} J_\parallel \boldsymbol{S}_j^{(s)} \cdot \boldsymbol{S}_{j+\delta_\parallel}^{(s)} - \sum_{s=1}^{N-1} \sum_{j\delta_\perp} 2J_\perp \boldsymbol{S}_j^{(s)} \cdot \boldsymbol{S}_{j+\delta_\perp}^{(s+1)}, \qquad (8.2.2)$$

where $\boldsymbol{S}_j^{(s)}$ is the spin located at the site $\boldsymbol{r}_j + s\mathbf{u}_z$. The absence of a factor 2 before the exchange integral $J_\parallel$ is due to a double counting coming from the sum over $j$ and $\delta_\parallel$. In the limit $T \to 0$, the equation of motion (8.1.7) splits into $N$ equations of motion for the Green functions $G_{n,n'}(\boldsymbol{k}_\parallel, \omega)$ with $n = 1, 2, \ldots, N$ given by

$$[\hbar(\omega - \omega_\parallel) - 2z_\perp J_\perp S] G_{1,n'}(\boldsymbol{k}_\parallel, \omega) + 2z_\perp J_\perp S\gamma_\perp G_{2,n'}(\boldsymbol{k}_\parallel, \omega) = 2\hbar S\delta_{1,n'} \qquad (8.2.3)$$

for $n = 1$ corresponding to the lower surface of the film,

$$[\hbar(\omega - \omega_\parallel) - 4z_\perp J_\perp S] G_{n,n'}(\boldsymbol{k}_\parallel, \omega) + 2z_\perp J_\perp S\gamma_\perp [G_{n-1,n'}(\boldsymbol{k}_\parallel, \omega) + G_{n+1,n'}(\boldsymbol{k}_\parallel, \omega)]$$
$$= 2\hbar S\delta_{n,n'} \qquad (8.2.4)$$

for $1 < n < N - 1$ corresponding to the internal layers (bulk) and

$$[\hbar(\omega - \omega_{\|}) - 2z_{\perp}J_{\perp}S]G_{N,n'}(\boldsymbol{k}_{\|},\omega) + 2z_{\perp}J_{\perp}S\gamma_{\perp}G_{N-1,n'}(\boldsymbol{k}_{\|},\omega) = 2\hbar S\delta_{N,n'} \tag{8.2.5}$$

for $n = N$ corresponding to the upper surface of the film. The frequency $\omega_{\|}$ and the structure factors occurring in Eqs. (8.2.3)–(8.2.5) are given by

$$\hbar\omega_{\|} = 2z_{\|}J_{\|}S(1 - \gamma_{\|}) \tag{8.2.6}$$

with

$$\gamma_{\|} = \frac{1}{z_{\|}}\sum_{\boldsymbol{\delta}_{\|}} e^{i\boldsymbol{k}_{\|}\cdot\boldsymbol{\delta}_{\|}} \tag{8.2.7}$$

and

$$\gamma_{\perp} = \frac{1}{z_{\perp}}\sum_{\boldsymbol{\delta}_{\perp}} e^{i\boldsymbol{k}_{\|}\cdot\boldsymbol{\delta}_{\perp}}, \tag{8.2.8}$$

respectively. In Eqs. (8.2.3)–(8.2.8), $z_{\|}$ and $z_{\perp}$ are the numbers of the NN spins in and out of $xy$-planes, respectively. Notice that for planes stacked in such a way that the spins along the $z$-direction lie on a straight line one has $z_{\perp} = 1$ and $\boldsymbol{k}_{\|}\cdot\boldsymbol{\delta}_{\perp} = 0$. This is the case for SC and H lattices, for instance. Defining

$$\hbar\omega = \hbar\omega_{\|} + 4z_{\perp}J_{\perp}S(1 - \gamma_{\perp}\cos\theta) \tag{8.2.9}$$

one may write the $N$ equations (8.2.3)–(8.2.5) in a matrix form as

$$\mathbf{A}\times\mathbf{G} = \frac{\hbar}{z_{\perp}J_{\perp}\gamma_{\perp}}\mathbf{1} \tag{8.2.10}$$

where

$$\mathbf{A} = \begin{pmatrix} \frac{1}{\gamma_{\perp}} - 2\cos\theta & 1 & 0 & \dots & 0 & 0 \\ 1 & -2\cos\theta & 1 & \dots & 0 & 0 \\ 0 & 1 & -2\cos\theta & \dots & 0 & 0 \\ \vdots & \vdots & \vdots & \ddots & \vdots & \vdots \\ 0 & 0 & 0 & \dots & -2\cos\theta & 1 \\ 0 & 0 & 0 & \dots & 1 & \frac{1}{\gamma_{\perp}} - 2\cos\theta \end{pmatrix}. \tag{8.2.11}$$

The matrix elements of $\mathbf{G}$ are $G_{n,n'}(\boldsymbol{k}_{\|},\omega)$ with $n, n' = 1, \dots, N$ and $\mathbf{1}$ is the unit matrix of dimension $N$. As one can see, the matrix $\mathbf{A}$ is a tridiagonal matrix symmetric with respect to its diagonals. The inverse matrix $\mathbf{A}^{-1}$ is no longer tridiagonal but it is yet symmetric with respect to its two diagonals so that the number of distinct $G_{n,n'}(\boldsymbol{k}_{\|},\omega)$ is drastically reduced. The solution of the matrix equation (8.2.10) is

$$G_{n,n'}(\boldsymbol{k}_{\|},\omega) = \frac{\hbar}{z_{\perp}J_{\perp}\gamma_{\perp}}(-1)^{n+n'}\frac{\det\mathbf{A}_{n'n}}{\det\mathbf{A}}, \tag{8.2.12}$$

where $\mathbf{A}_{n'n}$ is the matrix left after the $n'$th row and the $n$th column have been suppressed. Notice that the poles of the Green functions of Eq. (8.2.12) occur at frequencies $\omega_l$ (or $\theta_l$ according to Eq. (8.2.9)) for which the determinant of $\mathbf{A}$ vanishes. In order to evaluate such a determinant, we split the matrix $\mathbf{A}$ into two matrices like

$$\mathbf{A} = \boldsymbol{\Delta} - \mathbf{A}_0, \tag{8.2.13}$$

where $\boldsymbol{\Delta}$ is the square matrix of dimension $N$ whose elements are all zero except $\Delta_{1,1} = \Delta_{N,N} = \frac{1}{\gamma_\perp}$ and

$$\mathbf{A}_0 = \begin{pmatrix} 2\cos\theta & -1 & 0 & \dots & 0 & 0 \\ -1 & 2\cos\theta & -1 & \dots & 0 & 0 \\ 0 & -1 & 2\cos\theta & \dots & 0 & 0 \\ \vdots & \vdots & \vdots & \ddots & \vdots & \vdots \\ 0 & 0 & 0 & \dots & 2\cos\theta & -1 \\ 0 & 0 & 0 & \dots & -1 & 2\cos\theta \end{pmatrix}. \tag{8.2.14}$$

In this way, the determinant of $\mathbf{A}$ becomes

$$\det \mathbf{A} = (-1)^N D_N \ \det(\mathbf{1} - \mathbf{A}_0^{-1}\boldsymbol{\Delta}) \tag{8.2.15}$$

where $D_N = \det \mathbf{A}_0$, $\mathbf{A}_0^{-1}$ is the inverse matrix of $\mathbf{A}_0$ and $\mathbf{1}$ is the unit matrix of dimension $N$. A recursion formula for the determinant $D_N$ is easily obtained expanding the determinant itself along the first row. One obtains

$$D_N = 2\cos\theta D_{N-1} - D_{N-2}, \tag{8.2.16}$$

where $D_{N-1}$ and $D_{N-2}$ are determinants with the same structure of $D_N$ but with one and two rows and columns suppressed, respectively. It is direct to verify that the solution of Eq. (8.2.16) is

$$D_N = \frac{\sin(N+1)\theta}{\sin\theta} \tag{8.2.17}$$

and to obtain

$$\mathbf{A}_0^{-1}\boldsymbol{\Delta} = \begin{pmatrix} \dfrac{D_{N-1}}{\gamma_\perp D_N} & 0 & \dots & 0 & \dfrac{1}{\gamma_\perp D_N} \\ \dfrac{D_{N-2}}{\gamma_\perp D_N} & 0 & \dots & 0 & \dfrac{D_1}{\gamma_\perp D_N} \\ \vdots & \vdots & \ddots & \vdots & \vdots \\ \dfrac{D_1}{\gamma_\perp D_N} & 0 & \dots & 0 & \dfrac{D_{N-2}}{\gamma_\perp D_N} \\ \dfrac{1}{\gamma_\perp D_N} & 0 & \dots & 0 & \dfrac{D_{N-1}}{\gamma_\perp D_N} \end{pmatrix} \tag{8.2.18}$$

so that

$$\det(\mathbf{1} - \mathbf{A}_0^{-1}\boldsymbol{\Delta}) = \left(1 - \frac{D_{N-1}}{\gamma_\perp D_N}\right)^2 - \left(\frac{1}{\gamma_\perp D_N}\right)^2. \tag{8.2.19}$$

Replacing Eqs. (8.2.19) and (8.2.17) into Eq. (8.2.15), one obtains

$$\det \mathbf{A} = \frac{(-1)^N}{\gamma_\perp^2 D_N}[(\gamma_\perp D_N - D_{N-1})^2 - 1] \tag{8.2.20}$$

with $D_N$ given by Eq. (8.2.17).

The neutron scattering cross-section for a film can be deduced from Eq. (3.1.4). In order to simplify the formulas, we consider the neutron cross-section for a scattering wavevector directed perpendicularly to the surface of the film that is $\boldsymbol{K} = K_z\mathbf{u}_z$. For such a geometry, Eq. (3.1.4) reduces to

$$\frac{d^2\sigma}{d\Omega dE'} = r_0^2\frac{k'}{k}\left[\frac{1}{2}gF(\boldsymbol{K})\right]^2 e^{-2W(\boldsymbol{K})}\, 2S_\perp(\boldsymbol{K},\omega) \tag{8.2.21}$$

with

$$S_\perp(K_z,\omega) = -\frac{1}{4\pi\hbar}\sum_{n,n'} e^{-iK_z(n-n')}\Im\{[1+n(\omega)]G_{n,n'}(\omega) + n(-\omega)G_{n,n'}(-\omega)\}, \tag{8.2.22}$$

where $G_{n,n'}(\omega) = G_{n,n'}(\boldsymbol{K}_\parallel = 0,\omega)$. For $\boldsymbol{K}_\parallel = 0$, the solutions of the equation $\det \mathbf{A} = 0$ with $\det \mathbf{A}$ given by Eq. (8.2.20) are

$$\theta = \frac{\pi}{N}l, \quad l = 0,1,2,\ldots,N-1 \tag{8.2.23}$$

so that the Green functions $G_{n,n'}(\omega)$ appearing in Eq. (8.2.12) can be written in the form

$$G_{n,n'}(\omega) = \sum_{l=0}^{N-1}\frac{c_{n,n'}^{(l)}}{\omega - \omega_l} \tag{8.2.24}$$

with

$$\omega_l = \frac{4z_\perp J_\perp S}{\hbar}\left(1 - \cos\frac{\pi}{N}l\right). \tag{8.2.25}$$

By using the identity (3.2.12), one has

$$\Im G_{n,n'}(\pm\omega) = -\pi\sum_{l=0}^{N-1} c_{n,n'}^{(l)}\delta(\omega \mp \omega_l) \tag{8.2.26}$$

so that

$$S_\perp(K_z,\omega) = \sum_{l=0}^{N-1} I_l(K_z)\{[1+n(\omega_l)]\delta(\omega-\omega_l) + n(\omega_l)\delta(\omega+\omega_l)\} \tag{8.2.27}$$

with

$$I_l(K_z) = \frac{1}{4\hbar}\sum_{n,n'} c_{n,n'}^{(l)} e^{-iK_z(n-n')}. \tag{8.2.28}$$

The neutron cross-section (8.2.21) is a sum of $\delta$-peaks with intensity $I_l(K_z)$ whose number coincides with the number of layers of the film. Note that the existence of $\delta$-peaks in the neutron cross-section is directly connected to the RPA approximation that becomes exact as $T \to 0$. Higher-order decoupling has to be used at finite temperature and the related Green functions change the $\delta$-peaks into Lorentzian peaks, leading to the renormalization and damping of the excitations.

## 8.3. Bilayer

In order to show in detail the results illustrated in the previous section, we perform explicitly the calculations for a bilayer (two layers). For a bilayer, the matrix (8.2.11) becomes

$$\mathbf{A}_2 = \begin{pmatrix} \frac{1}{\gamma_\perp} - 2\cos\theta & 1 \\ 1 & \frac{1}{\gamma_\perp} - 2\cos\theta \end{pmatrix}. \tag{8.3.1}$$

The distinct Green functions given by Eq. (8.2.12) are

$$G_{1,1}(\boldsymbol{k}_\parallel, \omega) = \frac{\hbar}{z_\perp J_\perp \gamma_\perp} \frac{\frac{1}{\gamma_\perp} - 2\cos\theta}{\det \mathbf{A}_2} \tag{8.3.2}$$

and

$$G_{1,2}(\boldsymbol{k}_\parallel, \omega) = -\frac{\hbar}{z_\perp J_\perp \gamma_\perp} \frac{1}{\det \mathbf{A}_2} \tag{8.3.3}$$

where

$$\det \mathbf{A}_2 = \left(\frac{1}{\gamma_\perp} - 2\cos\theta\right)^2 - 1. \tag{8.3.4}$$

Owing to the properties of the matrix $\mathbf{A}_2$, one has $G_{2,1} = G_{1,2}$ (symmetry about the principal diagonal of the matrix) and $G_{2,2} = G_{1,1}$ (symmetry about the other diagonal of the matrix). Using the relationship (8.2.9), one can transform Eqs. (8.3.2) and (8.3.3) into

$$G_{1,1}(\boldsymbol{k}_\parallel, \omega) = S\left(\frac{1}{\omega - \omega_0} + \frac{1}{\omega - \omega_1}\right) \tag{8.3.5}$$

and

$$G_{1,2}(\boldsymbol{k}_\parallel, \omega) = S\left(\frac{1}{\omega - \omega_0} - \frac{1}{\omega - \omega_1}\right), \tag{8.3.6}$$

where

$$\omega_l = \omega_\parallel + \frac{2z_\perp J_\perp S}{\hbar}(1 - \gamma_\perp \cos \pi l) \tag{8.3.7}$$

with $l = 0, 1$. Using the relationship (3.2.12), the imaginary parts of the Green functions (8.3.5) and (8.3.6) are given by

$$\Im G_{1,1}(\boldsymbol{k}_\parallel, \pm\omega) = -\pi\ S[\delta(\omega \mp \omega_0) + \delta(\omega \mp \omega_1)] \tag{8.3.8}$$

and

$$\Im G_{1,2}(\boldsymbol{k}_\parallel, \pm\omega) = -\pi\ S[\delta(\omega \mp \omega_0) - \delta(\omega \mp \omega_1)]. \tag{8.3.9}$$

The neutron cross-section is given by (8.2.27) with $N = 2$ and

$$I_l(K_z) = \frac{S}{2\hbar}(1 + \cos K_z \cos \pi l) \tag{8.3.10}$$

with $l = 0, 1$.

## 8.4. Trilayer

In this section, we perform explicitly the calculations for a trilayer (three layers) for which the matrix (8.2.11) becomes

$$\mathbf{A}_3 = \begin{pmatrix} \frac{1}{\gamma_\perp} - 2\cos\theta & 1 & 0 \\ 1 & -2\cos\theta & 1 \\ 0 & 1 & \frac{1}{\gamma_\perp} - 2\cos\theta \end{pmatrix}. \tag{8.4.1}$$

The distinct Green functions given by Eq. (8.2.12) are

$$G_{1,1}(\boldsymbol{k}_\parallel, \omega) = \frac{\hbar}{z_\perp J_\perp \gamma_\perp} \frac{4\cos^2\theta - \frac{2}{\gamma_\perp}\cos\theta - 1}{\det \mathbf{A}_3}, \tag{8.4.2}$$

$$G_{1,2}(\boldsymbol{k}_\parallel, \omega) = -\frac{\hbar}{z_\perp J_\perp \gamma_\perp} \frac{\frac{1}{\gamma_\perp} - 2\cos\theta}{\det \mathbf{A}_3}, \tag{8.4.3}$$

$$G_{1,3}(\boldsymbol{k}_\parallel, \omega) = \frac{\hbar}{z_\perp J_\perp \gamma_\perp} \frac{1}{\det \mathbf{A}_3} \tag{8.4.4}$$

and

$$G_{2,2}(\boldsymbol{k}_\parallel, \omega) = \frac{\hbar}{z_\perp J_\perp \gamma_\perp} \frac{\left(\frac{1}{\gamma_\perp} - 2\cos\theta\right)^2}{\det \mathbf{A}_3}, \tag{8.4.5}$$

where

$$\det \mathbf{A}_3 = \left(\frac{1}{\gamma_\perp} - 2\cos\theta\right)\left(2\cos\theta - \frac{1 - \sqrt{1 + 8\gamma_\perp^2}}{2\gamma_\perp}\right)\left(2\cos\theta - \frac{1 + \sqrt{1 + 8\gamma_\perp^2}}{2\gamma_\perp}\right). \tag{8.4.6}$$

Owing to the properties of the matrix $\mathbf{A}_3$, one has $G_{2,1} = G_{1,2}$, $G_{1,3} = G_{3,1}$ (symmetry about the principal diagonal of the matrix) and $G_{3,3} = G_{1,1}$, $G_{2,3} = G_{3,2}$ (symmetry about the other diagonal of the matrix). Using the relationship (8.2.9), one can transform Eqs. (8.4.2)–(8.4.5) into

$$G_{1,1}(\boldsymbol{k}_\parallel, \omega) = \frac{S}{4}\left(\frac{\sqrt{1 + 8\gamma_\perp^2} + 1}{2\sqrt{1 + 8\gamma_\perp^2}} \frac{1}{\omega - \omega_0} + \frac{1}{\omega - \omega_1} + \frac{\sqrt{1 + 8\gamma_\perp^2} - 1}{2\sqrt{1 + 8\gamma_\perp^2}} \frac{1}{\omega - \omega_2}\right), \tag{8.4.7}$$

$$G_{1,2}(\boldsymbol{k}_{\|},\omega)=\frac{S}{2}\frac{\gamma_{\perp}}{\sqrt{1+8\gamma_{\perp}^2}}\left(\frac{1}{\omega-\omega_0}-\frac{1}{\omega-\omega_2}\right), \tag{8.4.8}$$

$$G_{1,3}(\boldsymbol{k}_{\|},\omega)=\frac{S}{4}\left(\frac{\sqrt{1+8\gamma_{\perp}^2}+1}{2\sqrt{1+8\gamma_{\perp}^2}}\frac{1}{\omega-\omega_0}-\frac{1}{\omega-\omega_1}\right.$$
$$\left.+\frac{\sqrt{1+8\gamma_{\perp}^2}-1}{2\sqrt{1+8\gamma_{\perp}^2}}\frac{1}{\omega-\omega_2}\right), \tag{8.4.9}$$

$$G_{2,2}(\boldsymbol{k}_{\|},\omega)=\frac{S}{4}\left(\frac{\sqrt{1+8\gamma_{\perp}^2}-1}{\sqrt{1+8\gamma_{\perp}^2}}\frac{1}{\omega-\omega_0}+\frac{\sqrt{1+8\gamma_{\perp}^2}+1}{\sqrt{1+8\gamma_{\perp}^2}}\frac{1}{\omega-\omega_2}\right), \tag{8.4.10}$$

where

$$\omega_0=\omega_{\|}+\frac{z_{\perp}J_{\perp}S}{\hbar}\left(3-\sqrt{1+8\gamma_{\perp}^2}\right), \tag{8.4.11}$$

$$\omega_1=\omega_{\|}+\frac{2z_{\perp}J_{\perp}S}{\hbar} \tag{8.4.12}$$

and

$$\omega_2=\omega_{\|}+\frac{z_{\perp}J_{\perp}S}{\hbar}\left(3+\sqrt{1+8\gamma_{\perp}^2}\right). \tag{8.4.13}$$

Using the relationship (3.2.12), the imaginary parts of the Green functions (8.4.7)–(8.4.10) are given by

$$\Im G_{1,1}(\boldsymbol{k}_{\|},\pm\omega)=-\frac{\pi S}{4}\left[\frac{\sqrt{1+8\gamma_{\perp}^2}+1}{2\sqrt{1+8\gamma_{\perp}^2}}\delta(\omega\mp\omega_0)+\delta(\omega\mp\omega_1)\right.$$
$$\left.+\frac{\sqrt{1+8\gamma_{\perp}^2}-1}{2\sqrt{1+8\gamma_{\perp}^2}}\delta(\omega\mp\omega_2)\right], \tag{8.4.14}$$

$$\Im G_{1,2}(\boldsymbol{k}_{\|},\pm\omega)=-\frac{\pi S}{2}\frac{\gamma_{\perp}}{\sqrt{1+8\gamma_{\perp}^2}}[\delta(\omega\mp\omega_0)-\delta(\omega\mp\omega_2)], \tag{8.4.15}$$

$$\Im G_{1,3}(\boldsymbol{k}_{\|},\pm\omega)=-\frac{\pi S}{4}\left[\frac{\sqrt{1+8\gamma_{\perp}^2}+1}{2\sqrt{1+8\gamma_{\perp}^2}}\delta(\omega\mp\omega_0)-\delta(\omega\mp\omega_1)\right.$$
$$\left.+\frac{\sqrt{1+8\gamma_{\perp}^2}-1}{2\sqrt{1+8\gamma_{\perp}^2}}\delta(\omega\mp\omega_2)\right], \tag{8.4.16}$$

$$\Im G_{2,2}(\boldsymbol{k}_{\|},\pm\omega)=-\frac{\pi S}{4}\left[\frac{\sqrt{1+8\gamma_{\perp}^2}-1}{\sqrt{1+8\gamma_{\perp}^2}}\delta(\omega\mp\omega_0)+\frac{\sqrt{1+8\gamma_{\perp}^2}+1}{\sqrt{1+8\gamma_{\perp}^2}}\delta(\omega\mp\omega_2)\right]. \tag{8.4.17}$$

The neutron cross-section is given by (8.2.27) with $N = 3$ and

$$I_0(K_z) = \frac{S}{3\hbar}\cos^4\frac{K_z}{2}, \tag{8.4.18}$$

$$I_1(K_z) = \frac{S}{4\hbar}\sin^2 K_z, \tag{8.4.19}$$

$$I_2(K_z) = \frac{S}{3\hbar}\sin^4\frac{K_z}{2} \tag{8.4.20}$$

and

$$\omega_l = \frac{4z_\perp J_\perp S}{\hbar}\left(1 - \cos\frac{\pi l}{3}\right) \tag{8.4.21}$$

with $l = 0, 1, 2$.

## 8.5. Classical Spin Waves in Multilayers

In order to get information about the spin dynamics in multilayers, we make use of the classical mechanics assuming the spins to be classical vectors. According to Section 1.6, the classical equations of motion for the spin components are given by the torque equations (1.6.5)–(1.6.7) the solution of which may be written

$$S^x_{j,n} = u_n \cos(\boldsymbol{k}_\parallel \cdot \boldsymbol{r}_j - \omega t), \tag{8.5.1}$$

$$S^y_{j,n} = u_n \sin(\boldsymbol{k}_\parallel \cdot \boldsymbol{r}_j - \omega t) \tag{8.5.2}$$

and

$$S^z_{j,n} = \sqrt{S^2 - u_n^2}, \tag{8.5.3}$$

where $\boldsymbol{r}_j$ is a 2D lattice vector in the layer and $n$ labels the $N$ layers. As one can see from Eqs. (8.5.1)–(8.5.3), each spin of the film rotates around the $z$-axis sweeping the surface of a cone of apex angle $\alpha_n = \arcsin\frac{u_n}{S} \simeq \frac{u_n}{S}$. While in the ferromagnetic "bulk" of Section 1.6, the apex angle $\alpha$ was the same at each lattice site, in the ferromagnetic "multilayer" of the present section the apex angle is a function of the number $n$ of the layer. In this way, the torque equation (1.6.4) leads to $N$ torque equations, one for each layer, and a homogeneous linear system of equations for the amplitudes $u_n$ is obtained. The matrix form of such a system is

$$\mathbf{A}_N \times \mathbf{U}_N = 0 \tag{8.5.4}$$

where the square matrix $\mathbf{A}_N$ of dimension $N$ is given by Eq. (8.2.11) and $\mathbf{U}_N$ is a vector whose components are the amplitude $u_n$ with $n = 1, 2, \ldots, N$, proportional to the apex angles of the cones in each plane. The frequency occurring in Eqs. (8.5.1) and (8.5.2) is the same that appears in Eq. (8.2.9). The values of $\theta$ that satisfy the determinant equation $\det \mathbf{A}_N = 0$ give the frequencies of the normal modes of the film. Because of the nature of the matrix $\mathbf{A}_N$, the vector $\mathbf{U}_N$ in Eq. (8.5.4) has a defined parity: its components are symmetric or antisymmetric with respect the centre of the multilayer. This fact allows us to reduce the number of independent equations of the linear system. Let us distinguish between the multilayers with an even or odd number of layers.

For $N$ *even*, the symmetric vector $\mathbf{U}_N^{(s)}$ has components satisfying the relationship $u_{\frac{N}{2}+l} = u_{\frac{N}{2}-l+1}$ with $l = 1, 2, \ldots, \frac{N}{2}$. Then Eq. (8.5.4) reduces to

$$\left[\frac{1}{\gamma_\perp} - 2\cos\theta^{(s)} + 1\right] u_1 = 0 \tag{8.5.5}$$

for $N = 2$ and

$$\mathbf{A}_{\frac{N}{2}}^{(s)} \times \mathbf{V}_{\frac{N}{2}} = 0 \tag{8.5.6}$$

for $N = 4$, 6, 8, etc. where

$$\mathbf{A}_{\frac{N}{2}}^{(s)} = \begin{pmatrix} \frac{1}{\gamma_\perp} - 2\cos\theta & 1 & 0 & \ldots & 0 & 0 \\ 1 & -2\cos\theta & 1 & \ldots & 0 & 0 \\ 0 & 1 & -2\cos\theta & \ldots & 0 & 0 \\ \vdots & \vdots & \vdots & \ddots & \vdots & \vdots \\ 0 & 0 & 0 & \ldots & 1 & (1 - 2\cos\theta) \end{pmatrix} \tag{8.5.7}$$

is a square matrix of dimension $\frac{N}{2}$ and $\mathbf{V}_{\frac{N}{2}}$ is a vector of components $u_1$, $u_2$, …, $u_{\frac{N}{2}}$. The determinant of the matrix (8.5.7) can be evaluated in a way similar to that of Section 8.2. One obtains

$$\begin{aligned} \det \mathbf{A}_{\frac{N}{2}}^{(s)} &= \frac{(-1)^{\frac{N}{2}}}{\gamma_\perp}\left[\left(\gamma_\perp - \frac{D_{\frac{N}{2}-1}}{D_{\frac{N}{2}}}\right)\left(D_{\frac{N}{2}} - D_{\frac{N}{2}-1}\right) - \frac{1}{D_{\frac{N}{2}}}\right] \\ &= \frac{(-1)^{\frac{N}{2}}}{\gamma_\perp \cos\frac{\theta^{(s)}}{2}}\left(\gamma_\perp \cos\frac{N+1}{2}\theta^{(s)} - \cos\frac{N-1}{2}\theta^{(s)}\right). \end{aligned} \tag{8.5.8}$$

The antisymmetric solution of Eq. (8.5.4) corresponds to a vector $\mathbf{U}_N^{(a)}$ whose components satisfy the relationship $u_{\frac{N}{2}+l} = -u_{\frac{N}{2}-l+1}$ with $l = 1, 2, \ldots, \frac{N}{2}$. Then Eq. (8.5.4) reduces to

$$\left[\frac{1}{\gamma_\perp} - 2\cos\theta^{(a)} - 1\right] u_1 = 0 \tag{8.5.9}$$

for $N = 2$ and

$$\mathbf{A}_{\frac{N}{2}}^{(a)} \times \mathbf{V}_{\frac{N}{2}} = 0 \tag{8.5.10}$$

for $N = 4$, 6, 8, etc. where

$$\mathbf{A}_{\frac{N}{2}}^{(a)} = \begin{pmatrix} \frac{1}{\gamma_\perp} - 2\cos\theta & 1 & 0 & \ldots & 0 & 0 \\ 1 & -2\cos\theta & 1 & \ldots & 0 & 0 \\ 0 & 1 & -2\cos\theta & \ldots & 0 & 0 \\ \vdots & \vdots & \vdots & \ddots & \vdots & \vdots \\ 0 & 0 & 0 & \ldots & 1 & -(1 + 2\cos\theta) \end{pmatrix}. \tag{8.5.11}$$

The determinant of the matrix (8.5.11) is given by

$$\det \mathbf{A}^{(a)}_{\frac{N}{2}} = \frac{(-1)^{\frac{N}{2}}}{\gamma_\perp}\left[\left(\gamma_\perp - \frac{D_{\frac{N}{2}-1}}{D_{\frac{N}{2}}}\right)\left(D_{\frac{N}{2}} - D_{\frac{N}{2}-1}\right) + \frac{1}{D_{\frac{N}{2}}}\right]$$

$$= \frac{(-1)^{\frac{N}{2}}}{\gamma_\perp \sin\frac{\theta^{(a)}}{2}}\left(\gamma_\perp \sin\frac{N+1}{2}\theta^{(a)} - \sin\frac{N-1}{2}\theta^{(a)}\right). \tag{8.5.12}$$

For $N$ *odd*, the symmetric vector $\mathbf{U}^{(s)}_N$ has components satisfying the relationship $u_{\frac{N+1}{2}+l} = u_{\frac{N+1}{2}-l}$ with $l = 1, 2, \ldots, \frac{N-1}{2}$. Then, Eq. (8.5.4) reduces to

$$\mathbf{A}^{(s)}_{\frac{N+1}{2}} \times \mathbf{V}_{\frac{N+1}{2}} = 0. \tag{8.5.13}$$

where

$$\mathbf{A}^{(s)}_{\frac{N+1}{2}} = \begin{pmatrix} \frac{1}{\gamma_\perp} - 2\cos\theta & 1 & 0 & \ldots & 0 & 0 \\ 1 & -2\cos\theta & 1 & \ldots & 0 & 0 \\ 0 & 1 & -2\cos\theta & \ldots & 0 & 0 \\ \vdots & \vdots & \vdots & \ddots & \vdots & \vdots \\ 0 & 0 & 0 & \ldots & 2 & -2\cos\theta \end{pmatrix} \tag{8.5.14}$$

is a square matrix of dimension $\frac{N+1}{2}$ and $\mathbf{V}_{\frac{N+1}{2}}$ is a vector of components $u_1$, $u_2, \ldots, u_{\frac{N+1}{2}}$. The determinant of the matrix (8.5.14) is given by

$$\det \mathbf{A}^{(s)}_{\frac{N+1}{2}} = \frac{(-1)^{\frac{N+1}{2}}}{\gamma_\perp}\left[\left(\gamma_\perp - \frac{D_{\frac{N-1}{2}}}{D_{\frac{N+1}{2}}}\right)\left(D_{\frac{N+1}{2}} - D_{\frac{N-3}{2}}\right) - \frac{2\cos\theta^{(s)}}{D_{\frac{N+1}{2}}}\right]$$

$$= (-1)^{\frac{N+1}{2}}\frac{2}{\gamma_\perp}\left(\gamma_\perp \cos\frac{N+1}{2}\theta^{(s)} - \cos\frac{N-1}{2}\theta^{(s)}\right). \tag{8.5.15}$$

The antisymmetric vector $\mathbf{U}^{(a)}_N$ has components satisfying the relationship $u_{\frac{N+1}{2}+l} = -u_{\frac{N+1}{2}-l}$ with $l = 1, 2, \ldots, \frac{N-1}{2}$ and $u_{\frac{N+1}{2}} = 0$. Then Eq. (8.5.4) reduces to Eq. (8.5.9) for $N = 3$ and to

$$\mathbf{A}^{(a)}_{\frac{N-1}{2}} \times \mathbf{V}_{\frac{N-1}{2}} = 0 \tag{8.5.16}$$

for $N = 5, 7, 9$, etc., where

$$\mathbf{A}^{(a)}_{\frac{N-1}{2}} = \begin{pmatrix} \frac{1}{\gamma_\perp} - 2\cos\theta & 1 & 0 & \ldots & 0 & 0 \\ 1 & -2\cos\theta & 1 & \ldots & 0 & 0 \\ 0 & 1 & -2\cos\theta & \ldots & 0 & 0 \\ \vdots & \vdots & \vdots & \ddots & \vdots & \vdots \\ 0 & 0 & 0 & \ldots & 1 & -2\cos\theta \end{pmatrix} \tag{8.5.17}$$

is a square matrix of dimension $\frac{N-1}{2}$ and $\mathbf{V}_{\frac{N-1}{2}}$ is a vector of components $u_1$, $u_2, \ldots, u_{\frac{N-1}{2}}$. The determinant of the matrix (8.5.17) is given by

$$\det \mathbf{A}^{(a)}_{\frac{N-1}{2}} = \frac{(-1)^{\frac{N-1}{2}}}{\gamma_\perp}\left(\gamma_\perp D_{\frac{N-1}{2}} - D_{\frac{N-3}{2}}\right)$$

$$= \frac{(-1)^{\frac{N}{2}}}{\gamma_\perp \sin\theta^{(a)}}\left(\gamma_\perp \sin\frac{N+1}{2}\theta^{(a)} - \sin\frac{N-1}{2}\theta^{(a)}\right). \qquad (8.5.18)$$

The frequencies of the normal modes of the film are obtained from Eq. (8.2.9) where $\theta$ are the solutions of the determinant equation $\det \mathbf{A} = 0$. Note that Eqs. (8.5.8) and (8.5.15) lead to the same symmetric spin wave frequencies. The same occurs for the antisymmetric spin wave frequencies of Eqs. (8.5.12) and (8.5.18).

Let us perform explicit calculations for small $N$. For $N = 2$ (bilayer), Eqs. (8.5.5) and (8.2.9) give

$$\omega^{(s)} = \omega_{\|} + \frac{2z_\perp J_\perp S}{\hbar}(1-\gamma_\perp), \quad u_1 = u_2 = u \qquad (8.5.19)$$

for the symmetric normal mode while Eqs. (8.5.9) and (8.2.9) give

$$\omega^{(a)} = \omega_{\|} + \frac{2z_\perp J_\perp S}{\hbar}(1+\gamma_\perp), \quad u_1 = -u_2 = u \qquad (8.5.20)$$

for the antisymmetric normal mode. The arbitrary constant $u$ appearing in Eqs. (8.5.19) and (8.5.20) can be fixed only giving the initial conditions. A snapshot of the spin configurations for the bilayer is shown in Fig. 8.1 where the index $n = 1, 2$ labels the two layers of the film and $\omega^{(s)}$ and $\omega^{(a)}$ indicate the normal mode frequencies corresponding to the symmetric and antisymmetric configurations. The apex angle of the cone is given by $\alpha = \arccos\frac{u}{S}$ and it is the same for each spin. The spins rotate about the $z$-axis with frequency $\omega^{(s)}$ or $\omega^{(a)}$ according to Eqs. (8.5.1)–(8.5.3). When the symmetric normal mode is excited, the spins in the two planes move in phase. When the antisymmetric normal mode is excited, the spins in the two planes move out of phase of an angle $\pi$. No qualitative difference exists between the spin waves in a bilayer and the ferromagnetic spin waves studied in Section 1.6 for a 3D lattice with PBC (see Eqs. (1.6.8)–(1.6.11)) except that in the bilayer, the allowed values of $k_z$, where the $z$-axis is chosen perpendicular to the film, have to be restricted to $k_z = 0$ and $k_z = \pi$.

For $N = 3$ (trilayer), the frequencies of the symmetric normal modes can be obtained from Eq. (8.5.15) that gives

$$2\gamma_\perp \cos^2\theta^{(s)} - \cos\theta^{(s)} - \gamma_\perp = 0. \qquad (8.5.21)$$

The two solutions of Eq. (8.5.21) lead to the frequencies

$$\omega_1^{(s)} = \omega_{\|} + \frac{z_\perp J_\perp S}{\hbar}\left(3 - \sqrt{1+8\gamma_\perp^2}\right) \qquad (8.5.22)$$

with

$$u_1 = u_3 = u, \quad u_2 = \frac{\sqrt{1+8\gamma_\perp^2}-1}{2\gamma_\perp}u \qquad (8.5.23)$$

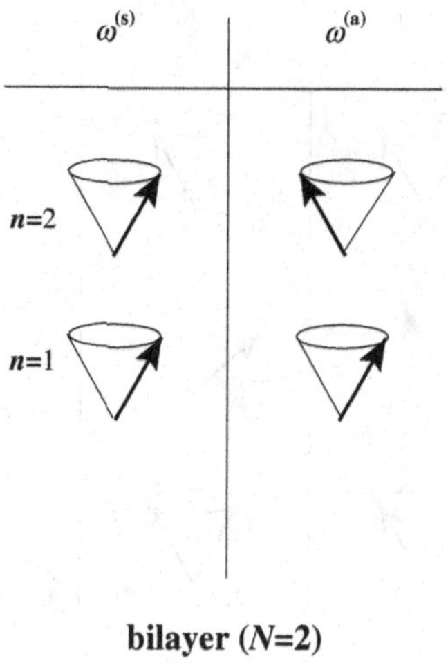


Fig. 8.1. Spin configuration of a bilayer.

and

$$\omega_2^{(s)} = \omega_\| + \frac{z_\perp J_\perp S}{\hbar}\left(3 + \sqrt{1 + 8\gamma_\perp^2}\right) \tag{8.5.24}$$

with

$$u_1 = u_3 = u, \quad u_2 = -\frac{\sqrt{1 + 8\gamma_\perp^2} + 1}{2\gamma_\perp} u. \tag{8.5.25}$$

Since $-1 \le \gamma_\perp \le 1$, in Eq. (8.5.23) $|u_2/u| \le 1$ so that for this mode the apex angle of the cone $\alpha_n = \arccos \frac{u_n}{S}$ *decreases* going from the surface $n = 1$ (or $n = 3$) to the internal layer $n = 2$: for this reason the corresponding normal mode is called[77] "monotonic surface magnon". On the contrary, in Eq. (8.5.25) $|u_2/u| \ge 1$ so that in this case the apex angle *increases* going from the surface to the centre of the film and the corresponding mode is called "bulk magnon". Equations (8.5.9) and (8.2.9) give the frequency of the antisymmetric normal mode

$$\omega^{(a)} = \omega_\| + \frac{2z_\perp J_\perp S}{\hbar}(1 + \gamma_\perp), \quad u_1 = -u_3 = u, \quad u_2 = 0. \tag{8.5.26}$$

The apex angle of the cone is the same on the two surfaces and is zero on the layer at the centre of the film: however, the spins on the two surfaces are out of phase of $\pi$. A snapshot of the spin configurations for the trilayer is shown in Fig. 8.2 where $n = 1, 2, 3$ labels the three layers of the film and $\omega_1^{(s)}$, $\omega_2^{(s)}$ and $\omega^{(a)}$ refer to the two symmetric and to the antisymmetric normal modes, respectively. In the trilayer, we have two monotonic surface magnons (one symmetric and one antisymmetric) and one (symmetric) bulk magnon. Note that the spin waves differ qualitatively from those of the 3D systems since the apex angle $\alpha_n$ of the cones swept by the spins changes from one layer to the other. For $\boldsymbol{k}_\| = 0$ ($\gamma_\perp = 1$), Eqs. (8.5.22) and (8.5.24)

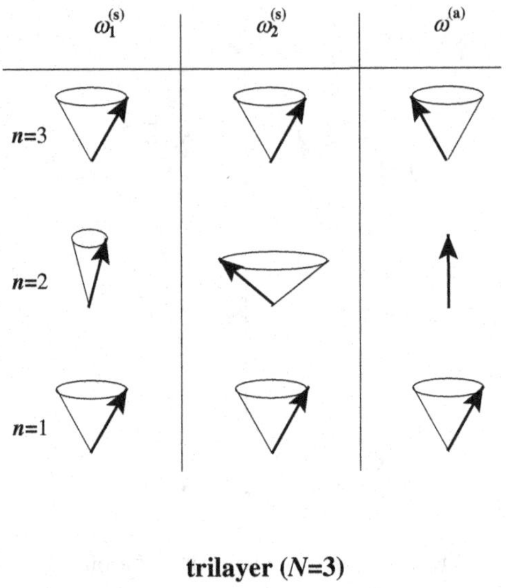


Fig. 8.2. Spin configuration of a trilayer.

and (8.5.26) reduce to

$$\omega_1^{(s)}(0) = 0, \quad u_1 = u_2 = u_3 = u, \tag{8.5.27}$$

$$\omega_2^{(s)}(0) = \frac{6z_\perp J_\perp S}{\hbar}, \quad u_1 = u_3 = u, \quad u_2 = -2u \tag{8.5.28}$$

and

$$\omega_1^{(a)}(0) = \frac{4z_\perp J_\perp S}{\hbar}, \quad u_1 = -u_3 = u, \quad u_2 = 0. \tag{8.5.29}$$

For $N = 4$ (four-layer), the frequencies of the symmetric normal modes can be obtained from Eq. (8.5.8) that gives

$$4\gamma_\perp \cos^2\theta^{(s)} - 2(1+\gamma_\perp)\cos\theta^{(s)} + 1 - \gamma_\perp = 0. \tag{8.5.30}$$

The two solutions of Eqs. (8.5.30) and (8.2.9) give the frequencies of the two symmetric spin waves

$$\omega_1^{(s)} = \omega_\parallel + \frac{z_\perp J_\perp S}{\hbar}\left(3 - \gamma_\perp - \sqrt{(1-\gamma_\perp)^2 + 4\gamma_\perp^2}\right) \tag{8.5.31}$$

with

$$u_1 = u_4 = u, \quad u_2 = u_3 = \frac{\sqrt{(1-\gamma_\perp)^2 + 4\gamma_\perp^2} - (1-\gamma_\perp)}{2\gamma_\perp} u \tag{8.5.32}$$

and

$$\omega_2^{(s)} = \omega_\parallel + \frac{z_\perp J_\perp S}{\hbar}\left(3 - \gamma_\perp + \sqrt{(1-\gamma_\perp)^2 + 4\gamma_\perp^2}\right) \tag{8.5.33}$$

with

$$u_1 = u_4 = u, \quad u_2 = u_3 = -\frac{\sqrt{(1-\gamma_\perp)^2 + 4\gamma_\perp^2} + (1-\gamma_\perp)}{2\gamma_\perp} u. \tag{8.5.34}$$

The frequencies of the antisymmetric normal modes are obtained from Eq. (8.5.12) that becomes

$$4\gamma_\perp \cos^2\theta^{(a)} - 2(1-\gamma_\perp)\cos\theta^{(a)} - (1+\gamma_\perp) = 0 \tag{8.5.35}$$

so that the two frequencies of the antisymmetric normal modes are

$$\omega_1^{(a)} = \omega_\parallel + \frac{z_\perp J_\perp S}{\hbar}\left(3 + \gamma_\perp - \sqrt{(1+\gamma_\perp)^2 + 4\gamma_\perp^2}\right) \tag{8.5.36}$$

with

$$u_1 = -u_4 = u, \quad u_2 = -u_3 = \frac{\sqrt{(1+\gamma_\perp)^2 + 4\gamma_\perp^2} - (1+\gamma_\perp)}{2\gamma_\perp} u \tag{8.5.37}$$

and

$$\omega_2^{(a)} = \omega_\parallel + \frac{z_\perp J_\perp S}{\hbar}\left(3 + \gamma_\perp + \sqrt{(1+\gamma_\perp)^2 + 4\gamma_\perp^2}\right) \tag{8.5.38}$$

with

$$u_1 = -u_4 = u, \quad u_2 = -u_3 = -\frac{\sqrt{(1+\gamma_\perp)^2 + 4\gamma_\perp^2} + (1+\gamma_\perp)}{2\gamma_\perp} u. \tag{8.5.39}$$

The spin configuration of the four-layer is shown in Fig. 8.3. Since $-1 \leq \gamma_\perp \leq 1$, in Eqs. (8.5.32) and (8.5.37) one has $|u_2/u| \leq 1$ and in Eqs. (8.5.34) and (8.5.39) one has $|u_2/u| \geq 1$ so that also for $N = 4$ we have two monotonic surface modes

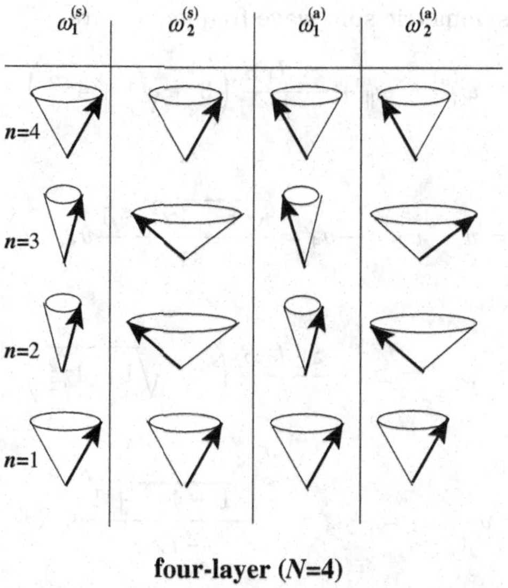


Fig. 8.3. Spin configuration of a four-layer.

(one symmetric and one antisymmetric) and two bulk magnons. Note that the frequencies of the surface magnons $\omega_1^{(s)}$ and $\omega_1^{(a)}$ are lower than the frequencies of the bulk magnons $\omega_2^{(s)}$ and $\omega_2^{(a)}$. For $\boldsymbol{k}_{\|} = 0$, Eqs. (8.5.31)–(8.5.34) and (8.5.36)–(8.5.39) reduce to

$$\omega_1^{(s)}(0) = 0, \quad u_1 = u_2 = u_3 = u_4 = u, \tag{8.5.40}$$

$$\omega_2^{(s)}(0) = \frac{4z_\perp J_\perp S}{\hbar}, \quad u_1 = u_4 = u, \quad u_2 = u_3 = -u, \tag{8.5.41}$$

$$\omega_1^{(a)}(0) = \frac{2z_\perp J_\perp S}{\hbar}(2 - \sqrt{2}), \quad u_1 = -u_4 = u, \quad u_2 = -u_3 = (\sqrt{2} - 1)u \tag{8.5.42}$$

and

$$\omega_2^{(a)}(0) = \frac{2z_\perp J_\perp S}{\hbar}(2 + \sqrt{2}), \quad u_1 = -u_4 = u, \quad u_2 = -u_3 = -(\sqrt{2} + 1)u. \tag{8.5.43}$$

For $N = 5$ (five-layer), the frequencies of the symmetric normal modes can be obtained from Eq. (8.5.15) that becomes

$$4\gamma_\perp \cos^3\theta^{(s)} - 2\cos^2\theta^{(s)} - 3\gamma_\perp \cos\theta^{(s)} + 1 = 0. \tag{8.5.44}$$

The solutions of the cubic equation (8.5.44) cannot be given in analytic form for a generic $\boldsymbol{k}_{\|}$ wavevector. Anyway, the numerical solution for several values of $\boldsymbol{k}_{\|}$ shows that the lowest frequency, say $\omega_1^{(s)}$, leads to a monotonic surface mode. The other two higher frequencies $\omega_2^{(s)}$ and $\omega_3^{(s)}$ correspond to bulk magnons. The frequencies of the antisymmetric normal modes are obtained from Eq. (8.5.18) that leads to

$$4\gamma_\perp \cos^2\theta^{(a)} - 2\cos\theta^{(a)} - \gamma_\perp = 0 \tag{8.5.45}$$

so that the two antisymmetric spin wave frequencies are

$$\omega_1^{(a)} = \omega_{\|} + \frac{z_\perp J_\perp S}{\hbar}\left(3 - \sqrt{1 + 4\gamma_\perp^2}\right) \tag{8.5.46}$$

with

$$u_1 = -u_5 = u, \quad u_2 = -u_4 = \frac{\sqrt{1 + 4\gamma_\perp^2} - 1}{2\gamma_\perp}u, \quad u_3 = 0 \tag{8.5.47}$$

and

$$\omega_2^{(a)} = \omega_{\|} + \frac{z_\perp J_\perp S}{\hbar}\left(3 + \sqrt{1 + 4\gamma_\perp^2}\right) \tag{8.5.48}$$

with

$$u_1 = -u_5 = u, \quad u_2 = -u_4 = -\frac{\sqrt{1 + 4\gamma_\perp^2} + 1}{2\gamma_\perp}u, \quad u_3 = 0. \tag{8.5.49}$$

Since $-1 \le \gamma_\perp \le 1$, in Eq. (8.5.47) one has $|u_2/u| \le 1$ so that the frequency $\omega_1^{(a)}$ corresponds to a monotonic surface magnon while in Eq. (8.5.49) one has $|u_2/u| \ge 1$

so that the frequency $\omega_2^{(a)}$ corresponds to a bulk magnon. For $\boldsymbol{k}_\parallel = 0$, Eq. (8.5.44) gives the solutions

$$\omega_1^{(s)}(0) = 0, \quad u_1 = u_2 = u_3 = u, \tag{8.5.50}$$

$$\omega_2^{(s)}(0) = \frac{z_\perp J_\perp S}{\hbar}(5 - \sqrt{5}), \quad u_1 = u, \quad u_2 = -\frac{3-\sqrt{5}}{2}u, \quad u_3 = -(\sqrt{5}-1)u \tag{8.5.51}$$

and

$$\omega_3^{(s)}(0) = \frac{z_\perp J_\perp S}{\hbar}(5 + \sqrt{5}), \quad u_1 = u, \quad u_2 = -\frac{3+\sqrt{5}}{2}u, \quad u_3 = (\sqrt{5}+1)u, \tag{8.5.52}$$

while Eqs. (8.5.46)–(8.5.49) reduce to

$$\omega_1^{(a)}(0) = \frac{z_\perp J_\perp S}{\hbar}(3 - \sqrt{5}), \quad u_1 = u, \quad u_2 = \frac{\sqrt{5}-1}{2}u, \quad u_3 = 0 \tag{8.5.53}$$

and

$$\omega_2^{(a)} = \frac{z_\perp J_\perp S}{\hbar}(3 + \sqrt{5}), \quad u_1 = u, \quad u_2 = -\frac{\sqrt{5}+1}{2}u, \quad u_3 = 0. \tag{8.5.54}$$

For a BCC lattice with the wavevector $\boldsymbol{k}_\parallel$ along the $(1,1,0)$ direction, one has $\gamma_\parallel = \cos k_\parallel$ and $\gamma_\perp = \cos^2 \frac{k_\parallel}{2}$. The spin wave frequencies are shown in the left panel of Fig. 8.4, assuming $J_\perp = J_\parallel$. There are three symmetric (s) spin waves frequencies (continuous curves) and two antisymmetric (a) spin waves frequencies (dashed curves). The lowest energy branches (s,1) and (a,1) are monotonic surface magnons. All other branches correspond to bulk spin waves. For $N = 6$ we obtain similar results. The spin wave frequencies are obtained from Eq. (8.5.8) (symmetric normal modes) and from Eq. (8.5.12) (antisymmetric normal modes). For a BCC lattice with the wavevector $\boldsymbol{k}_\parallel$ along the direction (1,1,0), the spin wave frequencies are shown in the right panel of Fig. 8.4 for $J_\perp = J_\parallel$. The symmetric normal mode frequencies are given by continuous curves and the antisymmetric normal mode frequencies are given by dashed curves. The lowest two branches (s,1) and (a,1) are monotonic surface magnons. For $\boldsymbol{k}_\parallel = 0$, the symmetric spin wave frequencies become

$$\omega_1^{(s)}(0) = 0, \quad u_1 = u_2 = u_3 = u, \tag{8.5.55}$$

$$\omega_2^{(s)}(0) = \frac{2z_\perp J_\perp S}{\hbar}, \quad u_1 = -u_3 = u, \quad u_2 = 0 \tag{8.5.56}$$

and

$$\omega_3^{(s)}(0) = \frac{3z_\perp J_\perp S}{\hbar}, \quad u_1 = u_3 = u, \quad u_2 = -2u. \tag{8.5.57}$$

The antisymmetric spin wave frequencies are

$$\omega_1^{(a)} = \frac{2z_\perp J_\perp S}{\hbar}(2 - \sqrt{3}), \quad u_1 = u, \quad u_2 = (\sqrt{3}-1)u, \quad u_3 = (2-\sqrt{3})u, \tag{8.5.58}$$

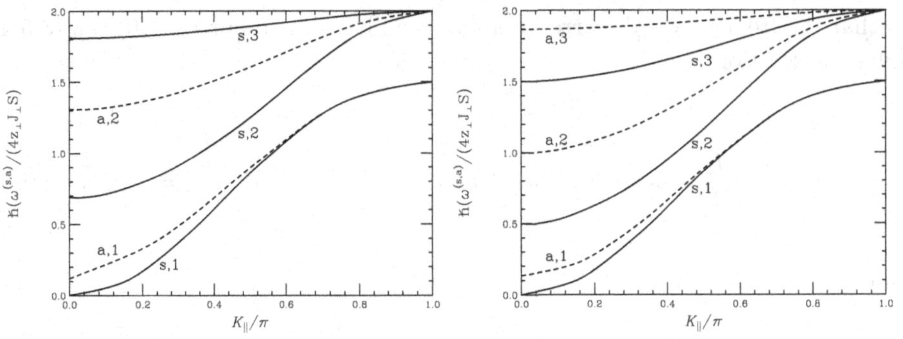


Fig. 8.4. Spin wave frequency in a film made up of five layers (left panel) and six layers (right panel) of a BCC structure with $J_\perp = J_\parallel$. The wavevector $\boldsymbol{k}_\parallel$ is along the (1,1,0) direction. The labels (s) and (a) indicate the symmetric (continuous curves) and antisymmetric (dashed curves) normal modes, respectively.

$$\omega_2^{(a)}(0) = \frac{4z_\perp J_\perp S}{\hbar}, \quad u_1 = -u_2 = -u_3 = u \tag{8.5.59}$$

and

$$\omega_3^{(a)}(0) = \frac{2z_\perp J_\perp S}{\hbar}(2+\sqrt{3}), \quad u_1 = u, \quad u_2 = -(\sqrt{3}+1)u, \quad u_3 = (2+\sqrt{3})u. \tag{8.5.60}$$

The existence of a soft mode at $\boldsymbol{k}_\parallel = 0$ for any film thickness reflects the rotation invariance of the Heisenberg Hamiltonian. This soft mode disappears, introducing a surface anisotropy as could be the case when the film is grown over a substrate of another material.

Now, we are able to investigate the spin wave frequencies for a film made up of an arbitrary number of layers. For $N$ even the spin wave frequencies are given from Eq. (8.2.9) where $\theta$ is obtained from Eq. (8.5.8) in the symmetric case and from Eq. (8.5.12) in the antisymmetric case. In particular, the lowest symmetric spin wave frequency is obtained from Eq. (8.2.9) where $\theta_1^{(s)}$ is the solution of the equation

$$\frac{\gamma_\perp \cosh \frac{N+1}{2}\theta_1^{(s)} - \cosh \frac{N-1}{2}\theta_1^{(s)}}{\cosh \frac{\theta_1^{(s)}}{2}} = 0. \tag{8.5.61}$$

The remaining $\frac{N}{2} - 1$ symmetric spin wave frequencies are still obtained from Eq. (8.2.9) where $\theta_l^{(s)}$ are the solutions of the equation

$$\frac{\gamma_\perp \cos \frac{N+1}{2}\theta_l^{(s)} - \cos \frac{N-1}{2}\theta_l^{(s)}}{\cos \frac{\theta_l^{(s)}}{2}} = 0. \tag{8.5.62}$$

In the antisymmetric case, one has to distinguish between the case

i) $\frac{N-1}{N+1} < \gamma_\perp < 1$ corresponding to magnons whose wavevectors belong to a region around the zone centre ($0 < k_\parallel < 2 \arccos \sqrt{\frac{N-1}{N+1}}$ for a BCC lattice along (1,1,0) direction) and the case

ii) $\gamma_\perp < \frac{N-1}{N+1}$ corresponding to magnons whose wavevectors are in a region around the ZC ($2\arccos\sqrt{\frac{N-1}{N+1}} < k_\| < \pi$ for a BCC lattice along (1,1,0) direction).

In the case i), the $\frac{N}{2}$ antisymmetric spin wave frequencies are obtained from Eq. (8.2.9) where $\theta_l^{(a)}$ are the solutions of the equation

$$\frac{\gamma_\perp \sin\frac{N+1}{2}\theta_l^{(a)} - \sin\frac{N-1}{2}\theta_l^{(a)}}{\sin\frac{\theta_l^{(a)}}{2}} = 0. \tag{8.5.63}$$

In the case ii), $\frac{N}{2} - 1$ spin wave frequencies are still given by Eq. (8.5.63); however, the remaining (lowest) frequency is obtained from Eq. (8.2.9) where $\theta_1^{(a)}$ is the solution of the equation

$$\frac{\gamma_\perp \sinh\frac{N+1}{2}\theta_1^{(a)} - \sinh\frac{N-1}{2}\theta_1^{(a)}}{\sinh\frac{\theta_1^{(a)}}{2}} = 0. \tag{8.5.64}$$

The existence of two distinct cases originates from the nature of the function in the left-hand side of Eq. (8.5.64). Indeed, it is an increasing monotonic function of $\theta$ whose value at $\theta = 0$ is positive in the case i) and negative in the case ii) implying that Eq. (8.5.64) has no solution in the former case and one solution in the latter case. It is worthwhile noticing that the eigenvectors belonging to $\theta_1^{(s)}$ and $\theta_1^{(a)}$ have components whose magnitude decreases going from $n = 1$ (surface) to the centre of the film $n = \frac{N}{2}$ so that these two modes are referred as[77] "monotonic surface magnons" in contrast with the remaining $N-2$ symmetric and antisymmetric modes that are classified as "bulk magnons".

For $N$ odd, the symmetric and antisymmetric spin wave frequencies are obtained from Eqs. (8.5.15) and (8.5.18), respectively. As for the symmetric case, $\theta_1^{(s)}$ is the solution of the equation

$$\gamma_\perp \cosh\frac{N+1}{2}\theta_1^{(s)} - \cosh\frac{N-1}{2}\theta_1^{(s)} = 0 \tag{8.5.65}$$

and the remaining $\frac{N-1}{2}$ spin wave frequencies are given by Eq. (8.2.9) where $\theta_l^{(s)}$ are the solutions of the equation

$$\gamma_\perp \cos\frac{N+1}{2}\theta_l^{(s)} - \cos\frac{N-1}{2}\theta_l^{(s)} = 0. \tag{8.5.66}$$

As before, in the antisymmetric case one has to distinguish between the two cases

i) $\frac{N-1}{N+1} < \gamma_\perp < 1$ and

ii) $\gamma_\perp < \frac{N-1}{N+1}$.

In the case i), all the $\frac{N-1}{2}$ antisymmetric spin wave frequencies are obtained from the $\theta_l^{(a)}$ that satisfy the equation

$$\frac{\gamma_\perp \sin \frac{N+1}{2}\theta_l^{(a)} - \sin \frac{N-1}{2}\theta_l^{(a)}}{\sin \theta_l^{(a)}} = 0. \tag{8.5.67}$$

In the case ii), $\frac{N-3}{2}$ spin wave frequencies are still obtained from Eq. (8.5.67) while the remaining (lowest) frequency $\theta_1^{(a)}$ is obtained from the equation

$$\frac{\gamma_\perp \sinh \frac{N+1}{2}\theta_1^{(a)} - \sinh \frac{N-1}{2}\theta_1^{(a)}}{\sinh \theta_1^{(a)}} = 0. \tag{8.5.68}$$

Also, for $N$ odd, the eigenvectors belonging to $\theta_1^{(s)}$ and $\theta_1^{(a)}$ have components decreasing with $n$ going from $n = 1$ (surface) to the centre of the film $n = \frac{N+1}{2}$ so that the existence of two "monotonic surface magnons" is a peculiarity of a BCC film with an arbitrary number of layers. For $\boldsymbol{k}_\| = 0$ ($\gamma_\perp = \gamma_\| = 1$), the spin wave frequencies are given by

$$\omega_l(0) = \frac{4z_\perp J_\perp S}{\hbar}(1 - \cos\theta_l) \tag{8.5.69}$$

with $\theta_l = \theta_l^{(s)} = \frac{2\pi l}{N}$ for the symmetric modes and $\theta_l = \theta_l^{(a)} = \frac{(2l+1)\pi}{N}$ for the antisymmetric modes. The soft mode corresponding to $l = 0$ in Eq. (8.5.69) leads to an eigenvector with components given by $u_n = u$: in this particular case, the monotonic surface magnon reduces to a conventional ferromagnetic spin wave with all spins parallel along a direction forming an angle $\alpha = \arccos \frac{u}{S}$ with the $z$-axis.

For SC and H lattices ($z_\perp = \gamma_\perp = 1$), an analytic result can be achieved. The symmetric spin wave frequencies are given by

$$\omega_l^{(s)} = \omega_\| + \frac{4J_\perp S}{\hbar}\left(1 - \cos\frac{2\pi l}{N}\right) \tag{8.5.70}$$

and the corresponding eigenvectors components $u_n^l$ are given by

$$u_n^l = u\,\frac{\cos\left[\frac{\pi l}{N}(N+1-2n)\right]}{\cos\frac{\pi l}{N}} \tag{8.5.71}$$

for $N$ even, where $l = 0, 1, \ldots, \frac{N}{2} - 1$ labels the $\frac{N}{2}$ symmetric normal modes and $u_{\frac{N}{2}+n} = u_{\frac{N}{2}-n+1}$. For $N$ odd, the frequencies are still given by Eq. (8.5.70) while the eigenvectors components are given by

$$u_n^l = u\,\cos\left[\frac{\pi l}{N}(N+1-2n)\right] \tag{8.5.72}$$

with $l = 0, 1, \ldots, \frac{N-1}{2}$ and $n = 1, 2, \ldots, \frac{N+1}{2}$.

The antisymmetric spin wave frequencies are given by

$$\omega_l^{(a)} = \omega_\parallel + \frac{4J_\perp S}{\hbar}\left[1 - \cos\frac{(2l+1)\pi}{N}\right] \tag{8.5.73}$$

and the corresponding eigenvectors components are given by

$$u_n^l = u\,\frac{\sin\left[\left(l+\frac{1}{2}\right)\frac{\pi}{N}(N+1-2n)\right]}{\sin\left(l+\frac{1}{2}\right)\frac{\pi}{N}} \tag{8.5.74}$$

for $N$ even, with $l = 0, 1, \ldots, \frac{N}{2}-1$, $n = 1, 2, \ldots, \frac{N}{2}$ and $u_{\frac{N}{2}+n} = -u_{\frac{N}{2}-n+1}$. For $N$ odd, the antisymmetric spin wave frequencies are still given by Eq. (8.5.73) while the eigenvectors components are given by

$$u_n^l = u\,\frac{\sin\left[\left(l+\frac{1}{2}\right)\frac{\pi}{N}(N+1-2n)\right]}{\sin\frac{(2l+1)\pi}{N}} \tag{8.5.75}$$

with $n = 1, \ldots, \frac{N-1}{2}$ and $l = 0, 1, \ldots, \frac{N-3}{2}$. Obviously, the remaining components are given by $u_{\frac{N+1}{2}+n} = -u_{\frac{N+1}{2}-n}$ and $u_{\frac{N+1}{2}} = 0$.

## 8.6. Classical Spin Waves in a Semi-Infinite Medium

In order to investigate how the presence of a surface modifies the spin waves of a 3D lattice, we study a model made up of a semi-infinite stacking of layers.[78] For a semi-infinite medium we can write a matrix equation like (8.5.4) where **A** and **U** are now an *infinite* square matrix and a vector with an infinite number of components, respectively. In particular, the linear system for the amplitudes $u_n$ becomes

$$(1 - 2\lambda\gamma_\perp)u_1 + \gamma_\perp u_2 = 0 \tag{8.6.1}$$

for the surface ($n = 1$) with

$$\omega = 2z_\parallel J_\parallel S(1-\gamma_\parallel) + 4z_\perp J_\perp S(1-\lambda\gamma_\perp) \tag{8.6.2}$$

and

$$-2\lambda\gamma_\perp u_n + u_{n-1} + u_{n+1} = 0 \tag{8.6.3}$$

for the bulk ($n > 1$) where $\gamma_\parallel$ and $\gamma_\perp$ are given by Eqs. (8.2.7) and (8.2.8), respectively. We point out that Eq. (8.6.3) represents a set of *infinite* linear equations at variance with the multilayer case where only a *finite* number of linear equations ($N-2$) for the bulk and one further equation for the second surface were required. The complication entered by the infinite number of equations, however, is balanced by the absence of the second surface equation so that a solution can be easily found. Indeed, for the surface mode we assume

$$u_n = u\,e^{-(n-1)\phi} \tag{8.6.4}$$

where $\phi$ is a parameter to be determined imposing that Eq. (8.6.4) is solution of the system (8.6.1)–(8.6.3). Replacing Eq. (8.6.4) into Eqs. (8.6.3) and (8.6.1), we find

$$\lambda = \frac{1}{2}(e^{\phi} + e^{-\phi}), \quad e^{\phi} = \frac{1}{\gamma_\perp} \tag{8.6.5}$$

so that the surface magnon frequency becomes

$$\omega = 2z_\parallel J_\parallel S(1 - \gamma_\parallel) + 2z_\perp J_\perp S(1 - \gamma_\perp^2) \tag{8.6.6}$$

and the amplitudes become $u_n = u(\gamma_\perp)^{n-1}$. Since $|\gamma_\perp| \leq 1$, the amplitudes $u_n$ decrease moving away from the surface, confirming that the spin wave with frequency (8.6.6) is really the "surface magnon" of the semi-infinite medium.

To recover the "bulk magnons", we look for the other solutions of the system (8.6.1)–(8.6.3). To do this, let us define $\lambda = \cos\theta$ and look for the zeros of the determinant of the infinite matrix $\mathbf{A}_\infty$ given by

$$\mathbf{A}_\infty = \begin{pmatrix} 2\cos\theta - \dfrac{1}{\gamma_\perp} & -1 & 0 & \dots \\ -1 & 2\cos\theta & -1 & \dots \\ 0 & -1 & -2\cos\theta & \dots \\ \vdots & \vdots & \vdots & \ddots \end{pmatrix}. \tag{8.6.7}$$

For finite $N$, the determinant of the square matrix $\mathbf{A}_N$ is obtained by expanding along the first row. One has

$$\det \mathbf{A}_N = \left(2\cos\theta - \frac{1}{\gamma_\perp}\right) D_{N-1} - D_{N-2} \tag{8.6.8}$$

with $D_N$ given by Eq. (8.2.17). Then the zeros of the determinant of $\mathbf{A}_N$ are given by the solutions of the equation

$$\tan N\theta = \frac{\gamma_\perp \sin\theta}{1 - \gamma_\perp \cos\theta} \tag{8.6.9}$$

with $0 < \theta < \pi$. The left-hand function of Eq. (8.6.9) is a function of $\theta$ with a series of asymptotes at $\theta = (l + \frac{1}{2})\frac{\pi}{N}$ while the right-hand function is a positive function with a maximum at $\theta_m = \arccos(\gamma_\perp)$ going to zero at $\theta = 0$ and $\pi$. Each solution of Eq. (8.6.9) is restricted to the interval $(l - \frac{1}{2})\frac{\pi}{N} < \theta_l < (l + \frac{1}{2})\frac{\pi}{N}$. For $N \to \infty$, the solutions become $\theta_l \to \frac{\pi}{N} l$ so that the bulk spin wave excitation energies of a semi-infinite system are given by

$$\hbar\omega = 2z_\parallel J_\parallel S(1 - \gamma_\parallel) + 4z_\perp J_\perp S\left(1 - \gamma_\perp \cos\frac{\pi l}{N}\right) \tag{8.6.10}$$

with $l$ integer and $N \to \infty$. The amplitudes can be obtained directly by the system (8.6.1)–(8.6.3), that is,

$$\begin{aligned} u_1 &= u, \quad u_2 = \frac{u}{\gamma_\perp}(2\gamma_\perp \cos\theta - 1), \\ u_3 &= \frac{u}{\gamma_\perp}(4\gamma_\perp \cos^2\theta - 2\cos\theta - \gamma_\perp), \\ u_4 &= \frac{u}{\gamma_\perp}(8\gamma_\perp \cos^3\theta - 4\cos^2\theta - 4\gamma_\perp \cos\theta + 1), \end{aligned} \tag{8.6.11}$$

and so on.

For the SC and H lattices ($\gamma_\perp = z_\perp = 1$), one has

$$u_1 = u, \quad u_n = \frac{\cos \frac{2n-1}{2}\theta}{\cos \frac{\theta}{2}} u \tag{8.6.12}$$

with $n > 1$. Notice that the well-known spin wave frequencies of the corresponding 3D lattice with PBC along all the three directions is not recovered. Indeed, the 3D spin waves for SC and H lattices with PBC are characterized by energies

$$\hbar\omega = 2z_\parallel J_\parallel S(1 - \gamma_\parallel) + 4J_\perp S(1 - \cos k_z) \tag{8.6.13}$$

with $k_z = \frac{2\pi l}{N}$ and constant amplitudes $u_n = u$. This result should have been expected since the presence of a surface (semi-infinite system) or two surfaces (film) is incompatible with the PBC that assure that each spin interacts in the same way with its NN. Indeed, the spins belonging to a surface interact only with their NN spins on the same surface and the NN spins of the bulk so that the equations of motion of the spins on the surfaces differ from the equation of motion of the spins in the bulk. Consequently, solutions like "surface magnons" appear and the "bulk magnons" are no longer characterized by uniform amplitudes. It is not surprising that the dynamics is deeply affected by the boundary conditions even though the thermal properties of macroscopic samples are not.

# Chapter 9

# SPIN WAVES IN SYSTEMS WITH LONG RANGE INTERACTION

## 9.1. Dipole–Dipole Interaction

In the previous chapters, we have studied magnetic systems where the spins are interacting via an exchange interaction that was not extended beyond a few shell of neighbours. This is well justified for the magnetic compounds with a high critical temperature indicating that the interaction between the spins originates from the exchange interaction. In these compounds, any other magnetic interaction can be neglected if one is interested to investigate their low temperature magnetic properties. However, the long range dipole–dipole interaction is always present in magnetic compounds and its effect becomes important in the vicinity of the critical point where the choice of the "universality class" in a continuous phase transition depends on the short or long range nature of the interaction. On the other hand, when the exchange interaction is very weak (comparable with the dipolar interaction) as for some quasi-1D or quasi-2D magnetic systems, the dipole–dipole interaction plays a crucial role in determining the magnetic LRO. Such magnetic systems are characterized by very low magnetic critical temperature as for some HTCS[61] like $RBa_2Cu_3O_{6+x}$ where[79] R is a rare earth (R = Dy, Er, Nd) that show an ordered magnetic phase at temperatures below $T_N \sim 1\,K$ depending on the oxygenation $x$. For instance,[79] the fully oxygenated $ErBa_2Cu_3O_7$ ($x = 1$) is a superconductor ($T_c = 92\,K$) with a Néel temperature $T_N = 0.62\,K$ while the deoxygenated $ErBa_2Cu_3O_6$ ($x = 0$) is an insulator without long range 3D magnetic order. In the fully oxygenated sample ($x = 1$), the magnetic $Er^{3+}$ ions crystallize in an orthorhombic lattice[79] with $a = 3.82\,Å$, $b = 3.88\,Å$ and $c = 11.66\,Å$ while in the deoxygenate sample ($x \simeq 0$) the $Er^{3+}$ ions crystallize[80] in a tetragonal lattice with $a = b = 3.85\,Å$ and $c = 11.79\,Å$. In the ordered magnetic phase of the oxygenated sample, the magnetic moments of the $Er^{3+}$ ions are aligned along the $b$-axis forming ferromagnetic chains along the $b$-axis and antiferromagnetic chains along the $a$-axis. Along the $c$-axis, both ferromagnetic and antiferromagnetic orders have been found.[79] Either the very low Néel temperature and the kind of order found in the ordered phase cannot be explained in terms of a NN exchange Hamiltonian.

As an example of quasi-1D ferromagnet, we point out the compound[81] $CsNiF_3$, a member of the large family of magnetic systems $ABX_3$ where A is an alkali metal, B a transition metal and X a halogen. The magnetic $Ni^{2+}$ ($S = 1$) ions crystallize in the H lattice with $a = b = 6.21\,\text{Å}$ and $c = 2.9\,\text{Å}$ and are ferromagnetically coupled along the $c$-axis (chain) by an exchange interaction $J/k_B = 11.8\,\text{K}$. The antiferromagnetic coupling between the chains $J'$ is about 500 times smaller than the intrachain coupling so that it is of the same order of magnitude of the dipolar interaction. The crucial role of the dipole–dipole interaction is pointed out by the very low transition temperature ($T_N = 2.67\,\text{K}$) and by the kind of order found in the ordered phase. Indeed, the antiferromagnetic order found in the $ab$-planes does not correspond to the expected 120°-phase of a NN Heisenberg antiferromagnet.

From a theoretical point of view, the dipolar interaction is interesting for two reasons:

i) the long range nature of the dipole–dipole interaction rules out the application of the Mermin and Wagner theorem[22] according to which the LRO is prevented at any finite temperature for Heisenberg or planar rotator models with short range interaction;
ii) the anisotropic nature of the dipole–dipole interaction changes the symmetry of the isotropic Hamiltonian choosing an easy-axis (Ising model) or an easy-plane (planar model) depending on the lattice structure.

The dipolar Hamiltonian is given by

$$\mathcal{H}_{\text{dip}} = -\frac{1}{2}\sum_{i\neq j}\frac{1}{r_{ij}^3}\left[3\frac{(\boldsymbol{\mu}_i\cdot\boldsymbol{r}_{ij})(\boldsymbol{\mu}_j\cdot\boldsymbol{r}_{ij})}{r_{ij}^2} - \boldsymbol{\mu}_i\cdot\boldsymbol{\mu}_j\right], \tag{9.1.1}$$

where $i, j$ run over all sites of the lattice, $\boldsymbol{r}_{ij}$ is the lattice vector joining the points $i$ and $j$ and $\boldsymbol{\mu}_i$ is the magnetic moment located at site $i$. Replacing the magnetic moments $\boldsymbol{\mu}_i$ by $g\mu_B\boldsymbol{S}_i$ where $g$ is the Landé factor and $\mu_B$ the Bohr magneton and writing the lattice site $\boldsymbol{r}_j$ as $\boldsymbol{r}_i + \boldsymbol{r}$, the Hamiltonian (9.1.1) becomes

$$\mathcal{H}_{\text{dip}} = -\frac{(g\mu_B)^2}{2}\sum_{\alpha,\beta}\sum_{i}\sum_{\boldsymbol{r}\neq 0}\frac{1}{r^3}\left[3\frac{r_\alpha r_\beta}{r^2} - \delta_{\alpha,\beta}\right] S_i^\alpha S_{i+\boldsymbol{r}}^\beta, \tag{9.1.2}$$

where $\alpha, \beta$ run over the cartesian crystallographic axes $x, y, z$. If the lattice is supposed to be infinite, the choice of the origin from which to measure the lattice vector $\boldsymbol{r}$ is independent of $i$ so that the sum over $\boldsymbol{r}$ is similar to the sum over $i$ except that $\boldsymbol{r} \neq 0$. In the following, we limit ourselves to account for spiral configurations. Then we introduce a reference system of local axes $\xi, \eta, \zeta$ in the same way as in Chapter 7. The ground-state spin configuration is then given by

$$\begin{aligned} S_i^x &= S\sin\theta\cos\,\phi_i, \\ S_i^y &= S\sin\theta\sin\,\phi_i, \\ S_i^z &= S\cos\theta \end{aligned} \tag{9.1.3}$$

with $\phi_i = \boldsymbol{Q}\cdot\boldsymbol{r}_i + \phi$ and the corresponding ground-state energy is

$$E_{\text{dip}} = -\frac{(g\mu_B S)^2}{2}\sum_i\sum_{\boldsymbol{r}\neq 0}\frac{1}{r^3}\left\{\left(3\frac{r_z^2}{r^2}-1\right)\cos^2\theta + \left[3\frac{r_x r_y}{r^2}\sin(\phi_i+\phi_{i+\boldsymbol{r}})\right.\right.$$
$$\left. + \left(3\frac{r_x^2}{r^2}-1\right)\cos\phi_i\cos\phi_{i+\boldsymbol{r}} + \left(3\frac{r_y^2}{r^2}-1\right)\sin\phi_i\sin\phi_{i+\boldsymbol{r}}\right]\sin^2\theta$$
$$\left. + \frac{3r_z}{r^2}[r_x\cos(\phi_i+\phi_{i+\boldsymbol{r}}) + r_y(\sin\phi_i+\sin\phi_{i+\boldsymbol{r}})]\sin\theta\cos\theta\right\}. \qquad (9.1.4)$$

By using the relationship

$$\sum_i e^{i\boldsymbol{Q}\cdot\boldsymbol{r}_i} = N\delta_{\boldsymbol{Q},\boldsymbol{G}}, \qquad (9.1.5)$$

where $\boldsymbol{G}$ is a reciprocal lattice vector, the sum over $i$ in Eq. (9.1.4) can be performed leading to

$$E_{\text{dip}}(\theta,\boldsymbol{Q},\phi) = -\frac{(g\mu_B S)^2}{2}N\left\{D^{zz}(0)\cos^2\theta + \left[\frac{1}{2}\left(D^{xx}\left(\frac{\boldsymbol{G}}{2}\right)\right.\right.\right.$$
$$\left. - D^{yy}\left(\frac{\boldsymbol{G}}{2}\right)\right)\delta_{\boldsymbol{Q},\frac{\boldsymbol{G}}{2}}\cos 2\phi + \frac{1}{2}(D^{xx}(\boldsymbol{Q}) + D^{yy}(\boldsymbol{Q}))$$
$$\left. + D^{xy}\left(\frac{\boldsymbol{G}}{2}\right)\delta_{\boldsymbol{Q},\frac{\boldsymbol{G}}{2}}\sin 2\phi\right]\sin^2\theta + 2\left(D^{xz}(0)\cos\phi\right.$$
$$\left.\left. + D^{yz}(0)\sin\phi\right)\delta_{\boldsymbol{Q},0}\sin\theta\cos\theta\right\}, \qquad (9.1.6)$$

where

$$D^{\alpha\beta}(\boldsymbol{Q}) = \sum_{\boldsymbol{r}\neq 0}\frac{e^{i\boldsymbol{Q}\cdot\boldsymbol{r}}}{r^3}\left(3\frac{r_\alpha r_\beta}{r^2} - \delta_{\alpha,\beta}\right). \qquad (9.1.7)$$

Equation (9.1.6) is correct for Bravais lattices for which $D^{\alpha\beta}(\boldsymbol{Q}) = D^{\alpha\beta}(-\boldsymbol{Q})$. From Eq. (9.1.7), it is direct to obtain the "sum rule"

$$D^{xx}(\boldsymbol{Q}) + D^{yy}(\boldsymbol{Q}) + D^{zz}(\boldsymbol{Q}) = 0 \qquad (9.1.8)$$

and $D^{\alpha\beta}(0) = 0$ for any $\alpha \neq \beta$. Using the sum rule (9.1.8), the dipolar energy (9.1.6) becomes

$$E_{\text{dip}}(\theta,\boldsymbol{Q},\phi) = -\frac{(g\mu_B S)^2}{2}N\left\{D^{zz}(0)\cos^2\theta - \frac{1}{2}D^{zz}(\boldsymbol{Q})\sin^2\theta\right.$$
$$+ \frac{1}{2}\left[\left(D^{xx}\left(\frac{\boldsymbol{G}}{2}\right) - D^{yy}\left(\frac{\boldsymbol{G}}{2}\right)\right)\cos 2\phi\right.$$
$$\left.\left. + D^{xy}\left(\frac{\boldsymbol{G}}{2}\right)\sin 2\phi\right]\delta_{\boldsymbol{Q},\frac{\boldsymbol{G}}{2}}\sin^2\theta\right\}, \qquad (9.1.9)$$

where the parameters $\theta$, $\boldsymbol{Q}$ and $\phi$ have to be determined minimizing the ground-state energy (9.1.9). Unlike the isotropic exchange Hamiltonian (7.1.4), in the ground-state energy (9.1.9), the angle $\phi$ does appear pointing out the anisotropic nature of the dipole–dipole interaction. The possible minima of the function (9.1.9) correspond to

i) ferromagnetic configuration with the spins directed along the $x$-axis ($\theta = \frac{\pi}{2}$, $\phi = 0$, $\boldsymbol{Q} = 0$) or along the $y$-axis ($\theta = \frac{\pi}{2}$, $\phi = \frac{\pi}{2}$, $\boldsymbol{Q} = 0$) or along the $z$-axis ($\theta = 0$) whose energy is given by

$$E^{\alpha}_{\rm dip}(0) = -\frac{(g\mu_B S)^2}{2} N D^{\alpha\alpha}(0) \tag{9.1.10}$$

with $\alpha = x, y, z$, respectively;

ii) antiferromagnetic configuration with the spins directed along the $x$-axis ($\theta = \frac{\pi}{2}$, $\phi = 0$, $\boldsymbol{Q} = \frac{\boldsymbol{G}}{2}$) or along the $y$-axis ($\theta = \frac{\pi}{2}$, $\phi = \frac{\pi}{2}$, $\boldsymbol{Q} = \frac{\boldsymbol{G}}{2}$) whose energy is given by

$$E^{\alpha}_{\rm dip}\left(\frac{\boldsymbol{G}}{2}\right) = -\frac{(g\mu_B S)^2}{2} N D^{\alpha\alpha}\left(\frac{\boldsymbol{G}}{2}\right) \tag{9.1.11}$$

with $\alpha = x, y$, respectively. As one can see from Eqs. (9.1.10) and (9.1.11), the dipolar interaction selects both the kind of magnetic order and the spin direction of the ground-state configuration. When $\boldsymbol{Q} \neq 0$, the long range character of the dipole–dipole interaction does not affect dramatically the sums (9.1.7) since the oscillating nature of the cosine function balances the long range of the dipolar interaction. On the contrary, for $\boldsymbol{Q} = 0$ the sums (9.1.7) are very slowly convergent: when they are restricted to a finite sample they depend on the shape of the sample. In the next section, we will evaluate explicitly the sums (9.1.7) with $\boldsymbol{Q} = 0$ for a spherical sample.

## 9.2. Dipolar Sums and Ewald's Method

Let us evaluate explicitly the sums (9.1.7) for an infinite 3D lattice following the Ewald's method.[82] The main problem of the sums (9.1.7) is that they are very slowly convergent. The Ewald's method consists on transforming such slowly convergent sums into quickly convergent sums entering generalized functions that decrease rapidly as their arguments increase. To do this, one splits the original sum in two parts: the former sum is extended to the direct lattice vectors while the latter sum is transformed in a sum over the reciprocal lattice vectors. We treat separately the two terms occurring in the sums (9.1.7). Let us begin with the last term of (9.1.7) and use the identity

$$\sum_{l\neq 0} \frac{e^{i\boldsymbol{Q}\cdot\boldsymbol{r}_l}}{r_l^3} = \frac{4}{\sqrt{\pi}} \sum_{l\neq 0} e^{i\boldsymbol{Q}\cdot\boldsymbol{r}_l} \int_0^{\infty} dz\ z^2 e^{-r_l^2 z^2} \equiv \frac{4}{\sqrt{\pi}}(A+B), \tag{9.2.1}$$

where

$$A = \sum_{l\neq 0} e^{i\boldsymbol{Q}\cdot\boldsymbol{r}_l} \int_0^{\eta} dz\ z^2 e^{-r_l^2 z^2} = -\frac{1}{3}\eta^3 + \sum_{l} e^{i\boldsymbol{Q}\cdot\boldsymbol{r}_l} \int_0^{\eta} dz\ z^2 e^{-r_l^2 z^2} \tag{9.2.2}$$

and

$$B = \sum_{l\neq 0} e^{i\boldsymbol{Q}\cdot\boldsymbol{r}_l} \int_\eta^\infty dz\, z^2 e^{-r_l^2 z^2} = \eta^3 \sum_{l\neq 0} e^{i\boldsymbol{Q}\cdot\boldsymbol{r}_l} \frac{1}{(\eta r_l)^3} \int_{\eta r_l}^\infty dt\; t^2 e^{-t^2}. \tag{9.2.3}$$

Taking advantage from the theta-function transformation[82]

$$\sum_l z^3 e^{i\boldsymbol{Q}\cdot\boldsymbol{r}_l} e^{-r_l^2 z^2} = \rho\pi^{3/2} \sum_l e^{-\frac{(\boldsymbol{G}_l+\boldsymbol{Q})^2}{4z^2}}, \tag{9.2.4}$$

where the first sum runs over the direct lattice vectors $\boldsymbol{r}_l$ while the second sum runs over the reciprocal lattice vectors $\boldsymbol{G}_l$ with $\rho = 1/v_c$, $v_c$ being the unit cell volume, Eq. (9.2.2) becomes

$$A = -\frac{1}{3}\eta^3 + \rho\pi^{3/2} \sum_l \int_0^\eta \frac{dz}{z} e^{-\frac{(\boldsymbol{G}_l+\boldsymbol{Q})^2}{4z^2}} = -\frac{1}{3}\eta^3 - \frac{1}{2}\rho\pi^{3/2} \sum_l \mathrm{Ei}\left[-\frac{(\boldsymbol{G}_l+\boldsymbol{Q})^2}{4\eta^2}\right], \tag{9.2.5}$$

where the change of variable $z = \frac{|\boldsymbol{G}_l+\boldsymbol{Q}|}{2\sqrt{t}}$ has been performed in Eq. (9.2.5) and

$$\mathrm{Ei}\left(-x^2\right) = -\int_{x^2}^\infty dt \frac{e^{-t}}{t} \tag{9.2.6}$$

is the exponential integral function.[4] Equation (9.2.3) may be written

$$B = \frac{1}{2}\eta^3 \sum_{l\neq 0} e^{i\boldsymbol{Q}\cdot\boldsymbol{r}_l} f_B(\eta r_l) \tag{9.2.7}$$

where

$$f_B(x) = \frac{2}{x^3}\int_x^\infty dz\; z^2 e^{-z^2} = \frac{e^{-x^2}}{x^2} + \frac{\sqrt{\pi}}{2x^3}\,\mathrm{erfc}(x) \tag{9.2.8}$$

and

$$\mathrm{erfc}(x) = \frac{2}{\sqrt{\pi}}\int_x^\infty dt\; e^{-t^2} \tag{9.2.9}$$

is the complementary error function.[4] Replacing Eqs. (9.2.5) and (9.2.7) into Eq. (9.2.1), one obtains

$$\sum_{l\neq 0} \frac{e^{i\boldsymbol{Q}\cdot\boldsymbol{r}_l}}{r_l^3} = -\frac{4\eta^3}{3\sqrt{\pi}} - 2\pi\rho \sum_l \mathrm{Ei}\left[-\frac{(\boldsymbol{G}_l+\boldsymbol{Q})^2}{4\eta^2}\right] + \frac{2\eta^3}{\sqrt{\pi}} \sum_{l\neq 0} e^{i\boldsymbol{Q}\cdot\boldsymbol{r}_l} f_B(\eta r_l). \tag{9.2.10}$$

In order to evaluate the first term of the sums (9.1.7), we use the relationship

$$\sum_{l\neq 0} \frac{e^{i\boldsymbol{Q}\cdot\boldsymbol{r}_l}}{r_l^5} r_l^\alpha r_l^\beta = -\frac{\partial^2}{\partial Q_\alpha \partial Q_\beta} \sum_{l\neq 0} \frac{e^{i\boldsymbol{Q}\cdot\boldsymbol{r}_l}}{r_l^5} \tag{9.2.11}$$

and the identity

$$\sum_{l\neq 0} \frac{e^{i\boldsymbol{Q}\cdot\boldsymbol{r}_l}}{r_l^5} = \frac{8}{3\sqrt{\pi}} \sum_{l\neq 0} e^{i\boldsymbol{Q}\cdot\boldsymbol{r}_l} \int_0^\infty dz\; z^4 e^{-r_l^2 z^2} \equiv \frac{8}{3\sqrt{\pi}}(C + D), \tag{9.2.12}$$

where

$$C = \sum_{l \neq 0} e^{i\boldsymbol{Q}\cdot\boldsymbol{r}_l} \int_0^\eta dz\, z^4 e^{-r_l^2 z^2} = -\frac{1}{5}\eta^5 + \sum_l e^{i\boldsymbol{Q}\cdot\boldsymbol{r}_l} \int_0^\eta dz z^4 e^{-r_l^2 z^2}$$

$$= -\frac{1}{5}\eta^5 + \frac{1}{2}\rho\pi^{3/2}\eta^2 \sum_l f_C\left[\frac{(\boldsymbol{G}_l + \boldsymbol{Q})^2}{4\eta^2}\right] \tag{9.2.13}$$

with

$$f_C(x^2) = x^2 \int_{x^2}^\infty \frac{dt}{t^2} e^{-t^2} = e^{-x^2} + x^2 \mathrm{Ei}(-x^2) \tag{9.2.14}$$

and

$$D = \sum_{l \neq 0} e^{i\boldsymbol{Q}\cdot\boldsymbol{r}_l} \int_\eta^\infty dz z^4 e^{-r_l^2 z^2} = \frac{1}{2}\eta^5 \sum_{l \neq 0} e^{i\boldsymbol{Q}\cdot\boldsymbol{r}_l} f_D(\eta r_l) \tag{9.2.15}$$

where

$$f_D(x) = \frac{2}{x^5} \int_x^\infty dt\, t^4 e^{-t^2} = \frac{3 + 2x^2}{2x^4} e^{-x^2} + \frac{3\sqrt{\pi}}{4x^5} \mathrm{erfc}(x). \tag{9.2.16}$$

Replacing Eqs. (9.2.13) and (9.2.15) into Eq. (9.2.12), one obtains

$$\sum_{l \neq 0} \frac{e^{i\boldsymbol{Q}\cdot\boldsymbol{r}_l}}{r_l^5} = -\frac{8\eta^5}{15\sqrt{\pi}} + \frac{4\pi}{3}\rho\eta^2 \sum_l f_C\left[-\frac{(\boldsymbol{G}_l + \boldsymbol{Q})^2}{4\eta^2}\right] + \frac{4\eta^5}{3\sqrt{\pi}} \sum_{l \neq 0} e^{i\boldsymbol{Q}\cdot\boldsymbol{r}_l} f_D(\eta r_l). \tag{9.2.17}$$

From the relationship (9.2.11), one obtains

$$\sum_{l \neq 0} \frac{e^{i\boldsymbol{Q}\cdot\boldsymbol{r}_l}}{r_l^5} r_l^\alpha r_l^\beta = -\frac{2\pi}{3}\rho\delta_{\alpha,\beta} \sum_l \mathrm{Ei}\left[\frac{-(\boldsymbol{G}_l + \boldsymbol{Q})^2}{4\eta^2}\right]$$

$$- \frac{4\pi}{3}\rho \sum_l \frac{(\boldsymbol{G}_l + \boldsymbol{Q})_\alpha (\boldsymbol{G}_l + \boldsymbol{Q})_\beta}{(\boldsymbol{G}_l + \boldsymbol{Q})^2} e^{-\frac{(\boldsymbol{G}_l + \boldsymbol{Q})^2}{4\eta^2}}$$

$$+ \frac{4\eta^5}{3\sqrt{\pi}} \sum_{l \neq 0} r_l^\alpha r_l^\beta e^{i\boldsymbol{Q}\cdot\boldsymbol{r}_l} f_D(\eta r_l). \tag{9.2.18}$$

Finally, replacing Eqs. (9.2.10) and (9.2.18) into the sums (9.1.7) one obtains

$$D_\infty^{\alpha\beta}(\boldsymbol{Q}) = -4\pi\rho \sum_l \frac{(\boldsymbol{G}_l + \boldsymbol{Q})_\alpha (\boldsymbol{G}_l + \boldsymbol{Q})_\beta}{(\boldsymbol{G}_l + \boldsymbol{Q})^2} e^{-\frac{(\boldsymbol{G}_l + \boldsymbol{Q})^2}{4\eta^2}} + \frac{4\eta^5}{\sqrt{\pi}} \sum_{l \neq 0} r_l^\alpha r_l^\beta e^{i\boldsymbol{Q}\cdot\boldsymbol{r}_l} f_D(\eta r_l)$$

$$+ \frac{2\eta^3}{\sqrt{\pi}}\delta_{\alpha,\beta}\left[\frac{2}{3} - \sum_{l \neq 0} e^{i\boldsymbol{Q}\cdot\boldsymbol{r}_l} f_B(\eta r_l)\right], \tag{9.2.19}$$

where the subscript $\infty$ means that the original dipolar sums on the left-hand side of (9.2.19) can be replaced by the generalized functions occurring in the right-hand side only for infinite lattices. Indeed, the theta-function transformation (9.2.4) implies an infinite sum over both the direct and the reciprocal lattice vectors. Equation (9.2.19) was given by Cohen and Keffer.[83] Notice the quick convergence of the series on the right-hand side of Eq. (9.2.19) due to the generalized functions $f_D(x) \simeq f_B(x) \simeq \frac{e^{-x^2}}{x^2}$ for $x \to \infty$. An interesting result is obtained for $\boldsymbol{Q} \to 0$. Indeed,

$$\lim_{\boldsymbol{Q}\to 0} D_\infty^{\alpha\beta}(\boldsymbol{Q}) = -4\pi\rho \left[ \frac{Q_\alpha Q_\beta}{Q^2} + \sum_{l\neq 0} \frac{G_l^\alpha G_l^\beta}{G_l^2} e^{-\frac{G_l^2}{4\eta^2}} \right] + \frac{4\eta^5}{3\sqrt{\pi}} \sum_{l\neq 0} r_l^\alpha r_l^\beta f_D(\eta r_l) + \frac{2\eta^3}{\sqrt{\pi}} \delta_{\alpha,\beta} \left[ \frac{2}{3} - \sum_{l\neq 0} f_B(\eta r_l) \right]. \quad (9.2.20)$$

For the lattices we are interested to (for instance, cubic, tetragonal, orthorhombic), one has

$$\sum_{l\neq 0} \frac{G_l^\alpha G_l^\beta}{G_l^2} e^{-\frac{G_l^2}{4\eta^2}} = \delta_{\alpha,\beta} \sum_{l\neq 0} \frac{(G_l^\alpha)^2}{G_l^2} e^{-\frac{G_l^2}{4\eta^2}} \quad (9.2.21)$$

and

$$\sum_{l\neq 0} r_l^\alpha r_l^\beta f_D(\eta r_l) = \delta_{\alpha,\beta} \sum_{l\neq 0} (r_l^\alpha)^2 f_D(\eta r_l) \quad (9.2.22)$$

so that Eq. (9.2.20) becomes

$$\lim_{\boldsymbol{Q}\to 0} D_\infty^{\alpha\beta}(\boldsymbol{Q}) = -4\pi\rho \frac{Q_\alpha Q_\beta}{Q^2} + \delta_{\alpha,\beta} d_\alpha \quad (9.2.23)$$

where

$$d_\alpha = \frac{4\eta^3}{3\sqrt{\pi}} - 4\pi\rho \sum_{l\neq 0} \frac{(G_l^\alpha)^2}{G_l^2} e^{-\frac{G_l^2}{4\eta^2}} + \frac{2\eta^3}{\sqrt{\pi}} \sum_{l\neq 0} [2(\eta r_l^\alpha)^2 f_D(\eta r_l) - f_B(\eta r_l)]. \quad (9.2.24)$$

Notice that the sum rule

$$d_x + d_y + d_z = 4\pi\rho \quad (9.2.25)$$

can be directly proved using Eq. (9.2.24) and the theta-function transformation (9.2.4). For a SC lattice, one has $d_x = d_y = d_z$ so that the sum rule (9.2.25) gives $d_\alpha = \frac{4}{3}\pi\rho$ and

$$\lim_{\boldsymbol{Q}\to 0} [D_\infty^{\alpha\beta}(\boldsymbol{Q})]_{\mathrm{SC}} = \frac{4\pi}{3a^3} \left( \delta_{\alpha,\beta} - 3\frac{Q_\alpha Q_\beta}{Q^2} \right). \quad (9.2.26)$$

As one can see, for an infinite system Eq. (9.2.23) does not lead to a unique value but to a range of values between $-\frac{8\pi}{3a^3}$ and $\frac{4\pi}{3a^3}$ depending on the direction along which the vector $\boldsymbol{Q}$ tends to zero. All the above results are obtained for an infinite system. Do these conclusions hold when a macroscopic sample of finite volume $V$ is considered? For a finite sample, one has the identity

$$D_V^{\alpha\beta}(\boldsymbol{Q}) = D_\infty^{\alpha\beta}(\boldsymbol{Q}) - D_{\infty-V}^{\alpha\beta}(\boldsymbol{Q}), \quad (9.2.27)$$

where $D_V^{\alpha\beta}(\boldsymbol{Q})$ is the dipolar sum (9.1.7) restricted to the lattice sites contained in the volume $V$ and $D_{\infty-V}^{\alpha\beta}(\boldsymbol{Q})$ is an integral extended to the whole space external to the volume $V$. The replacement of the discrete sum with an integral is justified by the fact that $V$ is a macroscopic volume so that the sum is expected to be insensitive to the lattice structure since it concerns spins far from the origin. To simplify the calculation of such an integral, let us suppose that the macroscopic sample is a sphere of radius $R$. In this case, we have

$$D_{\infty-V}^{\alpha\beta}(\boldsymbol{Q}) = \rho \iiint_{r^2>R^2} d^3\boldsymbol{r}\frac{1}{r^3}\left(3\frac{r_\alpha r_\beta}{r^2} - \delta_{\alpha,\beta}\right)e^{i\boldsymbol{Q}\cdot\boldsymbol{r}}. \tag{9.2.28}$$

To perform the calculation of the integral (9.2.28), we choose a new reference system $\xi\eta\zeta$ shown in Fig. 9.1 where the $\zeta$-axis is directed along the vector $\boldsymbol{Q}$ and $r_\xi = r\sin\vartheta\cos\varphi$, $r_\eta = r\sin\vartheta\sin\varphi$, $r_\zeta = r\cos\vartheta$ are the components of the vector $\boldsymbol{r}$ in the new frame. In the $\xi\eta\zeta$ frame one has

$$D_{\infty-V}^{\xi\xi}(0,0,Q) = \rho\int_R^\infty \frac{dr}{r}\int_{-1}^1 d\mu\int_0^{2\pi} d\varphi[3(1-\mu^2)\cos^2\varphi - 1]\,e^{iQr\mu}, \tag{9.2.29}$$

where the change of variable $\mu = \cos\vartheta$ has been done. The integration over $\varphi$ is direct and one obtains

$$\begin{aligned} D_{\infty-V}^{\xi\xi}(0,0,Q) &= \pi\rho\int_R^\infty \frac{dr}{r}\left[\int_{-1}^1 d\mu e^{iQr\mu} - 3\int_{-1}^1 d\mu\,\mu^2 e^{iQr\mu}\right] \\ &= \pi\rho\int_{QR}^\infty dt\left[-4\frac{\sin t}{t^2} - 12\frac{\cos t}{t^3} + 12\frac{\sin t}{t^4}\right], \end{aligned} \tag{9.2.30}$$

where an integration by parts of the second integral in the first row of Eq. (9.2.30) has been performed and the replacement $t = Qr$ has been done. A further integration by parts of the last two terms of the second row in Eq. (9.2.30) leads to

$$D_{\infty-V}^{\xi\xi}(0,0,Q) = 4\pi\rho f(QR), \tag{9.2.31}$$

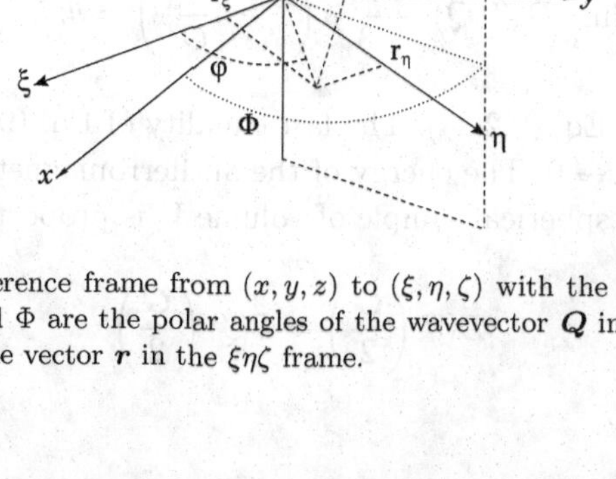

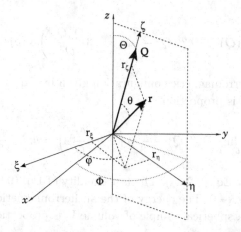

Fig. 9.1. Change of reference frame from $(x, y, z)$ to $(\xi, \eta, \zeta)$ with the $\zeta$-axis chosen parallel to the wavevector $\boldsymbol{Q}$. $\Theta$ and $\Phi$ are the polar angles of the wavevector $\boldsymbol{Q}$ in the $xyz$ frame; $\theta$ and $\phi$ are the polar angles of the vector $\boldsymbol{r}$ in the $\xi\eta\zeta$ frame.

where

$$f(x) = \frac{\sin x - x\cos x}{x^3}. \tag{9.2.32}$$

In a similar way, one obtains

$$\begin{gathered} D^{\eta\eta}_{\infty-V}(0,0,Q) = D^{\xi\xi}_{\infty-V}(0,0,Q), \quad D^{\zeta\zeta}_{\infty-V}(0,0,Q) = -2D^{\xi\xi}_{\infty-V}(0,0,Q), \\ D^{\xi\eta}_{\infty-V}(0,0,Q) = D^{\xi\zeta}_{\infty-V}(0,0,Q) = D^{\eta\zeta}_{\infty-V}(0,0,Q) = 0. \end{gathered} \tag{9.2.33}$$

Using Eqs. (9.2.33) and the relationship

$$\begin{pmatrix} r_x \\ r_y \\ r_z \end{pmatrix} = \begin{pmatrix} \sin\Phi & \cos\Theta\cos\Phi & \sin\Theta\cos\Phi \\ -\cos\Phi & \cos\Theta\sin\Phi & \sin\Theta\sin\Phi \\ 0 & -\sin\Theta & \cos\Theta \end{pmatrix} \begin{pmatrix} r_\xi \\ r_\eta \\ r_\zeta \end{pmatrix}, \tag{9.2.34}$$

one can write the $D^{\alpha\beta}_{\infty-V}(\boldsymbol{Q})$ with $\alpha,\beta = x,y,z$ given by Eq. (9.2.28) as functions of $D^{\rho\sigma}_{\infty-V}(0,0,Q)$ with $\rho,\sigma = \xi,\eta,\zeta$ given by Eq. (9.2.33). In particular,

$$D^{xx}_{\infty-V}(\boldsymbol{Q}) = (1 - 3\sin^2\Theta\cos^2\Phi) D^{\xi\xi}_{\infty-V}(0,0,Q) \equiv \left(1 - 3\frac{Q_x^2}{Q^2}\right) 4\pi\rho f(QR) \tag{9.2.35}$$

where $f(x)$ is given by Eq. (9.2.32). All the other terms can be obtained in a similar way and the generic term reads

$$D^{\alpha\beta}_{\infty-V}(\boldsymbol{Q}) = \left(\delta_{\alpha,\beta} - 3\frac{Q_\alpha Q_\beta}{Q^2}\right) 4\pi\rho f(QR) \tag{9.2.36}$$

so that

$$D^{\alpha\beta}_V(\boldsymbol{Q}) = D^{\alpha\beta}_\infty(\boldsymbol{Q}) - \left(\delta_{\alpha,\beta} - 3\frac{Q_\alpha Q_\beta}{Q^2}\right) 4\pi\rho f(QR). \tag{9.2.37}$$

The energy of the ferromagnetic configuration given by Eq. (9.1.10) for a spherical sample of volume $V$ is proportional to

$$D^{\alpha\alpha}_V(0) = \lim_{\boldsymbol{Q}\to 0} D^{\alpha\alpha}_\infty(\boldsymbol{Q}) - \frac{4\pi}{3}\rho\left(1 - 3\frac{Q_\alpha^2}{Q^2}\right) = d_\alpha - \frac{4\pi}{3}\rho, \tag{9.2.38}$$

where $d_\alpha$ is given by Eq. (9.2.24). The last equality of Eq. (9.2.38) comes from the limit $f(x) \to \frac{1}{3}$ for $x \to 0$. The energy of the antiferromagnetic configuration given by Eq. (9.1.11) for a spherical sample of volume $V$ is proportional to

$$D^{\alpha\alpha}_V\left(\frac{\boldsymbol{G}}{2}\right) = D^{\alpha\alpha}_\infty\left(\frac{\boldsymbol{G}}{2}\right) \tag{9.2.39}$$

since $f(x) \to 0$ for $x \to \infty$. Note that only for the ferromagnetic configuration, the finiteness of the sample enters a correction to the corresponding value for the infinite system. Moreover, the result for a finite sample has a unique value given by Eq. (9.2.38). For a non-spherical sample, the integral occurring in Eq. (9.2.28) cannot be performed in general. However, for ellipsoidic samples with semi-axis $A \geq B \geq C$ such integral may be expressed in terms of elliptic integrals[84] and the dipolar energy is given by Eq. (9.1.10) with

$$D_V^{\alpha\alpha}(0) = d_\alpha - \rho N_\alpha \tag{9.2.40}$$

where the factors $N_\alpha$ $(\alpha = x, y, z)$ are called "demagnetization factors". For a spherical sample, the demagnetization factors are given by $N_x = N_y = N_z = \frac{4\pi}{3}\pi$ so that Eq. (9.2.40) reduce to Eq. (9.2.38). For a "needle" shaped sample[84] $(A \gg B = C)$, the demagnetization factors reduce to

$$N_x = 4\pi \left(\frac{C}{A}\right)^2 \left(\ln \frac{2A}{C} - 1\right) + \cdots \to 0 \tag{9.2.41}$$

and

$$N_y = N_z = 4\pi \left[1 - \left(\frac{C}{A}\right)^2 \left(\ln \frac{2A}{C} - 1\right) + \cdots \right] \to 4\pi. \tag{9.2.42}$$

For a "slab" shaped sample[84] $(A = B \gg C)$, one obtains

$$N_x = N_y = \pi^2 \frac{C}{A} \left(1 - \frac{4}{\pi}\frac{C}{A} + \cdots \right) \to 0 \tag{9.2.43}$$

and

$$N_z = 4\pi \left(1 - \frac{\pi}{2}\frac{C}{A} + \cdots \right) \to 4\pi. \tag{9.2.44}$$

As one can see from Eqs. (9.2.40)–(9.2.44), the shape of the macroscopic sample determines the orientation of the magnetic moments even in a pure isotropic Heisenberg ferromagnet where the exchange interaction is much greater than the dipolar interaction. For instance, for a SC lattice one has $d_\alpha = \frac{4\pi}{3a^3}$ so that $D_V^{xx}(0) = D_V^{yy}(0) = D_V^{zz}(0) = 0$ for a spherical sample, $D_V^{xx}(0) = \frac{4\pi}{3a^3}$, $D_V^{yy}(0) = D_V^{zz}(0) = -\frac{8\pi}{3a^3}$, for a needle whose axis is directed along the $x$-axis and $D_V^{xx}(0) = D_V^{yy}(0) = \frac{4\pi}{3a^3}$, $D_V^{zz}(0) = -\frac{8\pi}{3a^3}$ for a slab with the surface parallel to the $xy$-plane. These values indicate that the magnetic moments may be parallel to any of the three NN directions for a spherical sample; on the contrary, the magnetic moments select the needle-axis direction for a needle-shaped sample and any of the two NN directions ($x$ or $y$) for a slab-shaped sample. In any case, the continuous symmetry of the isotropic Heisenberg Hamiltonian is broken. Obviously, the previous conclusions can be modified by the presence of an external magnetic field or accounting for the lattice anisotropy (crystalline electric field). In Table 9.1, we give

Table 9.1. Dipolar sums for SC lattice ($a = b = c$). The macroscopic sample is assumed to be a sphere of volume $V$.

| $Q = \frac{2\pi}{a}(q_1, q_2, q_3)$ | $a^3 D_V^{xx}(Q)$ | $a^3 D_V^{yy}(Q)$ | $a^3 D_V^{zz}(Q)$ |
|---|---|---|---|
| $(0, 0, 0)$ | 0 | 0 | 0 |
| $\left(\frac{1}{2}, 0, 0\right)$ | −9.6874430 | 4.8437215 | 4.8437215 |
| $\left(0, \frac{1}{2}, 0\right)$ | 4.8437215 | −9.6874430 | 4.8437215 |
| $\left(0, 0, \frac{1}{2}\right)$ | 4.8437215 | 4.8437215 | −9.6874430 |
| $\left(\frac{1}{2}, \frac{1}{2}, 0\right)$ | −2.6767887 | −2.6767887 | 5.3535774 |
| $\left(\frac{1}{2}, 0, \frac{1}{2}\right)$ | −2.6767887 | 5.3535774 | −2.6767887 |
| $\left(0, \frac{1}{2}, \frac{1}{2}\right)$ | 5.3535774 | −2.6767887 | −2.6767887 |
| $\left(\frac{1}{2}, \frac{1}{2}, \frac{1}{2}\right)$ | 0 | 0 | 0 |

the dipolar sums (9.2.38) and (9.2.39) of a SC lattice for a spherical sample: the generic lattice vector $\boldsymbol{r}_l$ is given by

$$\boldsymbol{r}_l = a\,(l_1\mathbf{u}_x + l_2\mathbf{u}_y + l_3\mathbf{u}_z) \tag{9.2.45}$$

and the generic reciprocal lattice vector $\boldsymbol{G}_l$ is given by

$$\boldsymbol{G}_l = \frac{2\pi}{a}(l_1\mathbf{u}_x + l_2\mathbf{u}_y + l_3\mathbf{u}_z). \tag{9.2.46}$$

The numbers given in Table 9.1 are obtained from Eq. (9.2.19) with $-10 < l_1, l_2, l_3 < 10$ and $\eta = \frac{2}{a}$: such numbers do not change within an error of $10^{-10}$ with increasing $l_1, l_2, l_3$. No difference between the values for the finite and infinite sample exists for $\boldsymbol{Q} \neq 0$. For $\boldsymbol{Q} = 0$ (first row), the result for the finite sample differs considerably from the result for the infinite system as one can see comparing the values of Table 9.1 with Eq. (9.2.26). Note that a direct evaluation of (9.1.7) summing over the lattice sites within a sphere of radius $R$ leads to values correct within $\frac{1}{(QR)^2}$ so that a sphere containing at least $\sim 10^{15}$ lattice sites is necessary to reach the same precision of the values obtained by the use of the Ewald's method!

From Table 9.1, one can see that the minimum energy configurations corresponding to the maxima of $D^{\alpha\alpha}(\boldsymbol{Q})$ occur for $\boldsymbol{Q} = (\frac{1}{2}, \frac{1}{2}, 0)$ (spins along $z$), $\boldsymbol{Q} = (\frac{1}{2}, 0, \frac{1}{2})$ (spins along $y$) and $\boldsymbol{Q} = (0, \frac{1}{2}, \frac{1}{2})$ (spins along $x$). Neither the ferromagnetic nor the Néel antiferromagnetic phase is supported by the dipolar interaction. The three degenerate ground-state configurations correspond to collinear antiferromagnetic phases in which the spins are directed along the ferromagnetic axis and the ferromagnetic chains are coupled antiferromagnetically to the NN ferromagnetic chains as shown in the upper part of Fig. 9.2. If a tetragonal distortion of the type $c > a = b$

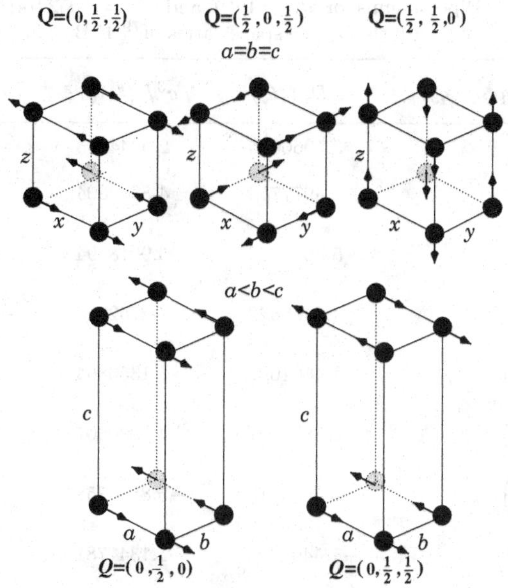


Fig. 9.2. Upper part: degenerate ground-state configurations of a SC lattice with dipole–dipole interactions. Lower part: quasi-degenerate ground-state configurations for an orthorhombic lattice with $b/a = 1.016$ and $c/a = 3.052$.

is entered, the three-fold degenerate ground state of the SC lattice is lifted into a two-fold degenerate ground state whose configurations correspond to $\boldsymbol{Q} = (0, \frac{1}{2}, \frac{1}{2})$ (spins along $x$) or $\boldsymbol{Q} = (\frac{1}{2}, 0, \frac{1}{2})$ (spins along $y$). The spin pattern of such configurations is shown in the first two cubes of the upper part of Fig. 9.2. On the contrary, for a tetragonal distortion of the type $c < a = b$, the SC three-fold degenerate ground state reduces to a single ground state corresponding to $\boldsymbol{Q} = (\frac{1}{2}, \frac{1}{2}, 0)$ (spins along $z$) whose spin pattern is shown in the third cube of the upper part of Fig. 9.2. The common feature of the ground-state configurations of a T lattice is that the spins are directed along the ferromagnetic axis and the NN ferromagnetic chains are coupled antiferromagnetically. An orthorhombic distortion ($a \neq b \neq c$) lifts any degeneracy and the ground-state energy corresponds to a configuration in which the spins are coupled ferromagnetically along the shortest lattice constant. For instance, when $b/a = 1.016$ and $c/a = 3.052$ (the ratios between the lattice constants suitable[79] for $ErBa_2Cu_3O_7$), the ground-state energy corresponds to $\boldsymbol{Q} = (0, \frac{1}{2}, \frac{1}{2})$ (spins along $x$) and the spin configuration is shown in the second parallelepiped of the lower part of Fig. 9.2.

## 9.3. Ground-State Configuration of $ErBa_2Cu_3O_{6+x}$

The aim of this section is to show how the experimental data on $ErBa_2Cu_3O_{6+x}$ can be used to select among the theoretical models of the previous section in order to check the experiment. Evaluating the dipolar sums given by Eqs. (9.2.38) and

Table 9.2. Dipolar sums for $b/a = 1.016$ and $c/a = 3.052$ (orthorhombic symmetry): these are the lattice space ratios of[79] $ErBa_2Cu_3O_7$.

| $\boldsymbol{Q} = \frac{2\pi}{a}(q_1, q_2, q_3)$ | $a^3 D_V^{xx}(\boldsymbol{Q})$ | $a^3 D_V^{yy}(\boldsymbol{Q})$ | $a^3 D_V^{zz}(\boldsymbol{Q})$ |
|---|---|---|---|
| $(0,0,0)$ | 3.1960006 | 2.9244265 | −6.1204271 |
| $\left(\frac{1}{2},0,0\right)$ | −5.8977753 | 4.8930604 | 1.0047149 |
| $\left(0,\frac{1}{2},0\right)$ | 5.0689676 | −5.9018694 | 0.8329018 |
| $\left(0,0,\frac{1}{2}\right)$ | 4.5468622 | 4.2752885 | −8.8221507 |
| $\left(\frac{1}{2},\frac{1}{2},0\right)$ | −1.4501090 | −1.1350864 | 2.5851954 |
| $\left(\frac{1}{2},0,\frac{1}{2}\right)$ | −5.8871227 | 4.8930607 | 0.9940620 |
| $\left(0,\frac{1}{2},\frac{1}{2}\right)$ | 5.0689679 | −5.8896758 | 0.8207079 |
| $\left(\frac{1}{2},\frac{1}{2},\frac{1}{2}\right)$ | −1.4497908 | −1.1347781 | 2.5845688 |

Table 9.3. Dipolar sums for $b/a = 1$ (tetragonal symmetry) and $c/a = 3.062$: these are the lattice space ratios of[80] $ErBa_2Cu_3O_6$.

| $\boldsymbol{Q} = \frac{2\pi}{a}(q_1, q_2, q_3)$ | $a^3 D_V^{xx}(\boldsymbol{Q})$ | $a^3 D_V^{yy}(\boldsymbol{Q})$ | $a^3 D_V^{zz}(\boldsymbol{Q})$ |
|---|---|---|---|
| $(0,0,0)$ | 3.1488185 | 3.1488185 | −6.2976372 |
| $\left(\frac{1}{2},0,0\right)$ | −6.0395796 | 5.0988729 | 0.9407067 |
| $\left(0,\frac{1}{2},0\right)$ | 5.0988729 | −6.0395796 | 0.9407067 |
| $\left(0,0,\frac{1}{2}\right)$ | 4.5168115 | 4.5168115 | −9.0336231 |
| $\left(\frac{1}{2},\frac{1}{2},0\right)$ | −1.3230812 | −1.3230812 | 2.6461624 |
| $\left(\frac{1}{2},0,\frac{1}{2}\right)$ | −6.0290914 | 5.0988731 | 0.9302183 |
| $\left(0,\frac{1}{2},\frac{1}{2}\right)$ | 5.0988731 | −6.0290914 | 0.9302183 |
| $\left(\frac{1}{2},\frac{1}{2},\frac{1}{2}\right)$ | −1.3228053 | −1.3228053 | 2.6456106 |

(9.2.39) for an orthorhombic lattice with the lattice constants of the oxygenate compound[79] ($x = 1$) and for a tetragonal lattice with the lattice constants of the deoxygenated compound[80] ($x \simeq 0$), we obtain Tables 9.2 and 9.3, respectively. The dipolar energy is given by

$$E^{\alpha}_{\text{dip}}(\boldsymbol{Q}) = -\frac{(g\mu_B S)^2}{2} N D_V^{\alpha\alpha}(\boldsymbol{Q}). \tag{9.3.1}$$

The Bohr magneton $\mu_B$ occurring in Eq. (9.3.1) is given by

$$\mu_B = \frac{e\hbar}{2m_e c} = 9.27401 \times 10^{-21}\,\mathrm{erg}^{\frac{1}{2}}\,\mathrm{cm}^{\frac{3}{2}} \qquad (9.3.2)$$

where $e = 4.8032 \times 10^{-10}$ stat C is the absolute value of the electron charge (1 stat C = 1 $\mathrm{erg}^{\frac{1}{2}}\mathrm{cm}^{\frac{1}{2}}$ from the Coulomb law in the CGS system), $\hbar = 1.05457 \times 10^{-27}$ erg × s is the reduced Planck's constant, $m_e = 9.10938 \times 10^{-28}$ g is the electron mass and $c = 2.99792 \times 10^{10}$ cm/s is the light speed. Using Eq. (9.3.2), one sees that the dipolar energy (9.3.1) measured in units of Boltzmann constant $k_B = 1.38065 \times 10^{-16}$ erg/K becomes

$$\frac{E^{\alpha}_{\mathrm{dip}}(\boldsymbol{Q})}{k_B N} = -0.62295\frac{(gS)^2}{2} D^{\alpha\alpha}_{V}(\boldsymbol{Q}) \quad (\mathrm{K\AA^3}). \qquad (9.3.3)$$

For[79] $gS \simeq 4.9$ and $a = 3.82$ Å, Eq. (9.3.3) gives

$$\frac{E^{\alpha}_{\mathrm{dip}}(\boldsymbol{Q})}{k_B N} = -0.1342\,[a^3 D^{\alpha\alpha}_{V}(\boldsymbol{Q})] \quad (\mathrm{K}), \qquad (9.3.4)$$

where $a^3 D^{\alpha\alpha}_V(\boldsymbol{Q})$ are given in Table 9.2 from which one can see that the ground-state energy corresponds to $D^{xx}_V(0, \frac{1}{2}, \frac{1}{2}) = 5.0689679/a^3$ even though the configuration corresponding to $D^{xx}_V(0, \frac{1}{2}, 0) = 5.0689676/a^3$ is very close in energy. The spin patterns of these two configurations are shown in the lower part of Fig. 9.2. In the ground-state configuration, the spins are aligned along the ferromagnetic $a$-axis and the ferromagnetic chains are coupled antiferromagnetically both along the $b$ and $c$-axes. In the quasi-degenerate configuration, the ferromagnetic chains are coupled ferromagnetically along the $c$-axis. The close energy values of the two configurations are explained by the very weak dipole–dipole coupling of the $Er^{3+}$ ions along the $c$-axis due to the large spacing $c \sim 3a$.

From Table 9.3, one can see that the ground state of the tetragonal $ErBa_2Cu_3O_6$ is degenerate corresponding to the two configurations $D^{xx}_V(0, \frac{1}{2}, \frac{1}{2}) = D^{yy}_V(\frac{1}{2}, 0, \frac{1}{2}) = 5.0988731/a^3$ with the spins directed along the ferromagnetic axis. Other two degenerate configurations corresponding to $D^{xx}_V(0, \frac{1}{2}, 0) = D^{yy}_V(\frac{1}{2}, 0, 0) = 5.0988729/a^3$ with the spins directed along the shortest ferromagnetic axis are very close in energy.

In a neutron scattering experiment,[79] both elastic Bragg peaks at $(\frac{1}{2}, 0, 0)$ and $(\frac{1}{2}, 0, \frac{1}{2})$ have been recorded for the orthorhombic $ErBa_2Cu_3O_7$ with the spins directed along the ferromagnetic $b$-axis. This contrasts with the values of the dipolar sums given in Table 9.2 where the configurations corresponding to the experiment are given by $D^{yy}_V(\frac{1}{2}, 0, 0) = 4.8930604/a^3$ and $D^{yy}_V(\frac{1}{2}, 0, \frac{1}{2}) = 4.8930607/a^3$, while the two lowest energies of the theoretical model correspond to $D^{xx}_V(0, \frac{1}{2}, \frac{1}{2}) = 5.0689679/a^3$ and $D^{xx}_V(0, \frac{1}{2}, 0) = 5.0689676/a^3$ with the spins aligned along the $a$-axis as shown in the lower part of Fig. 9.2. As for the tetragonal $ErBa_2Cu_3O_6$, no elastic Bragg peaks were recorded since the LRO is absent.

How to reconcile the theory with the experiment for the orthorhombic compound? In other words, which important features of the actual compound have been lost in the theoretical model? The theoretical model certainly neglects any antiferromagnetic exchange coupling between the NN erbium ions belonging to NN chains. However, it is expected that such interaction is very weak due to the indirect exchange through the bridges over the oxygen and copper ions. More importantly, such antiferromagnetic interaction should favour a Néel antiferromagnetic order in contrast to that observed in the experiment. The theoretical model also neglects any interaction between the $Er^{3+}$ ions and its surrounding oxygen or copper ions. The simplest way to account for such an interaction is to introduce the electrostatic potential between the electrons of the unfilled shell $4f$ of the $Er^{3+}$ ion and the surrounding oxygen ions represented as point-charges. This model is called the point-charge model[85] and the electrostatic potential is called the crystalline-electric-field (CEF) potential.

## 9.4. CEF Calculation for $ErBa_2Cu_3O_{6+x}$

In this section, we give a sketch of the CEF calculation[80] for $ErBa_2Cu_3O_{6+x}$. The eight NN oxygen ions are located on the vertices of a parallelepiped with edges $2a, 2b, 2c$ which do not coincide with the lattice constants of the orthorhombic lattice of the $Er^{3+}$ ions. The erbium ion location at the centre of the parallelepiped is assumed to be the origin of the reference coordinate system as shown in Fig. 9.3. Assuming that the oxygen ions behave like point-charges $q$, the crystalline field potential at a point $\boldsymbol{r} = (x, y, z)$ near the origin is given by[85]

$$V_{\mathrm{CEF}}(\boldsymbol{r}) = \sum_{i=1}^{8} \frac{q}{|\boldsymbol{r} - \boldsymbol{R}_i|} = \sum_{s=0}^{\infty} \frac{4\pi q}{2s+1} \frac{r^s}{R^{s+1}} \sum_{m=-s}^{s} A_s^{-m} Y_s^m \left(\frac{\boldsymbol{r}}{r}\right) \tag{9.4.1}$$

with $\boldsymbol{R}_i = (\pm a, \pm b, \pm c)$, $R = \sqrt{a^2 + b^2 + c^2}$ is the distance of each oxygen ion from the origin and

$$A_s^m = \sum_{i=1}^{8} (-1)^m Y_s^m \left(\frac{\boldsymbol{R}_i}{R}\right). \tag{9.4.2}$$

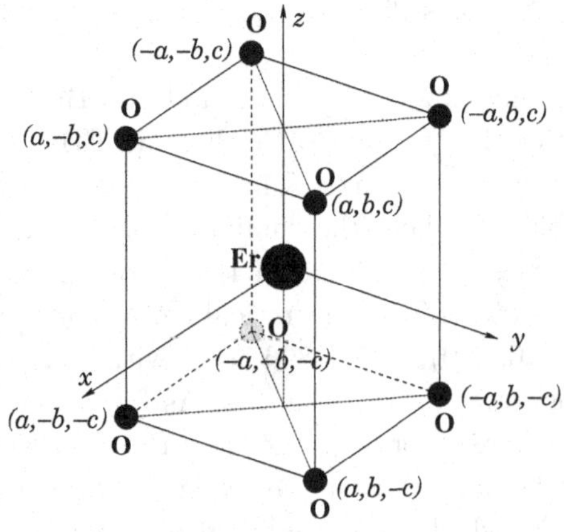


Fig. 9.3. Parallelepiped of oxygen ions O surrounding the $Er^{3+}$ ion.

The spherical harmonics expansion given in Eq. (9.4.1) is known as the "spherical harmonics addition theorem". Such an expansion consists of infinite terms each of which is a product of two spherical harmonics, one having as argument the polar angles of the vector $\boldsymbol{r}$ and the other one the polar angles of the vertices of the parallelepiped, that is, the coordinates of the oxygen point-charges. This model is referred as point-charge model.[85] The spherical harmonics written as a function of the cartesian coordinates instead of the polar angles are

$$Y_0^0 = \frac{1}{\sqrt{4\pi}}, \tag{9.4.3}$$

$$Y_1^0 = \frac{1}{2}\sqrt{\frac{3}{\pi}}\frac{z}{r}, \quad Y_1^{\mp 1} = \pm\frac{1}{2}\sqrt{\frac{3}{2\pi}}\frac{(x \mp iy)}{r}, \tag{9.4.4}$$

$$Y_2^0 = \frac{1}{4}\sqrt{\frac{5}{\pi}}\frac{(3z^2 - r^2)}{r^2}, \quad Y_2^{\mp 1} = \pm\frac{1}{2}\sqrt{\frac{15}{2\pi}}\frac{(x \mp iy)z}{r^2}, \quad Y_2^{\mp 2} = \frac{1}{4}\sqrt{\frac{15}{2\pi}}\frac{(x \mp iy)^2}{r^2}, \tag{9.4.5}$$

$$\begin{aligned} Y_3^0 &= \frac{1}{4}\sqrt{\frac{7}{\pi}}\frac{(5z^2 - 3r^2)z}{r^3}, \quad Y_3^{\mp 1} = \pm\frac{1}{8}\sqrt{\frac{21}{\pi}}\frac{(x \mp iy)(5z^2 - r^2)}{r^3}, \\ Y_3^{\mp 2} &= \frac{1}{4}\sqrt{\frac{105}{2\pi}}\frac{(x \mp iy)^2 z}{r^3}, \quad Y_3^{\mp 3} = \pm\frac{1}{8}\sqrt{\frac{35}{\pi}}\frac{(x \mp iy)^3}{r^3}, \end{aligned} \tag{9.4.6}$$

$$\begin{aligned} Y_4^0 &= \frac{3}{16}\sqrt{\frac{1}{\pi}}\frac{(35z^4 - 30z^2r^2 + 3r^4)}{r^4}, \quad Y_4^{\mp 1} = \pm\frac{3}{8}\sqrt{\frac{5}{\pi}}\frac{(x \mp iy)z(7z^2 - 3r^2)}{r^4}, \\ Y_4^{\mp 2} &= \frac{3}{8}\sqrt{\frac{5}{2\pi}}\frac{(x \mp iy)^2(7z^2 - r^2)}{r^4}, \quad Y_4^{\mp 3} = \pm\frac{3}{8}\sqrt{\frac{35}{\pi}}\frac{(x \mp iy)^3 z}{r^4}, \\ Y_4^{\mp 4} &= \frac{3}{16}\sqrt{\frac{35}{2\pi}}\frac{(x \mp iy)^4}{r^4}, \end{aligned} \tag{9.4.7}$$

$$\begin{aligned} Y_5^0 &= \frac{1}{16}\sqrt{\frac{11}{\pi}}\frac{(63z^4 - 70z^2r^2 + 15r^4)z}{r^5}, \\ Y_5^{\mp 1} &= \pm\frac{1}{16}\sqrt{\frac{165}{2\pi}}\frac{(x \mp iy)(21z^4 - 14z^2r^2 + r^4)}{r^5}, \\ Y_5^{\mp 2} &= \frac{1}{8}\sqrt{\frac{1155}{2\pi}}\frac{(x \mp iy)^2 z(3z^2 - r^2)}{r^5}, \quad Y_5^{\mp 3} = \pm\frac{1}{32}\sqrt{\frac{385}{\pi}}\frac{(x \mp iy)^3(9z^2 - r^2)}{r^5}, \\ Y_5^{\mp 4} &= \frac{3}{16}\sqrt{\frac{385}{2\pi}}\frac{(x \mp iy)^4 z}{r^5}, \quad Y_5^{\mp 5} = \pm\frac{3}{32}\sqrt{\frac{77}{\pi}}\frac{(x \mp iy)^5}{r^5}, \end{aligned} \tag{9.4.8}$$

$$Y_6^0 = \frac{1}{32}\sqrt{\frac{13}{\pi}}\frac{(231z^6 - 315z^4r^2 + 105z^2r^4 - 5r^6)}{r^6},$$

$$Y_6^{\mp 1} = \pm\frac{1}{16}\sqrt{\frac{273}{2\pi}}\frac{(x \mp iy)z(33z^4 - 30z^2r^2 + 5r^4)}{r^6},$$

$$
\begin{aligned}
Y_6^{\mp 2} &= \frac{1}{64}\sqrt{\frac{1365}{\pi}}\frac{(x \mp iy)^2(33z^4 - 18z^2r^2 + r^4)}{r^6}, \\
Y_6^{\mp 3} &= \pm\frac{1}{32}\sqrt{\frac{1365}{\pi}}\frac{(x \mp iy)^3 z(11z^2 - 3r^2)}{r^6}, \quad Y_6^{\mp 4} = \frac{3}{32}\sqrt{\frac{91}{2\pi}}\frac{(x \mp iy)^4(11z^2 - r^2)}{r^6}, \\
Y_6^{\mp 5} &= \pm\frac{3}{32}\sqrt{\frac{1001}{\pi}}\frac{(x \mp iy)^5 z}{r^6}, \quad Y_6^{\mp 6} = \frac{1}{64}\sqrt{\frac{3003}{\pi}}\frac{(x \mp iy)^6}{r^6}.
\end{aligned} \tag{9.4.9}
$$

Note that the sum in Eq. (9.4.2) can be performed easily using the spherical harmonics (9.4.3)–(9.4.9) evaluated at the points $\mathbf{R}_i = (\pm a, \pm b, \pm c)$. It is direct to see that due to the orthorhombic symmetry of the surrounding oxygen ions and to the nature of the spherical harmonics the only terms $A_s^m$ that give non-zero contribution are those with both $s$ and $m$ even. Then from Eq. (9.4.2), one obtains

$$A_0^0 = \frac{8}{\sqrt{4\pi}}, \tag{9.4.10}$$

$$A_2^0 = \frac{2}{R^2}\sqrt{\frac{5}{\pi}}(3c^2 - R^2), \tag{9.4.11}$$

$$A_2^{\pm 2} = \frac{2}{R^2}\sqrt{\frac{15}{2\pi}}(a^2 - b^2), \tag{9.4.12}$$

$$A_4^0 = \frac{3}{2R^4}\sqrt{\frac{1}{\pi}}(35c^4 - 30c^2R^2 + 3R^4), \tag{9.4.13}$$

$$A_4^{\pm 2} = \frac{3}{R^4}\sqrt{\frac{5}{2\pi}}(a^2 - b^2)(7c^2 - R^2), \tag{9.4.14}$$

$$A_4^{\pm 4} = \frac{3}{2R^4}\sqrt{\frac{35}{2\pi}}(a^4 - 6a^2b^2 + b^4), \tag{9.4.15}$$

$$A_6^0 = \frac{1}{4R^6}\sqrt{\frac{13}{\pi}}(231c^6 - 315c^4R^2 + 105c^2R^4 - 5R^6), \tag{9.4.16}$$

$$A_6^{\pm 2} = \frac{1}{8R^6}\sqrt{\frac{1365}{\pi}}(a^2 - b^2)(33c^4 - 18c^2R^2 + R^4), \tag{9.4.17}$$

$$A_6^{\pm 4} = \frac{3}{4R^6}\sqrt{\frac{91}{2\pi}}(a^4 - 6a^2b^2 + b^4)(11c^2 - R^2) \tag{9.4.18}$$

and

$$A_6^{\pm 6} = \frac{1}{8R^6}\sqrt{\frac{3003}{\pi}}(a^2 - b^2)(a^4 - 14a^2b^2 + b^4). \tag{9.4.19}$$

From Eqs. (9.4.1) and (9.4.10)–(9.4.19), one obtains the CEF potential. Then the CEF Hamiltonian is obtained multiplying the potential by the electron charge

$-e$ ($e = 4.8032 \times 10^{-10}$ stat C) and summing over the 11 electrons of the unfilled $4f$-shell of the $Er^{3+}$ ion. One obtains

$$\begin{aligned}\mathcal{H}_{\mathrm{CEF}} = &-\sum_i \frac{8eq}{R} - \frac{4\pi eq}{5R^3}\sum_i r_i^2\,[A_2^0 Y_2^0 + A_2^2(Y_2^2 + Y_2^{-2})] \\ &- \frac{4\pi eq}{9R^5}\sum_i r_i^4\,[A_4^0 Y_4^0 + A_4^2\,(Y_4^2 + Y_4^{-2}) + A_4^4\,(Y_4^4 + Y_4^{-4})] \\ &- \frac{4\pi eq}{13R^7}\sum_i r_i^6\,[A_6^0 Y_6^0 + A_6^2\,(Y_6^2 + Y_6^{-2}) \\ &+ A_6^4\,(Y_6^4 + Y_6^{-4}) + A_6^6\,(Y_6^6 + Y_6^{-6})], \end{aligned} \tag{9.4.20}$$

where the arguments of the spherical harmonics, omitted in Eq. (9.4.20), are the angular coordinates $(\theta_i, \phi_i)$ of the electrons. The free ion $Er^{3+}$ has an electronic configuration (Xe)$4f^{11}$ where Xe means the electronic configuration of the noble gas xenon. The spin and orbital quantum numbers of the ground-state multiplet are $S = \frac{3}{2}$ and $L = 6$, respectively. The total angular momentum is given by the Hund's rule $J = L + S = \frac{15}{2}$ so that the ground-state multiplet of the free $Er^{3+}$ ion is $|L = 6, S = \frac{3}{2}; J = \frac{15}{2}, M\rangle$ ($^4I_{15/2}$) with $M = -\frac{15}{2}, \ldots, \frac{15}{2}$ (16-fold degenerate). The first excited multiplet ($^4I_{13/2}$) has an energy 0.8 eV higher. Since the overall CEF splitting is of the order of $\sim$0.1 eV in a first approximation, one can neglect any mixing between different $J$ multiplets and one can use the first-order perturbation theory applied to the degenerate ground-state multiplet. Moreover, for electrons $4f$ the infinite terms of Hamiltonian (9.4.1) are reduced to the finite number of terms given in Eq. (9.4.20). Indeed, the matrix elements of the CEF Hamiltonian between the unperturbed states $|L = 6, S = \frac{3}{2}; J = \frac{15}{2}, M\rangle$ are combinations of Slater's determinants which are combinations of direct products of one-electron states $|\Psi_{nlm\sigma}\rangle = R_n(r) Y_l^m(\theta, \phi)\chi_\sigma$ where $n = 4$, $l = 3, m = -3, \ldots, 3$ and $\sigma = \pm\frac{1}{2}$. Then the matrix elements of the CEF Hamiltonian (9.4.20) $\langle\Psi_{4fm''\sigma''}|\mathcal{H}_{\mathrm{CEF}}|\Psi_{4fm'\sigma'}\rangle$ involve integrals like

$$(-1)^{m''}\delta_{\sigma',\sigma''}\langle r^s\rangle \int_0^\pi \sin\theta\, d\theta \int_0^{2\pi} d\phi Y_3^{-m''}(\theta,\phi) A_s^m Y_s^m(\theta,\phi) Y_3^{m'}(\theta,\phi), \tag{9.4.21}$$

where

$$\langle r^s\rangle = \int_0^\infty r^2 dr\, r^s R_4(r)^2 \tag{9.4.22}$$

with $m', m'' = -3, \ldots, 3$. The simultaneous occurrence of three spherical harmonics in the integral of Eq. (9.4.21) implies that it is not zero only if $s \le 6$ with $s$ an even integer and $|m'' - m'| = m$. In principle, one can obtain the matrix elements directly from Eq. (9.4.21) keeping the radial part (9.4.22) as a parameter since the radial function $R_4(r)$ of the free ion is not accurately known.

However, a much more convenient method of obtaining equivalent results is to replace the sums over the electron coordinates in Eq. (9.4.20) by the "Stevens' operator equivalents" in such a way that the matrix elements of the CEF Hamiltonian written in terms of the angular momentum operators $J_+, J_-, J_z$ between the states $|L,S;J,M\rangle$ are the same as the matrix elements of the original CEF Hamiltonian (9.4.20) between the same eigenfunctions. The correspondence between the sums over the eleven $4f$ electrons and the Stevens' operator equivalents is given by[85]

$$\sum_i r_i^2 Y_2^0\left(\frac{\boldsymbol{r}_i}{r_i}\right) = \frac{1}{4}\sqrt{\frac{5}{\pi}}\alpha_J\langle r^2\rangle O_2^0, \tag{9.4.23}$$

$$\sum_i r_i^2\left[Y_2^2\left(\frac{\boldsymbol{r}_i}{r_i}\right) + Y_2^{-2}\left(\frac{\boldsymbol{r}_i}{r_i}\right)\right] = \frac{1}{2}\sqrt{\frac{15}{2\pi}}\alpha_J\langle r^2\rangle O_2^2, \tag{9.4.24}$$

$$\sum_i r_i^4 Y_4^0\left(\frac{\boldsymbol{r}_i}{r_i}\right) = \frac{3}{16}\sqrt{\frac{1}{\pi}}\beta_J\langle r^4\rangle O_4^0, \tag{9.4.25}$$

$$\sum_i r_i^4\left[Y_4^2\left(\frac{\boldsymbol{r}_i}{r_i}\right) + Y_4^{-2}\left(\frac{\boldsymbol{r}_i}{r_i}\right)\right] = \frac{3}{4}\sqrt{\frac{5}{2\pi}}\beta_J\langle r^4\rangle O_4^2, \tag{9.4.26}$$

$$\sum_i r_i^4\left[Y_4^4\left(\frac{\boldsymbol{r}_i}{r_i}\right) + Y_4^{-4}\left(\frac{\boldsymbol{r}_i}{r_i}\right)\right] = \frac{3}{8}\sqrt{\frac{35}{2\pi}}\beta_J\langle r^4\rangle O_4^4, \tag{9.4.27}$$

$$\sum_i r_i^6 Y_6^0\left(\frac{\boldsymbol{r}_i}{r_i}\right) = \frac{1}{32}\sqrt{\frac{13}{\pi}}\gamma_J\langle r^6\rangle O_6^0, \tag{9.4.28}$$

$$\sum_i r_i^6\left[Y_6^2\left(\frac{\boldsymbol{r}_i}{r_i}\right) + Y_6^{-2}\left(\frac{\boldsymbol{r}_i}{r_i}\right)\right] = \frac{1}{32}\sqrt{\frac{1365}{\pi}}\gamma_J\langle r^6\rangle O_6^2, \tag{9.4.29}$$

$$\sum_i r_i^6\left[Y_6^4\left(\frac{\boldsymbol{r}_i}{r_i}\right) + Y_6^{-4}\left(\frac{\boldsymbol{r}_i}{r_i}\right)\right] = \frac{3}{16}\sqrt{\frac{91}{2\pi}}\gamma_J\langle r^6\rangle O_6^4 \tag{9.4.30}$$

and

$$\sum_i r_i^6\left[Y_6^6\left(\frac{\boldsymbol{r}_i}{r_i}\right) + Y_6^{-6}\left(\frac{\boldsymbol{r}_i}{r_i}\right)\right] = \frac{1}{32}\sqrt{\frac{3003}{\pi}}\gamma_J\langle r^6\rangle O_6^6, \tag{9.4.31}$$

where[85] $\alpha_J = \frac{4}{1575}$, $\beta_J = \frac{2}{45045}$ and $\gamma_J = \frac{8}{3864861}$ for $Er^{3+}$. For $J = \frac{15}{2}$, the Stevens' operator equivalents $O_n^m$ occurring in (9.4.23)–(9.4.31) are given by[85]

$$O_2^0 = 3J_z^2 - \frac{255}{4}, \tag{9.4.32}$$

$$O_4^0 = 35J_z^4 - \frac{3775}{2}J_z^2 + \frac{188955}{16}, \tag{9.4.33}$$

$$O_6^0 = 231J_z^6 - \frac{77385}{4}J_z^4 + \frac{6296829}{16}J_z^2 - \frac{72747675}{64}, \tag{9.4.34}$$

$$O_2^2 = \frac{1}{2}(J_+^2 + J_-^2), \tag{9.4.35}$$

$$O_4^2 = \frac{1}{4}\left[\left(7J_z^2 - \frac{275}{4}\right)(J_+^2 + J_-^2) + (J_+^2 + J_-^2)\left(7J_z^2 - \frac{275}{4}\right)\right], \tag{9.4.36}$$

$$O_6^2 = \frac{1}{4}\left[\left(33\,J_z^4 - \frac{2541}{2}J_z^2 + \frac{76857}{16}\right)(J_+^2 + J_-^2) + (J_+^2 + J_-^2)\left(33\,J_z^4 - \frac{2541}{2}J_z^2 + \frac{76857}{16}\right)\right], \tag{9.4.37}$$

$$O_4^4 = \frac{1}{2}(J_+^4 + J_-^4), \tag{9.4.38}$$

$$O_6^4 = \frac{1}{4}\left[\left(11\,J_z^2 - \frac{407}{4}\right)(J_+^4 + J_-^4) + (J_+^4 + J_-^4)\left(11\,J_z^2 - \frac{407}{4}\right)\right], \tag{9.4.39}$$

$$O_6^6 = \frac{1}{2}(J_+^6 + J_-^6) \tag{9.4.40}$$

and the CEF Hamiltonian (9.4.20) becomes[80]

$$\mathcal{H}_{\mathrm{CEF}} = B_2^0 O_2^0 + B_4^0 O_4^0 + B_6^0 O_6^0 + B_2^2 O_2^2 + B_4^2 O_4^2 + B_6^2 O_6^2 + B_4^4 O_4^4 + B_6^4 O_6^4 + B_6^6 O_6^6. \tag{9.4.41}$$

In Eq. (9.4.41), we have ignored the constant term of Eq. (9.4.20). Moreover, the coefficients $B_n^m$ occurring in Eq. (9.4.41) are proportional to the coefficients $A_n^m$ appearing in Eq. (9.4.20) through the averages $\langle r^2\rangle, \langle r^4\rangle$ and $\langle r^6\rangle$ which are not known. So the parameters $B_n^m$ have to be determined by a fitting with the experiment.[80] Indeed, the matrix elements of the CEF Hamiltonian (9.4.41)

$$\langle M|\mathcal{H}_{\mathrm{CEF}}|M'\rangle, \quad M, M' = \frac{15}{2}, \frac{13}{2}, \ldots, -\frac{13}{2}, -\frac{15}{2} \tag{9.4.42}$$

are evaluated using the matrix elements of the angular momentum operators $O_n^m$. The diagonal elements are given in Table 9.4. The non-zero off-diagonal terms of the

Table 9.4. Matrix elements of the diagonal Stevens' operator equivalents for $J = \frac{15}{2}$.

| $M$ | $\frac{1}{3}\langle M\|O_2^0\|M\rangle$ | $\frac{1}{60}\langle M\|O_4^0\|M\rangle$ | $\frac{1}{13860}\langle M\|O_6^0\|M\rangle$ |
|---|---|---|---|
| $\pm\frac{15}{2}$ | 35 | 273 | 65 |
| $\pm\frac{13}{2}$ | 21 | −91 | −117 |
| $\pm\frac{11}{2}$ | 9 | −221 | −39 |
| $\pm\frac{9}{2}$ | −1 | −201 | 59 |
| $\pm\frac{7}{2}$ | −9 | −101 | 87 |
| $\pm\frac{5}{2}$ | −15 | 23 | 45 |
| $\pm\frac{3}{2}$ | −19 | 129 | −25 |
| $\pm\frac{1}{2}$ | −21 | 189 | −75 |

Table 9.5. Matrix elements of the off-diagonal Stevens' operator equivalents for $J = \frac{15}{2}$.

| $M$ | $\langle \pm M \vert O_2^2 \vert \pm (M-2)\rangle$ | $\frac{1}{6}\langle \pm M \vert O_4^2 \vert \pm (M-2)\rangle$ | $\frac{1}{1320}\langle \pm M \vert O_6^2 \vert \pm (M-2)\rangle$ |
|---|---|---|---|
| $\frac{15}{2}$ | $\sqrt{105}$ | $39\sqrt{105}$ | $13\sqrt{105}$ |
| $\frac{13}{2}$ | $\sqrt{273}$ | $25\sqrt{273}$ | $\sqrt{273}$ |
| $\frac{11}{2}$ | $6\sqrt{13}$ | $80\sqrt{13}$ | $-21\sqrt{13}$ |
| $\frac{9}{2}$ | $2\sqrt{165}$ | $8\sqrt{165}$ | $-7\sqrt{165}$ |
| $\frac{7}{2}$ | $5\sqrt{33}$ | $-15\sqrt{33}$ | $-7\sqrt{33}$ |
| $\frac{5}{2}$ | $3\sqrt{105}$ | $-23\sqrt{105}$ | $3\sqrt{105}$ |
| $\frac{3}{2}$ | $12\sqrt{7}$ | $-120\sqrt{7}$ | $30\sqrt{7}$ |

Table 9.6. Matrix elements of the off-diagonal Stevens' operator equivalents for $J = \frac{15}{2}$.

| $M$ | $\frac{1}{12}\langle \pm M \vert O_4^4 \vert \pm (M-4)\rangle$ | $\frac{1}{660}\langle \pm M \vert O_4^6 \vert \pm (M-4)\rangle$ | $\frac{1}{360\sqrt{11}}\langle \pm M \vert O_6^6 \vert \pm (M-6)\rangle$ |
|---|---|---|---|
| $\frac{15}{2}$ | $\sqrt{1365}$ | $5\sqrt{1365}$ | $\sqrt{455}$ |
| $\frac{13}{2}$ | $\sqrt{5005}$ | $3\sqrt{5005}$ | $7\sqrt{39}$ |
| $\frac{11}{2}$ | $5\sqrt{429}$ | $7\sqrt{429}$ | $4\sqrt{273}$ |
| $\frac{9}{2}$ | $15\sqrt{77}$ | $3\sqrt{77}$ | 84 |
| $\frac{7}{2}$ | $10\sqrt{231}$ | $-6\sqrt{231}$ | $42\sqrt{5}$ |
| $\frac{5}{2}$ | $42\sqrt{15}$ | $-42\sqrt{15}$ | 0 |

CEF Hamiltonian are obtained from Tables 9.5 and 9.6. These tables reflect the properties of the raising and lowering angular momentum operators $J_\pm$. Indeed, the only non-zero terms of $O_n^2$ with $n = 2, 4, 6$ are those in which $M$ changes of $\pm 2$ since they contain $J_\pm^2$ operators. Similarly, the only non-zero matrix elements of the operators $O_n^4$ with $n = 4, 6$ and $O_6^6$ are those in which $M$ changes of $\pm 4$ and $\pm 6$, respectively. Note that the same matrix elements are obtained from the hermitian conjugate terms of CEF Hamiltonian since $(O_n^m)^+ = O_n^m$ so that the $16 \times 16$ matrix (9.4.42) is symmetric with respect to its main diagonal. The hermiticity of the CEF Hamiltonian assures that its eigenvalues are real. Using Tables 9.4–9.6 one can write all the elements of the CEF matrix for the ground-state multiplet. One can see that the CEF matrix is also symmetric with respect to its second diagonal, implying that all its eigenvalues are doubly degenerate. Obviously, this symmetry is broken by the introduction of an external magnetic field. The eigenvalues of the matrix (9.4.42) are functions of the parameters $B_n^m$ that are determined by fitting the differences between the matrix eigenvalues with the transition energies obtained from the INS. The best fit for $ErBa_2Cu_3O_7$ leads to the following values[80] of the $B_n^m$ (in meV)

$$\begin{aligned} &B_2^0 = 3.54 \times 10^{-2}, \quad B_4^0 = -1.43 \times 10^{-3}, \quad B_6^0 = 7.6 \times 10^{-6}, \\ &B_2^2 = 2.96 \times 10^{-2}, \quad B_4^2 = 4.56 \times 10^{-4}, \quad B_6^2 = -1.18 \times 10^{-6}, \\ &B_4^4 = 6.96 \times 10^{-3}, \quad B_6^4 = 2.16 \times 10^{-4}, \quad B_6^6 = 1.33 \times 10^{-6}. \end{aligned} \tag{9.4.43}$$

From the choice of parameters given in (9.4.43), one obtains the eigenvalues

$$\begin{aligned} E_1 &= E_2 = -41.33\,\text{meV} \Rightarrow E_{1,2}/k_B = 0\,\text{K}, \\ E_3 &= E_4 = -31.10\,\text{meV} \Rightarrow E_{3,4}/k_B = 119\,\text{K}, \\ E_5 &= E_6 = -30.46\,\text{meV} \Rightarrow E_{5,6}/k_B = 126\,\text{K}, \\ E_7 &= E_8 = -29.37\,\text{meV} \Rightarrow E_{7,8}/k_B = 139\,\text{K}, \\ E_9 &= E_{10} = 26.88\,\text{meV} \Rightarrow E_{9,10}/k_B = 792\,\text{K}, \\ E_{11} &= E_{12} = 30.74\,\text{meV} \Rightarrow E_{11,12}/k_B = 836\,\text{K}, \\ E_{13} &= E_{14} = 35.45\,\text{meV} \Rightarrow E_{13,14}/k_B = 891\,\text{K}, \\ E_{15} &= E_{16} = 39.19\,\text{meV} \Rightarrow E_{15,16}/k_B = 935\,\text{K}, \end{aligned} \tag{9.4.44}$$

where the last values of the energy levels are expressed in Kelvin degrees assuming the lowest energy doublet as the zero of the energy scale. The generic eigenvector belonging to the eigenvalue $E_n$ is given by

$$|\psi_n\rangle = \sum_{M=-15/2}^{M=15/2} a_M^n |M\rangle, \tag{9.4.45}$$

where the coefficients $a_M^n$ can be univocally determined using the normalization condition and breaking the symmetry of the CEF Hamiltonian entering an infinitesimal magnetic field directed along the $z$-axis. The infinitesimal magnetic field does not change the eigenvalues but eliminates the arbitrariness in the choice of the eigenstates of each doublet. The coefficients $a_M^n$ for the four doublets with the lowest energy are given in Table 9.7. The expectation values $\langle\psi_n|J_z|\psi_n\rangle$ between the eigenstates quoted in Table 9.7 are given by 1.903, −1.903, 5.995, −5.995, 0.586, −0.586, 0.028 and −0.028 for $n = 1, 2, \ldots$ and 8, respectively. Note that all the expectation values differ considerably from the saturation value $\frac{15}{2}$.

Now we evaluate the effective magnetic moments of the lowest energy doublet. Indeed, as one can see from Eq. (9.4.44), the excited doublets are more than 100 K higher in energy so that they do not contribute to the thermodynamic properties at temperatures $T \lesssim 1$ K where the ordered magnetic phase exists.[79] Using the second

Table 9.7. Eigenvectors belonging to the lower eigenvalues given in Eq. (9.4.44).

| $M$ | $a_M^1$ | $a_M^2$ | $a_M^3$ | $a_M^4$ | $a_M^5$ | $a_M^6$ | $a_M^7$ | $a_M^8$ |
|---|---|---|---|---|---|---|---|---|
| $\frac{15}{2}$ | 0 | 0.024 | 0.828 | 0 | 0 | 0.127 | −0.153 | 0 |
| $\frac{13}{2}$ | −0.542 | 0 | 0 | 0.082 | 0.135 | 0 | 0 | 0.461 |
| $\frac{11}{2}$ | 0 | −0.276 | −0.052 | 0 | 0 | −0.119 | −0.412 | 0 |
| $\frac{9}{2}$ | 0.003 | 0 | 0 | 0.048 | −0.254 | 0 | 0 | 0.059 |
| $\frac{7}{2}$ | 0 | −0.013 | −0.509 | 0 | 0 | −0.071 | 0.101 | 0 |
| $\frac{5}{2}$ | 0.584 | 0 | 0 | −0.067 | −0.096 | 0 | 0 | −0.347 |
| $\frac{3}{2}$ | 0 | 0.537 | 0.078 | 0 | 0 | 0.182 | −0.636 | 0 |
| $\frac{1}{2}$ | −0.005 | 0 | 0 | −0.182 | 0.917 | 0 | 0 | −0.238 |
| $-\frac{1}{2}$ | 0 | −0.005 | −0.182 | 0 | 0 | 0.917 | −0.238 | 0 |
| $-\frac{3}{2}$ | 0.537 | 0 | 0 | 0.078 | 0.182 | 0 | 0 | 0.636 |
| $-\frac{5}{2}$ | 0 | 0.584 | −0.067 | 0 | 0 | −0.096 | −0.347 | 0 |
| $-\frac{7}{2}$ | −0.013 | 0 | 0 | −0.509 | −0.071 | 0 | 0 | 0.101 |
| $-\frac{9}{2}$ | 0 | 0.003 | 0.048 | 0 | 0 | −0.254 | 0.059 | 0 |
| $-\frac{11}{2}$ | −0.276 | 0 | 0 | −0.052 | −0.119 | 0 | 0 | −0.412 |
| $-\frac{13}{2}$ | 0 | −0.542 | 0.082 | 0 | 0 | 0.135 | 0.461 | 0 |
| $-\frac{15}{2}$ | 0.024 | 0 | 0 | 0.828 | 0.127 | 0 | 0 | −0.153 |

and third column of Table 9.7, one finds

$$\langle\psi_1|J_z|\psi_1\rangle = -\langle\psi_2|J_z|\psi_2\rangle = 1.903, \quad \langle\psi_1|J_z|\psi_2\rangle = \langle\psi_2|J_z|\psi_1\rangle = 0 \tag{9.4.46}$$

so that one can write

$$J_z = 3.807\, S_z \Rightarrow \mu_z = g_J \mu_B J_z = 4.568 \mu_B S_z, \tag{9.4.47}$$

where $S_z = \frac{1}{2}\sigma_z$, $\sigma_z$ being one of the Pauli matrices, and the Landé factor $g_J = \frac{6}{5}$ is obtained using $J = \frac{15}{2}$, $L = 6$, $S = \frac{3}{2}$. The matrix elements of the raising and lowering angular momentum operators in the ground-state doublet are given by

$$\langle\psi_1|J_+|\psi_1\rangle = \langle\psi_2|J_+|\psi_2\rangle = 0, \quad \langle\psi_1|J_+|\psi_2\rangle = 6.442, \quad \langle\psi_2|J_+|\psi_1\rangle = -0.262 \tag{9.4.48}$$

and

$$\langle\psi_1|J_-|\psi_1\rangle = \langle\psi_2|J_-|\psi_2\rangle = 0, \quad \langle\psi_1|J_-|\psi_2\rangle = -0.262, \quad \langle\psi_2|J_-|\psi_1\rangle = 6.442 \tag{9.4.49}$$

leading to

$$J_x = \frac{1}{2}(J_+ + J_-) = 6.180\, S_x \Rightarrow \mu_x = g_J \mu_B J_x = 7.416\mu_B S_x \tag{9.4.50}$$

and

$$J_y = \frac{1}{2i}(J_+ - J_-) = 6.703\, S_y \Rightarrow \mu_x = g_J \mu_B J_y = 8.044\mu_B S_y, \tag{9.4.51}$$

where $S_x = \frac{1}{2}\sigma_x$ and $S_y = \frac{1}{2}\sigma_y$, $\sigma_x$ and $\sigma_y$ being the remaining two Pauli matrices. The principal information we receive from this calculation is that the presence of the surrounding oxygen ions removes the 16-fold degeneracy of the ground-state multiplet ($J = \frac{15}{2}$) of the "free" $Er^{3+}$ leading to a ground-state doublet ($S = \frac{1}{2}$) with anisotropic Landé factors[86] $g_{xx} = 7.416$, $g_{yy} = 8.044$ and $g_{zz} = 4.568$. As a consequence, the dipolar energy (9.1.11) becomes

$$E^{\alpha}_{\text{dip}}(\boldsymbol{G}/2) = -\frac{(g_{\alpha\alpha}\mu_B S)^2}{2} N D_V^{\alpha\alpha}(\boldsymbol{G}/2), \tag{9.4.52}$$

where $\alpha = x, y, z$, $S = \frac{1}{2}$ and $D_V^{\alpha\alpha}(\boldsymbol{G}/2)$ are given in Table 9.2 for $ErBa_2Cu_3O_7$. In particular, taking $a = 3.82$ Å and measuring the energy in Kelvin degrees one has

$$\begin{aligned} E^{x}_{\text{dip}}(\boldsymbol{G}/2)/(k_B N) &= -0.0768\,[a^3 D_V^{xx}(\boldsymbol{G}/2)], \\ E^{y}_{\text{dip}}(\boldsymbol{G}/2)/(k_B N) &= -0.0904\,[a^3 D_V^{yy}(\boldsymbol{G}/2)], \\ E^{z}_{\text{dip}}(\boldsymbol{G}/2)(k_B N) &= -0.0292[a^3 D_V^{zz}(\boldsymbol{G}/2)]. \end{aligned} \tag{9.4.53}$$

Looking at the values of $a^3 D_V^{\alpha\alpha}(\boldsymbol{G}/2)$ of Table 9.2 we conclude that the CEF due to the surrounding oxygen leads to a ground state corresponding to the configurations $(\frac{1}{2}, 0, \frac{1}{2})$ or $(\frac{1}{2}, 0, 0)$ (very close in energy) with the spins directed along the ferromagnetic $b$-axis in agreement with the experiment.[79] Note that these configurations become stable only for $g_{yy}/g_{xx} > 1.018$ as one can deduce from Eq. (9.4.53) and Table 9.2. For $g_{yy}/g_{xx} < 1.018$, the ground-state configurations are again those shown in Fig. 9.2. The lowest dipolar energy (9.4.53) becomes $E^{y}_{\text{dip}}(\frac{1}{2}, 0, \frac{1}{2})/(k_B N) = -0.442\,\text{K}$, of the order of magnitude of the transition temperature $T_N = 0.62\,\text{K}$. Moreover, the sublattice magnetization that corresponds to the square root of the intensity of the elastic peak at $(\frac{1}{2}, 0, \frac{1}{2})$ is seen to vanish at the transition following the power law $M_s \sim (T_N - T)^{1/8}$ that corresponds to the critical behaviour of a 2D, $S = \frac{1}{2}$ Ising model. This result is in agreement with the CEF calculation that leads to a ground state with an effective spin $S = \frac{1}{2}$ and uniaxial (Ising) anisotropy. The 2D character is explained by the large ratio $c/a \sim 3$ while the critical exponent of the order parameter $\beta = \frac{1}{8}$ is the exact result for a 2D, $S = \frac{1}{2}$ Ising model with NN exchange interaction.

The tetragonal symmetry ($a = b$) of the deoxygenated non-superconducting compound $ErBa_2Cu_3O_6$ reduces the number of terms of the CEF Hamiltonian (9.4.20). In particular, from Eqs. (9.4.12), (9.4.14), (9.4.17) and (9.4.19) one can see that $A_2^{\mp 2}{=}A_4^{\mp 2}{=}A_6^{\mp 2}{=}A_6^{\pm 6}{=}0$ and the best fit of the remaining non-zero parameters obtained from the INS[80] gives

$$\begin{gathered}B_2^0 = 1.6\times 10^{-2},\quad B_4^0 = -1.49\times 10^{-3},\quad B_6^0 = 7.39\times 10^{-6},\\ B_4^4 = 6.94\times 10^{-3},\quad B_6^4 = 2.16\times 10^{-4}.\end{gathered} \tag{9.4.54}$$

The energy levels of the CEF Hamiltonian become

$$\begin{aligned}
E_1 = E_2 &= -41.00\,\text{meV} \Rightarrow E_{1,2}/k_B = 0\,\text{K},\\
E_3 = E_4 &= -33.21\,\text{meV} \Rightarrow E_{3,4}/k_B = 90\,\text{K},\\
E_5 = E_6 &= -29.64\,\text{meV} \Rightarrow E_{5,6}/k_B = 132\,\text{K},\\
E_7 = E_8 &= -29.07\,\text{meV} \Rightarrow E_{7,8}/k_B = 138\,\text{K},\\
E_9 = E_{10} &= 28.03\,\text{meV} \Rightarrow E_{9,10}/k_B = 801\,\text{K},\\
E_{11} = E_{12} &= 31.32\,\text{meV} \Rightarrow E_{11,12}/k_B = 839\,\text{K},\\
E_{13} = E_{14} &= 34.62\,\text{meV} \Rightarrow E_{13,14}/k_B = 878\,\text{K},\\
E_{15} = E_{16} &= 38.95\,\text{meV} \Rightarrow E_{15,16}/k_B = 928\,\text{K}.
\end{aligned} \tag{9.4.55}$$

The two eigenstates belonging to the ground-state doublet are given by

$$|\psi_1\rangle = -0.558|13/2\rangle + 0.589|5/2\rangle + 0.520|-3/2\rangle - 0.266|-11/2\rangle \tag{9.4.56}$$

and

$$|\psi_2\rangle = -0.266|11/2\rangle + 0.520|3/2\rangle + 0.589|-5/2\rangle - 0.558|-13/2\rangle. \tag{9.4.57}$$

From the eigenstates (9.4.56) and (9.4.57), one obtains

$$J_z = 4.198\,S_z \Rightarrow \mu_z = g_J\mu_B J_z = 5.038\mu_B S_z, \tag{9.4.58}$$

$$J_x = 6.318\,S_x \Rightarrow \mu_x = g_J\mu_B J_x = 7.581\mu_B S_x \tag{9.4.59}$$

and

$$J_y = 6.318\,S_y \Rightarrow \mu_y = g_J\mu_B J_y = 7.581\mu_B S_y \tag{9.4.60}$$

so that $g_{zz} = g_{\parallel} = 5.038$, $g_{xx} = g_{yy} = g_{\perp} = 7.162$. The easy-plane character of the anisotropy does not modify the scenario obtained from the dipole–dipole interaction for a free ion. The two ground-state configurations deduced by Table 9.3 correspond to the wavevectors $(0, \frac{1}{2}, \frac{1}{2})$ and $(\frac{1}{2}, 0, \frac{1}{2})$ with the spins directed along the ferromagnetic axis ($a$ for the former, $b$ for the latter ground state). The INS experiment[79] on $ErBa_2Cu_3O_{6+x}$ with $x < 0.3$ rules out the existence of 3D LRO down to $T = 0.06\,\text{K}$. The 2D short range order corresponds to a ridge $\boldsymbol{Q} = (\frac{1}{2}, 0, q)$ with $0 < q < \frac{1}{2}$ with the spins in the $ab$-plane. The absence of LRO contrasts with the CEF and dipolar energy calculation that lead to a ground-state doublet that supports the existence of LRO. The specific-heat data[87] confirm the 2D, $S = \frac{1}{2}$ Ising-like behaviour for the (ordered) orthorhombic compound. On the contrary, the data indicate a 2D, planar rotator model behaviour for the (disordered) tetragonal compound.

## 9.5. Spin Waves in $ErBa_2Cu_3O_7$

To get the spin wave spectrum of $ErBa_2Cu_3O_7$, we make use of the transformation (7.1.2) assuming $\theta = \frac{\pi}{2}$, $\phi = \frac{\pi}{2}$, $\boldsymbol{Q} \equiv \frac{\boldsymbol{G}}{2} = (\frac{1}{2}, 0, \frac{1}{2})$, $g_{xx} = 7.416$, $g_{yy} = 8.044$ and $g_{zz} = 4.568$ corresponding to the ground-state configuration of $ErBa_2Cu_3O_7$. In this way, the transformation (7.1.2) reduces to

$$S_i^x = -S_i^\eta \cos\frac{\boldsymbol{G}\cdot\boldsymbol{r}_i}{2}, \quad S_i^y = S_i^\zeta \cos\frac{\boldsymbol{G}\cdot\boldsymbol{r}_i}{2}, \quad S_i^z = -S_i^\xi. \tag{9.5.1}$$

Then the spin-boson transformation

$$\begin{aligned} S_i^\xi &= \frac{\sqrt{2S}}{2}(a_i + a_i^+), \\ S_i^\eta &= \frac{\sqrt{2S}}{2i}(a_i - a_i^+), \\ S_i^\zeta &= S - a_i^+ a_i \end{aligned} \tag{9.5.2}$$

can be used to write the harmonic Hamiltonian

$$\begin{aligned} \mathcal{H}_2^{\text{dip}} = & E_{\text{dip}}^y\left(\frac{\boldsymbol{G}}{2}\right) + \sum_{\boldsymbol{q}} A_{\boldsymbol{q}}^{\text{dip}}\, a_{\boldsymbol{q}}^+ a_{\boldsymbol{q}} + \frac{1}{2}\sum_{\boldsymbol{q}} B_{\boldsymbol{q}}^{\text{dip}}\,(a_{\boldsymbol{q}} a_{-\boldsymbol{q}} + a_{\boldsymbol{q}}^+ a_{-\boldsymbol{q}}^+) \\ & + \frac{1}{2}\sum_{\boldsymbol{q}} iC_{\boldsymbol{q}}^{\text{dip}}\left(a_{\boldsymbol{q}} a_{-\boldsymbol{q}-\frac{\boldsymbol{G}}{2}} - a_{\boldsymbol{q}}^+ a_{-\boldsymbol{q}-\frac{\boldsymbol{G}}{2}}^+\right) + \frac{1}{2}\sum_{\boldsymbol{q}} iD_{\boldsymbol{q}}^{\text{dip}}\left(a_{\boldsymbol{q}}^+ a_{\boldsymbol{q}+\frac{\boldsymbol{G}}{2}} - a_{\boldsymbol{q}+\frac{\boldsymbol{G}}{2}}^+ a_{\boldsymbol{q}}\right), \end{aligned} \tag{9.5.3}$$

where

$$E_{\text{dip}}^y\left(\frac{\boldsymbol{G}}{2}\right) = -\frac{1}{2}(g_{yy}\mu_B S)^2 N D^{yy}\left(\frac{\boldsymbol{G}}{2}\right), \tag{9.5.4}$$

$$A_{\boldsymbol{q}}^{\text{dip}} = \frac{1}{2}\mu_B^2 S\left[2g_{yy}^2 D^{yy}\left(\frac{\boldsymbol{G}}{2}\right) - g_{xx}^2 D^{xx}\left(\frac{\boldsymbol{G}}{2} + \boldsymbol{q}\right) - g_{zz}^2 D^{zz}(\boldsymbol{q})\right], \tag{9.5.5}$$

$$B_{\boldsymbol{q}}^{\text{dip}} = \frac{1}{2}\mu_B^2 S\left[g_{xx}^2 D^{xx}\left(\frac{\boldsymbol{G}}{2} + \boldsymbol{q}\right) - g_{zz}^2 D^{zz}(\boldsymbol{q})\right], \tag{9.5.6}$$

$$C_{\boldsymbol{q}}^{\text{dip}} = \frac{1}{2}\mu_B^2 S g_{xx} g_{zz}\left[D^{xz}(\boldsymbol{q}) + D^{xz}\left(\boldsymbol{q} + \frac{\boldsymbol{G}}{2}\right)\right] \tag{9.5.7}$$

and

$$D_{\boldsymbol{q}}^{\text{dip}} = \frac{1}{2}\mu_B^2 S g_{xx} g_{zz}\left[D^{xz}\left(\boldsymbol{q} + \frac{\boldsymbol{G}}{2}\right) - D^{xz}(\boldsymbol{q})\right]. \tag{9.5.8}$$

To obtain Eq. (9.5.3), we have made use of the properties $C_{\boldsymbol{q}+\frac{\boldsymbol{G}}{2}}^{\text{dip}} = C_{\boldsymbol{q}}^{\text{dip}}$ and $D_{\boldsymbol{q}+\frac{\boldsymbol{G}}{2}}^{\text{dip}} = -D_{\boldsymbol{q}}^{\text{dip}}$ coming from Eqs. (9.5.7) and (9.5.8) and of the fact that $\boldsymbol{G}$ is a reciprocal lattice vector. The Hamiltonian (9.5.3) can be diagonalised using the same procedure of Section 7.9. The transformation

$$\alpha_{\boldsymbol{k}}^{(i)} = u_{\boldsymbol{k}}^{(i)} a_{\boldsymbol{k}} - l_{\boldsymbol{k}}^{(i)} a_{-\boldsymbol{k}}^{+} + v_{\boldsymbol{k}}^{(i)} a_{\boldsymbol{k}+\frac{\boldsymbol{G}}{2}} - m_{\boldsymbol{k}}^{(i)} a_{-\boldsymbol{k}-\frac{\boldsymbol{G}}{2}}^{+} \tag{9.5.9}$$

with

$$|u_{\boldsymbol{k}}^{(i)}|^2 - |l_{\boldsymbol{k}}^{(i)}|^2 + |v_{\boldsymbol{k}}^{(i)}|^2 - |m_{\boldsymbol{k}}^{(i)}|^2 = 1 \tag{9.5.10}$$

and

$$u_{-\boldsymbol{k}}^{(i)} = u_{\boldsymbol{k}}^{(i)*}, \quad l_{-\boldsymbol{k}}^{(i)} = l_{\boldsymbol{k}}^{(i)*}, \quad v_{-\boldsymbol{k}} = v_{\boldsymbol{k}}^{(i)*}, \quad m_{-\boldsymbol{k}}^{(i)} = m_{\boldsymbol{k}}^{(i)*} \tag{9.5.11}$$

brings the Hamiltonian (9.5.3) into the form

$$\mathcal{H}_2 = E_0' + \sum_i \sum_{\boldsymbol{k}} \hbar\omega_{\boldsymbol{k}}^{(i)*} \alpha_{\boldsymbol{k}}^{(i)+} \alpha_{\boldsymbol{k}}^{(i)} \tag{9.5.12}$$

where the frequencies $\omega_{\boldsymbol{k}}^{(i)}$ are the solutions of the determinant equation

$$\det \begin{pmatrix} A_{\boldsymbol{k}} - \hbar\omega_{\boldsymbol{k}}^{(i)} & B_{\boldsymbol{k}} & -iD_{\boldsymbol{k}} & iC_{\boldsymbol{k}} \\ B_{\boldsymbol{k}} & A_{\boldsymbol{k}} + \hbar\omega_{\boldsymbol{k}}^{(i)} & -iC_{\boldsymbol{k}} & iD_{\boldsymbol{k}} \\ iD_{\boldsymbol{k}} & iC_{\boldsymbol{k}} & A_{\boldsymbol{k}+\frac{\boldsymbol{G}}{2}} - \hbar\omega_{\boldsymbol{k}}^{(i)} & B_{\boldsymbol{k}+\frac{\boldsymbol{G}}{2}} \\ -iC_{\boldsymbol{k}} & -iD_{\boldsymbol{k}} & B_{\boldsymbol{k}+\frac{\boldsymbol{G}}{2}} & A_{\boldsymbol{k}+\frac{\boldsymbol{G}}{2}} + \hbar\omega_{\boldsymbol{k}}^{(i)} \end{pmatrix} = 0. \tag{9.5.13}$$

In Eq. (9.5.13) we have dropped the superscript "dip" occurring in Eq. (9.5.3). The determinant equation (9.5.13) leads to a biquadratic equation in the variable $\hbar\omega_{\boldsymbol{k}}^{(i)}$ whose positive (physical) solutions are[86]

$$\begin{aligned} \hbar\omega_{\boldsymbol{k}}^{(1,2)} = \Bigg\{ & \frac{1}{2}(A_{\boldsymbol{k}}^2 - B_{\boldsymbol{k}}^2 + A_{\boldsymbol{k}+\frac{\boldsymbol{G}}{2}}^2 - B_{\boldsymbol{k}+\frac{\boldsymbol{G}}{2}}^2) - C_{\boldsymbol{k}}^2 + D_{\boldsymbol{k}}^2 \\ & \pm \Bigg[ \frac{1}{4}(A_{\boldsymbol{k}}^2 - B_{\boldsymbol{k}}^2 - A_{\boldsymbol{k}+\frac{\boldsymbol{G}}{2}}^2 + B_{\boldsymbol{k}+\frac{\boldsymbol{G}}{2}}^2)^2 - C_{\boldsymbol{k}}^2 \Big( (A_{\boldsymbol{k}} - A_{\boldsymbol{k}+\frac{\boldsymbol{G}}{2}})^2 \\ & - (B_{\boldsymbol{k}} - B_{\boldsymbol{k}+\frac{\boldsymbol{G}}{2}})^2 \Big) + D_{\boldsymbol{k}}^2 \Big( (A_{\boldsymbol{k}} - A_{\boldsymbol{k}+\frac{\boldsymbol{G}}{2}})^2 - (B_{\boldsymbol{k}} - B_{\boldsymbol{k}+\frac{\boldsymbol{G}}{2}})^2 \Big) \\ & + 4C_{\boldsymbol{k}} D_{\boldsymbol{k}} \left( A_{\boldsymbol{k}} B_{\boldsymbol{k}+\frac{\boldsymbol{G}}{2}} - A_{\boldsymbol{k}+\frac{\boldsymbol{G}}{2}} B_{\boldsymbol{k}} \right) \Bigg]^{\frac{1}{2}} \Bigg\}^{\frac{1}{2}}. \end{aligned} \tag{9.5.14}$$

A considerable simplification of Eq. (9.5.14) is obtained by taking $C_{\boldsymbol{k}} = D_{\boldsymbol{k}} = 0$. This assumption is rigorously true for wavevectors $\boldsymbol{k}$ parallel to the high symmetry directions where $D^{xz}(\boldsymbol{k}) = 0$ (see Eqs. (9.5.7) and (9.5.8)). For $\boldsymbol{k} = (\frac{2\pi}{a}h, \frac{2\pi}{b}h, 1.3\frac{2\pi}{c}) \equiv (h, h, 1.3)$ (r.l.u), the scattering vectors used in the INS experiment[68], $D^{xz}(\boldsymbol{k})$ is not zero but its value is so small that the difference between

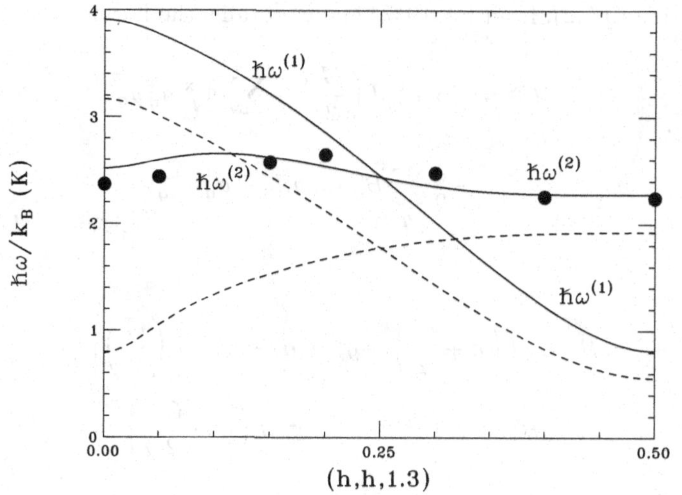


Fig. 9.4. Spin wave spectra $\hbar\omega^{(1)}/k_B$ (Eq. 9.5.15) and $\hbar\omega^{(2)}/k_B$ (Eq. 9.5.16) along $\boldsymbol{k} = (h, h, 1.3)$ (r.l.u.) for $ErBa_2Cu_3O_7$. The dashed curves are obtained for a pure dipolar model with $g_{xx} = 7.416$, $g_{yy} = 8.044$ and $g_{zz} = 4.568$. The continuous curves are obtained for a model with dipolar and exchange interaction ($J_{NN} = 0.0054$ K, $J'_{NN} = 0.05 J_{NN}$ and $J_{NNN} = 0.45 J_{NN}$). The full circles are the experimental data obtained from the INS[88] on $ErBa_2Cu_3O_7$.

the spin wave frequencies with and without $C_{\boldsymbol{k}}$ and $D_{\boldsymbol{k}}$ cannot be appreciated in Fig. 9.4 where the dash curves show the magnon spectra given by

$$\hbar\omega_{\boldsymbol{k}}^{(1)} = \sqrt{A_{\boldsymbol{k}}^2 - B_{\boldsymbol{k}}^2} \tag{9.5.15}$$

and

$$\hbar\omega_{\boldsymbol{k}}^{(2)} = \sqrt{A_{\boldsymbol{k}+\frac{\boldsymbol{G}}{2}}^2 - B_{\boldsymbol{k}+\frac{\boldsymbol{G}}{2}}^2} \tag{9.5.16}$$

obtained from Eq. (9.5.14) putting $C_{\boldsymbol{k}} = D_{\boldsymbol{k}} = 0$.

As one can see, the experimental data (full circles) are not recovered. We try to recover the experimental data by introducing a weak exchange interaction between the $Er^{3+}$ ions represented by the Hamiltonian

$$\mathcal{H}^{\text{ex}} = \sum_{i,\boldsymbol{\delta}_{\text{NN}}} \sum_{\alpha} g_{\alpha\alpha}^2 J_{\text{NN}} S_i^\alpha S_{i+\boldsymbol{\delta}_{\text{NN}}}^\alpha + \sum_{i,\boldsymbol{\delta}'_{\text{NN}}} \sum_{\alpha} g_{\alpha\alpha}^2 J'_{\text{NN}} S_i^\alpha S_{i+\boldsymbol{\delta}'_{\text{NN}}}^\alpha + \sum_{i,\boldsymbol{\delta}_{\text{NNN}}} \sum_{\alpha} g_{\alpha\alpha}^2 J_{\text{NNN}} S_i^\alpha S_{i+\boldsymbol{\delta}_{\text{NNN}}}^\alpha, \tag{9.5.17}$$

where $J_{\text{NN}}$ is the in-plane NN interaction with $\boldsymbol{\delta}_{\text{NN}} = (\pm a, 0, 0)$ and $(0, \pm b, 0)$, $J'_{\text{NN}}$ is the out-of-plane NN interaction with $\boldsymbol{\delta}'_{\text{NN}} = (0, 0, \pm c)$ and $J_{\text{NNN}}$ is the in-plane NNN interaction with $\boldsymbol{\delta}_{\text{NNN}} = (\pm a, \pm b, 0)$. At variance with the isotropic Heisenberg Hamiltonian, the factors $g_{\alpha\alpha}$ with $\alpha = x, y, z$ occur in Eq. (9.5.17) in agreement with the CEF calculation of the previous section. Using the spin-boson transformation (9.5.2), from Eq. (9.5.17) one obtains the harmonic Hamiltonian

$$\mathcal{H}^{\text{ex}} = NS^2 g_{yy}^2 J\left(\frac{\boldsymbol{G}}{2}\right) + \sum_{\boldsymbol{q}} A_{\boldsymbol{q}}^{\text{ex}}\, a_{\boldsymbol{q}}^+ a_{\boldsymbol{q}} + \frac{1}{2}\sum_{\boldsymbol{q}} B_{\boldsymbol{q}}^{\text{ex}}(a_{\boldsymbol{q}}a_{-\boldsymbol{q}} + a_{\boldsymbol{q}}^+ a_{-\boldsymbol{q}}^+) \tag{9.5.18}$$

where

$$A_{\boldsymbol{q}}^{\text{ex}} = S\left[g_{xx}^2 J\left(\boldsymbol{q} + \frac{\boldsymbol{G}}{2}\right) + g_{zz}^2 J(\boldsymbol{q}) - 2g_{yy}^2 J\left(\frac{\boldsymbol{G}}{2}\right)\right], \tag{9.5.19}$$

$$B_{\boldsymbol{q}}^{\text{ex}} = S\left[g_{zz}^2 J(\boldsymbol{q}) - g_{xx}^2 J\left(\boldsymbol{q} + \frac{\boldsymbol{G}}{2}\right)\right] \tag{9.5.20}$$

with

$$J(\boldsymbol{q}) = 2J_{\text{NN}}(\cos aq_x + \cos bq_y) + 2J_{\text{NN}}' \cos cq_z + 4J_{\text{NNN}} \cos aq_x \cos bq_y. \tag{9.5.21}$$

Note that from the lattice parameters of $ErBa_2Cu_3O_7$ ($c \sim 3a$ and $b \sim a$) one expects that $J_{\text{NN}} > J_{\text{NNN}} > J_{\text{NN}}'$. The only effect of a weak antiferromagnetic coupling $J_{\text{NN}}' > 0$ is to stabilize the antiferromagnetic ordering along the $c$-axis while the introduction of a NNN in-plane antiferromagnetic interaction $J_{\text{NNN}} > 0$ favours a columnar order, competing with the NN in-plane antiferromagnetic interaction $J_{\text{NN}} > 0$ that supports a Néel order. The ground-state configuration of $ErBa_2Cu_3O_7$, corresponding to $\boldsymbol{G} = (\frac{1}{2}, 0, \frac{1}{2})$, imposes an upper limit to $J_{\text{NN}}$ that can be obtained comparing the ground-state energy of the configuration really observed with the Néel configuration supported by the dominant antiferromagnetic exchange interaction. Indeed, imposing that $E_{\text{dip}}^y(\frac{1}{2}, 0, \frac{1}{2}) + NS^2 g_{yy}^2 J(\frac{1}{2}, 0, \frac{1}{2}) < E_{\text{dip}}^z(\frac{1}{2}, \frac{1}{2}, \frac{1}{2}) + NS^2 g_{zz}^2 J(\frac{1}{2}, \frac{1}{2}, \frac{1}{2})$ and using $g_{yy} = 8.044$, $g_{zz} = 4.568$, one obtains

$$J_{\text{NN}} < 0.0176 + 1.050\, J_{\text{NN}}' + 4.101\, J_{\text{NNN}} \ \text{(K)}. \tag{9.5.22}$$

The spin wave spectra including the exchange interaction are still given by Eqs. (9.5.15) and Eq. (9.5.14) with $A_{\boldsymbol{k}} = A_{\boldsymbol{k}}^{\text{dip}} + A_{\boldsymbol{k}}^{\text{ex}}$ and $B_{\boldsymbol{k}} = B_{\boldsymbol{k}}^{\text{dip}} + B_{\boldsymbol{k}}^{\text{ex}}$. The continuous curves of Fig. 9.4 show the spin wave spectra for $J_{\text{NN}} = 0.0054$ K, $J' = 0.05\, J_{\text{NN}}$ and $J_{\text{NNN}} = 0.45 J_{\text{NN}}$. The agreement of $\hbar\omega_{\boldsymbol{k}}^{(2)}$ with the experimental data is excellent. The choice of the exchange parameters is not unique: indeed, for $0.0045 \lesssim J_{\text{NN}} \lesssim 0.0063$ K, the agreement with the experimental data remains good provided that the value of $J_{\text{NNN}}$ is correspondingly changed.[86] The main discrepancy between the theory and the experiment pointed out in Fig. 9.4 is that the experimental data[88] fall on a unique branch. In order to understand why the INS cross-section leads to such a result, it becomes important to evaluate explicitly the neutron scattering cross-section for the model taken to fit experimental data. Using the spin wave spectra (9.5.15) and (9.5.16), the INS cross-section may be obtained in the same manner as in Section 7.10. One has

$$\left(\frac{d^2\sigma}{d\Omega\, dE'}\right)_{\text{inel}} = r_0^2 \frac{k'}{k}\left[\frac{1}{2}F(\boldsymbol{K})\right]^2 e^{-2W(\boldsymbol{K})}\frac{SN}{2\hbar}\Big\{I^{xx}(\boldsymbol{K})\Big[(1+n^{(2)})\delta(\omega-\omega_{\boldsymbol{K}}^{(2)}) + n^{(2)}\delta(\omega+\omega_{\boldsymbol{K}}^{(2)})\Big] + I^{zz}(\boldsymbol{K})\Big[(1+n^{(1)})\delta(\omega-\omega_{\boldsymbol{K}}^{(1)}) + n_1\delta(\omega+\omega_{\boldsymbol{K}}^{(1)})\Big]\Big\}, \tag{9.5.23}$$

where

$$I^{xx}(\boldsymbol{K}) = g_{xx}^2\left(1-\frac{K_x^2}{K^2}\right)\left[\frac{A_{\boldsymbol{K}+\frac{\boldsymbol{G}}{2}} - |B_{\boldsymbol{K}+\frac{\boldsymbol{G}}{2}}|}{A_{\boldsymbol{K}+\frac{\boldsymbol{G}}{2}} + |B_{\boldsymbol{K}+\frac{\boldsymbol{G}}{2}}|}\right]^{\frac{1}{2}}, \tag{9.5.24}$$

$$I^{zz}(\boldsymbol{K}) = g_{zz}^2\left(1-\frac{K_z^2}{K^2}\right)\left[\frac{A_{\boldsymbol{K}} + |B_{\boldsymbol{K}}|}{A_{\boldsymbol{K}} - |B_{\boldsymbol{K}}|}\right]^{\frac{1}{2}} \tag{9.5.25}$$

and

$$n^{(i)} = \frac{1}{e^{\beta\hbar\omega_{\boldsymbol{K}}^{(i)}} - 1}. \tag{9.5.26}$$

In Table 9.8, we give the peak intensities $I^{zz}(\boldsymbol{K})$ and $I^{xx}(\boldsymbol{K})$ along with the related frequencies $\omega_{\boldsymbol{K}}^{(1)}$ and $\omega_{\boldsymbol{K}}^{(2)}$ for $g_{xx} = 7.416, g_{zz} = 4.568, b/a = 1.016, c/a = 3.052$ and scattering vectors $\boldsymbol{K} = (h, h, 1.3)$. As one can see, the peak intensity $I^{xx}(\boldsymbol{K})$ corresponding to the branch $\omega_{\boldsymbol{K}}^{(2)}$ is greater than $I^{zz}(\boldsymbol{K})$ corresponding to the branch $\omega_{\boldsymbol{K}}^{(1)}$ for any $\boldsymbol{K}$ except near the zone boundary. This is due to the geometric factor $(1 - K_x^2/K^2)$ that is greater than $(1 - K_z^2/K^2)$ for any $h$ except $h \gtrsim 0.4$. Indeed, by using the parameters chosen in Table 9.8, one has

$$\frac{1-\frac{K_x^2}{K^2}}{1-\frac{K_z^2}{K^2}} = \frac{0.969h^2 + 0.181}{1.969h^2} \tag{9.5.27}$$

that goes to $\infty$ for $h \to 0$. In any case, this leads to a different weight of the two peaks in the neutron scattering cross-section even though it does not explain the

Table 9.8. Peak intensity $I^{xx}(\boldsymbol{K})$ and $I^{zz}(\boldsymbol{K})$ and related spin wave energies $\hbar\omega_{\boldsymbol{K}}^{(1)}$ frequencies and $\hbar\omega_{\boldsymbol{K}}^{(2)}$ for selected values of the scattering vector $\boldsymbol{K} = (h, h, 1.3)$ (r.l.u.).

| $h$ | $I^{zz}(\boldsymbol{K})$ | $\hbar\omega_{\boldsymbol{K}}^{(1)}/k_B$ (K) | $I^{xx}(\boldsymbol{K})$ | $\hbar\omega_{\boldsymbol{K}}^{(2)}/k_B$ (K) |
|---|---|---|---|---|
| 0 | 0 | 3.911 | 48.518 | 2.513 |
| 0.05 | 0.563 | 3.773 | 46.626 | 2.599 |
| 0.1 | 2.093 | 3.522 | 44.790 | 2.656 |
| 0.15 | 4.328 | 3.218 | 43.979 | 2.624 |
| 0.2 | 6.704 | 2.854 | 43.578 | 2.538 |
| 0.25 | 8.618 | 2.446 | 42.998 | 2.436 |
| 0.3 | 11.076 | 2.020 | 40.638 | 2.350 |
| 0.35 | 14.983 | 1.606 | 38.326 | 2.299 |
| 0.4 | 20.825 | 1.229 | 37.047 | 2.279 |
| 0.45 | 29.064 | 0.934 | 35.438 | 2.275 |
| 0.5 | 35.358 | 0.811 | 33.995 | 2.276 |

presence of a single peak in the experiment for $h \gtrsim 0.3$. It should be noted, however, that in any experiment a limited energy resolution prevents the occurrence of $\delta$-functions in the cross-section. In particular, in the neutron scattering cross-section of $ErBa_2Cu_3O_7$ the full width at half maximum (FWHM) of the data points[88] is $\frac{\hbar\Delta\omega}{k_B} \simeq 1.2$ K. Then a rough simulation of the neutron scattering peak profile at low temperature may be obtained from Eq. (9.5.23), neglecting all terms containing Bose factors and replacing the $\delta$-functions by Gaussians like

$$G^{(i)}(\boldsymbol{K},\omega) = \frac{1}{\sigma\sqrt{2\pi}} \exp^{-\frac{(\omega-\omega_{\boldsymbol{K}}^{(i)})^2}{2\sigma^2}}, \tag{9.5.28}$$

where $i = 1, 2$ and $\sigma = \frac{\hbar\Delta\omega}{2k_B\sqrt{\ln 4}} \simeq 0.51$ K according to the energy resolution of the experiment.[88] In Fig. 9.5, the dashed curves are the Gaussians $G^{(1)}$ and $G^{(2)}$ centred at the frequencies $\omega_{\boldsymbol{K}}^{(1)}$ and $\omega_{\boldsymbol{K}}^{(2)}$ multiplied by $I^{zz}(\boldsymbol{K})$ and $I^{xx}(\boldsymbol{K})$, respectively. The continuous curves are the sums of the two Gaussians. As one can see, the existence of a single peak for $h \lesssim 0.3$ is due to the ratio $I^{zz}/I^{xx} \ll 1$, so that the sums of the two Gaussians do not differ substantially from the Gaussian centred at $\omega_{\boldsymbol{K}}^{(2)}$. On the contrary, the existence of a single, even though asymmetric, peak for $h \gtrsim 0.3$ is due to the limited energy resolution comparable with the energy shift of the two spin wave branches.

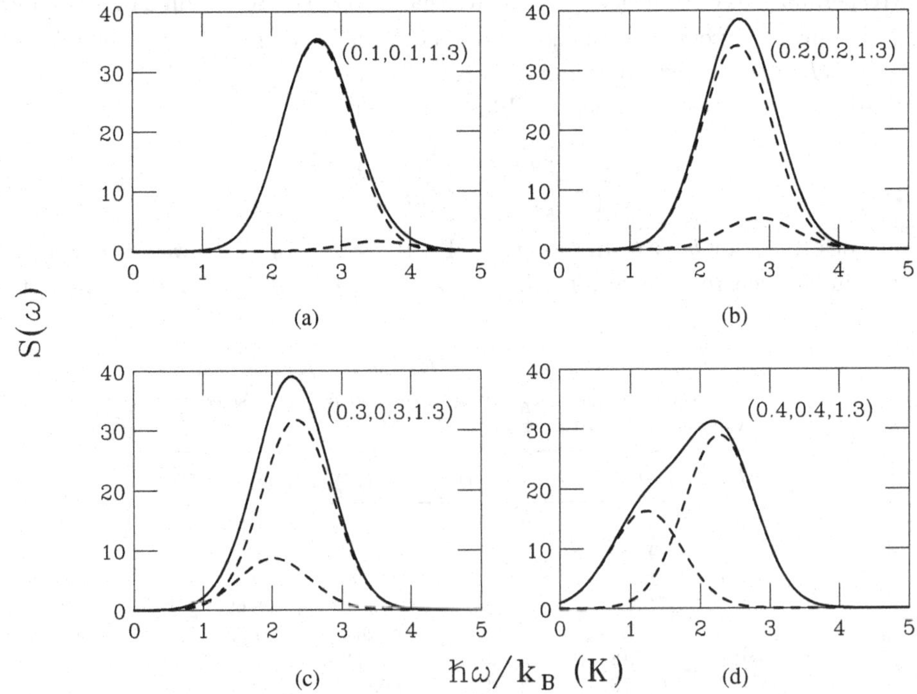


Fig. 9.5. Peak profiles (continuous curve) of the neutron scattering cross-section simulated by the sum of two Gaussians (dashed curves) with FWHM = 1.2 K centred at $\omega_{\boldsymbol{K}}^{(1)}$ and $\omega_{\boldsymbol{K}}^{(2)}$ for selected values of the scattering vector $\boldsymbol{K} = (h, h, 1.3)$: (a) $h = 0.1$; (b) $h = 0.2$; (c) $h = 0.3$; (d) $h = 0.4$.

To summarize, in this section we have tried to indicate what kind of problems must deal with a theoretical physicist when must explain the experimental data of a physical system. Taking $ErBa_2Cu_3O_7$ as an example, we have shown that the dipolar interactions are responsible of the antiferromagnetic configuration of the ground state. The crystalline electric field is essential to obtain at the correct spin orientation. The weak exchange interaction is crucial to recover the spin wave spectrum obtained from the INS cross-section. Finally, the limited energy resolution of the experiment is an important ingredient to describe the INS cross-section profile.

# Chapter 10

# LONG RANGE INTERACTIONS IN 2D SYSTEMS

## 10.1. Dipole–Dipole Interaction in 2D Systems

The study of 2D systems with dipole–dipole interaction is of interest for both theoretical and experimental reasons. The most intriguing question to which the statistical physics tries to give answer is: How the long range nature of the dipolar interaction affects the thermodynamic properties of a magnetic system? In other words, the LRO of a magnetic system can be recovered by the dipolar interaction in 2D system with continuous symmetry for which the Mermin and Wagner theorem[22] rules out any LRO in presence of short range interactions. Even though the 2D spin systems with continuous symmetry and short range interaction do not support LRO, different scenarios are found for the 2D Heisenberg model in which the spins are three-component vectors and the 2D planar rotator model in which the spins are two-component vectors. Indeed, the 2D Heisenberg model with NN exchange interaction is paramagnetic at any finite temperature while the 2D planar rotator model with NN exchange interaction shows[89] a Berezinskii–Kosterlitz–Thouless (BKT) phase characterized by divergent susceptibility and algebraic decaying of the spin–spin correlation function at temperatures[90] $T < T_{KT} = (0.895 \pm 0.005)J$. How is the BKT phase modified in replacing the NN exchange interaction by a long range interaction? When the spin–spin interaction is an *isotropic* long range interaction, for instance, assuming an exchange coupling decaying as $\frac{1}{r^3}$ where $r$ is the spin–spin distance, a variety of different behaviours are expected. A ferromagnetic long range interaction supports LRO as shown by renormalization group (RG) analysis[91] and confirmed by Monte Carlo (MC) simulation.[92] On the contrary, an antiferromagnetic long range coupling decaying with the same power law $\frac{1}{r^3}$ is expected to be unable to support LRO on the basis of MC simulation and confirmed by analytic calculations performed using the harmonic spin wave approximation.[93] What happens when the long range interaction is an *anisotropic* interaction as the dipole–dipole one? The ground state of a SQ planar rotator model with pure dipole–dipole interaction was found[94] to be infinitely many-fold degenerate. In particular, it consists of four sublattices[95,96] where the magnetic moments of each sublattice make an angle $\alpha$, $-\alpha$, $\pi + \alpha$ and $\pi - \alpha$, with the $x$-axis, $\alpha$ being arbitrary. Two configurations of this many-fold correspond to helix phases: indeed, the ground state with $\alpha = 0$ is recovered by a *columnar* phase with $\boldsymbol{Q} = \frac{2\pi}{a}(0, \frac{1}{2})$ and the ground state with $\alpha = \frac{\pi}{2}$ is recovered by another columnar configuration with $\boldsymbol{Q} = \frac{2\pi}{a}(\frac{1}{2}, 0)$. When the ground state is characterized by a continuous degeneracy, the harmonic spin wave approximation which is usually a reliable low temperature approach leads to a divergent mean-square angular displacement that destroys the LRO. However, the free energy obtained from the harmonic spin wave approximation is a function of the angle $\alpha$ so that its minimization selects a well-defined configuration,[95,96] lifting the continuous degeneracy at finite temperature and recovering the LRO. Such a phenomenology is called[97] *order by thermal disorder.* The recovering of the LRO cannot be obtained using the harmonic spin waves but one must use the *renormalized* spin waves as it will be illustrated in the next section.

The main numerical investigation on the statistical physics of the planar rotator model is obtained from the MC simulation. For a comprehensive book on this argument, we point out *A Guide to Monte Carlo Simulations in Statistical Physics* by David P. Landau and Kurt Binder (Cambridge University Press). In the next section, the analytic result obtained from the renormalized spin wave theory and the numerical result obtained from MC simulation will be compared at low temperature where the two approaches overlap.

Another important motivation to study 2D systems with dipole–dipole interaction comes from the technological interest in the ultrathin films (1–10 atomic layers) whose magnetization perpendicular to the film surface consists of *stripes* of spins aligned up and down. These unexpected configurations have been seen in ultrathin films[98] of Fe (2.5–3.5 atomic layers) grown on the $(1,0,0)$ surface of Cu and[99] Co (3–4.5 atomic layers) grown on the surface $(1,1,1)$ of Au. These ultrathin films can be represented by 2D Ising models with both short range exchange and long range dipolar interactions. The choice of the Ising model, in which the spins are one-component vectors, is justified by the strong anisotropy that favours the alignment of the magnetic moments perpendicular to the surface. Indeed, the experiments on the ultrathin films show a reorientation of the magnetization from the direction perpendicular to the surface to a direction parallel to the surface, increasing the number of layers, that is going from 2D to 3D systems. In Section 10.3, we will show the existence of stripes and checkerboard configurations by simply considering a 2D Ising model with NN ferromagnetic and dipole–dipole interaction and changing the weight of the short and long range interactions.

We conclude this section by giving the ground-state energy of a 2D system with dipole–dipole interaction and extending the Ewald's method of Section 9.2 to the 2D case. The key point is to change the 3D theta-function transformation given by Eq. (9.2.4) into the corresponding 2D theta-function transformation

$$\sum_l z^2 e^{i\boldsymbol{Q}\cdot\boldsymbol{r}_l} e^{-r_l^2 z^2} = \rho\pi \sum_l e^{-\frac{(\boldsymbol{G}_l+\boldsymbol{Q})^2}{4z^2}}. \tag{10.1.1}$$

By means of Eq. (10.1.1), Eq. (9.2.1) becomes

$$\sum_{l\neq 0}\frac{e^{i\boldsymbol{Q}\cdot\boldsymbol{r}_l}}{r_l^3} = -\frac{4\eta^3}{3\sqrt{\pi}} + 4\rho\eta\sqrt{\pi}\sum_l f_A\left(\frac{|\boldsymbol{G}_l+\boldsymbol{Q}|}{2\eta}\right) + \frac{2\eta^3}{\sqrt{\pi}}\sum_{l\neq 0} e^{i\boldsymbol{Q}\cdot\boldsymbol{r}_l} f_B(\eta r_l) \tag{10.1.2}$$

where

$$f_A(x) = \frac{x}{2}\int_{x^2}^{\infty} dz\,\frac{e^{-z}}{z^{3/2}} = e^{-x^2} - \sqrt{\pi}\,x\,\mathrm{erfc}(x) \tag{10.1.3}$$

and $f_B(x)$ is given by Eq. (9.2.8). Using again Eq. (10.1.1), the sum (9.2.18) becomes

$$\begin{aligned}\sum_{l\neq 0}\frac{e^{i\boldsymbol{Q}\cdot\boldsymbol{r}_l}}{r_l^5} r_l^\alpha r_l^\beta &= \frac{4}{3}\rho\eta\sqrt{\pi}\delta_{\alpha,\beta}\sum_l f_A\left(\frac{|\boldsymbol{G}_l+\boldsymbol{Q}|}{2\eta}\right)\\ &\quad -\frac{2}{3}\rho\pi\sum_l\frac{(\boldsymbol{G}_l+\boldsymbol{Q})_\alpha(\boldsymbol{G}_l+\boldsymbol{Q})_\beta}{|\boldsymbol{G}_l+\boldsymbol{Q}|}\,\mathrm{erfc}\left(\frac{|\boldsymbol{G}_l+\boldsymbol{Q}|}{2\eta}\right)\\ &\quad +\frac{4\eta^5}{3\sqrt{\pi}}\sum_{l\neq 0} r_l^\alpha r_l^\beta e^{i\boldsymbol{Q}\cdot\boldsymbol{r}_l} f_D(\eta r_l)\end{aligned} \tag{10.1.4}$$

where $f_D(x)$ is given by Eq. (9.2.16) and $\alpha,\beta = x,y$. Replacing Eqs. (10.1.2) and (10.1.4) into Eq. (9.2.19), the 2D dipolar sums become

$$\begin{aligned}D_\infty^{\alpha\beta}(\boldsymbol{Q}) &= -2\pi\rho\sum_l\frac{(\boldsymbol{G}_l+\boldsymbol{Q})_\alpha(\boldsymbol{G}_l+\boldsymbol{Q})_\beta}{|\boldsymbol{G}_l+\boldsymbol{Q}|}\,\mathrm{erfc}\left(\frac{|\boldsymbol{G}_l+\boldsymbol{Q}|}{2\eta}\right)\\ &\quad +\frac{4\eta^5}{\sqrt{\pi}}\sum_{l\neq 0} r_l^\alpha r_l^\beta e^{i\boldsymbol{Q}\cdot\boldsymbol{r}_l} f_D(\eta r_l) + \frac{2\eta^3}{\sqrt{\pi}}\delta_{\alpha,\beta}\left[\frac{2}{3}-\sum_{l\neq 0} e^{i\boldsymbol{Q}\cdot\boldsymbol{r}_l} f_B(\eta r_l)\right].\end{aligned} \tag{10.1.5}$$

In the limit $\boldsymbol{Q}\to 0$, one has

$$\lim_{\boldsymbol{Q}\to 0} D_\infty^{\alpha\beta}(\boldsymbol{Q}) = \delta_{\alpha,\beta}\, d_\alpha \tag{10.1.6}$$

where

$$d_\alpha = \frac{4\eta^3}{3\sqrt{\pi}} - 2\pi\rho\sum_{l\neq 0}\frac{(G_l^\alpha)^2}{G_l}\mathrm{erfc}\left(\frac{G_l}{2\eta}\right) + \frac{2\eta^3}{\sqrt{\pi}}\sum_{l\neq 0}\left[2(\eta r_l^\alpha)^2 f_D(\eta r_l) - f_B(\eta r_l)\right]. \tag{10.1.7}$$

For a SQ lattice, one has

$$\sum_{l\neq 0}\frac{(G_l^\alpha)^2}{G_l}\mathrm{erfc}\left(\frac{G_l}{2\eta}\right) = \frac{1}{2}\sum_{l\neq 0} G_l\,\mathrm{erfc}\left(\frac{G_l}{2\eta}\right) \tag{10.1.8}$$

and

$$\sum_{l\neq 0}(\eta r_l^{\alpha})^2 f_D(\eta r_l) = \frac{1}{2}\sum_{l\neq 0}(\eta r_l)^2 f_D(\eta r_l), \tag{10.1.9}$$

so that

$$\begin{aligned} d_x = d_y &= \frac{4\eta^3}{3\sqrt{\pi}} - \frac{\pi}{a^2}\sum_{l\neq 0} G_l \,\mathrm{erfc}\left(\frac{G_l}{2\eta}\right) \\ &+\frac{2\eta^3}{\sqrt{\pi}}\sum_{l\neq 0}\left[\frac{1+2(\eta r_l)^2}{2(\eta r_l)^2}e^{-\eta^2 r^2 l} + \frac{\sqrt{\pi}}{4(\eta r_l)^3}\mathrm{erfc}\,(\eta r_l)\right]. \\ &= \frac{4.5168108}{a^3}. \end{aligned} \tag{10.1.10}$$

The numerical value is evaluated by limiting the sums to $-5 < l_1, l_2 < 5$ and assuming $\eta = \frac{2}{a}$. As in the 3D case, the numerical accuracy is within $10^{-10}$. In 2D, the finite size effect is not as important as in 3D. Indeed, for a sample made up of a circle of radius $R$ ($S = \pi R^2$), the dipolar sums are given by

$$D_S^{\alpha\beta}(\boldsymbol{Q}) = D_{\infty}^{\alpha\beta}(\boldsymbol{Q}) - D_{\infty-S}^{\alpha\beta}(\boldsymbol{Q}) \tag{10.1.11}$$

where, choosing polar coordinates such that $\boldsymbol{Q} = Q(\cos\phi, \sin\phi)$ and $\boldsymbol{r} = r[\cos(\phi + \theta), \sin(\phi+\theta)]$, one has $\boldsymbol{Q}\cdot\boldsymbol{r} = Qr\cos\theta$ and

$$D_{\infty-S}^{xx}(\boldsymbol{Q}) = Q\frac{\pi}{a^2}\int_{QR}^{\infty}\frac{dx}{x^2}[J_0(x) - 3J_2(x)\cos 2\phi], \tag{10.1.12}$$

$$D_{\infty-S}^{yy}(\boldsymbol{Q}) = Q\frac{\pi}{a^2}\int_{QR}^{\infty}\frac{dx}{x^2}[J_0(x) + 3J_2(x)\cos 2\phi] \tag{10.1.13}$$

and

$$D_{\infty-S}^{xy}(\boldsymbol{Q}) = -Q\frac{3\pi}{a^2}\sin 2\phi\int_{QR}^{\infty}\frac{dx}{x^2}J_2(x) \tag{10.1.14}$$

where $J_n(x)$ are the Bessel functions of integer order[3] coming from the integration over $\theta$ and $x = Qr$. For $x \to 0$, one has

$$J_0(x) = 1 - \frac{1}{4}x^2 + \frac{1}{64}x^4 + \cdots \tag{10.1.15}$$

and

$$J_2(x) = \frac{1}{8}x^2 - \frac{1}{96}x^4 + \cdots \tag{10.1.16}$$

so that the integrals occurring in Eqs. (10.1.12) and (10.1.13) are singular for $Q \to 0$ while the integral occurring in Eq. (10.1.14) is finite for $Q = 0$. Then

$$\lim_{\boldsymbol{Q}\to 0} D_{\infty-S}^{xx}(\boldsymbol{Q}) = \lim_{\boldsymbol{Q}\to 0} D_{\infty-S}^{yy}(\boldsymbol{Q}) = \frac{\pi}{a^2 R} \tag{10.1.17}$$

and

$$\lim_{\boldsymbol{Q}\to 0} D_{\infty-S}^{xy}(\boldsymbol{Q}) = 0. \tag{10.1.18}$$

Equation (10.1.17) shows that the 2D dipolar sums for finite samples differ from the corresponding sums for infinite samples by terms of order $\frac{a}{R}$. For macroscopic

Table 10.1. Dipolar sums for SQ lattice for selected values of the wavevector $\boldsymbol{Q} = (q_1, q_2)$.

| $\boldsymbol{Q} = \frac{2\pi}{a}(q_1, q_2)$ | $a^3 D^{xx}(\boldsymbol{Q})$ | $a^3 D^{yy}(\boldsymbol{Q})$ | $a^3 D^{zz}(\boldsymbol{Q})$ |
|---|---|---|---|
| $(0, 0)$ | 4.5168108416 | 4.5168108416 | −9.0336216831 |
| $(\frac{1}{2}, 0)$ | −6.0343351486 | 5.0988729940 | 0.9354621546 |
| $(0, \frac{1}{2})$ | 5.0988729940 | −6.0343351486 | 0.9354621546 |
| $(\frac{1}{2}, \frac{1}{2})$ | −1.3229432662 | −1.3229432662 | 2.6458865323 |

samples, these corrections are of the order $10^{-7}$ or less. For this reason, in the following, we will ignore any distinction between finite and infinite systems and suppress any subscript in the dipolar sums. In Table 10.1, we give the dipolar sums $D^{\alpha\alpha}(\boldsymbol{Q})$ for a SQ lattice obtained from Eq. (10.1.5) for selected values of $\boldsymbol{Q} = \frac{2\pi}{a}(q_1, q_2)$. The last column is obtained from the first and second ones by means of the sum rule

$$D^{xx}(\boldsymbol{Q}) + D^{yy}(\boldsymbol{Q}) + D^{zz}(\boldsymbol{Q}) = 0. \tag{10.1.19}$$

As one can see from Table 10.1, the ground-state energy of a planar rotator model is

$$E_0^x\left(0, \frac{1}{2}\right) = E_0^y\left(\frac{1}{2}, 0\right) = -5.0988730\frac{(\mu S)^2}{2a^3}N, \tag{10.1.20}$$

where the superscripts $x$ and $y$ indicate the spin direction. The corresponding configurations are *columnar* configurations characterized by alternating ferromagnetic rows or columns with the spins directed along the ferromagnetic chains. The dipolar sum for a 2D Ising model with the spins directed along the $z$-axis assumed to be perpendicular to the plane is

$$D^{zz}(\boldsymbol{Q}) = -\sum_{l \neq 0} \frac{e^{i\boldsymbol{Q}\cdot\boldsymbol{r}_l}}{r_l^3}. \tag{10.1.21}$$

The last column of Table 10.1 gives the values of $D^{zz}(\boldsymbol{Q})$ for selected wavevectors $\boldsymbol{Q}$. The ground state of the 2D Ising model with pure dipole–dipole interaction and spins perpendicular to the plane is

$$E_0^z\left(\frac{1}{2}, \frac{1}{2}\right) = -2.6458865\frac{(\mu S)^2}{2a^3}N \tag{10.1.22}$$

corresponding to a Néel antiferromagnetic configuration.

## 10.2. Planar Rotator Model with Long Range Interactions

The Hamiltonian of the planar rotator model with pure dipolar interaction reads

$$\mathcal{H} = -\frac{1}{2}\mu^2 \sum_i \sum_{\boldsymbol{r}\neq 0} \sum_{\alpha\beta} \frac{1}{r^3}\left(3\frac{r^\alpha r^\beta}{r^2} - \delta_{\alpha,\beta}\right) S_i^\alpha S_{i+\boldsymbol{r}}^\beta, \tag{10.2.1}$$

where $\mu$ is the magnetic moment; $\alpha, \beta = x, y$ label the two spin components; $i$ labels the $N$ sites of a SQ lattice; $\boldsymbol{r} = a(l_1\mathbf{u}_x + l_2\mathbf{u}_y)$, with $l_1, l_2$ integers, is the generic lattice vector; $a$ is the lattice constant. The ground-state spin configuration selected by thermal fluctuations is a columnar configuration[95,96] whose energy is given by Eq. (10.1.20). To study the elementary excitations, we assume that the spin components of the planar rotator model are

$$S_i^x = S\cos(\boldsymbol{Q}\cdot\boldsymbol{r}_i + \psi_i), \quad S_i^y = S\sin(\boldsymbol{Q}\cdot\boldsymbol{r}_i + \psi_i), \tag{10.2.2}$$

where $\boldsymbol{Q} = \frac{2\pi}{a}(0, \frac{1}{2})$ is the order wavevector characterizing the columnar configuration and $\psi_i$ is the angular displacement from the ground-state configuration at site $i$. The elementary excitations correspond to small oscillations around the ground-state configuration so that when we expand Hamiltonian (10.2.1) in powers of the angular displacements $\psi_i$, keeping contributions up to fourth order, and use the Fourier transforms of the angular displacements

$$\psi_i = \frac{1}{\sqrt{N}}\sum_{\boldsymbol{q}} \psi_{\boldsymbol{q}}\, e^{i\boldsymbol{q}\cdot\boldsymbol{r}_i}, \tag{10.2.3}$$

the Hamiltonian (10.2.1) becomes

$$\mathcal{H} = E_0 + \mathcal{H}_2 + \mathcal{H}_3 + \mathcal{H}_4 + \cdots \tag{10.2.4}$$

where

$$E_0 = -\frac{1}{2}N\mu^2S^2D^{xx}(\boldsymbol{Q}) = -5.0988730\,\frac{(\mu S)^2}{2a^3}N, \tag{10.2.5}$$

$$\mathcal{H}_2 = \frac{1}{2}\mu^2S^2\sum_{\boldsymbol{q}}[D^{xx}(\boldsymbol{Q}) - D^{yy}(\boldsymbol{Q}-\boldsymbol{q})]\psi_{\boldsymbol{q}}\psi_{-\boldsymbol{q}}, \tag{10.2.6}$$

$$\mathcal{H}_3 = -\frac{1}{2}\mu^2S^2\frac{1}{\sqrt{N}}\sum_{\boldsymbol{q}_1,\boldsymbol{q}_2,\boldsymbol{q}_3}\delta_{\boldsymbol{q}_1+\boldsymbol{q}_2+\boldsymbol{q}_3,0}D^{xy}(\boldsymbol{Q}-\boldsymbol{q}_1)\psi_{\boldsymbol{q}_1}\psi_{\boldsymbol{q}_2}\psi_{\boldsymbol{q}_3}, \tag{10.2.7}$$

$$\begin{aligned}\mathcal{H}_4 = -\frac{1}{24}\mu^2S^2\frac{1}{N}\sum_{\boldsymbol{q}_1,\boldsymbol{q}_2,\boldsymbol{q}_3,\boldsymbol{q}_4}&\delta_{\boldsymbol{q}_1+\boldsymbol{q}_2+\boldsymbol{q}_3+\boldsymbol{q}_4,0}[D^{xx}(\boldsymbol{Q}) + 3D^{xx}(\boldsymbol{Q}-\boldsymbol{q}_1-\boldsymbol{q}_2)\\ &- 4D^{yy}(\boldsymbol{Q}-\boldsymbol{q}_1)]\psi_{\boldsymbol{q}_1}\psi_{\boldsymbol{q}_2}\psi_{\boldsymbol{q}_3}\psi_{\boldsymbol{q}_4}.\end{aligned} \tag{10.2.8}$$

The harmonic spin wave spectrum is

$$\hbar\omega_2(\boldsymbol{q}) = \frac{\mu^2S^2}{2a^3}\,\epsilon_{\boldsymbol{q}} \tag{10.2.9}$$

where

$$\epsilon_{\boldsymbol{q}} = a^3[D^{xx}(\boldsymbol{Q}) - D^{yy}(\boldsymbol{Q}-\boldsymbol{q})]. \tag{10.2.10}$$

Note that the spectrum (10.2.9) vanishes at $\boldsymbol{q}_0 = \frac{2\pi}{a}(\frac{1}{2}, \frac{1}{2})$ as one can see from Table 10.1. Indeed, for $\boldsymbol{q} = \boldsymbol{q}_0$, one has $\boldsymbol{Q} - \boldsymbol{q}_0 = -\frac{2\pi}{a}(\frac{1}{2}, 0) \equiv -\boldsymbol{Q}'$ and from Table 10.1, one sees that $a^3D^{xx}(\boldsymbol{Q}) = a^3D^{yy}(\boldsymbol{Q}') = 5.0988730$. The order parameter in harmonic approximation is zero for any finite temperature. Indeed,

$$\langle\psi_i^2\rangle = \frac{1}{N}\sum_{\boldsymbol{q}}\langle\psi_{\boldsymbol{q}}\psi_{-\boldsymbol{q}}\rangle_0 = \frac{1}{N}\sum_{\boldsymbol{q}}\frac{\int_0^\infty d|\psi_{\boldsymbol{q}}|\,|\psi_{\boldsymbol{q}}|^2\,e^{-\beta\hbar\omega_2(\boldsymbol{q})|\psi_{\boldsymbol{q}}|^2}}{\int_0^\infty d|\psi_{\boldsymbol{q}}|\,e^{-\beta\hbar\omega_2(\boldsymbol{q})|\psi_{\boldsymbol{q}}|^2}} = \frac{1}{N}\sum_{\boldsymbol{q}}\frac{t}{\epsilon_{\boldsymbol{q}}} \tag{10.2.11}$$

where

$$t = \frac{k_B T a^3}{\mu^2 S^2} \tag{10.2.12}$$

is a reduced temperature. The expansion of $\epsilon_{\boldsymbol{q}}$ about $\boldsymbol{q}_0$ gives

$$\epsilon_{\boldsymbol{q}\to\boldsymbol{q}_0} = 0.14474(aq_x - \pi)^2 + 1.7873(aq_y - \pi)^2. \tag{10.2.13}$$

Equation (10.2.13) is obtained using the series expansions of Eq. (10.1.5) about $\boldsymbol{Q}' = (\frac{1}{2}, 0)$ for $\alpha = \beta = y$. The vanishing of the harmonic spin wave spectrum about $\boldsymbol{q}_0$ leads to a logarithmic divergence of the sum (10.2.11) so that $\langle\psi_i^2\rangle \to \infty$ and the order parameter

$$\langle\cos\psi_i\rangle = \cos\boldsymbol{Q}\cdot\boldsymbol{r}_i\;e^{-\frac{1}{2}\langle\psi_i^2\rangle} \tag{10.2.14}$$

goes to zero at any finite temperature. This event claims for a more accurate treatment of the elementary excitations spectrum going beyond the harmonic approximation. To this end, we make use of the "classical" equivalent of the Dyson's equation of Section 3.5. To obtain the perturbation expansion of the classical propagator (cumulant expansion), we split the Hamiltonian (10.2.4) in two parts: the unperturbed (harmonic) Hamiltonian $\mathcal{H}_2$ and the perturbation $V = \mathcal{H} - \mathcal{H}_2$. Then we expand the exponential function containing $V$ in a power series of $\beta V$. By doing so, one obtains

$$\langle\psi_{\boldsymbol{q}}\psi_{-\boldsymbol{q}}\rangle = \frac{\sum_{n=0}^\infty(-1)^n\frac{\beta^n}{n!}\langle\psi_{\boldsymbol{q}}\psi_{-\boldsymbol{q}}V^n\rangle_0}{\sum_{n=0}^\infty(-1)^n\frac{\beta^n}{n!}\langle V^n\rangle_0} \tag{10.2.15}$$

where the subscript 0 means a thermal average over the ensemble with the Hamiltonian $\mathcal{H}_2$. By means of the linked cluster theorem and the diagrams expansion of Chapter 4, Eq. (10.2.15) becomes

$$\langle\psi_{\boldsymbol{q}}\psi_{-\boldsymbol{q}}\rangle = \sum_{n=0}^\infty(-1)^n\frac{\beta^n}{n!}\langle\psi_{\boldsymbol{q}}\psi_{-\boldsymbol{q}}V^n\rangle_0^c \tag{10.2.16}$$

where the superscript $c$ means "connected" diagrams. Restricting to the second-order perturbation theory, the expansion (10.2.16) for the Hamiltonian (10.2.4) reads

$$\langle\psi_{\boldsymbol{q}}\psi_{-\boldsymbol{q}}\rangle = \langle\psi_{\boldsymbol{q}}\psi_{-\boldsymbol{q}}\rangle_0 - \beta\langle\psi_{\boldsymbol{q}}\psi_{-\boldsymbol{q}}\mathcal{H}_4\rangle_0^c + \frac{1}{2}\beta^2\langle\psi_{\boldsymbol{q}}\psi_{-\boldsymbol{q}}\mathcal{H}_3^2\rangle_0^c \tag{10.2.17}$$

where

$$\langle\psi_{\boldsymbol{q}}\psi_{-\boldsymbol{q}}\rangle_0 = \frac{t}{\epsilon_{\boldsymbol{q}}} \tag{10.2.18}$$

is the harmonic classical propagator of the non-interacting system. By means of Eqs. (10.2.8) and (10.2.7), one obtains

$$\langle\psi_{\boldsymbol{q}}\psi_{-\boldsymbol{q}}\mathcal{H}_4\rangle_0^c = -\mu^2 S^2 \langle\psi_{\boldsymbol{q}}\psi_{-\boldsymbol{q}}\rangle_0^2 \frac{1}{N}\sum_{\boldsymbol{k}}[D^{xx}(\boldsymbol{Q}) + D^{xx}(\boldsymbol{Q}+\boldsymbol{q}-\boldsymbol{k}) - D^{yy}(\boldsymbol{Q}-\boldsymbol{q}) - D^{yy}(\boldsymbol{Q}-\boldsymbol{k})]\langle\psi_{\boldsymbol{k}}\psi_{-\boldsymbol{k}}\rangle_0 \tag{10.2.19}$$

and

$$\langle\psi_{\boldsymbol{q}}\psi_{-\boldsymbol{q}}\mathcal{H}_3^2\rangle_0^c = \mu^4 S^4 \langle\psi_{\boldsymbol{q}}\psi_{-\boldsymbol{q}}\rangle_0^2 \frac{1}{N}\sum_{\boldsymbol{k}}[D^{xy}(\boldsymbol{Q}-\boldsymbol{q})^2 + 4D^{xy}(\boldsymbol{Q}-\boldsymbol{q})D^{xy}(\boldsymbol{Q}-\boldsymbol{k}) + 2D^{xy}(\boldsymbol{Q}-\boldsymbol{k})^2 + 2D^{xy}(\boldsymbol{Q}-\boldsymbol{k})D^{xy}(\boldsymbol{Q}-\boldsymbol{k}+\boldsymbol{q})] \times \langle\psi_{\boldsymbol{k}}\psi_{-\boldsymbol{k}}\rangle_0\langle\psi_{\boldsymbol{k}-\boldsymbol{q}}\psi_{-\boldsymbol{k}+\boldsymbol{q}}\rangle_0. \tag{10.2.20}$$

Then the renormalized classical propagator of the interacting system reads

$$\langle\psi_{\boldsymbol{q}}\psi_{-\boldsymbol{q}}\rangle = \frac{1}{\langle\psi_{\boldsymbol{q}}\psi_{-\boldsymbol{q}}\rangle_0^{-1} - \Sigma_{\boldsymbol{q}}} = \frac{t}{\epsilon_{\boldsymbol{q}} - t\,\Sigma_{\boldsymbol{q}}}, \tag{10.2.21}$$

where the proper self-energy $\Sigma_{\boldsymbol{q}}$, obtained from Eqs. (10.2.19) and (10.2.20), is given by

$$\Sigma_{\boldsymbol{q}} = \frac{1}{N}\sum_{\boldsymbol{k}}[D^{xx}(\boldsymbol{Q}) + D^{xx}(\boldsymbol{Q}+\boldsymbol{q}-\boldsymbol{k}) - D^{yy}(\boldsymbol{Q}-\boldsymbol{q}) - D^{yy}(\boldsymbol{Q}-\boldsymbol{k})]\frac{a^3}{\epsilon_{\boldsymbol{k}}} + \frac{1}{2N}\sum_{\boldsymbol{k}}[D^{xy}(\boldsymbol{Q}-\boldsymbol{q})^2 + 4D^{xy}(\boldsymbol{Q}-\boldsymbol{q})D^{xy}(\boldsymbol{Q}-\boldsymbol{k}) + 2D^{xy}(\boldsymbol{Q}-\boldsymbol{k})^2 + 2D^{xy}(\boldsymbol{Q}-\boldsymbol{k})D^{xy}(\boldsymbol{Q}-\boldsymbol{k}+\boldsymbol{q})]\frac{a^6}{\epsilon_{\boldsymbol{k}}\,\epsilon_{\boldsymbol{k}-\boldsymbol{q}}}. \tag{10.2.22}$$

The renormalized spectrum is then

$$\hbar\omega(\boldsymbol{q}) = \frac{\mu^2 S^2}{2a^3}(\epsilon_{\boldsymbol{q}} - t\,\Sigma_{\boldsymbol{q}}). \tag{10.2.23}$$

The self-energy (10.2.22) evaluated at $\boldsymbol{q} = \boldsymbol{q}_0$ gives

$$\Sigma_{\boldsymbol{q}_0} = \frac{1}{N}\sum_{\boldsymbol{k}}[D^{xx}(\boldsymbol{Q}+\boldsymbol{q}_0-\boldsymbol{k}) - D^{yy}(\boldsymbol{Q}-\boldsymbol{k})]\frac{a^3}{\epsilon_{\boldsymbol{k}}} + \frac{1}{N}\sum_{\boldsymbol{k}} D^{xy}(\boldsymbol{Q}-\boldsymbol{k})[D^{xy}(\boldsymbol{Q}-\boldsymbol{k}) + D^{xy}(\boldsymbol{Q}-\boldsymbol{k}+\boldsymbol{q}_0)]\frac{a^6}{\epsilon_{\boldsymbol{k}}\,\epsilon_{\boldsymbol{k}-\boldsymbol{q}_0}} = -0.575 \tag{10.2.24}$$

so that

$$\hbar\omega(\boldsymbol{q}_0) = \frac{\mu^2 S^2}{2a^3}0.575\,t. \tag{10.2.25}$$

The gap at $\boldsymbol{q} = \boldsymbol{q}_0$ originates from thermal fluctuations and increases as temperature increases. This behaviour[100] is very peculiar since, usually, the energy gap is *reduced*

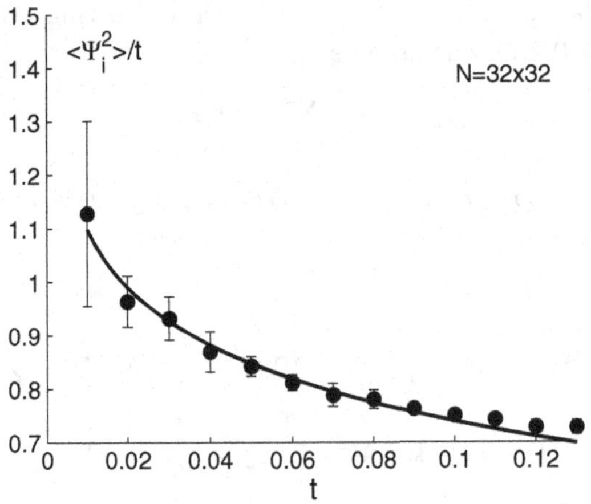


Fig. 10.1. Mean-square angular displacement versus temperature. The continuous curve is the analytic result (10.2.27) with $O(t) = 0.38t$. Full circles are MC data for a 32×32 SQ lattice.[100]

rather than *increased* by thermal fluctuations. A calculation of the renormalized spectrum in the neighbourhood of $\boldsymbol{q}_0$ gives

$$\hbar\omega(\boldsymbol{q} \to \boldsymbol{q}_0) = \frac{\mu^2 S^2}{2a^3}[0.575\, t + 0.14474(\pi - aq_x)^2 + 1.7873(\pi - aq_y)^2] \qquad (10.2.26)$$

and the mean-square angular displacement (10.2.4), making use of Eq. (10.2.23), becomes

$$\langle \psi_i^2 \rangle = \frac{1}{N}\sum_{\boldsymbol{k}} \frac{t}{\epsilon_{\boldsymbol{k}} - t\,\Sigma_{\boldsymbol{k}}} \simeq -0.156\, t \ln t + O(t), \qquad (10.2.27)$$

where the last equality points out the singular contribution to the sum coming from the renormalized spectrum about $\boldsymbol{q} = \boldsymbol{q}_0$. As one can see from Eq. (10.2.27), at low temperature, $\langle \psi_i^2 \rangle / t$ diverges logarithmically as shown by the continuous curve in Fig. 10.1, assuming $O(t) = 0.38t$. Note that the same quantity evaluated in systems with conventional LRO is a constant since in this case, only the regular contribution $O(t)$ in the sum (10.2.27) is present. Using the renormalized spectrum, the order parameter (10.2.4) becomes

$$\langle \cos \psi_i \rangle = \cos(l_2 \pi)\, e^{0.0882\, t \ln t}. \qquad (10.2.28)$$

Note the infinite slope of the order parameter for $t \to 0$. In this limit, the energy cost to create a spin wave of wavevector $\boldsymbol{q} \sim \boldsymbol{q}_0$ is vanishing so that a large number of spin waves can be excited. However, the interaction between the spin waves creates a gap in the spectrum that increases at increasing temperature and the LRO is restored.

For comparison, we consider the isotropic long range planar antiferromagnet in which the spin–spin coupling decays as $\frac{1}{r^3}$ where $r$ is the spin–spin distance. The

Hamiltonian of such a model reads

$$\mathcal{H} = \frac{1}{2}\sum_i \sum_{\boldsymbol{r}\neq 0} J(r)\boldsymbol{S}_i \cdot \boldsymbol{S}_{i+\boldsymbol{r}} = \frac{1}{2}Ja^3S^2 \sum_i \sum_{\boldsymbol{r}\neq 0} \frac{1}{r^3}\cos(\boldsymbol{Q}\cdot\boldsymbol{r} + \psi_i - \psi_{i+\boldsymbol{r}}). \tag{10.2.29}$$

The expansion of the Hamiltonian (10.2.29) for small angular displacements gives

$$\mathcal{H} = E_0 + \mathcal{H}_2 + \mathcal{H}_4 + \cdots \tag{10.2.30}$$

where

$$E_0 = \frac{1}{2}NS^2J(\boldsymbol{Q}), \tag{10.2.31}$$

$$\mathcal{H}_2 = \frac{1}{2}S^2\sum_{\boldsymbol{q}}[J(\boldsymbol{Q}-\boldsymbol{q}) - J(\boldsymbol{Q})]\psi_{\boldsymbol{q}}\psi_{-\boldsymbol{q}}, \tag{10.2.32}$$

$$\mathcal{H}_4 = \frac{1}{24}S^2\frac{1}{N}\sum_{\boldsymbol{q}_1,\boldsymbol{q}_2,\boldsymbol{q}_3,\boldsymbol{q}_4} \delta_{\boldsymbol{q}_1+\boldsymbol{q}_2+\boldsymbol{q}_3+\boldsymbol{q}_4,0}[J(\boldsymbol{Q}) - 2J(\boldsymbol{Q}+\boldsymbol{q}_4) + 3J(\boldsymbol{Q}+\boldsymbol{q}_3+\boldsymbol{q}_4) - 2J(\boldsymbol{Q}-\boldsymbol{q}_1)]\psi_{\boldsymbol{q}_1}\psi_{\boldsymbol{q}_2}\psi_{\boldsymbol{q}_3}\psi_{\boldsymbol{q}_4} \tag{10.2.33}$$

with

$$J(\boldsymbol{q}) = Ja^3\sum_{\boldsymbol{r}\neq 0}\frac{1}{r^3}e^{i\boldsymbol{q}\cdot\boldsymbol{r}}. \tag{10.2.34}$$

The minimum of $E_0(\boldsymbol{Q})$ in Eq. (10.2.31) is obtained for $\boldsymbol{Q} = \frac{2\pi}{a}(\frac{1}{2},\frac{1}{2})$ corresponding to a Néel antiferromagnetic configuration. Note the absence of cubic terms in the angular displacements in the Hamiltonian (10.2.30). The harmonic spin wave approximation gives an elementary excitation energy

$$\hbar\omega_2(\boldsymbol{q}) = \frac{1}{2}JS^2\,\epsilon_{\boldsymbol{q}} \tag{10.2.35}$$

where

$$\epsilon_{\boldsymbol{q}} = \frac{1}{J}[J(\boldsymbol{Q}-\boldsymbol{q}) - J(\boldsymbol{Q})] \tag{10.2.36}$$

that vanishes quadratically as $q \to 0$. Indeed, expanding Eq. (10.2.36) about $\boldsymbol{q} = 0$, one obtains

$$\epsilon_{\boldsymbol{q}\to 0} = 0.40388\,(aq)^2. \tag{10.2.37}$$

The first-order renormalized spin wave spectrum is given by

$$\hbar\omega(\boldsymbol{q}) = \frac{1}{2}JS^2\,(\epsilon_{\boldsymbol{q}} - t\,\Sigma_{\boldsymbol{q}}) \tag{10.2.38}$$

where $t = \frac{k_BT}{JS^2}$ and

$$\Sigma_{\boldsymbol{q}} = -\frac{1}{JN}\sum_{\boldsymbol{k}}\left\{J(\boldsymbol{Q}) - J(\boldsymbol{Q}-\boldsymbol{k}) - J(\boldsymbol{Q}-\boldsymbol{q}) + \frac{1}{2}[J(\boldsymbol{Q}-\boldsymbol{q}-\boldsymbol{k}) + J(\boldsymbol{Q}+\boldsymbol{q}-\boldsymbol{k})]\right\}\frac{1}{\epsilon_{\boldsymbol{k}}} \tag{10.2.39}$$

that for $q \to 0$ vanishes as $\Sigma_{q\to 0} = 0.101(aq)^2$. In the long wavelength limit, both the harmonic and the renormalized spectra vanish quadratically as $q \to 0$, leading to a divergence of the mean-square angular displacement

$$\langle \psi_i^2 \rangle \simeq \langle \psi_i^2 \rangle_0 = \frac{1}{N} \sum_{\boldsymbol{q}} \frac{t}{\epsilon_{\boldsymbol{q}}} \simeq -0.39406\, t \lim_{q\to 0} \left[ \ln \left( \frac{aq}{2\pi} \right) \right] \to +\infty \qquad (10.2.40)$$

and therefore to the absence of LRO at any finite temperature. For a finite system, one may replace the argument of the logarithmic function in Eq. (10.2.40) by $\frac{1}{\sqrt{N}}$ where $N$ is the total number of the spins obtaining

$$\langle \psi_i^2 \rangle_N \simeq 0.19703\, t \ln N + O(t) \qquad (10.2.41)$$

where $O(t)$ is the regular term independent of $N$. Equation (10.2.41) shows that the mean-square angular displacement diverges logarithmically in the thermodynamic limit as shown by the upper continuous curve of Fig. 10.2 for $t = 0.1$. The divergence for $N \to \infty$ is related to the soft mode at $\boldsymbol{q} = 0$ that is not destroyed by thermal fluctuations in the isotropic long range antiferromagnetic planar rotator. This fact could be expected since the rotational invariance of Hamiltonian (10.2.29) implies the existence of a soft mode at $\boldsymbol{q} = 0$.

In order to check the analytic calculation at low temperature and to extend the study of the model at higher temperature, one can perform MC simulations. In particular, the temperature dependence of the mean-square angular displacement can be evaluated and compared with the analytic result at low temperature.[100] As shown in Fig. 10.1, MC data (full circles) fit very well the analytic result (10.2.27) obtained by the renormalized spin wave theory (continuous curve). MC data[100] are taken from a simulation on a sample $32 \times 32$ averaging over $10^3$ configurations after having disregarded $10^5$ steps for equilibration. An approach of periodic "images" is

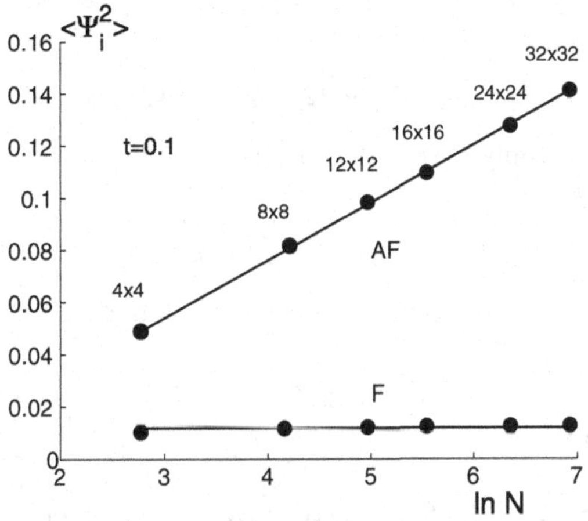


Fig. 10.2. Mean-square angular displacement at $t = 0.1$ versus $\ln N$ for SQ planar model with ferromagnetic (F) and antiferromagnetic (AF) isotropic long range interactions decaying as $\frac{1}{r^3}$. The full circles are obtained from MC simulations.[100] The continuous curves are the analytic results.

used.[101] Such an approach, which is based on a periodic arrangement of MC cells, seems to be the most convenient method to treat systems with long range interactions. Note that the singular behaviour of the mean-square angular displacement is captured by MC simulation only if the region of low temperature ($0.01 < t < 0.1$) is carefully investigated.

In Fig. 10.2, the size dependence of the mean-square angular displacement at $t = 0.1$ is shown for the ferromagnetic and antiferromagnetic SQ planar model with isotropic long range interaction decaying as $1/r^3$ (full circles). The existence of LRO in the ferromagnetic model is supported by the size independence of the mean-square angular displacement (lower curve). On the contrary, the absence of LRO in the antiferromagnetic model is clearly pointed out by the size dependence of the mean-square angular displacement which agrees very well with the analytic result (10.2.41). The existence of LRO in the isotropic long range ferromagnetic planar rotator model can be justified analytically by evaluating the mean-square angular displacement by using the harmonic spin wave approximation. In this case, the ground state corresponds to a ferromagnetic configuration with $\boldsymbol{Q} = 0$ and the harmonic spin wave spectrum is given by Eq. (10.2.35) with

$$\epsilon_{\boldsymbol{q}} = \frac{1}{J}[J(0) - J(\boldsymbol{q})]. \tag{10.2.42}$$

Expanding Eq. (10.2.42) about $\boldsymbol{q} \to 0$, one obtains

$$\epsilon_{\boldsymbol{q}\to 0} = 2\pi(aq) - 0.97507\,(aq)^2. \tag{10.2.43}$$

The linear increment of the harmonic spectrum as a function of the wavevector $q$ in the ferromagnetic model prevents any divergence of the mean-square angular displacement. Indeed, one obtains

$$\langle \psi_i^2 \rangle_0 = \frac{1}{N}\sum_{\boldsymbol{q}} \frac{t}{\epsilon_{\boldsymbol{q}}} = 0.1275\,t \tag{10.2.44}$$

and the LRO is recovered. In Fig. 10.2, the lower continuous curve is the analytic result (10.2.44) for $t = 0.1$. The agreement between the MC simulation and the harmonic spin wave theory is very good.

## 10.3. Stripes and Checkerboards in 2D Ising Model

The $S = \frac{1}{2}$, SQ Ising model with ferromagnetic NN exchange interaction ($J > 0$) and dipole–dipole ($g$) interaction has been extensively investigated[102–107] because of its theoretical interest in studying the effect of long range interactions in the critical region and in view of its possible application in describing the qualitative features of the spin configurations observed in ultrathin films of magnetic atoms on metal substrates. The Hamiltonian of the model is

$$\mathcal{H} = -J\sum_{\langle i,j\rangle} \sigma_i\sigma_j + g\sum_{i\neq j} \frac{1}{r_{ij}^3}\sigma_i\sigma_j, \tag{10.3.1}$$

where the first sum is restricted to distinct pairs of NN spins and a ferromagnetic interaction $J > 0$ is assumed; in the second sum, $i$ and $j$ run over all the sites of a SQ lattice with $L \times L = N$ and $g = \frac{\mu_B S^2}{a^3}$. The spins $\sigma_i = \pm 1$ are supposed to be aligned out-of-plane. The analytic investigation of the zero temperature energy of Eq. (10.3.1) leads to a variety of configurations depending on the ratio $J/g$. The *stripe* and *checkerboard* configurations can be obtained assuming a double periodic spin configuration of the ground state. In particular, let us assume a periodic spin configuration with period $p_1 = 2h_1$ along the $x$-axis and $p_2 = 2h_2$ along the $y$-axis of a SQ lattice of size $L$ where $L$ is an even number and label each lattice site by a couple of integer numbers $(n, m)$ with $n, m = 1, \ldots, L$. The double periodicity implies that the spin $\sigma_{m,n}$ at site $\boldsymbol{r}_i = (m, n)$ is the same as the spin $\sigma_{m+s_1 p_1, n+s_2 p_2}$ at site $(m + s_1 p_1, n + s_2 p_2)$ with $s_1 = 0, 1, \ldots, L/p_1 - 1$ and $s_2 = 0, 1, \ldots, L/p_2 - 1$. The *finite* Fourier transform of the spin $\sigma_{m,n}$ is given by

$$\begin{aligned}\sigma_{\boldsymbol{Q}} \equiv \sigma_{q_1,q_2} &= \frac{1}{L} \sum_{m=1}^{L} \sum_{n=1}^{L} \sigma_{m,n} e^{-i\frac{2\pi}{L} q_1 m} e^{-i\frac{2\pi}{L} q_2 n} \\ &= \frac{1}{L} \sum_{m=1}^{p_1} \sum_{n=1}^{p_2} \sigma_{m,n} e^{-i\frac{2\pi}{L} q_1 m} e^{-i\frac{2\pi}{L} q_2 n} \sum_{s_1=0}^{L/p_1-1} \sum_{s_2=1}^{L/p_2-1} e^{-i\frac{2\pi}{L} q_1 s_1 p_1} e^{-i\frac{2\pi}{L} q_2 s_2 p_2}\end{aligned} \tag{10.3.2}$$

where $\boldsymbol{Q} = 2\pi(\frac{q_1}{L}, \frac{q_2}{L})$ with $q_1, q_2 = 0, 1, \ldots, L-1$. The last two sums occurring in Eq. (10.3.2) can be easily performed. Indeed,

$$\sum_{s_1=0}^{L/p_1-1} e^{-i\frac{2\pi}{L} q_1 s_1 p_1} = \frac{L}{p_1} \delta_{\frac{q_1}{L}, \frac{l_1}{p_1}}, \tag{10.3.3}$$

where $l_1 = 0, 1, \ldots, p_1 - 1$. Then Eq. (10.3.2) reduces to

$$\sigma_{q_1,q_2} = \frac{L}{p_1 p_2} \delta_{\frac{q_1}{L}, \frac{l_1}{p_1}} \delta_{\frac{q_2}{L}, \frac{l_2}{p_2}} \sum_{m=1}^{p_1} \sum_{n=1}^{p_2} \sigma_{m,n} e^{-i\frac{2\pi}{L} q_1 m} e^{-i\frac{2\pi}{L} q_2 n}. \tag{10.3.4}$$

To perform the sums occurring in Eq. (10.3.4), one has to account for the checkerboard structure of the ground state. Indeed, one takes $\sigma_{m,n} = 1$ for $1 \le m \le h_1$, $1 \le n \le h_2$ and $h_1 + 1 \le m \le 2h_1$, $h_2 + 1 \le n \le 2h_2$ and $\sigma_{m,n} = -1$ for $1 \le m \le h_1$, $h_2 + 1 \le n \le 2h_2$ and $h_1 + 1 \le m \le 2h_1$, $1 \le n \le h_2$. Then the sums in Eq. (10.3.4) become

$$\begin{aligned}\sum_{m=1}^{p_1} \sum_{n=1}^{p_2} \sigma_{m,n} e^{-i\frac{2\pi}{L} q_1 m} e^{-i\frac{2\pi}{L} q_2 n} &= \sum_{m=1}^{h_1} \sum_{n=1}^{h_2} e^{-i\frac{2\pi}{L} q_1 m} e^{-i\frac{2\pi}{L} q_2 n} \\ &\quad \times (1 - e^{-i\frac{2\pi}{L} q_1 h_1})(1 - e^{-i\frac{2\pi}{L} q_2 h_2}).\end{aligned} \tag{10.3.5}$$

Accounting for the $\delta$-functions occurring in Eq. (10.3.4), from Eq. (10.3.5) one obtains the relationship

$$\delta_{\frac{q_1}{L},\frac{l_1}{p_1}}(1-e^{-i\frac{2\pi}{L}q_1h_1})\sum_{m=1}^{h_1} e^{-i\frac{2\pi}{L}q_1m} = \begin{cases} 0 & \text{for } \frac{q_1}{L}=\frac{s_1}{h_1}, \\ \frac{4e^{-i\frac{\pi}{h_1}(2s_1+1)}}{1-e^{-i\frac{\pi}{h_1}(2s_1+1)}} & \text{for } \frac{q_1}{L}=\frac{s_1+\frac{1}{2}}{h_1} \end{cases} \tag{10.3.6}$$

with $s_1 = 0, 1, \ldots, h_1 - 1$. Replacing the result (10.3.6) into Eq. (10.3.4), one has

$$\sigma_{q_1,q_2} = \frac{4L}{h_1h_2}\delta_{\frac{q_1}{L},\frac{1}{h_1}(s_1+\frac{1}{2})}\delta_{\frac{q_2}{L},\frac{1}{h_2}(s_2+\frac{1}{2})}\frac{e^{-i\frac{\pi}{h_1}(2s_1+1)}}{1-e^{-i\frac{\pi}{h_1}(2s_1+1)}} \times \frac{e^{-i\frac{\pi}{h_2}(2s_2+1)}}{1-e^{-i\frac{\pi}{h_2}(2s_2+1)}} \tag{10.3.7}$$

with $s_1 = 0, 1, \ldots, h_1-1$ and $s_2 = 0, 1, \ldots, h_2-1$. The square modulus of Eq. (10.3.7) is given by

$$|\sigma_{q_1,q_2}|^2 = \frac{L^2}{h_1^2h_2^2}\delta_{\frac{q_1}{L},\frac{1}{h_1}(s_1+\frac{1}{2})}\delta_{\frac{q_2}{L},\frac{1}{h_2}(s_2+\frac{1}{2})}\frac{1}{\sin^2\left[\frac{\pi}{h_1}(s_1+\frac{1}{2})\right]\sin^2\left[\frac{\pi}{h_2}(s_2+\frac{1}{2})\right]} \tag{10.3.8}$$

and the general formula for the ground-state energy of a checkerboard configuration with ferromagnetic NN exchange and dipole–dipole interaction becomes

$$\begin{aligned} E_{\langle h_1,h_2\rangle} &= -\sum_{q_1,q_2}\left[J\left(\cos\frac{2\pi q_1}{L}+\cos\frac{2\pi q_2}{L}\right)+gD^{zz}\left(\frac{q_1}{L},\frac{q_2}{L}\right)\right]|\sigma_{q_1,q_2}|^2 \\ &= -\frac{L^2}{h_1^2h_2^2}\sum_{s_1=0}^{h_1-1}\sum_{s_2=0}^{h_2-1}\left\{J\left[\cos\frac{2\pi(s_1+\frac{1}{2})}{h_1}+\cos\frac{2\pi(s_2+\frac{1}{2})}{h_2}\right]\right. \\ &\quad \left. +gD^{zz}\left(\frac{s_1+\frac{1}{2}}{h_1},\frac{s_2+\frac{1}{2}}{h_2}\right)\right\}\frac{1}{\sin^2\left[\frac{\pi}{h_1}(s_1+\frac{1}{2})\right]\sin^2\left[\frac{\pi}{h_2}(s_2+\frac{1}{2})\right]}, \end{aligned} \tag{10.3.9}$$

where $D^{zz}(\mathbf{Q}) \equiv D^{zz}(\frac{q_1}{L},\frac{q_2}{L})$ is given by Eq. (10.1.21). From the relationships

$$\frac{1}{h^2}\sum_{s=0}^{h-1}\frac{1}{\sin^2\left[\frac{\pi}{h}(s+\frac{1}{2})\right]} = 1 \tag{10.3.10}$$

and

$$\frac{1}{h^2}\sum_{s=0}^{h-1}\frac{\cos\left[\frac{2\pi}{h}(s+\frac{1}{2})\right]}{\sin^2\left[\frac{\pi}{h}(s+\frac{1}{2})\right]} = 1-\frac{2}{h}, \tag{10.3.11}$$

one obtains

$$\begin{aligned} \frac{E_{\langle h_1,h_2\rangle}}{gL^2} &= -2\frac{J}{g}\left(1-\frac{1}{h_1}-\frac{1}{h_2}\right) \\ &\quad -\frac{1}{h_1^2h_2^2}\sum_{s_1=0}^{h_1-1}\sum_{s_2=0}^{h_2-1}\frac{D^{zz}\left(\frac{s_1+\frac{1}{2}}{h_1},\frac{s_2+\frac{1}{2}}{h_2}\right)}{\sin^2\left[\frac{\pi}{h_1}(s_1+\frac{1}{2})\right]\sin^2\left[\frac{\pi}{h_2}(s_2+\frac{1}{2})\right]}. \end{aligned} \tag{10.3.12}$$

Equation (10.3.12) is the general zero temperature energy of a *checkerboard* configuration where blocks of $h_1 \times h_2$ spins up and down alternate in the SQ lattice. A study has to be done to determine the regions of stability of the different checkerboard configurations as function of the ratio $J/g$. The checkerboard configuration was proposed as possible ground-state configuration of an Ising model with NN exchange and dipolar interactions by Czech and Villain[108] in the limit of large $J/g$. They arrived at the conclusion that for $\langle h_1, h_2 \rangle$ configurations with $h_1, h_2 \to \infty$, the ground-state energy was characterized by a *regular* checkerboard configuration with $h_1 = h_2$. On the other hand, *stripe* configurations corresponding to $\langle h_1, h_2 \to \infty \rangle$ or $\langle h_1 \to \infty, h_2 \rangle$ were proven to have a lower energy with respect to the checkerboard configurations,[109] at least for small $J/g$. A careful investigation of the minima of the function (10.3.12) varying the ratio $J/g$ shows[107] that the checkerboard configuration is never stable for generic $\langle h_1, h_2 \rangle$. Only the checkerboard phases corresponding to $h_1 = 1$ and $h_2 \equiv n = 1, 2, \ldots$ become stable in a restricted region of $J/g$. For the checkerboard phase $\langle 1, n \rangle$, the zero temperature energy (10.3.12) becomes

$$\frac{E_{\langle 1,n \rangle}}{gL^2} = \frac{2}{n}\left(\frac{J}{g}\right) - \frac{1}{n^2}\sum_{s=0}^{n-1} \frac{D^{zz}\left(\frac{1}{2}, \frac{1}{n}\left(s+\frac{1}{2}\right)\right]}{\sin^2\left[\frac{\pi}{n}\left(s+\frac{1}{2}\right)\right]} \equiv a\frac{J}{g} + b. \tag{10.3.13}$$

In Table 10.2, we give the coefficients $a$ and $b$ and the region of stability for several $\langle 1, n \rangle$ phases obtained from Eq. (10.3.13). For instance, the Néel antiferromagnet, corresponding to the checkerboard configuration $\langle 1, 1 \rangle$ ($n = 1$ in Table 10.2), is stable for $J/g < 0.83040611$ in agreement with the result (10.1.22) obtained for the Ising model with pure dipole–dipole interaction. In the narrow range $0.830406111750 < J/g < 0.880624471981$, the sequence of $\langle 1, n \rangle$-phases, with $n = 2, 3, \ldots, 9$, appears. Each phase is characterized by $n$ identical antiferromagnetic rows followed by the same number of antiferromagnetic rows with overturned spins. For $n \geq 10$, all the checkerboard phases collapse to the multiphase point $J/g = 0.880624471981$. Note that for $n \to \infty$, the checkerboard phase reduces to a stripe configuration corresponding to a "columnar" phase in which ferromagnetic

Table 10.2. Coefficients $a$ and $b$ of the zero temperature energy $E_{\langle 1,n \rangle}/(gL^2) = a(J/g) + b$ for several $\langle 1, n \rangle$ phases.

| $n$ | $a$ | $b$ | Region of stability |
|---|---|---|---|
| 1 | 2 | −2.6458865323 | (0, 0.830406111750) |
| 2 | 1 | −1.8154804206 | (0.830406111750, 0.878839805732) |
| 3 | 2/3 | −1.5225338186 | (0.878839805732, 0.880558140666) |
| 4 | 1/2 | −1.3757741285 | (0.880558140666, 0.880621920068) |
| 5 | 2/5 | −1.2877119365 | (0.880621920068, 0.880624371628) |
| 6 | 1/3 | −1.2290036451 | (0.880624371628, 0.880624467975) |
| 7 | 2/7 | −1.1870691466 | (0.880624467975, 0.880624471820) |
| 8 | 1/4 | −1.1556182726 | (0.880624471820, 0.880624471975) |
| 9 | 2/9 | −1.1311564817 | (0.880624471975, 0.880624471981) |
| $n \geq 10$ | $2/n$ | $-0.9354621546 - \dfrac{1.761248944}{n}$ | $\dfrac{J}{g} = 0.880624471981$ |

columns of spin up and down alternate. As shown in Table 10.2, the width $n$ of the bands that are made up of repeated antiferromagnetic rows becomes larger and larger as $J/g$ increases. For $n \geq 10$, one can see that the coefficient $b$ occurring in Eq. (10.3.13) is given by

$$b = D^{zz}\left(\frac{1}{2}, 0\right) + \frac{1.761248944}{n} \tag{10.3.14}$$

where $D^{zz}(\frac{1}{2}, 0)$ is given in Table 10.1. Then for $n \geq 10$, the checkerboard energy (10.3.13) becomes

$$\frac{E_{\langle 1,n\rangle}}{gL^2} = \frac{2}{n}\left(\frac{J}{g}\right) - 0.9354621546 - \frac{1.761248944}{n}. \tag{10.3.15}$$

Note that for $J/g = 0.8808624471981$, the energy of the $\langle 1, n\rangle$ phase is independent of $n$ and an "accumulation point" occurs, similar to the *multiphase point* ($T = 0, J_2/J_1 = -1/2$) present in the anisotropic next nearest neighbour Ising (ANNNI) model[110] in 2D and 3D where the competition between the NN interaction $J_1$ and the NNN interaction $J_2$ is restricted to one direction (i.e. the $x$-axis) of a SQ or SC lattice. In Fig. 10.3, we show the checkerboard configurations $\langle 1, 1\rangle$, $\langle 1, 2\rangle$, $\langle 1, 3\rangle$ and the stripe configuration $\langle 1, \infty\rangle$ for a SQ lattice of

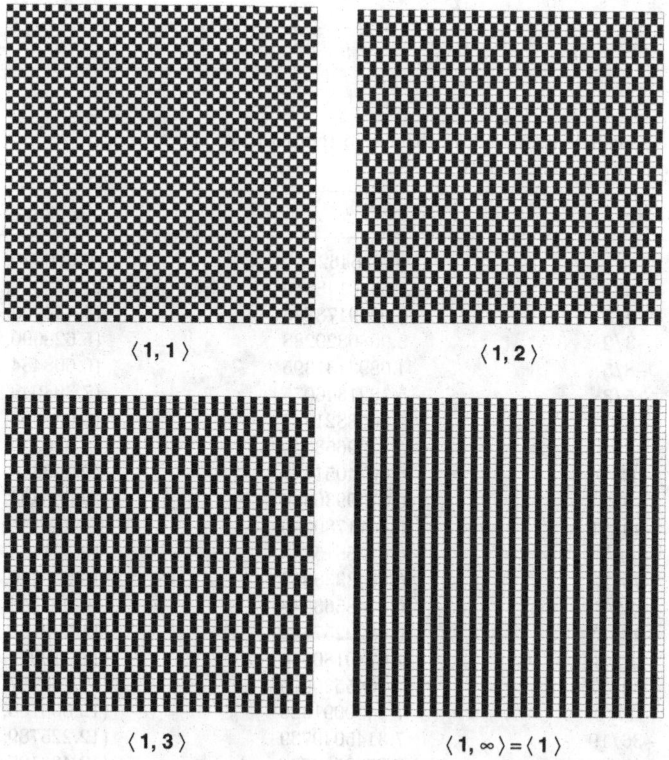


Fig. 10.3. Zero temperature configuration of the phases $\langle 1, 1\rangle$ (upper left), $\langle 1, 2\rangle$ (upper right), $\langle 1, 3\rangle$ (lower left) and $\langle 1, \infty\rangle$ (lower right).

side $L = 48$. The white and black squares correspond to spins of opposite orientation. As one can see, the $\langle 1, n\rangle$ phases correspond to irregular checkerboard. We stress that the checkerboard phases have energy lower than the stripe phases in a very narrow region of $J/g$ located between the Néel phase and the columnar phase. Any other checkerboard configuration $\langle m, n\rangle$ with $m \neq 1$ is *never* stable for any $J/g$.

As we have seen, the *stripe* configuration can be obtained from the checkerboard configuration taking the limit $h_1 \to \infty$ (horizontal stripes of width $h_2$) or $h_2 \to \infty$ (vertical stripes of width $h_1$). The energies of the stripe configurations are obtained from Eq. (10.3.12), assuming $h_1 = h$, $h_2 \to \infty$ or $h_1 \to \infty$, $h_2 = h$ and making use of Eq. (10.3.10). The general stripe configuration energy becomes

$$\frac{E_{\langle h\rangle}}{gL^2} = -2\frac{J}{g}\left(1 - \frac{1}{h}\right) - \frac{1}{h^2}\sum_{s=0}^{h-1} \frac{D^{zz}(0, \frac{s+\frac{1}{2}}{h})}{\sin^2\left[\frac{\pi}{h}\left(s + \frac{1}{2}\right)\right]}. \tag{10.3.16}$$

The energies of several stripe configurations are given in Table 10.3. The zero temperature energy of the ferromagnetic configuration is obtained taking the limit $h \to \infty$ of Eq. (10.3.16), that is,

$$\frac{E_{\mathrm{F}}}{gL^2} = -2\frac{J}{g} - D^{zz}(0,0) \tag{10.3.17}$$

Table 10.3. Coefficients $a$ and $b$ of the zero temperature energy $E_{\langle h\rangle}/(gL^2) = a(J/g) + b$ for stripe configurations of width $h$.

| $h$ | $a$ | $b$ | Region of stability |
|---|---|---|---|
| 1 | 0 | −0.9354621546 | (0.880625, 2.517077) |
| 2 | −1 | 1.5816148819 | (2.517077, 4.344909) |
| 3 | −4/3 | 3.0299178788 | (4.344909, 5.628990) |
| 4 | −3/2 | 3.9680829288 | (5.628990, 6.608484) |
| 5 | −8/5 | 4.6893131395 | (6.608484, 7.398049) |
| 6 | −5/3 | 5.1221346075 | (7.398049, 8.058698) |
| 7 | −12/7 | 5.5058821312 | (8.058698, 8.626345) |
| 8 | −7/4 | 5.8139658640 | (8.626345, 9.123814) |
| 9 | −16/9 | 6.0674051477 | (9.123814, 9.566483) |
| 10 | −9/5 | 6.2799936626 | (9.566483, 9.965190) |
| 11 | −20/11 | 6.4611789296 | (9.965190, 10.327855) |
| 12 | −11/6 | 6.6176615768 | (10.327855, 10.660440) |
| 13 | −24/13 | 6.7543338863 | (10.660440, 10.967544) |
| 14 | −13/7 | 6.8748563518 | (10.967544, 11.252788) |
| 15 | −28/15 | 6.9820257605 | (11.252788, 11.519072) |
| 16 | −15/8 | 7.0780180265 | (11.519072, 11.768757) |
| 17 | −32/17 | 7.1645530020 | (11.768757, 12.003788) |
| 18 | −17/9 | 7.2430091326 | (12.003788, 12.225789) |
| 19 | −36/19 | 7.3145049739 | (12.225789, 12.436127) |
| 20 | −19/10 | 7.3799582736 | (12.436127, 12.635966) |
| $h \geq 20$ | $-2(1-\frac{1}{h})$ | $9.0336216831 - \frac{8}{h}\ln h - \frac{9.10467}{h} - \frac{1.09642}{h^3}$ | |

where $D^{zz}(0,0)$ is given in Table 10.1. For stripes of large width ($h \geq 20$), Eq. (10.3.16) is well represented by the formula

$$\frac{E_{\langle h\rangle}}{gL^2} = -2\frac{J}{g}\left(1-\frac{1}{h}\right) - D^{zz}(0,0) - \frac{8}{h}\ln h - \frac{9.10467}{h} - \frac{1.09642}{h^3}. \qquad (10.3.18)$$

The energies obtained from Eq. (10.3.18) differ from those obtained by direct (numerical) calculation by less than $10^{-6}$ for $20 < h < 10^5$. The logarithmic term in the asymptotic expansion (10.3.18) is reminiscent of the linear dependence on the magnitude $Q = |\boldsymbol{Q}|$ of $D^{zz}(\boldsymbol{Q})$ for $\boldsymbol{Q} \to 0$, that is

$$a^3 D^{zz}(\boldsymbol{Q}) = -9.0336216831 + 2\pi(aQ) - 0.975066230(aQ)^2 \cdots. \qquad (10.3.19)$$

In Fig. 10.4, we show the stripe configurations $\langle 2\rangle$, $\langle 3\rangle$, $\langle 4\rangle$ and $\langle 8\rangle$ for a SQ lattice of side $L = 48$. For $J/g$ increasing from the multiphase point ($J/g = 0.880624471981$), a sequence of stripe configurations $\langle h\rangle$ with $h = 1, 2, \ldots$ occurs until the ferromagnetic phase F corresponding to $h \to \infty$ is reached. The transition between the F phase and the widest stripe phase $\langle L/2\rangle$ occurs at $J/g = 1.79975 +$

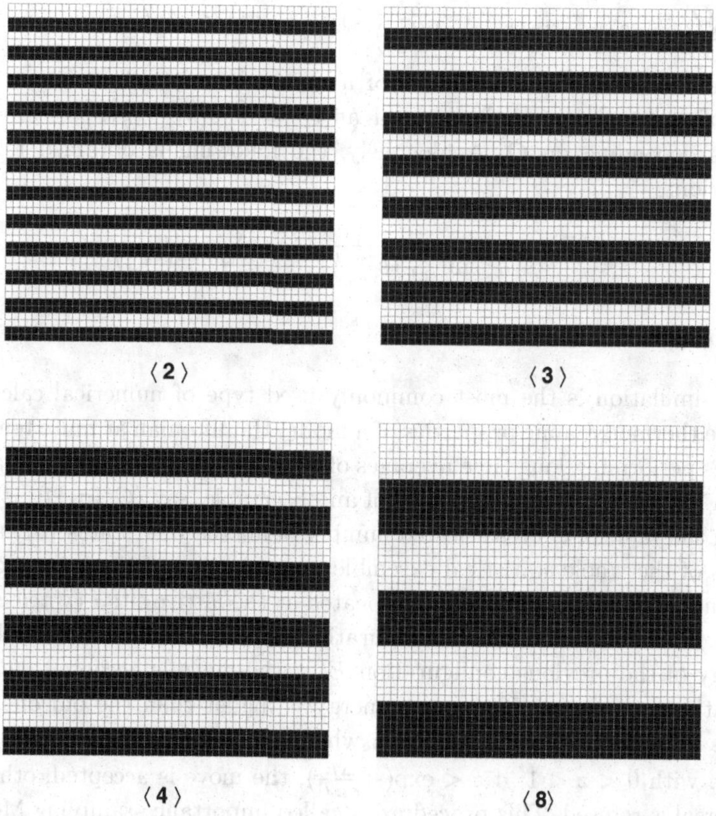


Fig. 10.4. Zero temperature configuration of the stripe phases $\langle 2\rangle$ (upper left), $\langle 3\rangle$ (upper right), $\langle 4\rangle$ (lower left) and $\langle 8\rangle$ (lower right).

$\ln L^4$ as it can be obtained by comparing the energy of the F phase with that of the stripe phase with $h = L/2$ given by the asymptotic expansion (10.3.18).

The structure factor of a checkerboard configuration can be obtained from Eq. (3.1.2). Noticing that the Ising Hamiltonian (10.3.1) commutes with each local spin observable $\sigma_i$, one deduces that *all* $\sigma_i$ are constants of motion, that is $\sigma_i(t) = \sigma_i$. This property reduces Eq. (3.1.2) to

$$S^{zz}(\boldsymbol{Q}, \omega) = \sum_{i,j} e^{i\boldsymbol{Q}\cdot(\boldsymbol{r}_i - \boldsymbol{r}_j)} \langle \sigma_i \sigma_j \rangle \delta(\omega). \tag{10.3.20}$$

Using Eq. (10.3.2), one obtains

$$S^{zz}(\boldsymbol{Q}, \omega) = N^2 \delta(\omega) \langle S(\boldsymbol{Q}) \rangle \tag{10.3.21}$$

where $N = L^2$ and

$$S(\boldsymbol{Q}) = \frac{|\sigma_{\boldsymbol{Q}}|^2}{L^2} \tag{10.3.22}$$

is the static structure factor. At zero temperature, the static structure factor for a checkerboard configuration can be obtained directly from Eq. (10.3.8):

$$S(\boldsymbol{Q}) = \frac{1}{h_1^2 h_2^2 \sin^2\left(\frac{Q_x}{2}\right) \sin^2\left(\frac{Q_y}{2}\right)} \delta_{Q_x, \frac{2\pi}{h_1}\left(s_1 + \frac{1}{2}\right)} \delta_{Q_y, \frac{2\pi}{h_2}\left(s_2 + \frac{1}{2}\right)}. \tag{10.3.23}$$

As one can see, the structure factor of a checkerboard configuration consists of a series of $\delta$-like elastic peaks located at $\boldsymbol{Q} = \left[\frac{2\pi}{h_1}\left(s_1 + \frac{1}{2}\right), \frac{2\pi}{h_2}\left(s_2 + \frac{1}{2}\right)\right]$ with intensities depending on $\boldsymbol{Q}$. Analogously, the following is obtained for a stripe configuration:

$$S(\boldsymbol{Q}) = \frac{1}{h^2 \sin^2\left[\frac{\pi}{h}\left(s + \frac{1}{2}\right)\right]} \delta_{Q_x, 0} \delta_{Q_y, \frac{2\pi}{h}\left(s + \frac{1}{2}\right)}. \tag{10.3.24}$$

## 10.4. Monte Carlo Simulation

The MC simulation is the most commonly used type of numerical calculation to obtain the thermodynamic properties of a model Hamiltonian at finite temperature. It involves performing long time averages of the physical observables. In particular, for a spin lattice, the thermal average of an observable, say $\langle \mathcal{O}(\{\sigma_i\}) \rangle$, is evaluated by taking the value of $\mathcal{O}$ on a selected number of microscopic states $\{\sigma_i\}$ among the $2^N$ states of the whole statistical ensemble: the selection of the microscopic states is performed by reversing a given spin located at the lattice site $i$ (*single move*) and evaluating the energy of the new configuration $E_{\text{new}}$: if the new energy is lower than the energy of the previous configuration $E_{\text{old}}$, the move is accepted, and the new configuration is assumed. If the new energy is higher than the old one, then one compares the quantity $\exp(-\frac{\Delta E}{k_B T}) < 1$ where $\Delta E = E_{\text{new}} - E_{\text{old}}$ with a random number $z$ with $0 < z < 1$: if $z < \exp(-\frac{\Delta E}{k_B T})$, the move is accepted, otherwise the spin reversal is refused. This procedure is called importance-sampling Monte Carlo calculation.[111] A MC *step* (MCS) consists of $L^2$ spin-flip trials, one for every spin of the lattice. The statistical average $\langle \mathcal{O} \rangle$ reduces to the arithmetic average of $\mathcal{O}$

over the selected configurations (*snapshots*). To obtain a reliable statistics, one has to perform long MC *runs* ($10^5$ MCSs in the present case) recording one configuration every 10 MCSs and disregarding the initial $10^4$ MCSs for equilibration at each temperature. In order to improve the statistics, an average over eight independent runs was considered.[107] A convenient procedure that allow us to reduce the equilibration time consists of starting from the low temperature configuration (ground state) which is known exactly, increasing the temperature by small steps $\Delta T = 0.01$ (in units of $\frac{g}{k_B}$) and assuming the starting configuration of the next temperature as the final configuration of the previous one. In this way, the system is always kept near the equilibrium even though near a first-order phase transition, a small uncertainty about the transition temperature between the two coexisting phases is entered because of the reluctance of the system to change its configuration due to the use of single moves. However, the approach based on a gradual rise of the temperature gives accurate results everywhere except in the proximity of a first-order phase transition. Moreover, this approach saves computing time with respect to the approach that assumes a random configuration as initial configuration at any temperature so increasing the equilibration time.[106] Indeed, a serious problem of computing time occurs due to the long range nature of the dipole–dipole interaction: the computing time is proportional to $L^2$ for a system with short range (exchange) interaction while it grows as $L^4$ for systems with long range (dipolar) interaction. With the actually available computers the time required for one MCS on a SQ lattice with $L = 48$ is of the order of $10^{-2}$ s, so that it is easy to evaluate the time required for a MC simulation. Another problem implicit in the long range interactions is how to cut them on a *finite* lattice: Kretschmer and Binder[101] showed that in MC simulations, the most effective approach is to think of the finite lattice of side $L$ as a "cell" belonging to an infinite lattice made up of infinitely repeated "images" of the original cell. Then the true dipole–dipole interaction between the spins $\sigma_i$ and $\sigma_j$ belonging to the finite lattice under investigation is replaced by an effective interaction that accounts for the interaction of the spin $\sigma_i$ with *all* the images of the spin $\sigma_j$ belonging to the infinitely repeated cells. Then we evaluate such an interaction by means of the Ewald's method of Section 10.1. Obviously, this approach results in an excess of "order" due to the assumption of the periodic image arrangement. Finally, the usual PBC are imposed on the finite lattice.

Among the several thermodynamic quantities that one can evaluate by means of the MC simulation, we will consider the specific heat

$$C = \frac{\langle \mathcal{H}^2 \rangle - \langle \mathcal{H} \rangle^2}{L^2 k_B T^2}, \tag{10.4.1}$$

the internal energy per spin in units of dipolar interaction

$$E = \frac{\langle \mathcal{H} \rangle}{L^2 g}, \tag{10.4.2}$$

the static structure factor $S(\boldsymbol{Q})$ given by Eq. (10.3.22) and the order parameterbooth95

$$O_{hv} = \left\langle \left| \frac{n_h - n_v}{n_h + n_v} \right| \right\rangle, \tag{10.4.3}$$

where $n_h$ (respectively, $n_v$) is the number of horizontal (respectively, vertical) pairs of NN antiparallel spins. At $T = 0$, this unconventional order parameter is 1 in the stripe configuration since $n_h = 0$, $n_v = \frac{L}{h}$ for horizontal stripes of width $h$ and $n_h = \frac{L}{h}$, $n_v = 0$ for vertical stripes of width $h$; it is $\frac{n-1}{n+1}$ in the checkerboard phase $\langle 1, n \rangle$; it vanishes in the Néel (N) configuration ($n_h = n_v = L$) and it is indeterminate in the ferromagnetic (F) configuration ($n_h = n_v = 0$). In the last two cases, the order parameter (10.4.3) is replaced by the staggered or sublattice magnetization in the N phase and by the magnetization in the F phase, respectively.

Other quantities that we will investigate in the MC simulation are the energy density distribution $P(E)$ and the order parameter density distribution $P(O_{hv})$. These quantities are particularly useful to establish the order of the phase transition: indeed, a two-peak structure in $P(E)$ is expected if the phase transition is of the first-order while a single peak that broadens at the transition is expected in a second-order phase transition.[113,114] The two peaks in $P(E)$ occurring at the first-order phase transition are related to the energies of the two coexisting phases and the transition temperature will be determined as the temperature at which the two peaks have the same area. Indeed, for a finite system, the existence of the two peaks, even though with different areas, is recorded over a narrow but finite range of temperatures at variance with a macroscopic system ($L \to \infty$) where the coexistence is restricted to a single temperature: the transition temperature of the first-order phase transition. As for the order parameter density distribution[104] $P(O_{hv})$, it is more convenient to neglect the absolute value in Eq. (10.4.3): in this way, we are able to distinguish between horizontal ($O_{hv} = 1$ at $T = 0$) and vertical ($O_{hv} = -1$ at $T = 0$) stripe phases and a three-peak structure in $P(O_{hv})$ can be observed at a first-order phase transition: the central peak corresponds to the paramagnetic phase while the two symmetric peaks around the origin correspond to horizontal and vertical stripes respectively. The coexistence of these two peaks is traced back to the occurrence of a "tetragonal" phase[112] where the orientational order is lost because the stripes rotate freely, the paramagnetic phase entering only at higher temperature. The tetragonal phase differs from the conventional paramagnetic phase because it is characterized by a two-peak profile in $P(O_{hv})$ without the central peak. In both tetragonal and paramagnetic phase, the LRO is absent: it is the short range order that differs in the two phases. Strictly speaking, no phase transition is expected between the tetragonal and paramagnetic phase.

Any MC simulation is performed on a SQ lattice $48 \times 48$ with PBC.[107] The size of the lattice is suggested by the fact that it is the smallest one supporting stripes of width $h = 1, 2, 3$ and 4. The remaining stripes consistent with $L = 48$ are 6, 8, 12 and 24. In Table 10.4, we give the zero temperature energies and the stability regions of the $\langle 1, n \rangle$ and $\langle h \rangle$ configurations which are *consistent* with the PBC of the lattice $48 \times 48$. Comparing Table 10.4 with Tables 10.2 and 10.3, one can check

Table 10.4. Coefficients $a$ and $b$ of the zero temperature energy for $\langle 1, n\rangle$ and $\langle h\rangle$ configurations which are consistent with a SQ lattice of side $L = 48$ and PBC.

| $\langle 1, n\rangle$, $\langle h\rangle$ | $a$ | $b$ | Region of stability |
|---|---|---|---|
| $\langle 1, 1\rangle = \mathrm{N}$ | 2 | −2.6458865323 | (0, 0.830406) |
| $\langle 1, 2\rangle$ | 1 | −1.8154804206 | (0.830406, 0.878840) |
| $\langle 1, 3\rangle$ | 2/3 | −1.5225338186 | (0.878840, 0.880558) |
| $\langle 1, 4\rangle$ | 1/2 | −1.3757741285 | (0.880558, 0.880622) |
| $\langle 1, 6\rangle$ | 1/3 | −1.2290036451 | (0.880622, 0.880624) |
| $\langle 1, 8\rangle$ | 1/4 | −1.1556182726 | $\frac{J}{g} = 0.880624$ |
| $\langle 1, \infty\rangle = \langle 1\rangle$ | 0 | −0.9354621546 | (0.880624, 2.517077) |
| $\langle 2\rangle$ | −1 | 1.5816148819 | (2.517077, 4.344909) |
| $\langle 3\rangle$ | −4/3 | 3.0299178788 | (4.344909, 5.628990) |
| $\langle 4\rangle$ | −3/2 | 3.9680829288 | (5.628990, 6.924310) |
| $\langle 6\rangle$ | −5/3 | 5.1221346075 | (6.924310, 8.301975) |
| $\langle 8\rangle$ | −7/4 | 5.8139658640 | (8.301975, 9.644349) |
| $\langle 12\rangle$ | −11/6 | 6.6176615768 | (9.644349, 11.726019) |
| $\langle 24\rangle$ | −23/12 | 7.5948298428 | (11.726019, 17.265502) |
| $\langle \infty\rangle = \mathrm{F}$ | −2 | 9.0336216831 | $\frac{J}{g} > 17.265502$ |

the merits and the faults of the finite-size lattice. In the finite-size lattice, some checkerboard and stripe phases disappear from the sequence of the infinite lattice. Indeed, checkerboard $\langle 1, n\rangle$ and stripe $\langle h\rangle$ phases with $n$ and $h$ such that $L/(2n)$ and $L/(2h)$ are *not* integer numbers are prevented by the PBC. By construction, the periodic images assumption implies that the checkerboard and stripe configurations that are consistent with the finite lattice, have the same zero temperature energy $E$ of the corresponding configurations of the infinite lattice. The lattice $48 \times 48$, however, prevents the occurrence of phases with $n$ and $h = 5, 7, 9, 10$, etc. Indeed, the existence regions of the "regular" stripe phases $\langle 4\rangle$, $\langle 6\rangle$ and $\langle 8\rangle$ become wider at the expense of the stripe phases $\langle 5\rangle$ and $\langle 7\rangle$. However, some "irregular" stripe phases reminiscent of the regular stripe phases suppressed by the finite-size effect reappear.[107] Indeed, a configuration consisting of two stripes of width $h = 4$ followed by eight stripes of width $h = 5$ ($\langle 4^2 5^8\rangle$) becomes stable in the range $6.63 < J/g < 7.23$ in the lattice $48 \times 48$, while it is *never* stable in the infinite lattice where the phase $\langle 5\rangle$ has a lower energy. Let us define the phase $\langle 4^2 5^8\rangle$ as $\langle \sim 5\rangle$ because of its vague recollection to the stripe phase $\langle 5\rangle$ of the infinite lattice. An analogous phase $\langle \sim 7\rangle$ should be expected between the stripe phases $\langle 6\rangle$ and $\langle 8\rangle$ but any attempt to find it was unsuccessful. Indeed, phases like $\langle 5^4 7^4\rangle$, $\langle 5^3 6^3 7^2\rangle$ and $\langle 10^2 7^4\rangle$ have a zero temperature energy higher than, even though very close to, the regular stripe phases $\langle 6\rangle$ or $\langle 8\rangle$ in the lattice $48 \times 48$: one has to conclude that the size of the lattice under investigation is too *small* to support irregular stripe configurations of the type $\langle \sim h\rangle$ with $h \geq 7$. To check this statement, an investigation on a SQ lattice $32 \times 32$ for which the consistent regular stripe phases are characterized by $h = 1, 2, 4, 8, 16$ was performed: a phase $\langle 3^9 5\rangle = \langle \sim 3\rangle$ becomes stable in the region

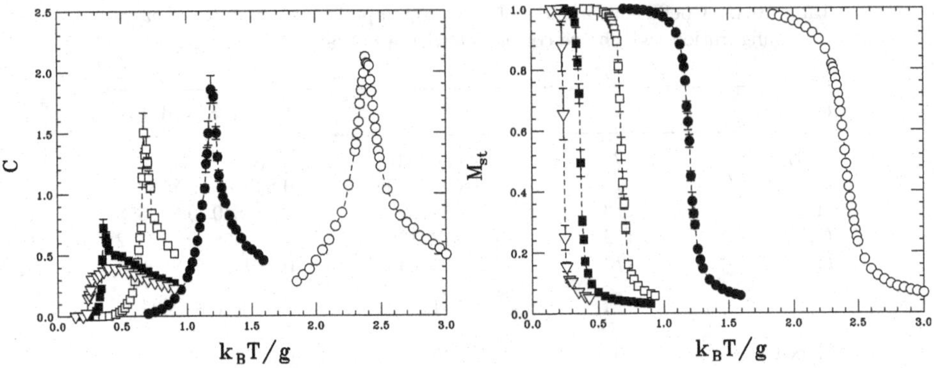


Fig. 10.5. Specific heat (left) and staggered magnetization (right) versus temperature for the Néel (N) antiferromagnet on a SQ lattice 48 × 48: $J/g = 0$ (open circles), $J/g = 0.5$ (full circles), $J/g = 0.7$ (open squares), $J/g = 0.8$ (full squares) and $J/g = 0.83$ (open downward triangles). All MC data is the result of an average over eight independent runs.

$4.56 < J/g < 5.41$ between the stripe phases $\langle 2\rangle$ and $\langle 4\rangle$ and a phase $\langle 5^4 6^2\rangle$ becomes stable for $6.77 < J/g < 7.99$ between the stripe phases $\langle 4\rangle$ and $\langle 8\rangle$. We stress that all irregular stripe phases are unstable in the infinite lattice. For this reason, the detailed investigation of the checkerboard and the stripe phases at finite temperature will be restricted to $n \leq 3$ and $h \leq 4$ respectively. Note that for checkerboard phases $\langle 1, n\rangle$ with $n > 8$, the region of stability is so narrow that it cannot be shown within the numerical precision of Table 10.4. The "largest" stripe configuration at zero temperature supported by the lattice 48 × 48 is the stripe phase $\langle 24\rangle$ and the phase boundary between it and the F phase is given in Table 10.4: $J/g = 17.27$.

In Fig. 10.5, we show the specific heat $C$ (left panel) and the staggered magnetization $M_{\text{st}}$ (right panel) versus temperature for selected values of $J/g$ in the range $0 \leq J/g \leq 0.83$ (N phase). The peak in the specific heat and the drop in the order parameter point out the existence of a phase transition. As one can see from the left panel of Fig. 10.5, both the height of the peak of the specific heat and the transition temperature decrease as $J/g$ increases: the Néel temperature is $k_BT_N/g = 2.38, 1.19, 0.67, 0.36$ for $J/g = 0, 0.5, 0.7, 0.8$, respectively. Moreover, the shape of the peak becomes increasingly sharper when approaching the boundary between the N and the $\langle 1, 2\rangle$ phase ($J/g = 0.83$). Correspondingly, the slope of the staggered magnetization curves at $T_N$ increases going from $J/g = 0$ to 0.83 as shown in the right panel of Fig. 10.5. All the MC data presented in Fig. 10.5 are obtained averaging over eight independent MC runs. The sharpening of the peak and the increment of the error bars of the data around the transition temperature approaching the phase boundary N$-\langle 1, 2\rangle$, seems to point out a change in the order of the phase transition. Indeed, the peak of the specific heat grows as $L^2$ in a first-order phase transition while it grows "only" as $\ln L$ in a second-order (continuous) phase transition of a 2D Ising model.[113] Moreover, the "critical region" is absent in a first-order phase transition, leading to a narrowing of the peak itself. Even though a finite-size scaling is not presented here, the second-order character of the transition can be checked for $0 < J/g \lesssim 0.5$. The large error bars for $J/g \gtrsim 0.5$ are due to the small change in the location of the peak of the specific heat going from one run to another. Indeed, one finds a sharp peak in the specific heat in *each run* whose location and height, however, fluctuate from one run to another, indicating that MC runs of $10^5$ MCSs are too "short" to reach the equilibrium. Incidentally, much longer runs ($10^6$ MCSs) also do not give significantly better results. The fluctuation in the location of the peak causes a drastic reduction of the height of the peak itself when an average over eight runs is taken. This drawback does not affect a second-order phase transition as illustrated in Fig. 10.5 looking at the specific heat of the models with $J/g = 0$ and 0.5. Because of the impossibility of performing much longer MC runs, we investigate the order of the phase transition by looking at the energy and the order parameter density distribution obtained from a single MC run.

In Fig. 10.6, we show the specific heat $C$, the staggered magnetization $M_{\rm st}$ versus temperature for $J/g = 0.83$ together with the energy density distribution $P(E)$ and the order parameter density distribution $P(M_{\rm st})$ for selected temperatures. A sharp peak in the specific heat and a sudden drop in the staggered magnetization is seen at $k_BT/g = 0.27$. Correspondingly, a two-peak profile in $P(E)$ and an intricate structure in $P(M_{\rm st})$ are seen at the same temperature. Out of $k_BT/g = 0.27$, a single

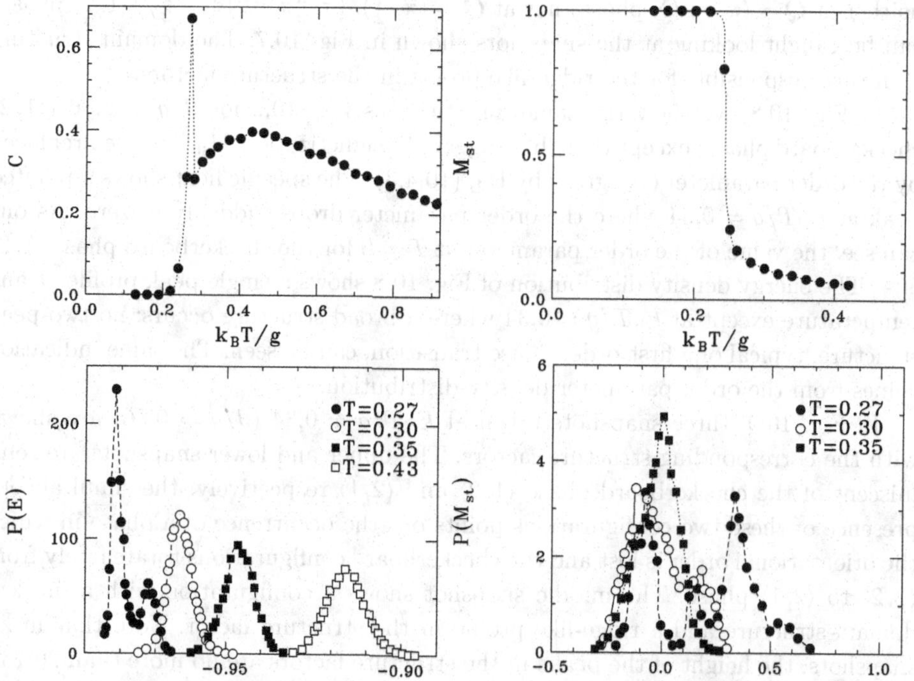


Fig. 10.6. Specific heat (upper left) and order parameter (upper right) versus temperature for $J/g = 0.83$ (Néel phase). Energy density distribution $P(E)$ (lower left) and order parameter density distribution $P(M_{\rm st})$ (lower right) for selected temperatures obtained from a single MC run.

peak structure is recorded. In particular, for $k_BT/g \leq 0.23$, the single $\delta$-like peak in $P(M_{\rm st})$ is located at $E = -0.986$ and the single peak in $P(M_{\rm st})$ is located at $M_{\rm st} = 1$. These values are indistinguishable from their homologous at $T = 0$. For $k_BT/g \geq 0.30$, a single peak structure reappears: by increasing the temperature, the peak in $P(E)$ moves smoothly towards higher energy and broadens while the peak in $P(M_{\rm st})$ broadens without moving away from the origin. For $k_BT/g \leq 0.23$, the structure factor $\langle S(\boldsymbol{Q})\rangle$ shows a $\delta$-like peak of intensity $\simeq 1$ located at $\boldsymbol{Q} = (\pi, \pi)$ as expected for the N antiferromagnet. At $k_BT/g = 0.24$, the height of the peak reduces to 0.65 without changing location. In the narrow range $0.25 < k_BT/g < 0.27$, a ridge-like structure appears around the peak at $\boldsymbol{Q} = (\pi, \pi)$ whose intensity is suddenly reduced to 0.09. Both the snapshots of Fig. 10.7 are taken at $k_BT/g = 0.27$. The microscopic spin configurations are shown on the left while the corresponding structure factor $S(\boldsymbol{Q})$ is shown on the right. The value and the location of the maximum of $S(\boldsymbol{Q})$, the order parameter $O_{hv}$, the staggered magnetization $M_{\rm st}$ and the energy per spin in units of dipolar interaction $E$ are recorded over each snapshot. The energy and the staggered magnetization of the upper snapshot correspond approximately to the location of first peak on the left of both $P(E)$ and $P(M_{\rm st})$ at $k_BT/g = 0.27$ as shown in Fig. 10.6. The energy and the staggered magnetization of the lower snapshot correspond approximately to the second peak of $P(E)$ and $P(M_{\rm st})$ of Fig. 10.6. A mixing of $\langle 1, 1\rangle$ and $\langle 1, 2\rangle$ domains leading to "incommensurate peaks" located neither at $\boldsymbol{Q} = (\pi, \pi)$ (N phase) nor at $\boldsymbol{Q} = (\pi, \frac{\pi}{2})$, $(\pi, \frac{3}{2}\pi) \equiv (\pi, -\frac{\pi}{2})$ ($\langle 1, 2\rangle$ phase) can be caught looking at the snapshots shown in Fig. 10.7. The domain structure is in turn responsible for the ridge-like profile in the structure factors.

In Fig. 10.8, we show the same quantities as Fig. 10.6 for $J/g = 0.86$ ($\langle 1, 2\rangle$ checkerboard phase) except that the staggered magnetization $M_{\rm st}$ has been replaced by the order parameter $O_{hv}$ given by Eq. (10.4.3). The specific heat shows a pointed peak at $k_BT/g = 0.34$ where the order parameter drops suddenly to zero. As one can see, the value of the order parameter at $T = 0$ for the checkerboard phase $\langle 1, 2\rangle$ is $\frac{1}{3}$. The energy density distribution of Fig. 10.8 shows a single-peak profile at any temperature except at $k_BT/g = 0.34$ where a broad structure occurs: no two-peak structure, typical of a first-order phase transition, can be seen. The same indication comes from the order parameter density distribution.

In Fig. 10.9, three snapshots taken at $k_BT/g = 0.34$ ($J/g = 0.86$) are shown with the corresponding structure factors. The upper and lower snapshots are reminiscent of the checkerboard phase $\langle 1, 2\rangle$ and $\langle 2, 1\rangle$ respectively: the simultaneous presence of these two configurations points out the occurrence of a phase in which the orientational order is lost and the checkerboard configurations rotate freely from $\langle 1, 2\rangle$ to $\langle 2, 1\rangle$ phase. The middle snapshot shows a configuration with a marked domain-structure and a ridge-like profile in the structure factor. Note that in all snapshots, the height of the peaks in the structure factors are no more than 20% of the corresponding peaks at $T = 0$.

Figure 10.10 shows the specific heat (left) and the order parameter (right) versus temperature for several $J/g$ in the range $0.88 < J/g < 2.52$ ($\langle 1\rangle$ phase). For $J/g$

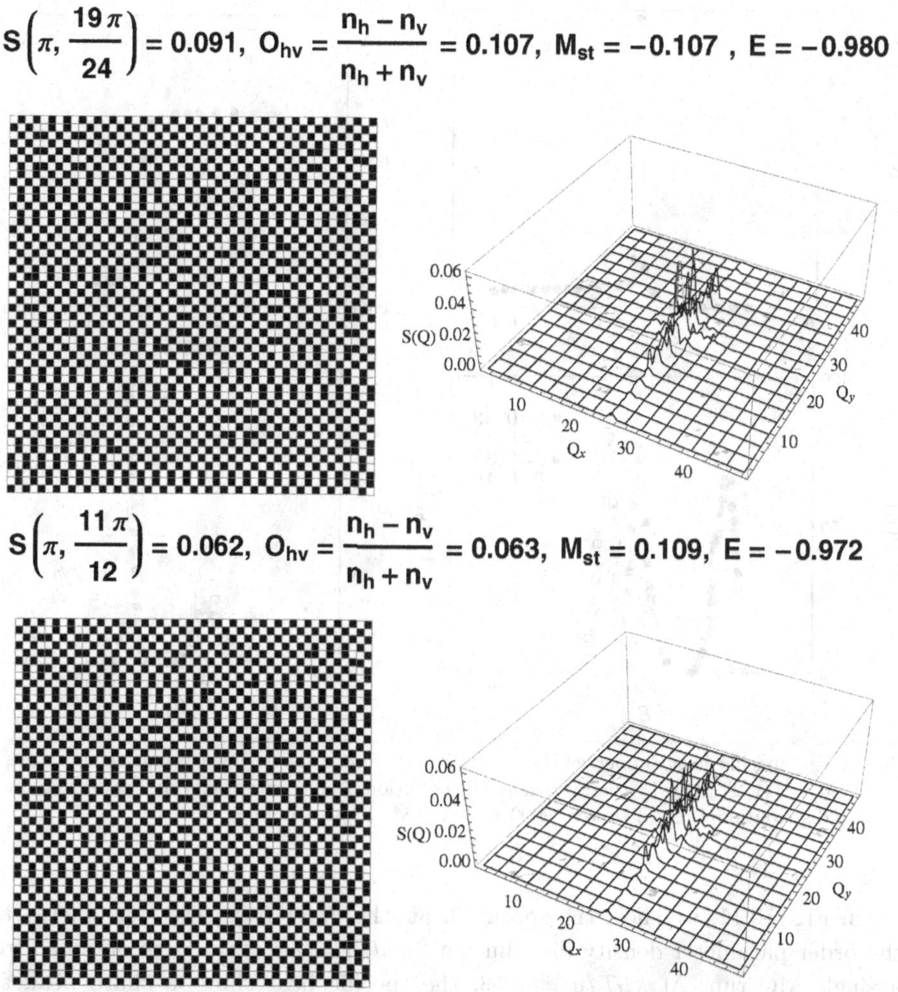


Fig. 10.7. Snapshots and corresponding structure factors for $J/g = 0.83$ (Néel phase) and $k_BT/g = 0.27$. The values of the maximum of the structure factor and of $O_{hv}$, $M_{\rm st}$ and $E$ are indicated over each snapshot. $Q = (\frac{2m\pi}{48}, \frac{2n\pi}{48})$ with $n, m = 0, 1, \ldots, 47$.

close to the phase boundary between the stripe $\langle 1 \rangle$ and the checkerboard $\langle 1, 2 \rangle$ (full circle curves) and for $J/g$ near to the boundary between the stripe phases $\langle 1 \rangle$ and $\langle 2 \rangle$ (open upward triangle curves), the specific heat shows a narrow peak and the order parameter a sudden drop. Moving away from the boundaries towards the centre of the existence region of the phase $\langle 1 \rangle$, the peak in the specific heat broadens and the slope of the drop in the order parameter decreases. Note that the error bars entered by the average over eight independent runs grow when approaching the boundaries.

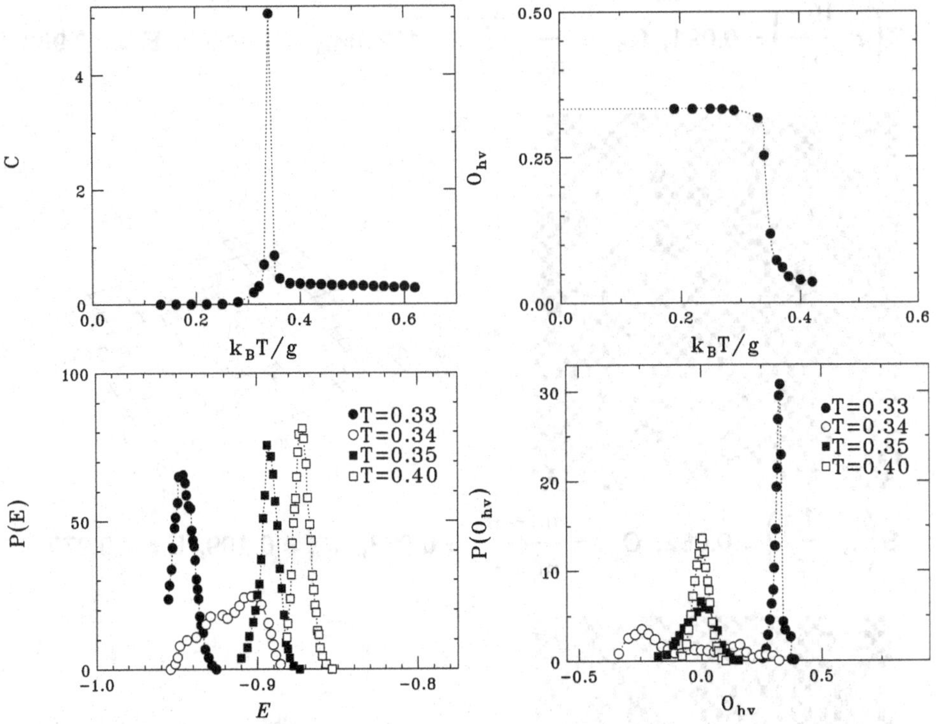


Fig. 10.8. Specific heat (upper left) and order parameter (upper right) versus temperature for $J/g = 0.86$ ($\langle 1, 2\rangle$ phase). Energy density distribution $P(E)$ (lower left) and order parameter density distribution $P(O_{hv})$ (lower right) for selected temperatures.

In Fig. 10.11, we show the specific heat, the order parameter, the energy and the order parameter density distribution for $J/g = 2.5$ ($h = 1$) as obtained from a single MC run. At $k_BT/g = 0.53$, the specific heat shows a sharp peak and the order parameter a single large step from $O_{hv} \simeq 1$ corresponding to the low temperature stripe configuration to $O_{hv} \sim 0$ corresponding to the high temperature paramagnetic phase. At the same time, the energy density distribution (lower left panel of Fig. 10.11) shows a shift in the location of the peak going from $k_BT/g = 0.52$ (full circles) to 0.54 (full squares). However, at $k_BT/g = 0.53$ (open circles), any structure is hard to be identified. Notice the smooth displacement of the peak towards higher energy when increasing the temperature from $k_BT/g = 0.54$. A similar behaviour is found in the order parameter density distribution $P(O_{hv})$ (lower right panel of Fig. 10.11): a shift from the sharp peak centred at $O_{hv} = -1$ at $k_BT/g = 0.52$ to the broad peak centred at $O_{hv} = 0$ at $k_BT/g = 0.6$ is clearly seen. A very flat profile is observed at $k_BT/g = 0.54$.

In Fig. 10.12, we show two snapshots taken at $k_BT/g = 0.53$. The upper snapshot corresponds to a stripe configuration $\langle 1\rangle$ while the lower snapshot corresponds to a "tetragonal" phase[112] where the orientational order is lost. As one can see looking at the corresponding structure factor profiles, the stripe phase $\langle 1\rangle$ is characterized

$S\left(\pi, \dfrac{\pi}{2}\right) = 0.122$, $O_{hv} = \dfrac{n_h - n_v}{n_h + n_v} = 0.301$, $E = -0.940$

$S\left(\dfrac{9\pi}{24}, \pi\right) = 0.048$, $O_{hv} = \dfrac{n_h - n_v}{n_h + n_v} = -0.200$, $E = -0.910$

$S\left(\dfrac{11\pi}{24}, \pi\right) = 0.114$, $O_{hv} = \dfrac{n_h - n_v}{n_h + n_v} = -0.303$, $E = -0.931$

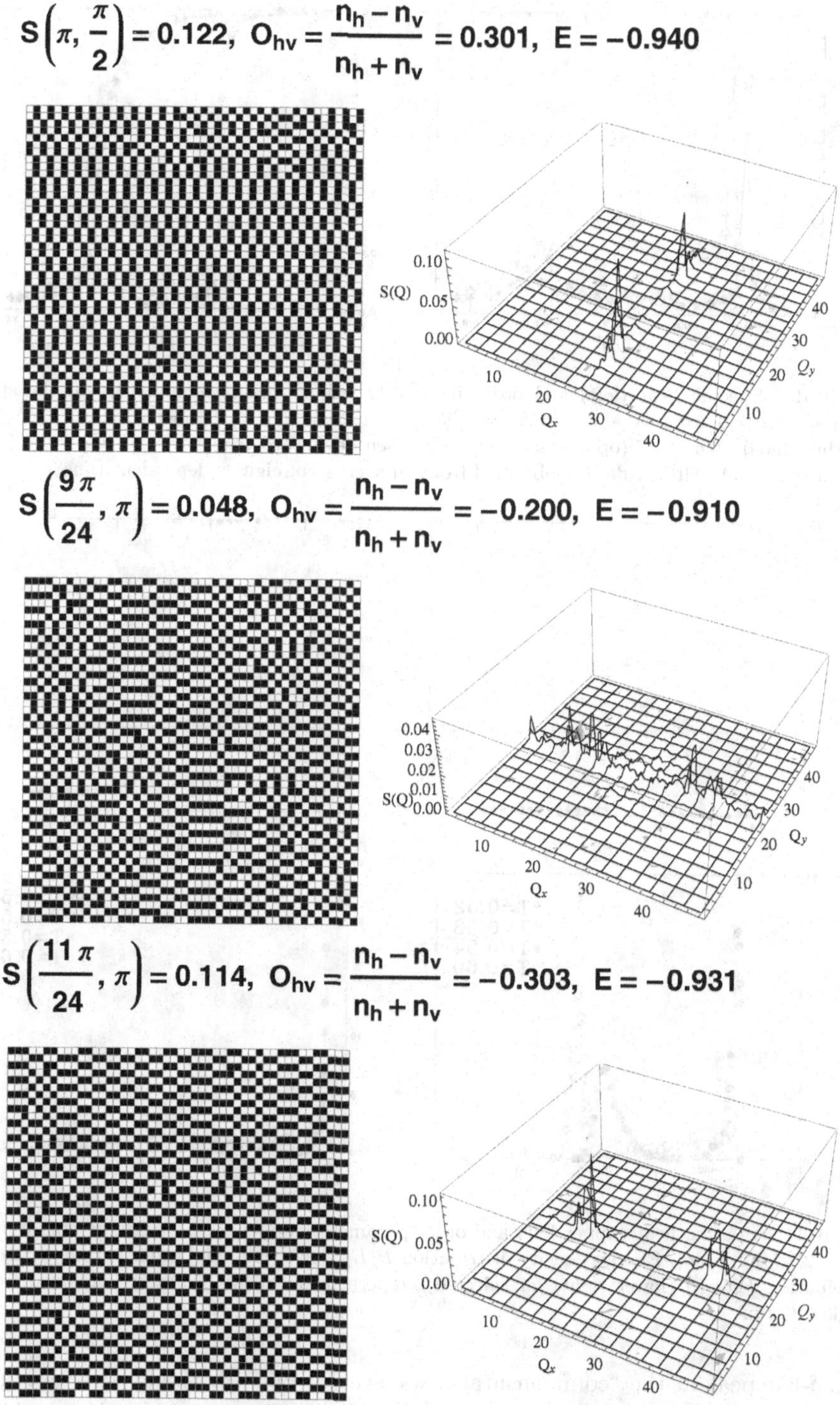

Fig. 10.9. Snapshots and corresponding structure factors for $J/g = 0.86$ ($\langle 1, 2 \rangle$ checkerboard phase) and $k_B T/g = 0.34$. The values of the maximum of the structure factor and of $O_{hv}$ and $E$ are indicated over each snapshot.

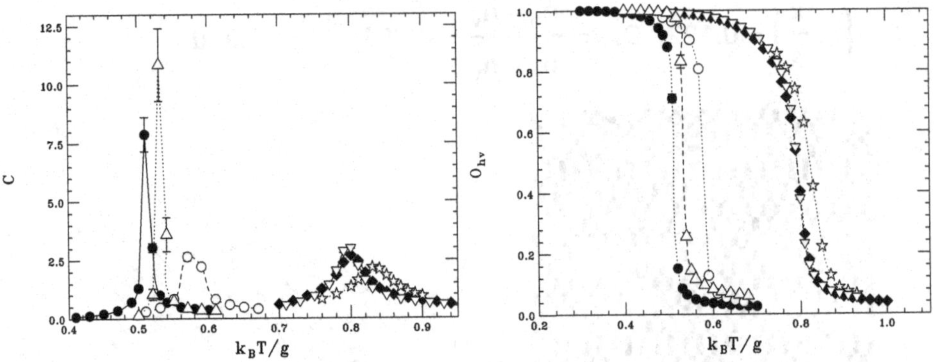


Fig. 10.10. Specific heat (left) and order parameter (right) versus temperature in the stripe phase with $h = 1$ for several $J/g$: $J/g = 0.95$ (full circles), $J/g = 1$ (open circles), $J/g = 1.4$ (full diamonds), $J/g = 1.7$ (open stars), $J/g = 2$ (open downward triangles) and $J/g = 2.5$ (open upward triangles). All MC data is obtained from an average of eight independent runs.

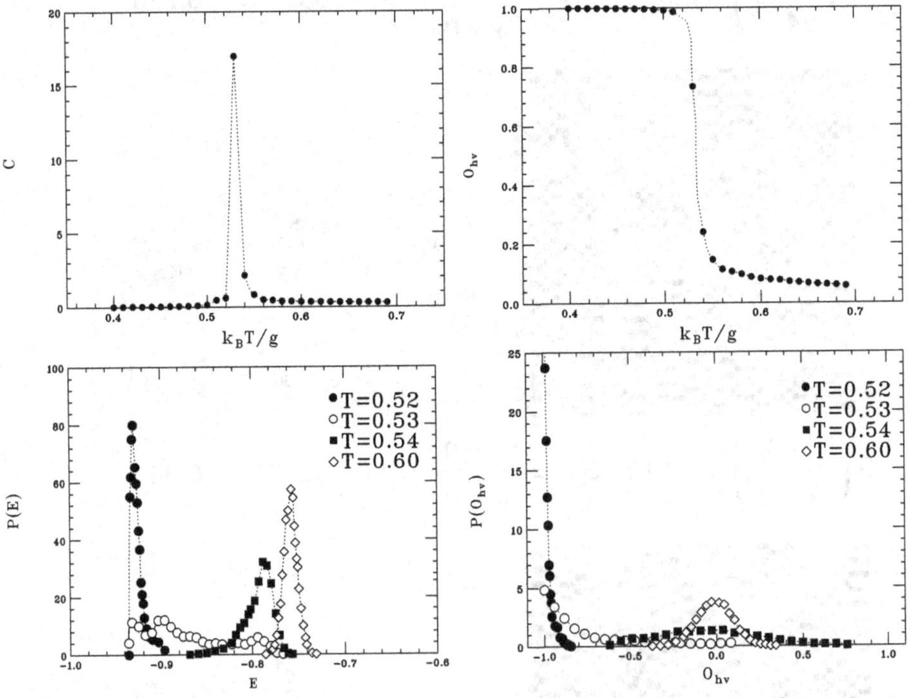


Fig. 10.11. Specific heat (upper left) and order parameter (upper right) versus temperature for $J/g = 2.5$ ($\langle 1 \rangle$ phase). Energy density distribution $P(E)$ (lower left) and order parameter density distribution $P(O_{hv})$ (lower right) for selected temperatures. All MC data are obtained from a single MC run.

by a $\delta$-like peak at the "commensurate" wavevector $\boldsymbol{Q} = (0, \pi)$ while the tetragonal phase is characterized by a crown-shaped profile.[102] These two snapshots, taken at $k_BT/g = 0.53$, point out the coexistence of the stripe configuration $\langle 1 \rangle$ with the disordered phase. Their energies correspond approximately to the left and right limit

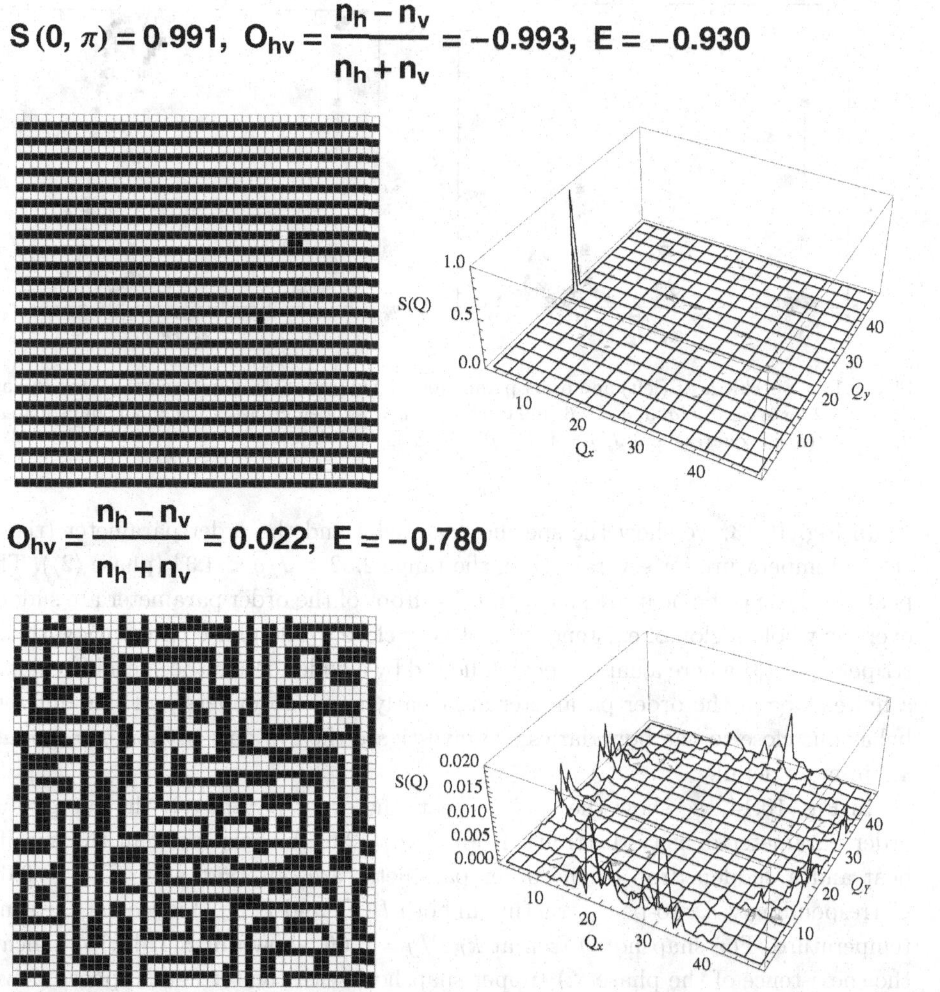


Fig. 10.12. Snapshots and corresponding structure factors for $J/g = 2.5$ (stripe phase with $h = 1$) and $k_BT/g = 0.53$. The values of the maximum of the structure factor and of the observables $O_{hv}$ and $E$ are indicated over each snapshot.

of the flat profile seen in the lower left panel of Fig. 10.11 for $k_BT/g = 0.53$. Many other snapshots were taken: those with an energy $-0.93 \lesssim E \lesssim -0.81$ are similar to the upper snapshot even though some labyrinth disorder begins to appear, leading to a decrease of the height of the commensurate peak from 1 to ~0.3; on the contrary, those with an energy $-0.81 \lesssim E \lesssim -0.78$ are very similar to the lower snapshot since the labyrinth disorder spreads on the whole lattice (tetragonal phase). The crown-shape of the structure factor remains for higher temperatures: the peak in $P(E)$ of Fig. 10.11 that moves smoothly to higher energy for $k_BT/g \geq 0.54$ corresponds to the tetragonal phase. The labyrinth shape of the tetragonal configuration differs from the pattern of the usual paramagnetic configuration which is recovered, however, at higher temperatures.

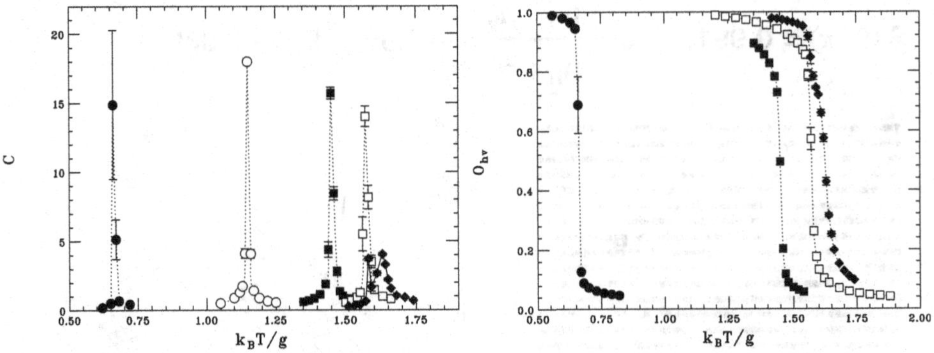


Fig. 10.13. Specific heat (left) and order parameter (right) versus temperature in the stripe phase with $h = 2$ for several $J/g$: $J/g = 2.6$ (full circles), $J/g = 3$ (open circles), $J/g = 3.4$ (full squares), $J/g = 3.7$ (open squares) and $J/g = 4$ (full diamonds).

In Fig. 10.13, we show the specific heat (left) and the order parameter (right) versus temperature for several $J/g$ in the range $2.52 < J/g < 4.34$ (phase $\langle 2 \rangle$). The peaks in the specific heat are sharp and the drops of the order parameter are sudden over the whole region of existence except very close to the phase boundary with the stripe phase $\langle 3 \rangle$ where a narrow peak followed by a wider one in the specific heat and a finite slope in the order parameter are clearly seen. To understand this different behaviour close to the boundaries, we investigate the systems with $J/g = 2.6$ and 4.3 in more detail.

In Fig. 10.14, we show the specific heat, the order parameter, the energy and order parameter density distribution for $J/g = 2.6$. A sharp peak in the specific heat and a sudden drop of the order parameter are observed at $k_BT/g = 0.66$. Correspondingly, a two-peak structure in both $P(E)$ and $P(O_{hv})$ is seen at the same temperature. The snapshots taken at $k_BT/g = 0.66$, shown in Fig. 10.15, indicate the coexistence of the phase $\langle 2 \rangle$ (upper snapshot) with the tetragonal phase (lower snapshot) which looks very similar to the lower snapshot of Fig. 10.12.

The MC simulation for $J/g = 4.3$ leads to a different scenario as shown in Fig. 10.16. The specific heat shows two narrows peaks at $k_BT/g = 1.42$ and 1.48 and a broad one around $k_BT/g = 1.65$. The order parameter shows two small steps in correspondence of the two sharp peaks of the specific heat and a smooth decrease around $k_BT/g = 1.65$. The energy and the order parameter density distributions show a two-peak structure at $k_BT/g = 1.42$ and 1.48. The two snapshots of Fig. 10.17 taken at $k_BT/g = 1.42$ show the coexistence of the stripe phase $\langle 2 \rangle$ with peaks in the structure factor at $\boldsymbol{Q} = (0, \pm\frac{\pi}{2})$ and a *modulated* phase with peaks at the incommensurate wavevectors $\boldsymbol{Q} = (0, \pm\frac{11}{24}\pi)$. As previously explained, the occurrence of peaks in the structure factor in correspondence of incommensurate wavevectors reflects a mixing of two defected stripe phases with different $h$: in the present case, the stripe phases $\langle 2 \rangle$ and $\langle 3 \rangle$. Analogously, the two snapshots of Fig. 10.18 taken at $k_BT/g = 1.48$ point out the coexistence of two different incommensurate phases with peaks at $\boldsymbol{Q} = (0, \pm\frac{11}{24}\pi)$ (upper snapshot) and

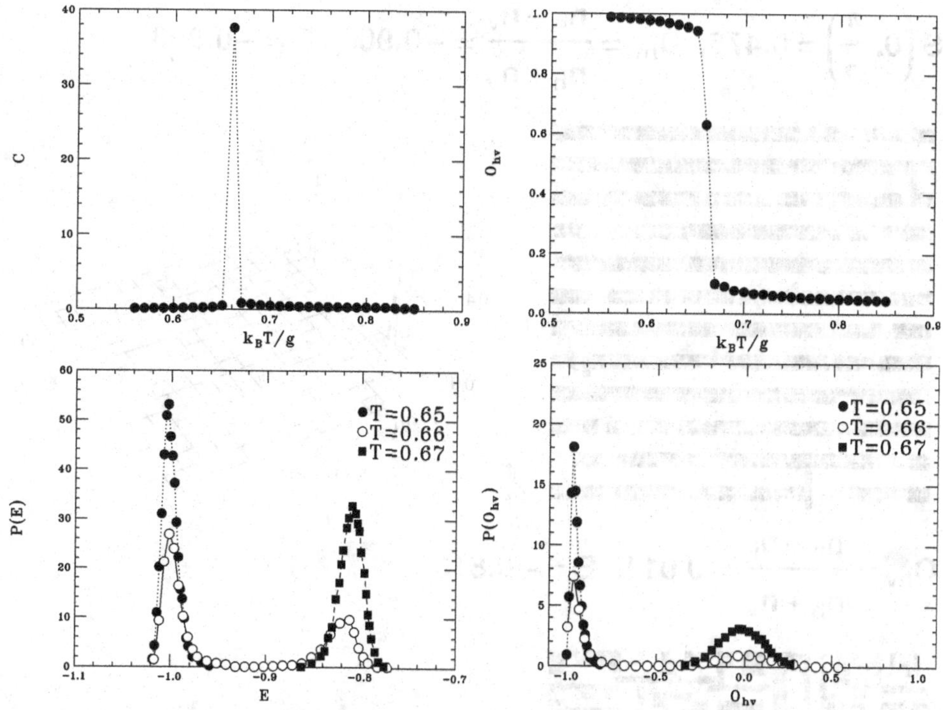


Fig. 10.14. Specific heat (upper left) and order parameter (upper right) versus temperature for $J/g = 2.6$ ($\langle 2 \rangle$ phase). Energy density distribution $P(E)$ (lower left) and order parameter density distribution $P(O_{hv})$ (lower right) for selected temperatures.

$\boldsymbol{Q} = (0, \pm\frac{10}{24}\pi)$ (lower snapshot), respectively. The weight of the phase $\langle 3 \rangle$ increases and the incommensurate wavevectors move towards those characterizing the stripe $\langle 3 \rangle$: $\boldsymbol{Q} = (0, \pm\frac{\pi}{3})$. Finally, the three snapshots of Fig. 10.19 taken at $k_BT/g = 1.65$ indicate the existence of a tetragonal phase without orientational order[102] in which the stripes rotate freely (upper and lower snapshots). The labyrinth shape of the middle snapshot and the corresponding crown-shaped structure factor profile is similar to the analogous snapshots of Figs. 10.12 and 10.15. The snapshots of Figs. 10.17–10.19 allow us to understand the density distributions of Fig. 10.16. The two-peak structure of both the energy and the order parameter density distribution at $k_BT/g = 1.42$ points out a first-order phase transition between the commensurate (C) stripe phase $\langle 2 \rangle$ and the incommensurate phase $I_1$ characterized by $\boldsymbol{Q} = (0, \pm\frac{11}{24}\pi)$. Similarly, the two-peak structure at $k_BT/g = 1.48$ points out a transition between the incommensurate phases $I_1$ and $I_2$, the latter being characterized by $\boldsymbol{Q} = (0, \pm\frac{10}{24}\pi)$. The transition C$-I_1$ is almost certainly a true phase transition whereas the transition $I_1-I_2$ is almost certainly a finite-size effect: indeed, for any finite lattice, any change in the wavevector $\boldsymbol{Q}$ may occur only by discrete steps $\Delta Q = \frac{2\pi}{L}$ as well as any change in the corresponding energy (leading to repeated narrow peaks in the specific heat) and in the order parameter (repeated small steps in $O_{hv}$). Most probably, the modulated phase is a genuine one but the

$$S\left(0, \frac{\pi}{2}\right) = 0.475, \quad O_{hv} = \frac{n_h - n_v}{n_h + n_v} = -0.906, \quad E = -0.993$$

$$O_{hv} = \frac{n_h - n_v}{n_h + n_v} = 0.011, \quad E = -0.815$$

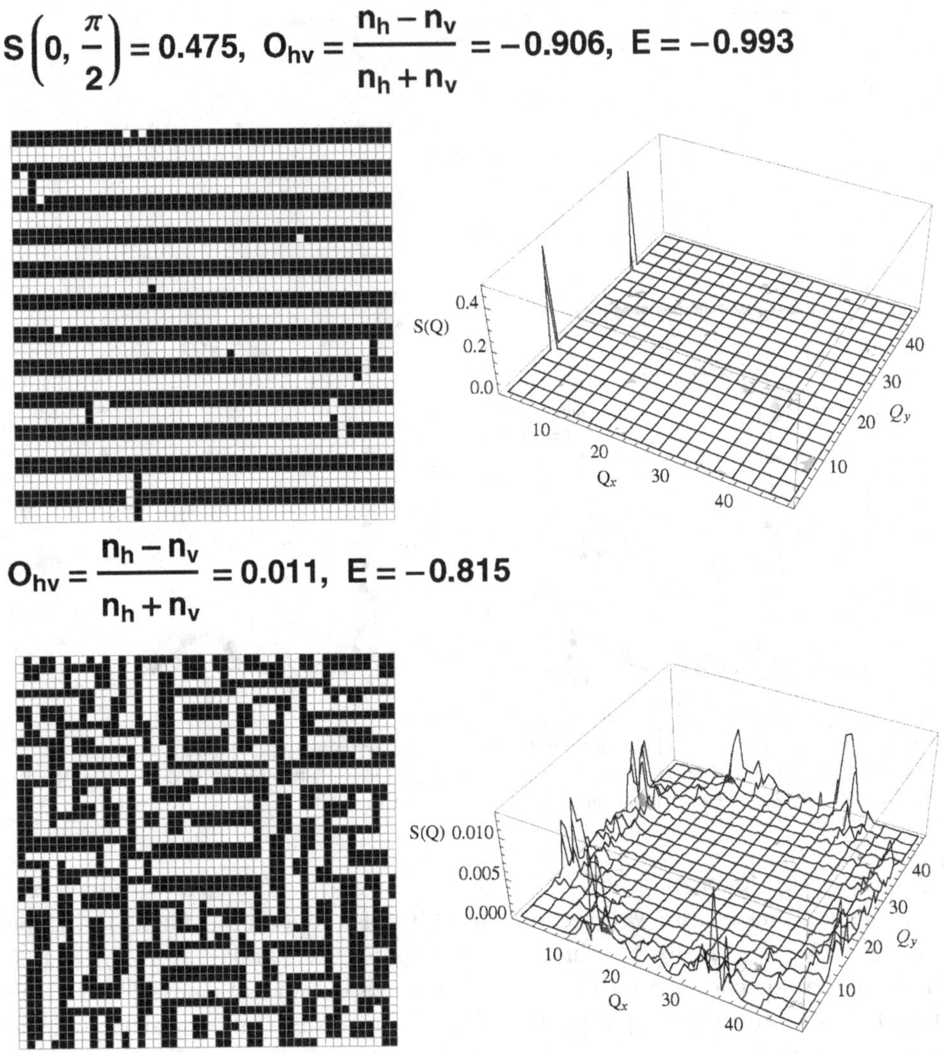

Fig. 10.15. Snapshots and corresponding structure factors for $J/g = 2.6$ (stripe phase with $h = 2$) and $k_B T/g = 0.66$. The values of the maximum of the structure factor and of the observables $O_{hv}$ and $E$ are indicated over each snapshot.

incommensurate wavevector $Q$ changes continuously and not by steps for a lattice with $L \to \infty$.

What about the existence of LRO in the modulated phase and the order of the phase transition? No proof exists for the existence of LRO in the modulated phase: if the LRO is absent, the true phase transition occurs at the first peak of the specific heat, otherwise the first peak corresponds to a transition between two ordered phases and only in the proximity of the broad peak, an order–disorder transition occurs. This second hypothesis seems less realistic than the first one since no peak is seen at the "second" transition. A more realistic scenario is to think of the

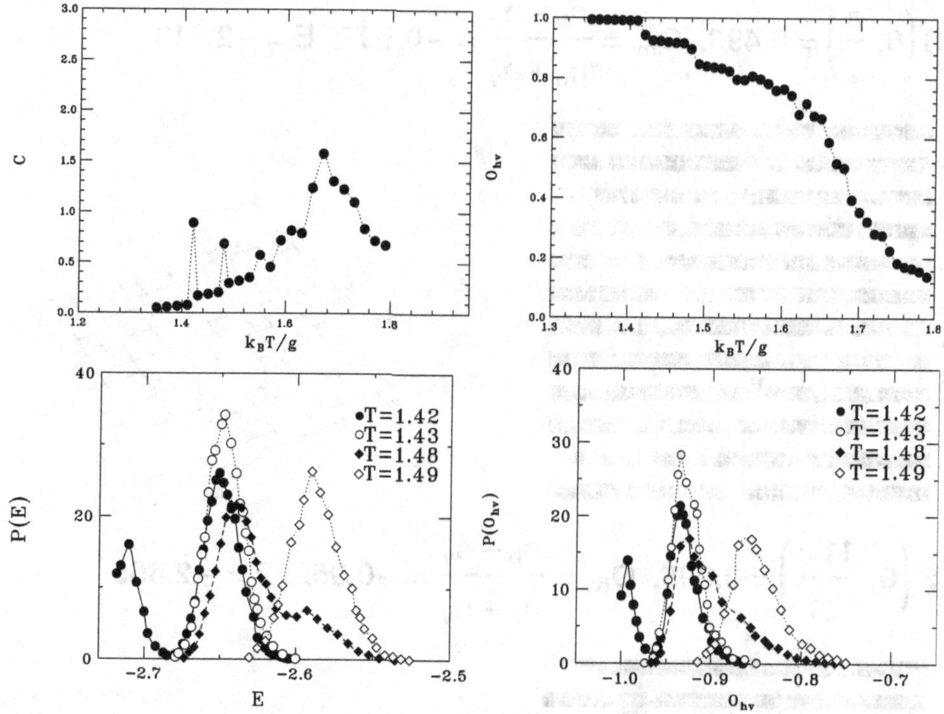


Fig. 10.16. Specific heat (upper left) and order parameter (lower left) versus temperature for $J/g = 4.3$ ($\langle 2 \rangle$ phase). Energy density distribution $P(E)$ (upper right) and order parameter density distribution $P(O_{hv})$ (lower right) for selected temperatures.

modulated phase as a BKT phase[89] in which the LRO is absent but the correlation functions decay algebraically with distance, in contrast with the paramagnetic phase where the correlations decay exponentially. In this case, two transitions occur: C–I (stripe-modulated phase) and I–P (modulated-paramagnetic phase). The transition between the BKT and the paramagnetic phase does not imply the existence of a sharp peak in the specific heat. In any case, the occurrence of the modulated phase seems well established.

The specific heat and the order parameter corresponding to the stripe phase $\langle 3 \rangle$ ($4.34 < J/g < 5.63$) are shown in Fig. 10.20. The two peaks in the specific heat are seen only near to the phase boundaries $\langle 2 \rangle - \langle 3 \rangle$ and $\langle 3 \rangle - \langle 4 \rangle$ respectively.

In Fig. 10.21, the specific heat, the order parameter and their density distributions are shown for $J/g = 4.4$. A sharp peak in the specific heat and a small step in the order parameter are clearly seen at $k_BT/g = 1.52$. Moreover, one can see a broad peak in the specific heat at $k_BT/g = 1.69$ and a corresponding smooth decrease of the order parameter around the same temperature. The energy density distribution shows a peak followed by a shoulder at $k_BT/g = 1.52$ and a well-defined single peak at higher temperature. The order parameter density distribution shows a peak with an imperceptible shoulder at $k_BT/g = 1.52$ and two symmetric peaks at $O_{hv} \simeq \pm 0.7$ at $k_BT/g = 1.69$, indicating the free rotation of the

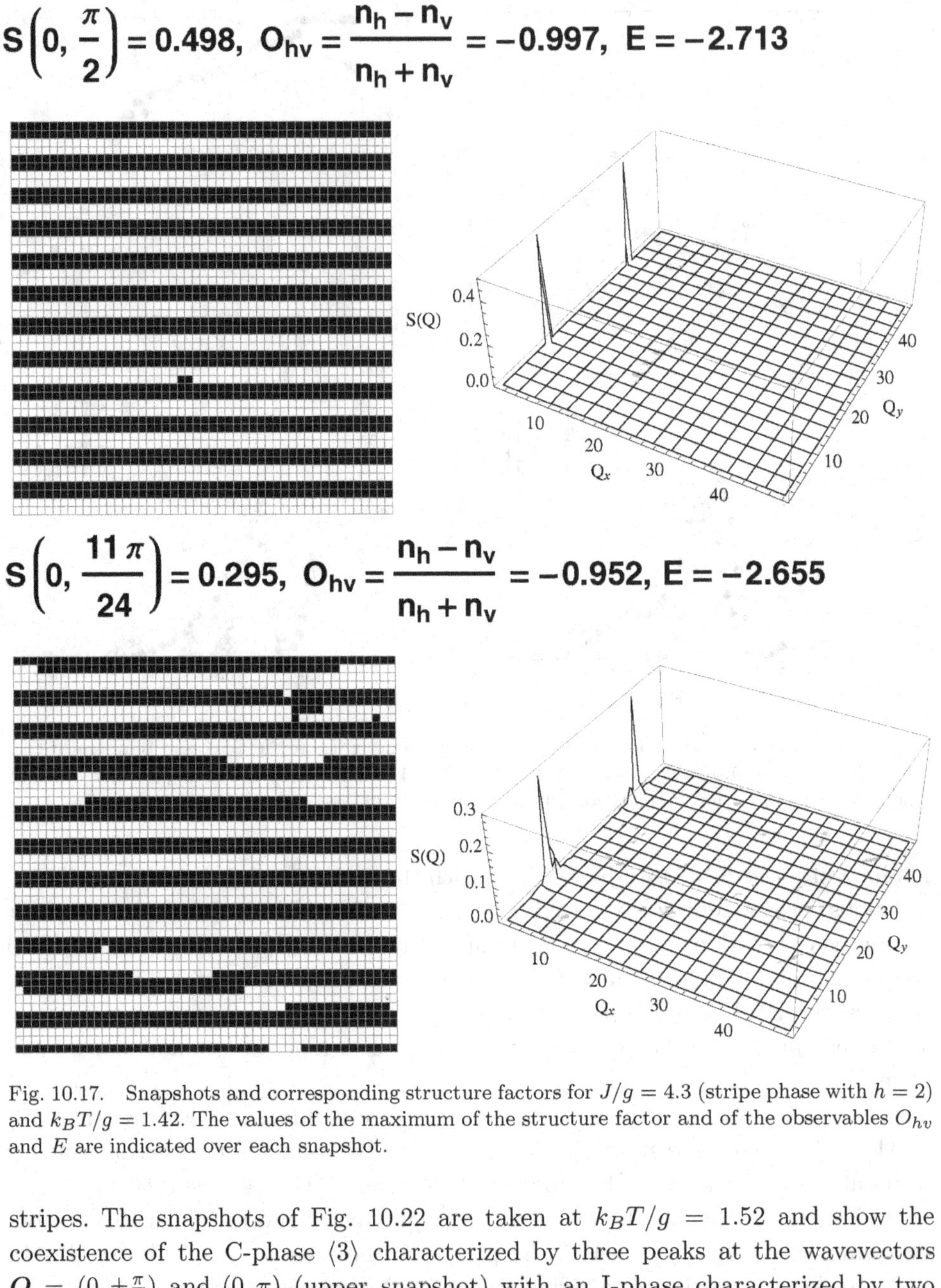


Fig. 10.17. Snapshots and corresponding structure factors for $J/g = 4.3$ (stripe phase with $h = 2$) and $k_BT/g = 1.42$. The values of the maximum of the structure factor and of the observables $O_{hv}$ and $E$ are indicated over each snapshot.

stripes. The snapshots of Fig. 10.22 are taken at $k_BT/g = 1.52$ and show the coexistence of the C-phase $\langle 3 \rangle$ characterized by three peaks at the wavevectors $\boldsymbol{Q} = (0, \pm\frac{\pi}{3})$ and $(0, \pi)$ (upper snapshot) with an I-phase characterized by two peaks at wavevectors $\boldsymbol{Q} = (0, \pm\frac{3}{8}\pi)$ (lower snapshot). Note that in the C-phase $\langle 3 \rangle$, the location of the three $\delta$-peaks in the profile of the structure factor along with their corresponding heights can be obtained directly from Eq. (10.3.24) at $T = 0$. The upper snapshot of Fig. 10.22, even though taken at $k_BT/g = 1.52$, looks very similar to the ground-state configuration. The snapshots of Fig. 10.23 are taken at $k_BT/g = 1.69$ and show the free rotation of the stripes (upper and lower snapshot)

$$S\left(0, \frac{11\pi}{24}\right) = 0.252, \quad O_{hv} = \frac{n_h - n_v}{n_h + n_v} = -0.934, \quad E = -2.642$$



$$S\left(0, \frac{10\pi}{24}\right) = 0.176, \quad O_{hv} = \frac{n_h - n_v}{n_h + n_v} = -0.863, \quad E = -2.599$$

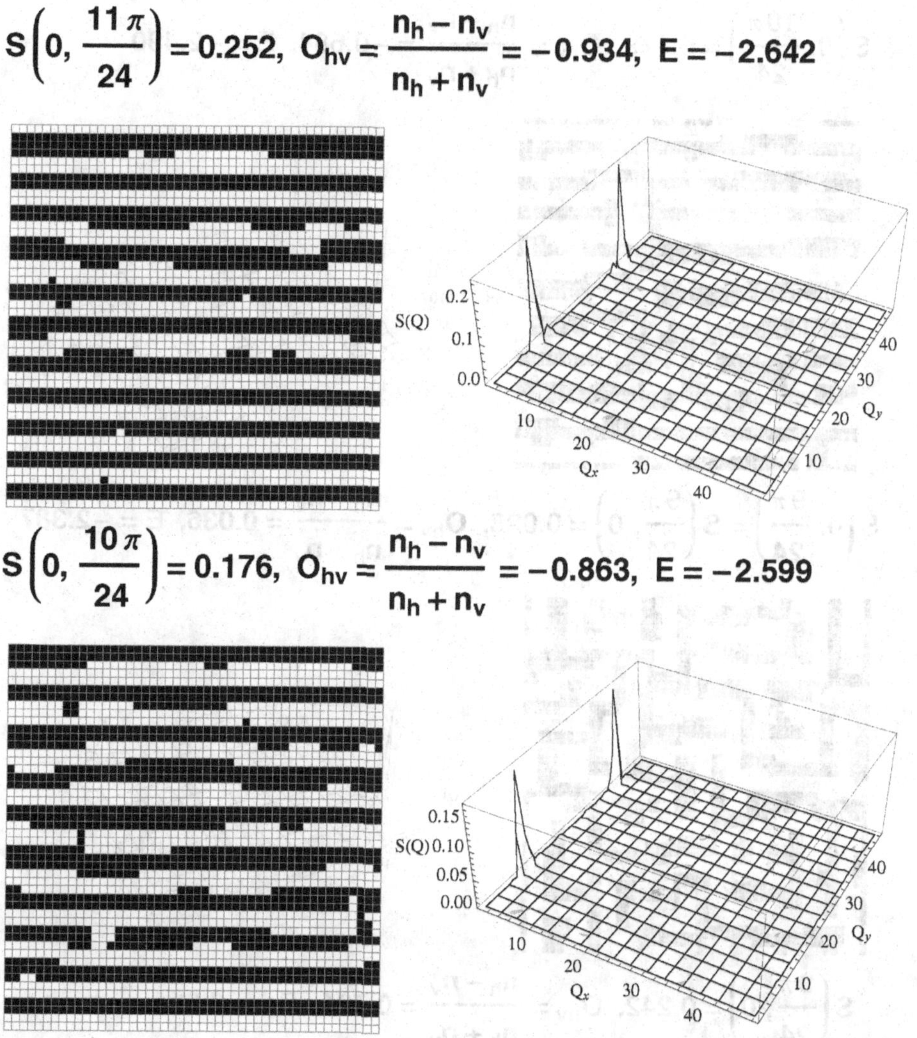

Fig. 10.18. Snapshots and corresponding structure factors for $J/g = 4.3$ (stripe phase with $h = 2$) and $k_B T/g = 1.48$. The values of the maximum of the structure factor and of the observables $O_{hv}$ and $E$ are indicated over each snapshot.

along with the labyrinth configuration (middle snapshot). All these configurations are typical of the tetragonal phase.[112]

The specific heat and the order parameter corresponding to the stripe phases with $h = 4$ ($5.63 < J/g < 6.92$) are shown in Fig. 10.24. For $J/g > 17.27$, where the stable phase is the ferromagnetic (F) one, an interesting behaviour of the specific heat versus temperature is shown in Fig. 10.25 where a two-peak structure is clearly seen for specific heats with $20 < J/g < 50$: the low temperature peak is very narrow while the high temperature peak is rather broad. For $60 < J/g < 100$, a single peak is seen. For $J/g = 20$, the error bars on the height of the first peak are very large because of the uncertainty of the exact location of the temperature at which the

$$\mathrm{S}\left(0, \frac{10\,\pi}{24}\right) = 0.179, \quad \mathrm{O_{hv}} = \frac{\mathrm{n_h} - \mathrm{n_v}}{\mathrm{n_h} + \mathrm{n_v}} = -0.684, \quad \mathrm{E} = -2.490$$

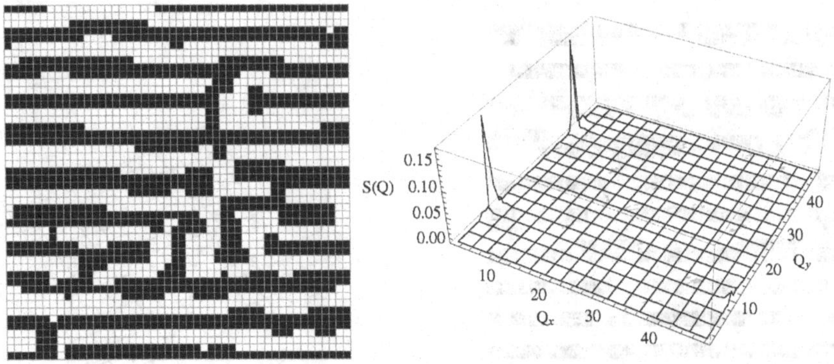


$$\mathrm{S}\left(0, \frac{9\,\pi}{24}\right) = \mathrm{S}\left(\frac{9\,\pi}{24}, 0\right) = 0.028, \quad \mathrm{O_{hv}} = \frac{\mathrm{n_h} - \mathrm{n_v}}{\mathrm{n_h} + \mathrm{n_v}} = 0.036, \quad \mathrm{E} = -2.387$$

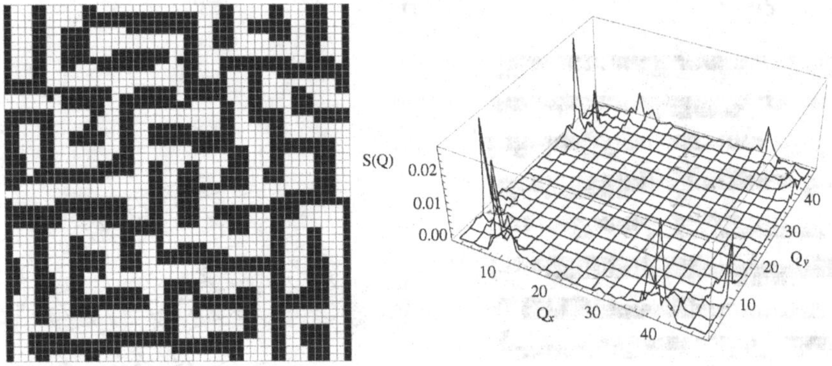


$$\mathrm{S}\left(\frac{9\,\pi}{24}, 0\right) = 0.242, \quad \mathrm{O_{hv}} = \frac{\mathrm{n_h} - \mathrm{n_v}}{\mathrm{n_h} + \mathrm{n_v}} = 0.731, \quad \mathrm{E} = -2.514$$

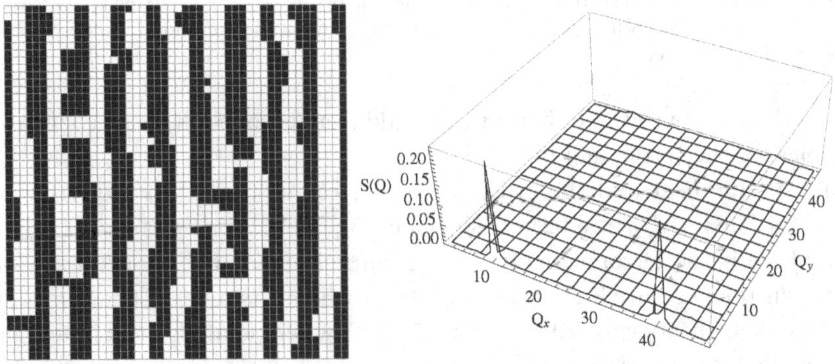


Fig. 10.19. Snapshots and corresponding structure factors for $J/g = 4.3$ (stripe phase with $h = 2$) and $k_BT/g = 1.65$. The values of the maximum of the structure factor and of the observables $O_{hv}$ and $E$ are indicated over each snapshot.

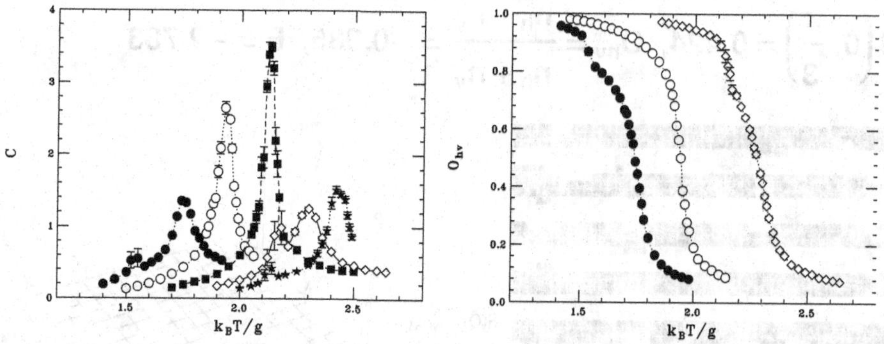


Fig. 10.20. Specific heat versus temperature in the stripe phase with $h = 3$ for several $J/g$: $J/g = 4.4$ (full circles), $J/g = 4.7$ (open circles), $J/g = 5$ (full squares), $J/g = 5.3$ (open diamonds) and $J/g = 5.6$ (full stars).

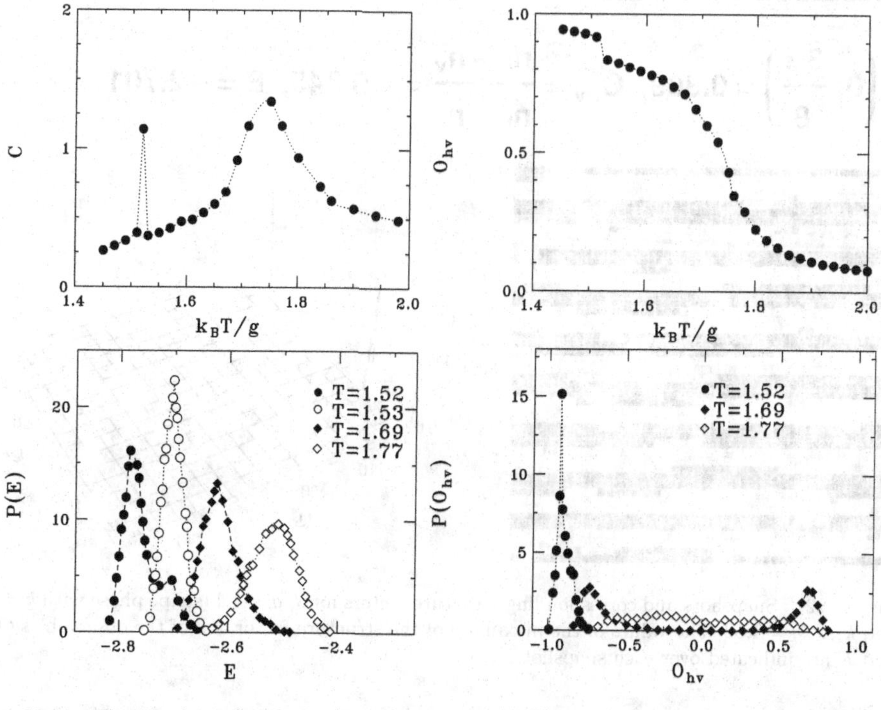


Fig. 10.21. Specific heat (upper left) and order parameter (lower left) versus temperature for $J/g = 4.4$ ($\langle 3 \rangle$ phase). Energy density distribution $P(E)$ (upper right) and order parameter density distribution $P(O_{hv})$ (lower right) for selected temperatures.

peak occurs. The error bars on the narrow peaks gradually reduce with increasing $J/g$. These facts seem to point out the existence of a first-order phase transition between the F and the paramagnetic phase for $J/g < 50$. On the contrary, for $J/g > 50$, the narrow peak merges into the broad one and a second-order phase transition seems to occur. A least square linear fit of the temperature at which the

$$S\left(0, \frac{\pi}{3}\right) = 0.424, \quad O_{hv} = \frac{n_h - n_v}{n_h + n_v} = -0.935, \quad E = -2.783$$

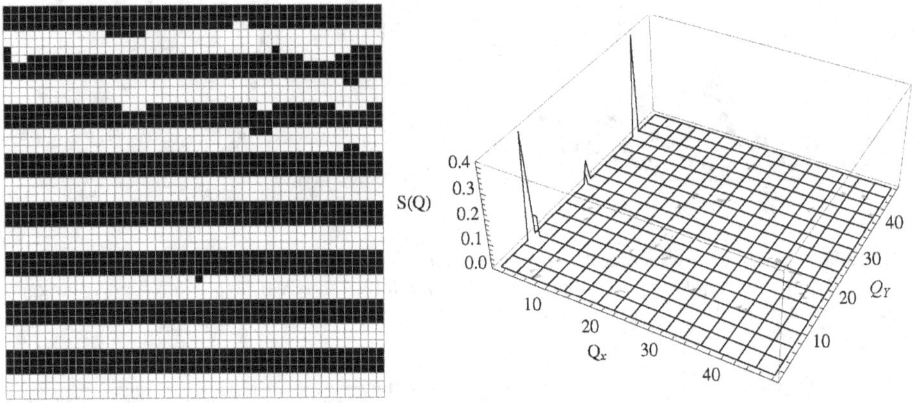


$$S\left(0, \frac{3\pi}{8}\right) = 0.309, \quad O_{hv} = \frac{n_h - n_v}{n_h + n_v} = -0.845, \quad E = -2.701$$

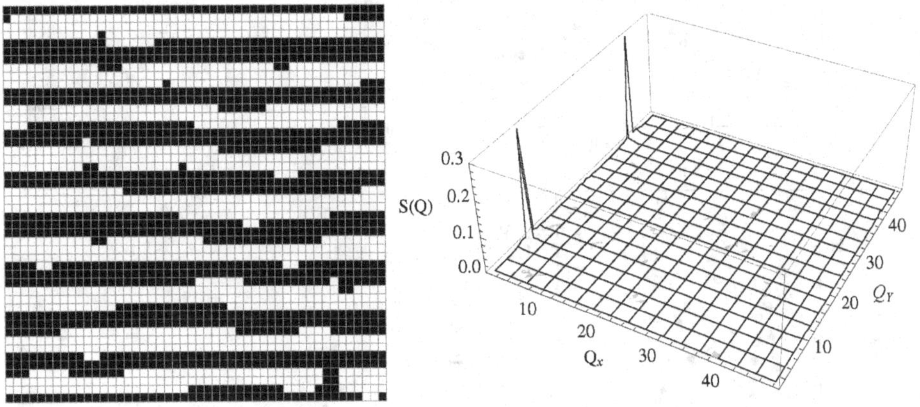


Fig. 10.22. Snapshots and corresponding structure factors for $J/g = 4.4$ (stripe phase with $h = 3$) and $k_BT/g = 1.52$. The values of the maximum of the structure factor and of the observables $O_{hv}$ and $E$ are indicated over each snapshot.

maximum of the specific heat occurs gives

$$k_BT_c = 2.285J - (24.1 \pm 0.4)g. \tag{10.4.4}$$

The finite-size correction to the critical temperature for a SQ Ising model with NN exchange interaction was given by Ferdinand and Fisher[115]:

$$k_BT_c(L)/J = 2.269185 + 0.818/L \tag{10.4.5}$$

$$S\left(0,\frac{3\pi}{8}\right)=0.264,\quad O_{hv}=\frac{n_h-n_v}{n_h+n_v}=-0.701,\quad E=-2.618$$

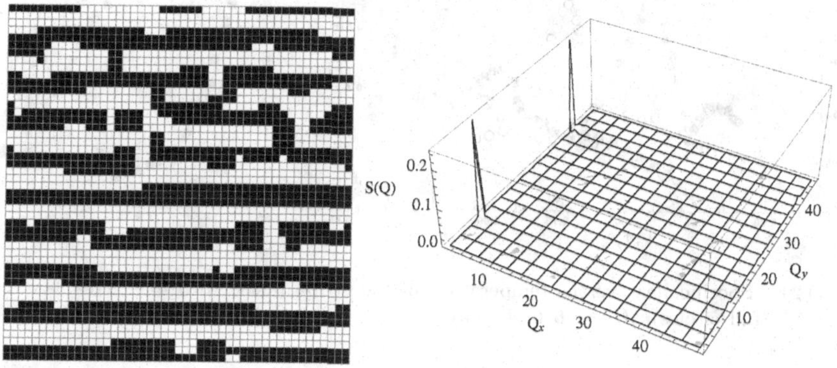


$$S\left(\frac{3\pi}{8},0\right)=S\left(0,\frac{3\pi}{8}\right)=0.040,\quad O_{hv}=\frac{n_h-n_v}{n_h+n_v}=0.034,\quad E=-2.537$$

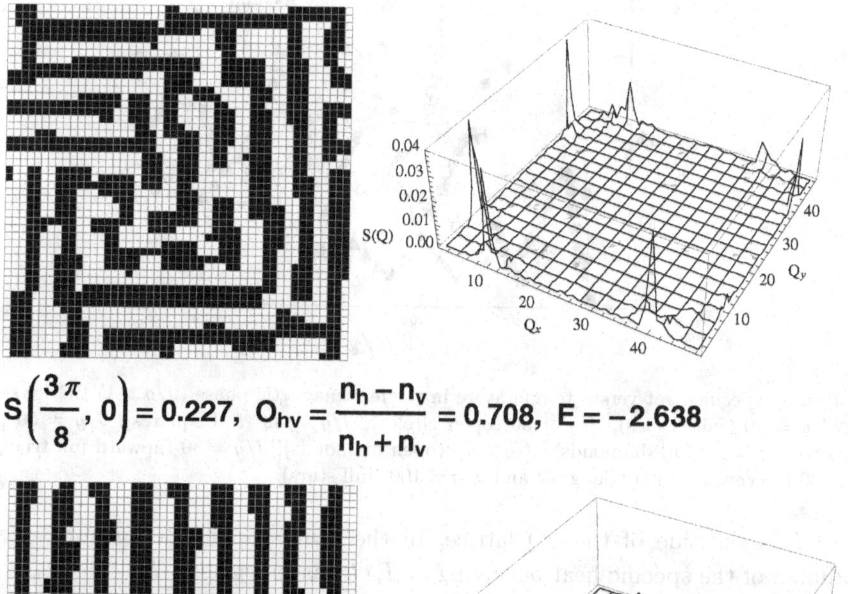


$$S\left(\frac{3\pi}{8},0\right)=0.227,\quad O_{hv}=\frac{n_h-n_v}{n_h+n_v}=0.708,\quad E=-2.638$$

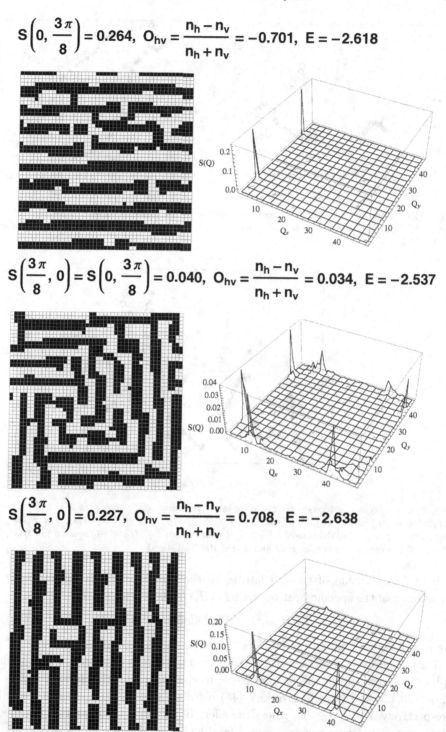

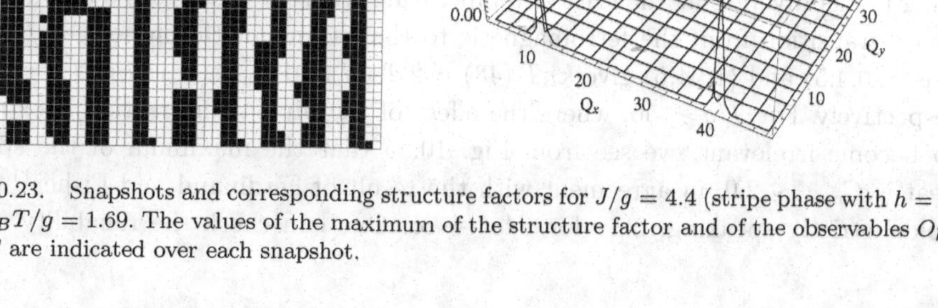


Fig. 10.23. Snapshots and corresponding structure factors for $J/g = 4.4$ (stripe phase with $h = 3$) and $k_BT/g = 1.69$. The values of the maximum of the structure factor and of the observables $O_{hv}$ and $E$ are indicated over each snapshot.

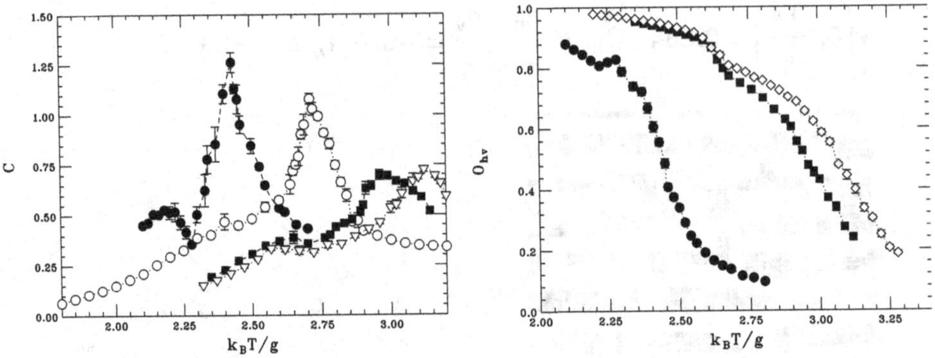


Fig. 10.24. Specific heat versus temperature in the stripe phase with $h = 4$ for several $J/g$: $J/g = 5.7$ (full circles), $J/g = 6$ (open circles), $J/g = 6.5$ (full squares) and $J/g = 6.6$ (open diamonds).

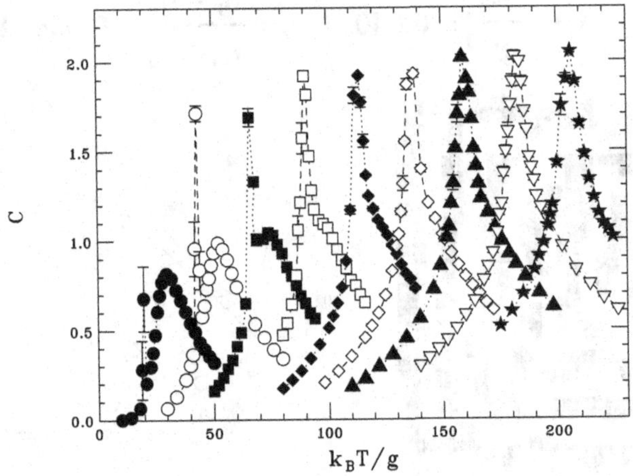


Fig. 10.25. Specific heat versus temperature in the ferromagnetic phase ($J/g > 17.27$) for several $J/g$: $J/g = 20$ (full circles), $J/g = 30$ (open circles), $J/g = 40$ (full squares), $J/g = 50$ (open squares), $J/g = 60$ (full diamonds), $J/g = 70$ (open diamonds), $J/g = 80$ (upward full triangles), $J/g = 90$ (downward open triangles) and $J/g = 100$ (full stars).

where $L$ is the side of the SQ lattice. In the same paper, the authors give the maximum of the specific heat occurring at $T_c(L)$ as function of $L$:

$$C_{\max}(L)/(k_B L^2) = 0.4945386 \ln L + 0.201359. \tag{10.4.6}$$

It is well known that in the thermodynamic limit, the specific heat of the 2D NN Ising model diverges at the critical temperature where it undergoes a second-order phase transition from the ferromagnetic to the paramagnetic phase. For $L = 48$, Eqs. (10.4.5) and (10.4.6) give $k_B T_c(48) = 2.286J$ and $C_{\max}(48)/(k_B 48^2) = 2.116$, respectively. For $J/g \gtrsim 50$, where the effect of the dipolar interaction is expected to become irrelevant, we see from Fig. 10.25 that the maximum of the specific heat is $C_{\max} \sim 2.0$, in agreement with the result of Ferdinand and Fisher[115] that holds, strictly speaking, only for $J/g \to \infty$. One can check the reliability of the

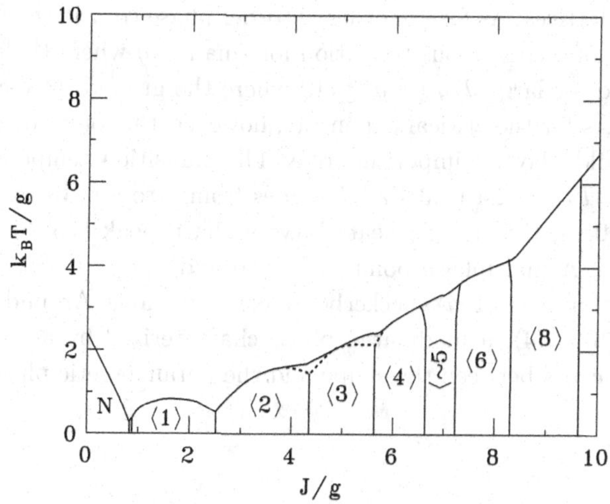


Fig. 10.26. Phase diagram of a SQ lattice with $L = 48$. The region between the dotted and continuous line is the region where the modulated phase is expected to occur. The checkerboard phases are restricted to the very narrow region between $N$ and $\langle 1 \rangle$ phase.

MC simulation by performing a finite-size scaling analysis on a SQ NN Ising model ($g = 0$): one obtains $k_BT_c/J = 2.269 \pm 0.001 + (0.76 \pm 0.04)/L$ and $C_{\max}/(k_BL^2) = (0.493 \pm 0.008) \ln L + (0.20 \pm 0.03)$ in very good agreement with the exact result of Ferdinand and Fisher.[115] Incidentally, the shape of the specific heat of the NN model is indistinguishable from that shown in Fig. 10.25 for $J/g = 100$. The checked reliability of the MC simulations for $g = 0$ gives confidence in the anomalous specific heat behaviour observed for $J/g < 50$: a change of the order of the phase transition may occur around $J/g \simeq 50$.

In Fig. 10.26, we give the phase diagram as obtained from analytic calculations at $T = 0$ and MC simulations on a lattice $48 \times 48$ at finite temperature. The continuous line in Fig. 10.26 was obtained by the location of the specific heat peak characterizing the transition to the tetragonal or paramagnetic phase. The dotted curve is the transition between the stripe and the modulated phase. As one can see, the phase diagram shows a variety of configurations depending on the ratio $J/g$. Indeed, for $J = 0$ (pure dipole interaction), the ground state corresponds to an antiferromagnetic N configuration. As the ratio $J/g$ increases, the ground state is characterized by $\langle 1, n \rangle$ configurations in the narrow range $0.8304 - 0.8806$ and by the stripe configurations $\langle 1 \rangle$, $\langle 2 \rangle$, $\langle 3 \rangle$, $\langle 4 \rangle$, $\langle {\sim}5 \rangle$, $\langle 6 \rangle$, $\langle 8 \rangle$, $\langle 12 \rangle$, $\langle 24 \rangle$ (the last two not shown in the phase diagram) for $0.8806 < J/g < 17.27$. Finally, for $J/g > 17.27$, the F configuration becomes stable. An important difference between the phase diagram shown in Fig. 10.26 and a previous phase diagram[109] obtained for the same model using a much smaller lattice of side $L = 16$, is the existence of a modulated phase in proximity of the phase boundary $\langle 2 \rangle - \langle 3 \rangle$ and $\langle 3 \rangle - \langle 4 \rangle$. The strong reduction of the existence region of the modulated phase away from the boundaries and its disappearance in proximity of the boundary $\langle 1 \rangle - \langle 2 \rangle$ indicate the importance of using large-size lattices. As for the order–disorder phase transition, we are in favour of the existence of a continuous transition for small $J/g$ where the ground state is a Néel state and for large $J/g$ ($J/g \gtrsim 50$) where the ground-state configuration is ferromagnetic. As for the critical exponents, however, the long range dipole–dipole interaction should play an important role. The transition temperature decreases linearly from $k_BT/g = 2.4$ to 0.4 as $J/g$ goes from 0 to 0.8. In the narrow region $0.83 < J/g < 0.88$, the specific heat shows a sharp peak typical of a first-order phase transition. A multiphase point at $T = 0$ and $J/g = 0.8806$ corresponds to the confluence of infinite $\langle 1, n \rangle$ checkerboard configurations. Around the boundaries $\langle 2 \rangle - \langle 3 \rangle$ and $\langle 3 \rangle - \langle 4 \rangle$, a modulated phase characterized by an incommensurate wavevector $\boldsymbol{Q}$ occurs between the stripe and the paramagnetic phase.

# REFERENCES


1. W. Heitler and F. London, *Z. Phys.* **44**, 455 (1927).
2. Y. Sugiura, *Z. Phys.* **45**, 484 (1927).
3. I. S. Gradshteyn and I. M. Ryzhik, *Table of Integrals, Series and Products*, Academic, London, 1965.
4. M. Abramowitz and I. A. Stegun, *Handbook of Mathematical Functions*, Dover Publications, New York, 1965.
5. L. I. Shiff, *Quantum Mechanics*, McGraw-Hill, London, 1968.
6. N. Fukuda and M. Wortis, *J. Phys. Chem. Solids* **24**, 1675 (1963).
7. D. C. Mattis, *The Theory of Magnetism Made Simple*, World Scientific, London, 2006.
8. M. A. Rudermann and C. Kittel, *Phys. Rev.* **96**, 99 (1954); T. Kasuya, *Prog. Theor. Phys.* **16**, 45 (1956); K.Yosida, *Phys. Rev.* **106**, 893 (1957).
9. J. Jensen and A. R. Mackintosh, *Rare Earth Magnetism*, Clarendon Press, Oxford, 1991.
10. H. Bethe, *Z. Phys.* **71**, 205 (1931).
11. L. Hulthén, *Ark. Mat. Astron. Fys.* **26**A, 11 (1938).
12. J. Des Cloizeaux and J. J. Pearson, *Phys. Rev.* **128**, 2131 (1962).
13. A. L. Fetter and J. D. Walecka, *Quantum Theory of Many-Particle Systems*, McGraw-Hill, London, 1971.
14. F. Holstein and H. Primakoff, *Phys. Rev.* **58**, 1048 (1940).
15. F. J. Dyson, *Phys. Rev.* **102**, 1217, 1230 (1956).
16. S. V. Maleev, *Sov. Phys. JETP* **6**, 776 (1958).
17. J. I. Davis, *Ann. Phys.* **58**, 529 (1970).
18. E. Rastelli and A. P. Lindgård, *J. Phys. C* **12**, 1899 (1979).
19. D. Forster, *Hydrodynamic Fluctuations, Broken Symmetry, and Correlation Functions*, Benjamin, London, 1975.
20. K. Huang, *Statistical Mechanics*, Wiley, London, 1963.
21. F. Bloch, *Z. Phys.* **61**, 206 (1930); **74**, 295 (1932).
22. N. D. Mermin and H. Wagner, *Phys. Rev. Lett.* **17**, 1133 (1966).
23. S. W. Lovesey, *Theory of Neutron Scattering from Condensed Matter*, Clarendon, Oxford, 1984.
24. D. N. Zubarev, *Sov. Phys. Uspekhi* **3**, 320 (1960).
25. R. Silberglitt and A. B. Harris, *Phys. Rev.* **174**, 640 (1968).
26. F. Leoni and C. R. Natoli, *J. Phys. C* **3**,1462 (1970).
27. T. Morita, *J. Phys. Soc. Japan* **29**, 850 (1970); T. Morita and T. Tanaka, *Phys. Rev.* **137**, A648 (1965); **138**, A1395 (1965).
28. R. A. Tahir-Kheli and D. Ter Haar, *Phys. Rev.* **127**, 95 (1962).
29. M. Bloch, *Phys. Rev. Lett.* **9**, 286 (1962).

30. G. S. Rushbrook and P. J. Wood, *Mol. Phys.* **1**, 257 (1962).
31. A. B. Harris, *Phys. Rev.* **175**, 674 (1968); **184**, 606 (1969).
32. G. C. Wick, *Phys. Rev.* **80**, 268 (1950).
33. T. Matsubara, *Prog. Theor. Phys.* **14**, 351 (1955).
34. J. Goldstone, *Proc. Royal Soc. A* **239**, 267 (1957).
35. Murray R. Spiegel, *Theory and Problems of Complex Variable*, Schaum's Outline Series, McGraw-Hill, New York 1964.
36. R. Silberglitt and A. B. Harris, *Phys. Rev. Lett.* **19**, 30 (1967).
37. S. W. Lovesey, *Phys. Lett. A* **86**, 43 (1981); S. W. Lovesey and M. Hood, *Z. Phys. B* **47**, 327 (1982).
38. T. Oguchi, *J. Phys. Soc. Japan* **31**, 394 (1971).
39. M. Wortis, *Phys. Rev.* **132**, 85 (1963).
40. E. Rastelli, S. Sedazzari and A. Tassi, *J. Phys.: Condens. Matter* **4**, 6283 (1992).
41. A. M. Bonnot and J. Hanus, *Phys. Rev.* **7**, 2207 (1973).
42. R. Silberglitt and J. Torrance, *Phys. Rev. B* **2**, 772 (1970).
43. T. Tonegawa, *Prog. Theor. Phys.* **46**, 61 (1970).
44. I. Ono, S. Mikado and T. Oguchi *J. Phys. Soc. Japan* **30**, 358 (1971).
45. J. E. Van Himbergen and J, A. Tjon, *Physica A* **82**, 389 (1976).
46. E. Rastelli and A. Tassi, *J. Phys.* **C 17**, 727 (1984).
47. S. T. Beliaev, *Sov. Phys. JETP* **7**, 289 (1958).
48. J. Goldstone, *Nuovo Cimento* **19**, 154 (1961).
49. Y. Nambu, *Phys. Rev.* **117**, 648 (1960).
50. A. B. Harris, D. Kumar, B. I. Halpering and P. C. Hohenberg, *Phys. Rev. B* **3**, 961 (1971).
51. U. Balucani, M. G. Pini, A. Rettori and V. Tognetti, *J. Phys. C* **13**, 3895 (1980).
52. P. A. Lindgård and O. Danielsen, *J. Phys. C* **7**, 1523 (1974).
53. E. Rastelli and A. Tassi, *Phys. Rev. B* **11**, 4711 (1975).
54. J. Villain, *J. Phys. Chem. Solids* **11**, 303 (1959); T. A. Kaplan, *Phys. Rev.* **124**, 329 (1961); R. J. Elliott, *Phys. Rev.* **124**, 346 (1961); T. Nagamiya, K. Nagata and Y. Kitano, *Prog. Theor. Phys.* **27**, 1253 (1962).
55. E. Rastelli, L. Reatto and A. Tassi, *J. Phys. C* **18**, 353 (1985).
56. B. Coqblin, *The Electronic Structure of Rare-Earth Metals and Alloys: The Magnetic Heavy Rare-Earths*, Academic Press, London, 1977.
57. P. Day, M. W. Moore, T. E. Wood, D. McK. Paul, K. R. A. Ziebeck, L. P. Regnault and J. Rossat-Mignod, *Sol. State Commun.* **51**, 627 (1984).
58. S. R. Kuindersma, J. P. Sanchez and C. Haas, *Physica B* **111**, 231 (1981).
59. F. Keffer, *Handbuch der Physik*, Vol. XVIII/2, Springer-Verlag, Heidelberg, 1966.
60. H. Kadowaki, K. Ubukoshi and K. Hirakawa, *J. Phys. Soc. Japan* **54**, 363 (1985).
61. G. Bednorz and K. A. Müller, *Z. Phys. B* **64**, 189 (1986).
62. G. H. Wannier, *Phys. Rev.* **79**, 357 (1950).
63. A. Danielian, *Phys. Rev.* **133**, A1344 (1964).
64. B. C. Frazer, G. Shirane, D. E. Cox and C. E. Olsen, *Phys. Rev.* **140**, A1418 (1965).
65. E. Rastelli and A. Tassi, *J. Phys. C* **20**, L303 (1987).
66. E. Rastelli and A. Tassi, *J. Phys. C* **19**, L423 (1986).
67. G. C. De Fotis, *Phys. Rev. B* **23**, 4714 (1981).
68. M. F. Collins and O. A. Petrenko, *Can. J. Phys.* **75**, 605 (1997).
69. E. Rastelli, L. Reatto and A. Tassi, *Physica B* **97**, 1 (1979).
70. E. Rastelli and A. Tassi, *Physica B* **156–157**, 115 (1989).
71. R. J. Birgeneau, H. J. Guggenheim and G. Shirane, *Phys. Rev. Lett.* **22**, 720 (1969); J. Skalyo, Jr, G. Shirane, R. J. Birgeneau and H. J. Guggenheim, *Phys. Rev. Lett.* **23**, 1394 (1969).

72. P. W. Stephens, R. J. Birgeneau, C. F. Majkrzak and G. Shirane, *Phys. Rev. B* **28**, 452 (1983).
73. A. Chahid, F. J. Bermejo, A. Criado, J. L. Martinez and M. Garcia-Hernandez, *J. Phys.: Condens. Matter* **5**, 6295 (1993).
74. E. Rastelli and A. Tassi, *J. Phys. C* **21**, 1003 (1988).
75. H. Kadowaki, H. Kikuchi and Y. Ajiro, *J. Phys.: Condens. Matter* **2**, 4485 (1990).
76. I. Harada and O. Nagai, *J. Phys. Soc. Japan* **42**, 738 (1930).
77. P. Politi, M. G. Pini and A. Rettori, *Phys. Rev. B* **46**, 8312 (1992).
78. R. E. De Wames and T. Wolfram, *Phys. Rev.* **185** 762, (1969).
79. T. W. Clinton, L. W. Lynn, J. Z. Liu, X. Y. Jia, T. J. Goodwin, R. N. Shelton, B. W. Lee, M. Buchgeister, M. B. Maple and J. L. Peng, *Phys. Rev. B* **51**, 15429 (1995).
80. J. Mesot, P. A. Allenspach, U. Staub, A. Furrer, H. Mutka, R. Osborn and A. Taylor, *Phys. Rev. B* **47**, 6027 (1993).
81. M. Baher, M. Winkelmann, P. Voderwisch, M. Steiner, C. Pich and F. Schwabl, *Phys. Rev. B* **54**, 12932 (1996).
82. P. Ewald, *Ann. Phys.* **369**, 253 (1921).
83. M. H. Cohen and F. Keffer, *Phys. Rev.* **99**.
84. J. A. Osborn, *Phys. Rev.* **67**, 351 (1945).
85. M. T. Hutchings, *Point-Charge Calculations of Energy Levels of Magnetic Ions in Crystalline Electric Fields*, Solid State Physics, eds. F. Seitz and D. Turnbull, Vol. 16, Academic Press, New York, 1965, p. 227.
86. E. Rastelli and A. Tassi, Eur. *Phys. J. B* **4**, 285 (1998).
87. S. Simizu, G. H. Bellesis, J. L. Lukin, S. A. Friedberg, H. S. Lessure, S. M. Fine and M Greenblatt, *Phys. Rev. B* **39**, 9099 (1989).
88. S. Skanthakumar, J. W. Lynn and F. Dogan, *J. Appl. Phys.* **81**, 4934 (1997).
89. V. L. Berezinskii, *Zh. Exsp. Teor. Fiz.* **59**, 907 (1970) [*Sov. Phys. JETP* **32**, 493 (1971)]; **61**, 1144 (1971) [**34**, 610 (1973)]; J. M. Kosterlitz and D. J. Thouless, *J. Phys. C* **6**, 1181 (1973); J. M. Kosterlitz, *J. Phys. C* **7**, 1046 (1974).
90. R. Gupta, J. DeLapp, G. G. Batrouni, G. C. Fox, C. F. Baillie and J. Apostolakis, *Phys. Rev. Lett.* **61**, 1996 (1988); R. Gupta and C. F. Baillie, *Phys. Rev. B* **45**, 2883 (1992); P. Olson, *Phys. Rev. Lett.* **73**, 3339 (1994); J. K. Kim, *Phys. Lett. A* **223**, 261 (1996).
91. M. E. Fisher, S. Ma and B. G. Nickel, *Phys. Rev. Lett.* **29**, 917 (1972).
92. S. Romano, *Il Nuovo Cimento B* **100**, 447 (1987).
93. S. Romano, *Phys. Rev. B* **44**, 7066 (1991).
94. P. I. Belobrov, R. S. Gekht and V. A. Ignatchenko, *Sov. Phys. JETP* **57**, 636 (1983).
95. K. De'Bell, A. B. MacIsaac, I. N. Booth and J. P. Whitehead, *Phys. Rev. B* **55**, 15108 (1997).
96. E. Rastelli, A. Carbognani, S. Regina and A. Tassi, *Eur. Phys. J. B* **9**, 641 (1999).
97. J. Villain, R. Bideaux, J. P. Carton and R. Conte, J. Phys. (*France*) **41**, 1263 (1980); E. Rastelli, L. Reatto and A. Tassi, *J. Phys. C* **16**, L331 (1983); C. L. Henley, *Phys. Rev. Lett.* **62**, 2056 (1989).
98. D. P. Pappas, K. A. Kämper and H. Hopster, *Phys. Rev. Lett.* **64**, 3179 (1990).
99. R. Allenspach, M. Stampanoni and A. Bischof, *Phys. Rev. Lett.* **65**, 3344 (1990).
100. A. Carbognani, E. Rastelli, S. Regina and A. Tassi, *Phys. Rev. B* **62**, 1015 (2000).
101. R. Kretschmer and K. Binder, *Z. Phys. B* **34**, 375 (1979).
102. K. De'Bell, A. B. MacIsaac and J. P. Whitehead, *Rev. Mod. Phys.* **72**, 225 (2000).
103. S. A. Cannas, D. A. Stariolo and F.A. Tamarit, *Phys. Rev. B* **69**, 092409 (2004).
104. E. Rastelli, S. Regina and A. Tassi, *J. Appl. Phys.* **99**, 08F708-1 (2006).
105. E. Rastelli, S. Regina and A. Tassi, *Phys. Rev. B* **73**, 144418 (2006).

106. S. A. Cannas, M. F. Michelon, D. A. Stariolo and F. A. Tamarit, *Phys. Rev. B* **73**, 184425 (2006).
107. E. Rastelli, S. Regina and A. Tassi, *Phys. Rev. B* **76**, 054438-1 (2007).
108. R. Czech and J. Villain, *J. Phys.: Condens. Matter B* **1**, 619 (1989).
109. A. B. MacIsaac, J. P. Whitehead, M. C. Robinson and K. De'Bell, *Phys. Rev. B* **51**, 16033 (1995).
110. W. Selke, *Phys. Rep.* **170**, 213 (1988).
111. D. P. Landau, *Phys. Rev. B* **13**, 2997 (1976).
112. I. Booth, A. B. MacIsaac, J. P. Whitehead and K. De'Bell, *Phys. Rev. Lett.* **75**, 950 (1995).
113. M. S. S. Challa, D. P. Landau and K. Binder, *Phys. Rev. B* **34**, 1841 (1986).
114. K. Binder, *Rep. Prog. Phys.* **50**, 783 (1987).
115. A. E. Ferdinand and M. E. Fisher, *Phys. Rev.* **135**, 832 (1969).

# INDEX